20th International Conference on Ground Penetrating Radar 2024

Journal of Physics: Conference Series Volume 2887

Online
23-27 June 2024

ISBN: 979-8-3313-1023-3
ISSN: 1742-6588

Printed from e-media with permission by:

Curran Associates, Inc.
57 Morehouse Lane
Red Hook, NY 12571

CURRAN ASSOCIATES INC.
proceedings
.com

Some format issues inherent in the e-media version may also appear in this print version.

Institute of Physics
Dirac House, Temple Back
Bristol BS1 6BE UK

Phone: 44 1 17 929 7481
Fax: 44 1 17 920 0979

techtracking@iop.org

Additional copies of this publication are available from:

Curran Associates, Inc.
57 Morehouse Lane
Red Hook, NY 12571 USA
Phone: 845-758-0400
Fax: 845-758-2633
Email: curran@proceedings.com
Web: www.proceedings.com

TABLE OF CONTENTS

Journal of Physics: Conference Series **2887** (2024) 011001 doi:10.1088/1742-6596/2887/1/011001

Preface

It is with great pleasure that we present the proceedings of the 20th International Conference on Ground Penetrating Radar (GPR, 2024), held from June 23-27, 2024, at Jilin University in Changchun, China. This conference, organized by Jilin University and the Chinese Geophysical Society, it is the second time this prestigious event has been hosted in mainland China.

The theme of this year's conference, "GPR for Better City," reflects our focus on leveraging GPR technology to enhance urban life and development. The conference was remarkable, bringing together over 330 participants from 21 countries and regions, including 64 international delegates. The conference featured 104 oral presentations, 55 poster sessions, and specialized short courses on Gprmax GPR numerical simulation software and Geotilx GPR data processing, attracting over 80 experts and scholars. The conference also hosted an exhibition showcasing the latest GPR instruments and software, with participation from 14 organizations and manufacturers from China, Russia, the United States, Canada, South Korea, and Latvia.

We were honored to have seven distinguished scholars, including experts from Japan, the Netherlands, Belgium, and China, deliver keynote and plenary speeches. Their presentations covered various topics, such as advancements in electromagnetic theory for understanding GPR signals, progress in underground structure detection using navigation and communication satellite signals, and new developments in UAV-mounted GPR systems. Additionally, the latest achievements in China's "Chang'e 6" lunar mission and "Tianwen-2" Mars mission were shared, sparking lively discussions among participants.

We extend our heartfelt thanks to all the authors who contributed their work, the keynote speakers who shared their valuable insights, and the reviewers who ensured the high quality of the contributions. Special appreciation goes to the organizing and technical program committees for their dedication and hard work in bringing this conference to fruition.

The papers included in this volume will serve as a significant resource for GPR researchers, practitioners, and enthusiasts, contributing to the continued advancement of GPR technology. We also hope that GPR 2024 has fostered new collaborations and research initiatives that will drive innovation in this field for years.

We look forward to welcoming you to future editions of the GPR International Conference, where we will continue to explore and develop this vital technology for society's benefit.

The Committee of GPR 2024

Journal of Physics: Conference Series **2887** (2024) 011001 doi:10.1088/1742-6596/2887/1/011001

Organizing Committees

Local Organizing Committee

Honorary Chair: Zhaofa Zeng, Jilin University

General Chair: Xuan Feng, Jilin University

Members:

Sixin Liu, Jilin University

Jing Li, Jilin University

Qi Lu, Jilin University

Wenjing Liang, Jilin University

Haoqiu Zhou, Jilin University

Zejun Dong, Jilin University

International Advisory Committee

Erica Utsi, EMC Radar Consulting, Haddenham, United Kingdom

Evert Slob, Delft University of Technology, Delft, The Netherlands

Mary Collins, University of Florida (professor emeritus)

Motoyuki Sato, Tohoku University, Sendai, Japan

Raffaele Persico, National Research Council-IBAM, Italy

Richard Yelf, Georadar Pty Ltd., Coffs Harbour, NSW, Australia

Sebastien Lambot, Université Catholique de Louvain, Belgium

Steven Koppenjan, Bechtel Nevada/Special Technologies Laboratory, UC Santa Barbara, USA

Timo Saarenketo, Roadscanners, Rovaniemi, Finland

Wallace Lai, The Hong Kong Polytechnic University, Hong Kong

Xiongyao Xie, Tongji University, Shanghai, China

Johannes Hugenschmi, Eastern Switzerland University of Applied Science, Switzerlanddt

John Bradford Colorado School of Mines, Golden, Colorado, USA

Local Committee

Suping Peng China University of Mining and Technology(Beijing)

Jun Lin Jilin University

Guangyou Fang Aerospace Information Research Institute, Chinese Academy of Sciences

Jinhai Zhang Institute of Geology and Geophysics, Chinese Academy of Sciences

Yibo Wang Institute of Geology and Geophysics, Chinese Academy of Sciences

Rongyi Qian China University of Geoscience(Beijing)

Deshan Feng Central South University

Hai Liu Guangzhou University

Xiaopeng Yang Beijing Institute of Technology

Tong Hao Tongji University

Hui Zhou China University of Petroleum(Beijing)

Honghua Wang Guilin University of Technology

Tian Jin National University of Defence Technology

Wenzhe Fa Peking University

Yan Su National Astronomical Observatories, Chinese Academy of Sciences

Yi Xu Macau University of Science and Technology

Yonghui Zhao Tongji University

Feng Zhou China University of Geoscience(Wuhan)

Bin Liu Shandong University

Yuelin Xu Key Laboratory of Near Surface Detection Technology

Jie Chen ZOROY Technology and Development Co., Ltd.

Chunyu Ding Shenzhen University

Ling Zhang Sun Yat-sen University

Lei Fu China University of Geoscience(Wuhan)

Journal of Physics: Conference Series **2887** (2024) 011002 doi:10.1088/1742-6596/2887/1/011002

Peer Review Statement

All papers published in this volume have been reviewed through processes administered by the Editors. Reviews were conducted by expert referees to the professional and scientific standards expected of a proceedings journal published by IOP Publishing.

- **Type of peer review:** Undefined
- **Conference submission management system:** Morressier
- **Number of submissions received:** 77
- **Number of submissions sent for review:** 77
- **Number of submissions accepted:** 77
- **Acceptance Rate (Submissions Accepted / Submissions Received × 100):** 100
- **Average number of reviews per paper:** 3
- **Total number of reviewers involved:** 230
- **Contact person for queries:**
  Name: Jing Li
  Email: inter_lijing@jlu.edu.cn
  Affiliation: Jilin University - Jilin University College of Geoexploration Science and Technology

Tunnel Quality Inspection Based on GPR and Machine Learning

Shan Shan Huang[1 2 3],Jun Chao Wang[1], Dan Dan Cheng[1],Hong Yan Geng[1],Li Yong Zhang[2 *],Yu Qi Yao[2],and Pei Wen Yao[2]

[1]The 22nd Research Institute of China Electronics Technology Group Corporatipon,Qingdao,Shandong,China.
[2]XiDian University,Xi'an, Shaanxi,China.
[3]Email:huangss@crirp.ac.cn.
*Corresponding author,email:zhangliyong@xidian.edu.cn.

Abstract. The development of non-destructive testing, particularly ground penetrating radar (GPR) technology, has gained popularity in underground detection due to its high resolution and non-destructive capabilities. However, the lack of intuitive imaging methods in GPR data analysis leads to poor interpretability, requiring skilled interpreters. To address this, the article suggests a machine learning-based algorithm for tunnel defect detection using GPR data. The YOLOv5 model network is used to train GPR data from actual road and railway tunnel inspections, and a CBAM attention mechanism is added to the original model to achieve high-precision radar scanning image defect recognition and localization. Compared with the original recognition model, the optimized recognition algorithm significantly improves recognition accuracy and greatly improves the efficiency of defect recognition compared to manual interpretation.

1. Introduction

Ground penetrating radar data preprocessing demands prior knowledge and manual screening, being time-consuming. Skilled professionals interpret and process data for precise target positioning, aiding subsequent analysis and report generation. Most image processing based detection technologies in China are surface defect detection, and ground penetrating radar is an important non-destructive testing method. There is relatively little research on radar defect detection. This article achieves higher precision defect recognition of radar echo images by data augmentation and expansion of images, training with YOLOv5 model, and combining CBAM attention mechanism.

2. Dataset construction

In the field of image recognition, there are various open-source high-resolution image datasets, but there is very little information about ground penetrating radar, and the vast majority are private and non-public data. The dataset used in this experiment is based on over 1500 ground penetrating radar images collected by the LTD series ground penetrating radar developed by the China Radio Wave Propagation Research Institute. The majority of this dataset consists of echo images of urban subways and urban tunnels detected by ground penetrating radar. The main detection target of the entire dataset is the initial lining or secondary lining steel mesh on the inner wall of the tunnel.

Ground-penetrating radar's frequent data acquisition leads to limited accuracy with B-Scan imaging due to noise and clutter, affecting radar analysis. In order to improve this situation, this article applies the IDSP7 post-processing software developed by the China Research Institute of Radio Wave Propagation to the preprocessing of radar images, effectively improving their signal-to-noise ratio and helping to accurately reveal the details of underground structures.

This article improves the geometric features of images through geometric transformations including Mixup, Cutmix[1], and Mosaic, thereby significantly improving the accuracy and precision of training. Therefore, based on the distribution characteristics of underground steel bars, this article applies geometric transformations to annotated data to increase the capacity of the dataset.

3. Experiment

3.1. YOLOv5 network

YOLOv5 is an open-source project announced by Ultralytics LLC on GitHub in June 2020[2][3][4]. The YOLOv5 model mainly consists of three core parts: Backbone, Neck, and Prediction. Among these components, Backbone can effectively perform feature extraction. Use FPN+PANet in the Neck section to fuse the image features of this stage. The Prediction section is used for target prediction and outputs through prediction. The network structure of YOLOv5 is shown in figure 1:

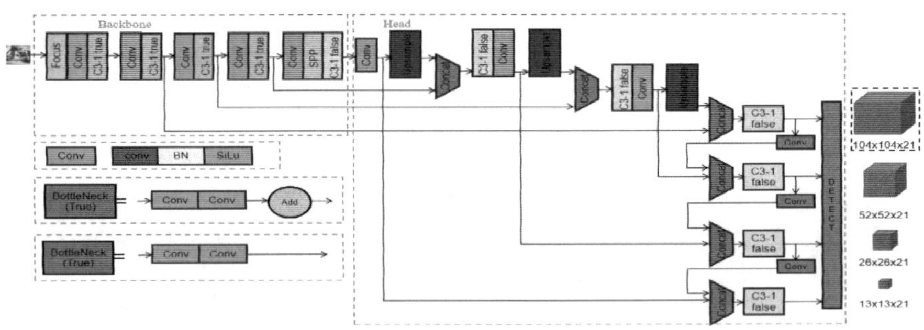

Figure 1. Schematic diagram of YOLOv5 network structure.

3.2. CBAM attention mechanism

Neural networks use attention mechanisms to manage computer resources efficiently, handle information overload, and focus on essential task content. This approach filters out irrelevant data, improving task completion accuracy. Two attention types are channel attention, assessing channel importance, and spatial attention, prioritizing pixel importance. In 2018, Bahdanau[5] et al. combined the two and processed the input feature layers using channel attention module and spatial attention module, abbreviated as CBAM. This article introduces the CBAM attention mechanism on the original YOLOv5 network structure. In this research, we replaced the C3 module in the original network with the CBAM attention mechanism to enhance the network's extraction of steel bar features.

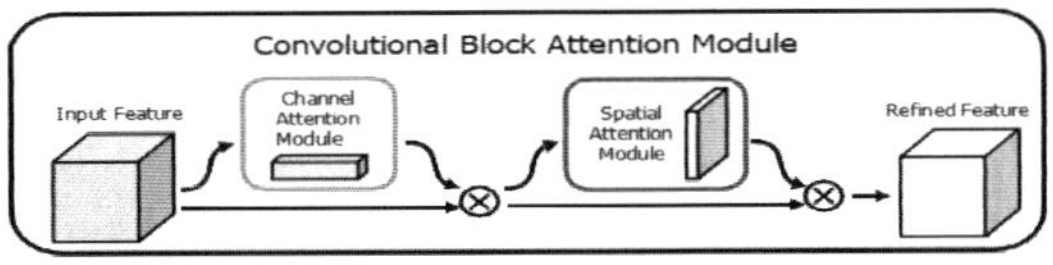

Figure 2. Implementation diagram of CBAM.

4. Result

The experiment in this article was conducted in a Windows environment, and we used Anaconda3 to deploy a deep learning environment. The specific configuration details of the environment are shown in table 1：

Table 1. Details of experimental environment configuration.

Device	Version
OS	Intel(R) Core(TM) i7-9750H CPU @ 2.60GHz 2.59 GHz
GPU	GeForce GTX 1660 Ti
CUDA	11.7
Python	3.9.13

After introducing the attention mechanism in YOLOv5, it can be seen from figure 3 that although the convergence speed of the loss function curve is slower, it is more stable. Before adding the CBAM attention mechanism, although obj_loss converges, there are significant fluctuations. After joining this mechanism, it is very stable without significant fluctuations, which means that after our improvement, the stability and robustness of YOLOv5 have been significantly improved, showing extremely high stability with almost no fluctuations. The comparison of model size recognition accuracy before and after improvement is detailed in table 2:

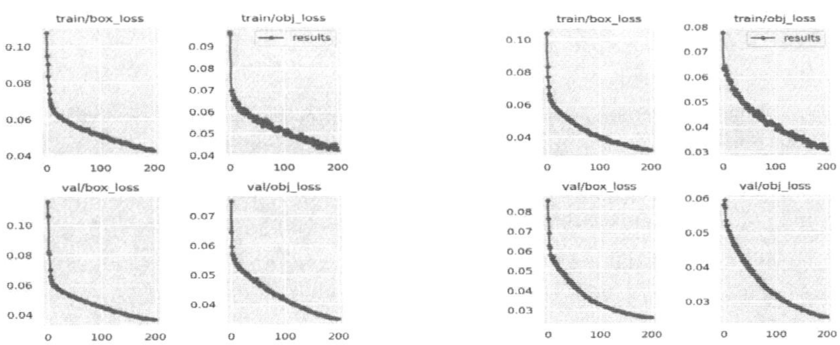

(a) Not incorporating attention mechanism (b) Add attention mechanism

Figure 3. Comparison chart of loss functions before and after optimization.

Table 2. Comparison of results before and after model improvement

	P	R	memory
Original	0.965	0.96	20.29
Improved	0.984	0.982	11.6

Figures 4 and 5 show the detection effect of the YOLOv5 model on steel bars before and after optimization. Compared to the basic YOLOv5 model, the optimized model not only has higher positioning accuracy of the detection box, but also can avoid the occurrence of missed detection when multiple steel bars overlap.

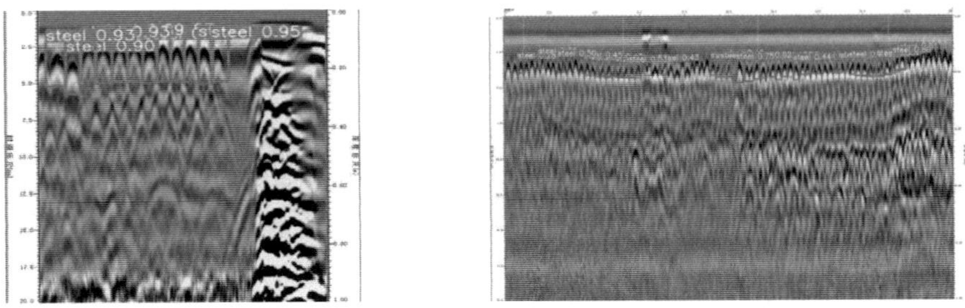

(a) (b)

Figure 4. Effect diagram of steel bar detection before model optimization.

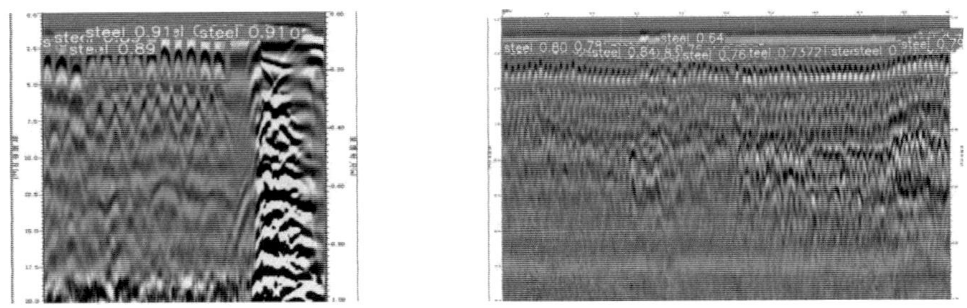

(a) (b)

Figure 5. Effect diagram of steel bar detection after model optimization.

5. Conclusion

This article aims to solve the problems of high false alarm rate and slow detection speed in detecting underground steel bar images using artificial ground penetrating radar. We analyzed the obtained data and generated corresponding datasets. Finally, after optimization, The optimization method of embedding CBAM attention mechanism into YOLOv5 structure can improve the detection performance of YOLOv5 model on ground penetrating radar steel bar dataset. After improvement, the model parameters were reduced to 11.6MB, the model size was reduced to 75%, and the recognition accuracy was significantly improved, reaching 98.4%. Compared with the original recognition model, it increased by 2%, laying a solid foundation for the intelligent deployment of future detection radar systems.

References

[1] Yun S , Han D , Oh S J ,et al. 2019 CutMix: Regularization Strategy to Train Strong Classifiers with Localizable Features.

[2] Redmon, Joseph , and A. Farhadi .2018 "YOLOv3: An Incremental Improvement." *arXiv e-prints.*

[3] Redmon, Joseph , et al. 2016 "You Only Look Once: Unified, Real-Time Object Detection." *Computer Vision & Pattern Recognition IEEE.*

[4] Redmon, Joseph , and A. Farhadi .2017 "YOLO9000: Better, Faster, Stronger." *IEEE Conference on Computer Vision & Pattern Recognition IEEE.*

[5] Woo, Sanghyun , et al. 2018 "CBAM: Convolutional Block Attention Module." *Springer, Cham.*

Monitoring Technology of Urban Underground Space Infrastructure Structure Based on Fully Polarized Ground Penetrating Radar

Yue Yu[1], XuanFeng[2], Zejun Dong[2], Enhedelihai Alex Nilot[2]

1: Changchun Institute of Technology, Changchun 130026, China, 894146687@qq.com
2: Jilin University, Changchun 130026, China, fengxuan@jlu.edu.cn, dongzejun@jlu.edu.cn, alexnilot123@gmail.com

Abstract. In order to maintain the safety of underground infrastructures, this paper applies fully polarized ground penetrating radar to develop non-destructive technology for monitoring underground infrastructures such as underground utility tunnels. More specifically, representative numerical models of concrete tunnels were created using GPRMAX to simulate fully polarized microwave scattering simulation data for different degrees of structural changes, pipeline displacements, and concrete wall moisture contents. Algorithms for extracting different physical parameters and scattering polarization characteristics were developed to establish a qualitative relationship between these extracted parameters and actual changes. Monitoring criteria and formulas were subsequently developed based on these relationships. These simulation data will be used to train and optimize the parameters of the monitoring criteria in future study.

Index Terms—**Ground Penetrating Radar, GPRMax, Tunnel, Dielectric Constant**

1. Introduction

These Urban underground space facilities mainly include urban lifeline projects such as gas pipes, water pipes, sewage pipes, electricity lines, communication cables, and heating pipes, as well as urban public living spaces such as rail transit, underground commercial streets, underground shopping malls, and underground parking lots. The underground comprehensive pipe gallery will integrate the scattered and disorderly underground pipeline facilities of the past into the lifeline of modern cities. How to effectively monitor the potential safety threats of these vital arteries of modern cities, such as subways and underground tunnels, is an urgent issue that needs to be addressed.

At present, advanced countries are lagging behind in their underground development and construction due to concerns about the safety and management of underground comprehensive pipe galleries [1] and a lack of comprehensive planning for supporting regulations [2]. At present, these concerns can be alleviated through intelligent management and installation of monitoring instruments and firefighting equipment. For example, Germany has increased its underground pipe gallery ventilation system and fire alarm and automatic fire extinguishing devices. China has accumulated rich experience in the planning, construction, and management of modern underground comprehensive pipe galleries, which can be used as a reference for other countries [3]. For example, the comprehensive pipe gallery in Pudong is equipped with advanced monitoring and alarm systems, making it the first modern urban underground comprehensive pipe gallery in China. Currently, the safety maintenance of underground comprehensive pipe galleries and tunnels mainly relies on intelligent management and monitoring of local areas such as sound and light, temperature and humidity, air quality, gas, smoke, and vibration [4][5][6] conducted a survey report on some existing tunnel structure safety monitoring technologies. Although optical fibers have also been used to monitor the deformation and temperature of the entire tunnel structure [7], their installation is not easy. Tan et al. [8] also studied how to strengthen the security and maintenance management of underground commercial areas by combining the Internet of Things.

2. 2D GPRMax Tunnel Simulation Setup

The GPRMax tunnel model consists of antenna at center if the tunnel, frees-pace, circle-concrete and dry sand ambient medium, as shown Figure 1. The radius of a circular tunnel is 0.2m. A 250 MHz Ricker pulse waveforms were adopted for all simulations discussed in this paper. To obtain polarimetric scattering data in (1), separate simulations using x-polarized (point to the right) and z-polarized (out of paper) transmitting dipoles were used as follows.
Ezz: The Ez output using the Z hertzian dipole source.
Exz The Ex output using the Z hertzian dipole source.
Exx: The Ex output using the X hertzian dipole source.

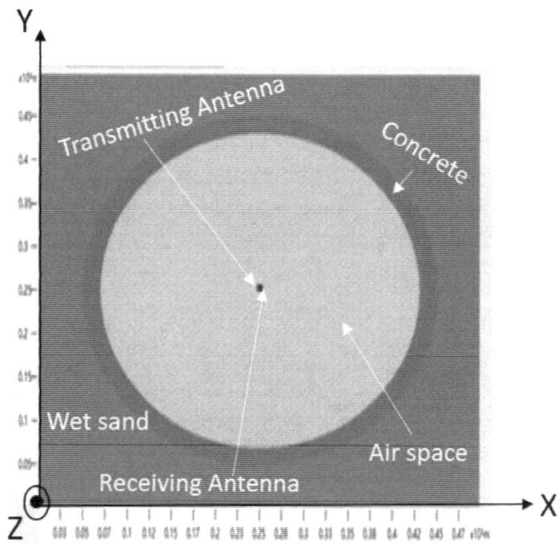

Figure 1. Observation system of GPRMax

$$\begin{bmatrix} E_{zz} & E_{xz} \\ E_{xz} & E_{xx} \end{bmatrix} \quad (1)$$

3. Monitoring technology for infrastructure structures

A. Concrete Wall Dielectric Constant Monitoring

For a radar source located at the center of a tunnel. Concrete walls in a tunnel will produce co-polarized reflection whose intensity will increase with the increase of moisture content inside the concrete due to decrease in its wave impedance. We monitored the degree of water seepage into tunnel wall by measuring the reflection coefficient of the wall's front surface. The dielectric constant of the wall was then calculated from the surface reflection using (1)-(9). The calibration was done by comparing measured reflection data to that obtained from a metallic wall. The radius of a circular

doi:10.1088/1742-6596/2887/1/012002

tunnel is 0.2m. The antenna is located at midpoint x=2.5m, y=2.5m. In order to remove the undesired antenna self-coupling, we also conducted background simulation in of antenna along in free space.

Figure 1 shows the geometry of concrete tunnel used for measurement simulations. The measured reflection coefficient of dry concrete surface was approximately 7dB. To study the effect of wet concrete surface on the reflection coefficient, we subsequently extracted the dielectric constant from the simulated reflection data.

By comparing the true dielectric constant of the tunnel to 7, 9, 10.3, 11.6, 13.08, 15.06, 16.65, 21.56, 24.22 and the extracted dielectric constants at the center frequency: 6.6, 8.4, 9.5, 11.3, 12.7, 14.6, 16.1, 20.6, 23.04, we observed good agreements between them, as shown in Figure2.

$$R = \frac{Z_2 - Z_1}{Z_2 + Z_1} \tag{2}$$

$$Z_1 = \sqrt{\frac{\mu_1}{\varepsilon_1}} \tag{3}$$

$$Z_2 = \sqrt{\frac{\mu_2}{\varepsilon_2}} \tag{4}$$

$$Z_1 = \sqrt{\frac{\mu_0}{\varepsilon_0 \varepsilon_{1r}}} \tag{5}$$

$$Z_2 = \sqrt{\frac{\mu_0}{\varepsilon_0 \varepsilon_{2r}}} \tag{6}$$

$$R = \frac{\sqrt{\varepsilon_{1r}} - \sqrt{\varepsilon_{2r}}}{\sqrt{\varepsilon_{2r}} + \sqrt{\varepsilon_{1r}}} \tag{7}$$

$$\varepsilon_{1r} = 1 \tag{8}$$

$$R = \frac{1 - \sqrt{\varepsilon_{2r}}}{\sqrt{\varepsilon_{2r}} + 1} \tag{9}$$

$$\varepsilon_{2r} = \left(\frac{1 - R}{1 + R} \right)^2 \tag{10}$$

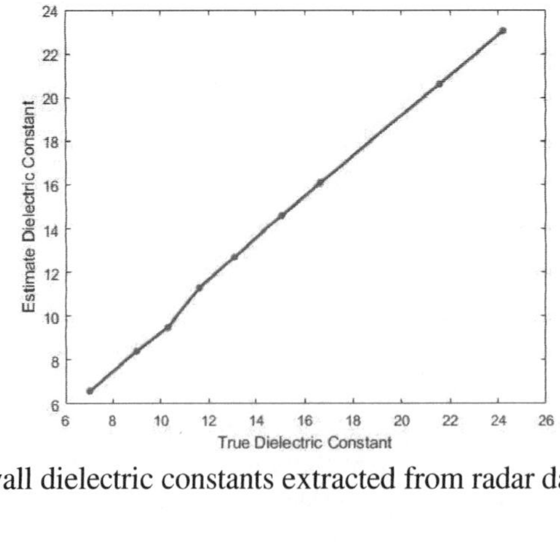

Figure 2. Tunnel wall dielectric constants extracted from radar data versus true values.

doi:10.1088/1742-6596/2887/1/012002

B. Tunnel Wall Displacement Monitoring

The inner surface of the tunnel wall can be determined from the time of flight of the first radar echo. Figure 3 compares the extracted inner surface radius of the Figure 1 tunnel for different tunnel radii with the same wall thickness and true radii to show that the radar can accurately monitor the wall displacement.

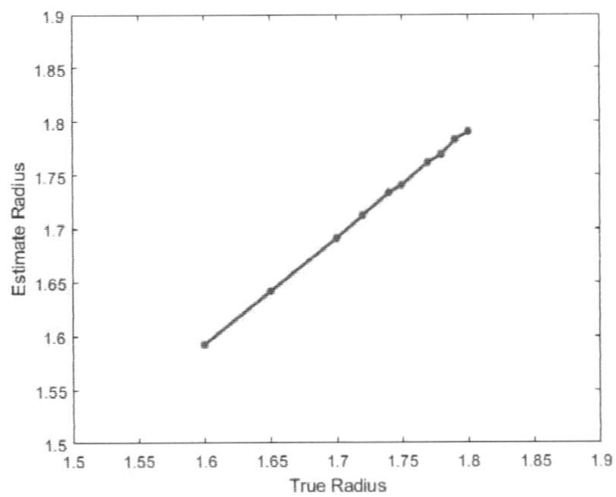

Figure 3. Tunnel radius (Figure 1) extracted from radar data verses true radius.

C. Tunnel Floor Waterlogging Monitoring

Figure 4 shows the simulation setup of water accumulation on the floor of a square concrete tunnel. The radar is able to detect the presence of such water accumulation when compared the data collected when there was no-water case. Using the time of the flight of the water surface reflection, radar is able to tell how high if the accumulated water on the floor by measuring the distance from the radar to the water surface for different water heights maybe from 1cm to 20cm, as demonstrated in Figure 5.

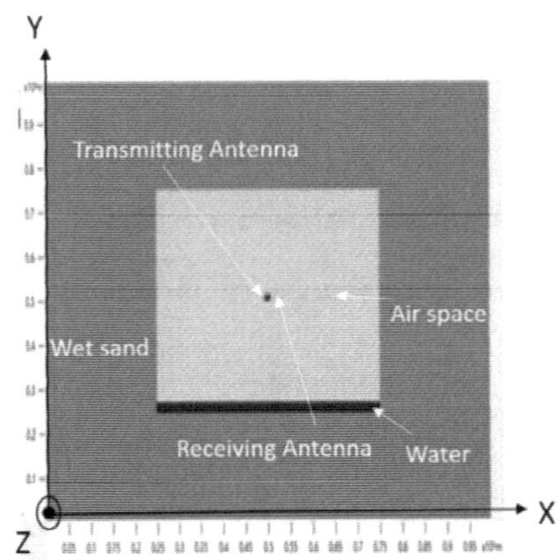

Figure 4. 2D GPRMax model of waterlogging in a square tunnel

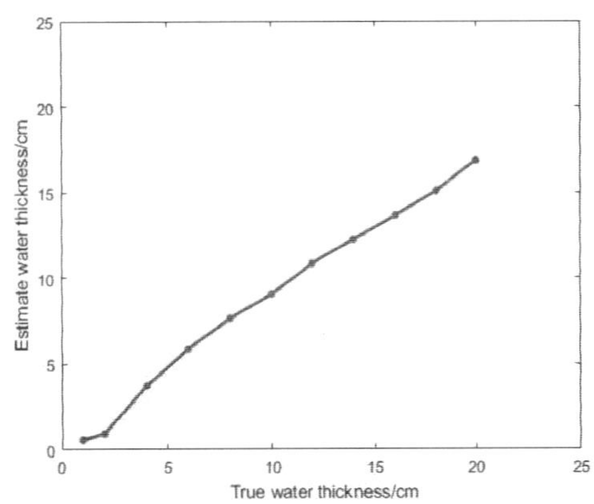

Figure 5. Estimated floor water depth versus true water depth.

D. Metal Pipe/Cable Displacement Monitoring

We also studied the scattering of a solid PEC cylinder near the ceiling of the tunnel with different diameters from 1cm, 2cm, 5 cm to 10cm. For each diameter case, the cylinder was placed 30cm below the tunnel ceiling and then change the height from 20cm to 40 cm at 2cm increments to see how accurate can radar track the position changes, as demonstrated in Figure 6. Solid cylinders must be used because EM waves cannot penetrate PEC. Cables or utility lines are usually suspended from the ceiling or sidewalls to avoid water on the ground and allow maintenance personnel to walk along the tunnel. The Simulate results as shown Figure 7.

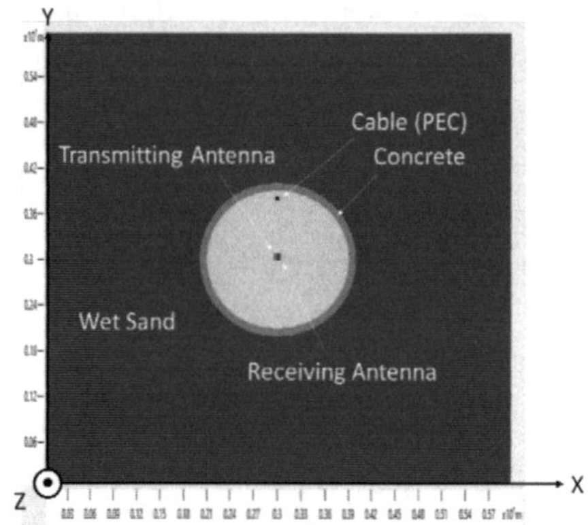

Figure 6. Metal Pipe/Cable Displacement Monitoring model

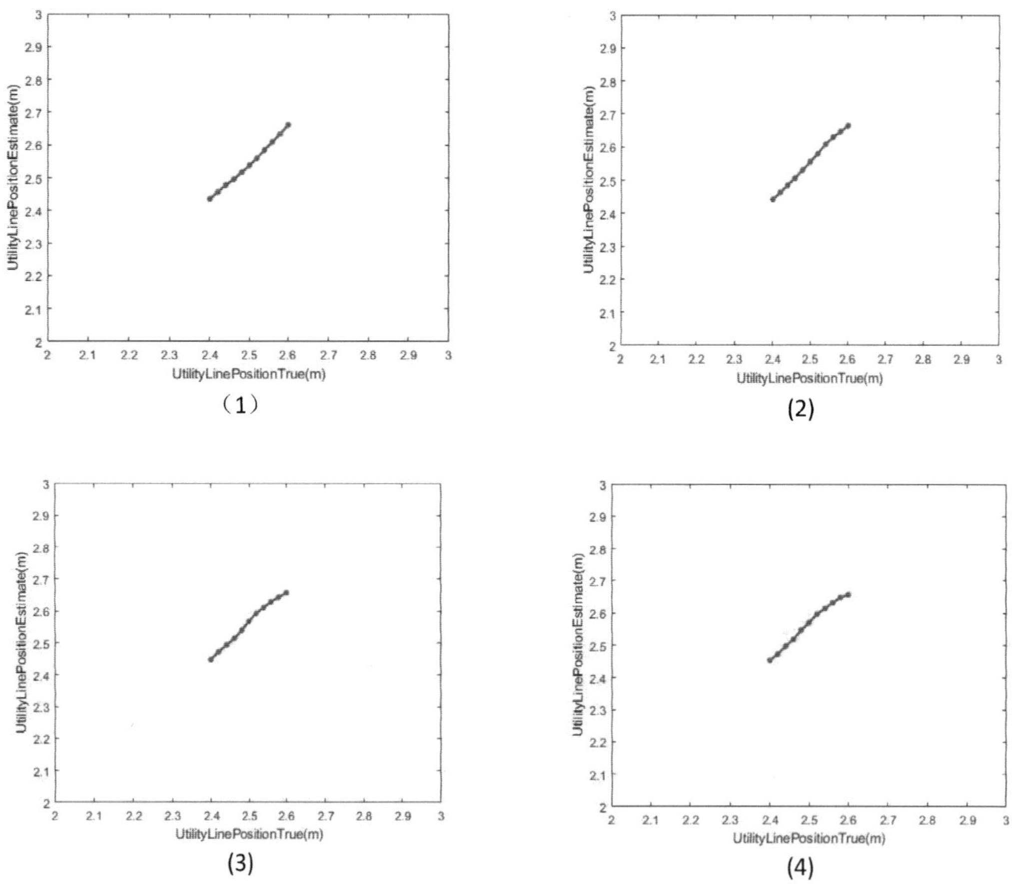

Figure 7. Metal pipe position estimation from radar data versus true position. (1) Pipe Radius = 10 cm (2) Pipe Radius = 5cm (3) Pipe Radius = 1cm (4) Pipe Radius = 0.5cm

4. Conclusion

A theoretical study of using 250 MHz pulsed radar for monitoring some key physical conditions of an underground tunnel with concrete walls. These conditions included tunnel radius, tunnel wall dielectric constant, tunnel floor water accumulation, and metal cable/wire position. These studied were conducted using 2D GPRMax Finite-Difference Time-Domain (FDTD) simulation software for efficient proof-of-concept simulations. The results of these study were presented and discussed in this paper. These results clearly showed that in-tunnel radar sensors can be used as an effective non-contacting method for monitoring underground tunnel conditions, and thus preventing disastrous accidents from happening.

5. Acknowledgment

This work was supported by Science and Technology Development Plan Project of Jilin Province, China (Grant No.20220203059SF)

References

[1] 2013 US National Research Council Report.
[2] Q.Zhen "Development and Utilization of Urban Underground Space" 4th International Conference on Energry Equipment Science and Engineering.pp1-3.
[3] Y.Luo,,C.Zhang, A. Hammad "Recent Development Of Multi-purpose Utility Tunnels In China",CSCE Annual Conference, June 12-15,2019.
[4] Isam Shahrour, Hanbing Bian, Xiongyao Xie and Zixin Zhang, "Use Of Smart Technology To Improve Management Of Utility Tunnels" Applied Sciences,pp.1-14,2020.
[5] Rui Hana,, Dan Shaob, and Xian Luc, "Maintenance Engineering For Urban Utility Tunnel using 3D Simulation", International Journal of Performability Engineering,Vol.16,No.7,pp.1118-1129.July 2020.
[6] Jung-Ryul Kim, Hyun-Suk Yoo, Soon-Wook Kwon,.et, "Integrated Tunnel Monitoring System Using Wireless Automated Data Collection Technology",The 25th International Symposium On Automation and Robotics in Construction, June 26-29,2008.
[7] Marco Domaneschi ,Sara Casciati,et. "Structural Health Monitoring Of In-service Tunnels", Int.J.Sustainable Materials and Structural Systems, Vol.4 , pp.268-291,2020.
[8] Tan Zhang lu,Zhang Chang lu, "Construction Of The Safety Management System For Urban Underground Business District With The Application Of IOT",2013 IEEE.

Journal of Physics: Conference Series **2887** (2024) 012003 doi:10.1088/1742-6596/2887/1/012003

A compact ultra-wideband circular bowtie antenna for ground penetrating radar

Shenzhao Lv[1], Chunlin Huang[1], Xiaoji Song[1*], Zhihua He[1], and Tao Liu[1]

[1] National University of Defense Technology, Changsha, China.

Abstract. A compact Ultra-Wideband (UWB) circular bowtie antenna for Ground Penetrating Radar (GPR) is presented in this paper. The bowtie antenna is composed of two circular arms with parallel slots at the edges to obtain UWB performance while maintaining compact in size ($0.31\lambda_{max} \times 0.19\lambda_{max} \times 0.16\lambda_{max}$). Each slot has a resistor loading at the edges to minimize the size of antenna. In addition, a microstrip balun is designed for impedance matching. The performance of the proposed antenna is investigated by simulations and the results show a UWB operating bandwidth ranging from 0.4 to 3.52 GHz, a relative bandwidth of 159%, a maximum gain of 2 dBi, and lower late-time ringing level compare to the basic antenna.

1. Introduction

Thanks to the ability of non-destructive detection capability and high efficiency, GPR has been used in various fields, such as geological exploration [1], architectural engineering [2], archaeology [3], and unexploded ordnance detection [4]. As key components of GPR systems, antennas have a significant impact on system performance. GPR antennas are desired to be low-frequency, wide-bandwidth, and have good time-domain characteristics to emit UWB impulses [5, 6]. In addition, lightweight and compact size are important for GPR systems, especially for those intended to be deployed on portable equipment.

There are different types of GPR antennas, including spiral antennas, resistor-loading dipoles, bowtie antennas, and Vivaldi antennas. The spiral antenna has a compact size and a wide bandwidth, but it is a dispersive antenna, which leads to serious signal distortion that cannot satisfy the time-domain requirements of the impulse GPR system [7]. Classic antennas, such as dipoles, have large dimensions that to large to used small GPR design [8, 9]. Many techniques have been adopted to reduce the antenna weight and size. In [10], a lightweight cage dipole was proposed, but it is huge in size. In [11], a folded structure and RC-loading was applied to reduce the size of the antenna. In [12], An UWB antipodal Vivaldi antenna is designed using triangle-shaped slots. In [13], V-shaped slots and resistors were added to broaden the low frequency.

In this paper, a miniaturization technique is applied to the design of a UWB bowtie antenna. Unlike common bowtie antenna, this design adopts circular arms to achieve a wider bandwidth. Parallel slots are designed on arms side with resistor loading at the edges of the arms to extend the current path. This design helps to reduce the dimension of the antenna while preserving its low-frequency characteristics.

* songxj@nudt.edu.cn

Furthermore, a balun is designed to convert unbalanced signals to balanced signals for feeding and impedance matching. The feasibility of the proposed design is demonstrated through computer simulation.

This paper is organized as follows. The design principles, antenna miniaturization techniques and configuration of the proposed antenna are provided Section II. The simulation results are presented in Section III. Conclusions are provided in Section IV.

2. Antenna Design

The design of the proposed antenna is shown in Figure.1. The antenna consists of a bowtie antenna and a microstrip balun. The size of the antenna measured $0.31\lambda_{max}\times0.19\lambda_{max}\times0.14\lambda_{max}$. All the parameters of the proposed antenna are listed in Table 1.

As shown in Figure.1(a), the bowtie antenna consists of two circular arms, each containing parallel slots and loading resistors on edge. All resistors have equal resistance values. The bowtie antenna size measured 230mm×140mm×2mm. The substrate of the bowtie antenna is FR-4 with a thickness of 2 mm, a relative permittivity of 4.4, and a loss tangent of 0.025.

As shown in Figure.1(b), we designed an exponentially gradual microstrip line to double parallel line balun to achieve impedance matching. The balun dimensions were 30mm×100mm×2mm, and the substrate was Teflon ZYF255DA, with a thickness of 1.52mm, a relative permittivity of 2.55, and a loss tangent of 0.0018.

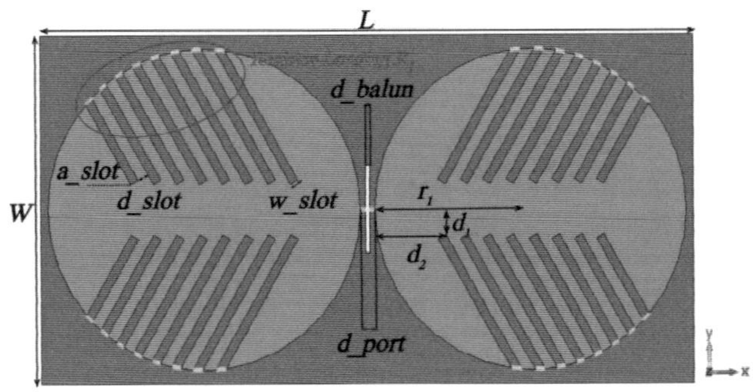

(a) Top view of the proposed antenna.

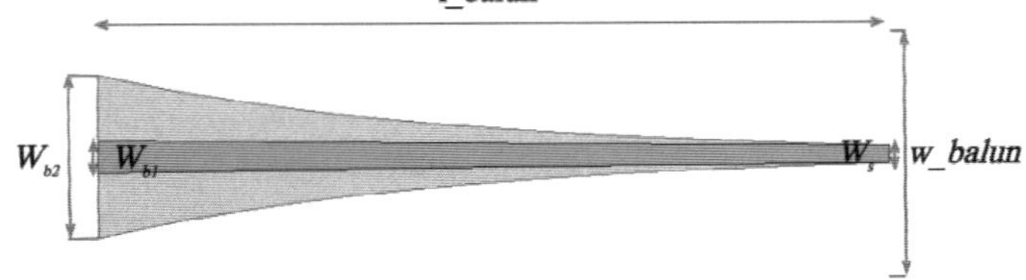

(b) Balun.

Figure 1. Configuration of the proposed bowtie and balun.

Table 1. The parameter of the proposed antenna.

Parameter	Value	Parameter	Value	Parameter	Value
L	230mm	d_1	12mm	*w_balun*	30mm
W	140mm	d_2	21.5mm	*h_balun*	1.52mm
r_1	55mm	*d_port*	5mm	*l_balun*	100mm
a_slot	30°	R_1	50Ω	W_s	2.2mm
w_slot	3mm	W_{b1}	4mm		
d_slot	5mm	W_{b2}	20mm		

2.1. Circle Bowtie

The arm of traditional bowtie antenna is usually either triangular or rectangular in shape, which leads to a limited operating impedance bandwidth. Figure.2 shows the impedance characteristics of the triangular-armed bowtie and circular-armed bowtie. The real and imaginary parts of the impedance of the circular arm are smoother than those of the triangular arm. Therefore, better impedance matching over operating bandwidth can be achieved.

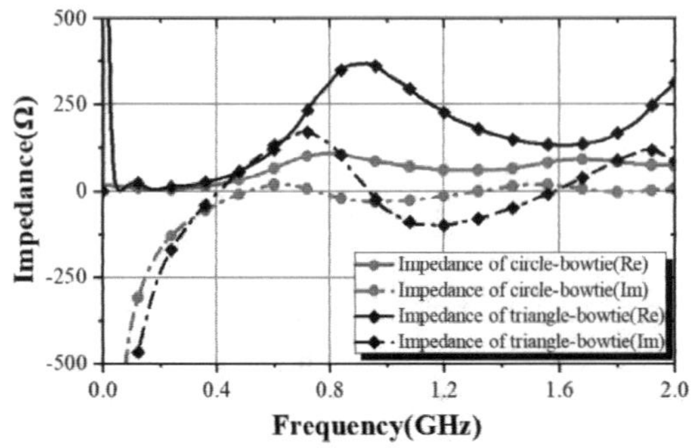

Figure 2. Impedance characteristics of triangle and circle bowtie.

2.2. Slots and Resistor-Loading

In this paper, a method of slots and resistor-loading is proposed. The slots extend the effective current length, and the resistor-loading eliminates oscillating currents. In order to demonstrate the effect of slots and resistor-loading, Figure.3 shows the surface current distribution with and without slots and resistor-loading at 0.5 GHz. Figure.3 shows that the slot structure effectively constrains the flow of current. Nevertheless, the closure of the slot structure results in current reflection within it, compromising the antenna's performance. Therefore, resistor-loading is applied to the edge of the slots to dissipate the reflected current.

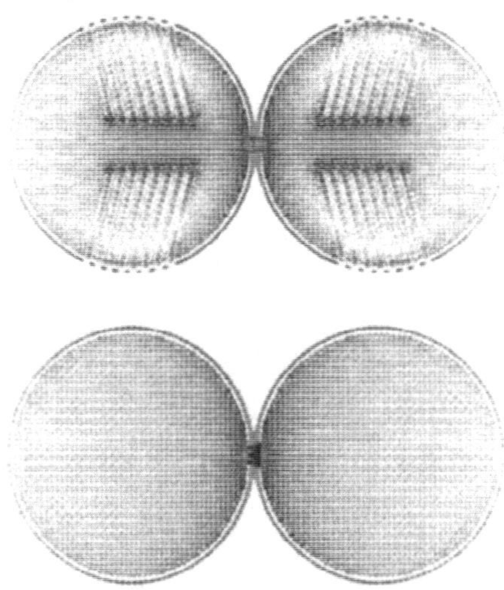

Figure 3. Surface current of with and without slots and resistor-loading.

3. Results And Discussion

In this section, $|S_{11}|$, VSWR, realized gain, far-field radiation patterns, and time-domain waveform are presented. The simulation experiments are conducted using CST, and the baluns in these experiments are identical.

3.1. $|S_{11}|$ *and VSWR*

Figure.4 and Figure.5 demonstrate the $|S_{11}|$ and VSWR of the antenna with and without slots and resistor-loading. Figure.4 demonstrate the current distribution at 0.5 GHz. The antenna with the loading exhibits a longer electrical length and a lowered cut-off frequency of 100 MHz. The proposed antenna's operating bandwidth ranging from 0.40 to 3.51 GHz, a relative bandwidth of 159% with the basic circle bowtie bandwidth ranges from 0.52 to 3.50 GHz, a relative bandwidth of 148%.

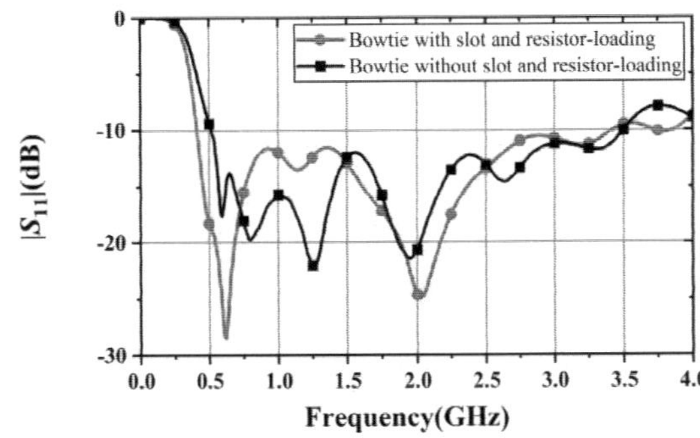

Figure 4. Simulated $|S_{11}|$ of the proposed antenna.

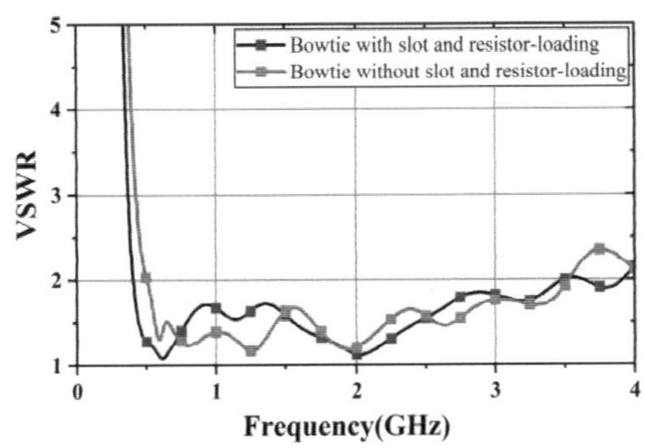

Figure 5. Simulated VSWR of the proposed antenna.

3.2. Realized Gain

Figure 6 illustrates the realized gain of the proposed antenna with and without slots and resistor-loading. The realized gain in the operating band range varies from -1 to 2 dBi. Nevertheless, the antenna without loading can achieve up to 4 dBi. The observed reduction in gain can be attributed to the presence of absorbing resistors, which have the effect of reducing the overall radiation efficiency of the antenna. At a frequency of 1.2 GHz, the reduction in radiation efficiency is approximately 40%.

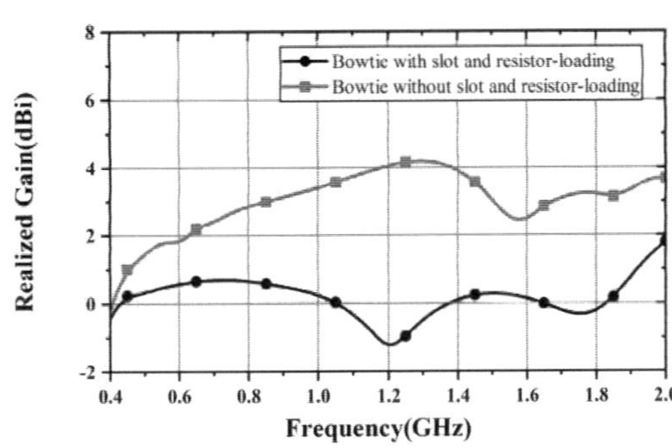

Figure 6. Simulated Realized Gain of the proposed antenna.

3.3. Far-field Pattern

The far-field pattern of the antenna at the frequencies of 400 MHz, 1.2 GHz, and 2 GHz is shown in Figure.7. The radiation patterns shown in Figure.7 are similar to the pattern of a dipole, exhibiting a primary radiation direction along the symmetry plane of the two arms. As illustrated in Figures 7(b) and 7(c), both patterns of the antenna exhibit slight distortion. The antenna employing the loading technique shows the main beam splitting at lower frequencies compared to the not employing. However, this kind of main beam splitting is usually difficult to avoid in UWB antenna design due to the wide range of frequency change.

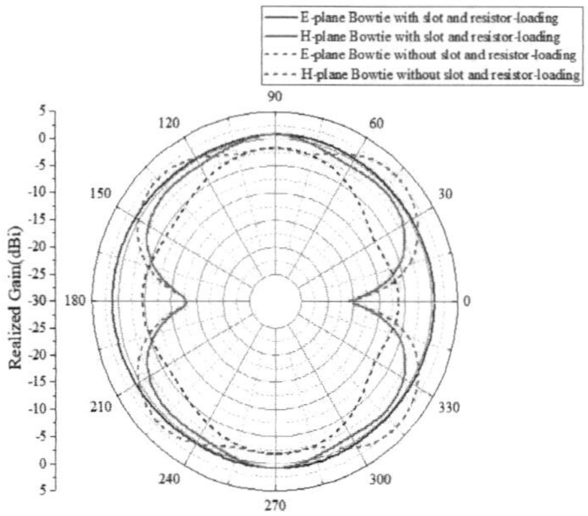

(a)0.4 GHz

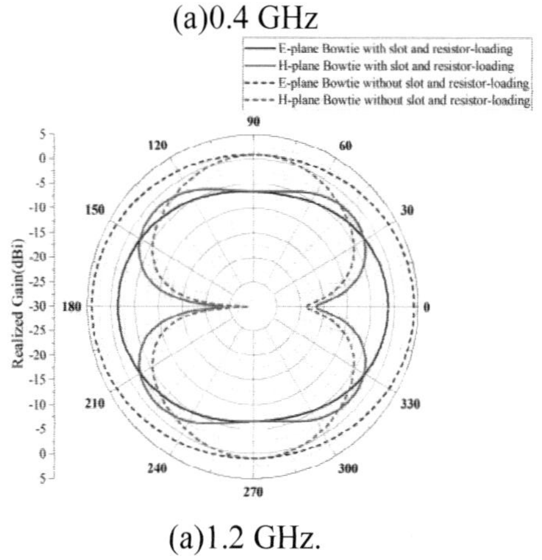

(a)1.2 GHz.

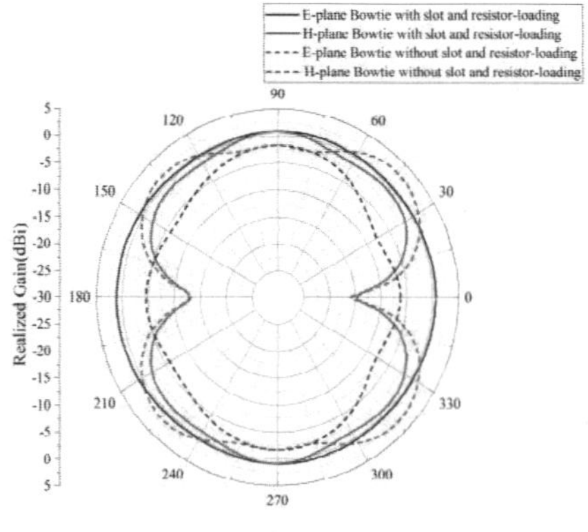

(a)2 GHz.

Figure 7. Far-Field Pattern..

3.4. Time-domain Waveform

In simulation experiments, the excitation waveform is a Gaussian pulse with a center frequency of 500 MHz, and the probe is placed 1 m directly in front of the antenna. As shown in Figure.8, in absence of resistor-loading, the closed structure of the narrow slot induces internal energy vibration, leading to serious late-time ringing effects that degrade the antenna's time-domain performance. Therefore, the resistor is loaded to reduce current reflections within the enclosed structure, reduces the reflection to a similar level and enhance the time-domain performance of the antenna.

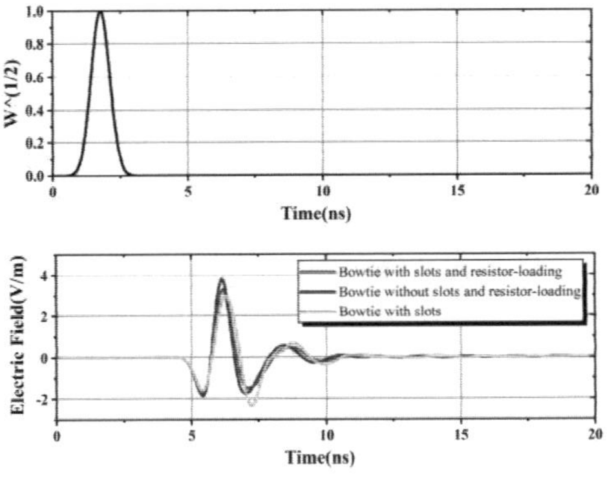

Figure 8. Time-Domin Waveform.

4. Conclusion

In this paper, we propose a novel compact UWB bowtie antenna operating in the frequency bandwidth of 0.40 to 3.52 GHz with a compact size of $0.31\lambda_{max}\times0.19\lambda_{max}\times0.16\lambda_{max}$. The design achieving UWB by incorporating a circular arm with slots and resistor-loading. The antenna's far-field pattern resembles that of a dipole, with radiation concentrated in the plane of symmetry of the antenna arm. The late-time ringing effect of the antenna is acceptable. However, the antenna gain is limited due to resistor-loading, resulting in a minimum gain of -1 dBi, maximum gain of 2 dBi. In future work, we aim to enhance the efficiency of the antenna by considering incorporating a back cavity to achieve higher gain.

5. References

[1] Chunyu Ding, Jing Li and Rong Hu 2023 Moon-Based Ground-Penetrating Radar Observation of the Latest Volcanic Activity at the Chang'E-4 Landing Site. *IEEE Transactions on Geoscience and Remote Sensing*, **61** 1.

[2] Yan Xu, Dongwen Hua, Jian Wang and Jinbin Li 2021 Research on application of Ground Penetrating Radar in road Inspection. *IOP conference series: Earth and environmental science*, **781(2)** 022019.

[3] Zayed Mohammad and Andrew M. Chrysler 2023 Airborne Reflector Based Ground Penetrating Radar for Environmental and Archaeological Studies. *IEEE Open Journal of Antennas and Propagation*, **4** 1.

[4] Yu, Yue and Chen, Chi-Chih 2023 Automatic subsurface unexploded ordnance detection using a wideband full-polarization ground penetrating radar and entropy-based polarimetric signatures. *Journal of Applied Geophysics*, **209** 104913

[5] Mohammed Mahmoud Mohanna, Esmat A. F. Abdallah, Hadia El-Hennawy and Magdy Ahmed Attia 2018 A novel high directive Willis-Sinha tapered slot antenna for GPR application in detecting landmine. *Progress In Electromagnetics Research C*, **80** 181.

[6] Liu, Lanbo and Liu, Sixin 2014 Remote Detection of Human Vital Sign With Stepped-Frequency Continuous Wave Radar. *IEEE Journal of Selected Topics in Applied Earth Observations and Remote Sensing*, **7(3)** 775.

[7] Mohamed A. Elmansouri and Dejan S. Filipovic 2012 Low-dispersion spiral antennas. *IEEE Transactions on Antennas and Propagation*, **60(12)** 5522.

[8] Guangyao Yang, Shengbo Ye, Yicai Ji, Xiaojuan Zhang and Guangyou Fang 2020 Radiation Enhancement of an Ultrawideband Unidirectional Folded Bowtie Antenna for GPR Applications. *IEEE Access*, **8** 182218.

[9] María García Fernández, Guillermo Álvarez Narciandi, Ana Arboleya, Carlos Vázquez Antuña, Fernando Las-Heras Andrés and Yuri Álvarez López 2021 Development of an Airborne-based GPR system for landmine and IED detection: antenna analysis and intercomparison. *IEEE Access*, **9** 1.

[10] Gaurav Kumar, Pratik Mevada, Soumyabrata C, Milind B. Mahajan 2022 Ultrawide band cage dipole antenna forultra high frequency band ground penetrating radar system. *International Journal of RF and Microwave Computer-Aided Engineering*, **32(6)** 23139.

[11] Guangyao Yang, Shengbo Ye, Yicai Ji, Xiaojuan Zhang and Guangyou Fang 2020 Radiation Enhancement of an Ultrawideband Unidirectional Folded Bowtie Antenna for GPR Applications. *IEEE Access,* **8** 182218.

[12] Li J., Liu L., Zeng, Z. , & Liu, F. . (2014). Advanced signal processing for vital sign extraction with applications in UWB radar detection of trapped victims in complex environments. *IEEE Journal of Selected Topics in Applied Earth Observations & Remote Sensing*, 7(3), 783-791.

[13] Li J., Zeng Z., Sun J., Liu F., (2012) Through-wall detection of human being's movement by UWB radar. *IEEE Geoscience and Remote Sensing Letter,* 9(6), 1079-1083.

Enhancing GPR-based tunnel lining delamination detection: An unsupervised domain-adaptive YOLO network

Yuanzheng Wang[1], Hui Qin[1,*], Jinbo Liu[2], Youjian Chen[2], Lipeng Wang[3], Hongyong Ren[3], Xianbiao Meng[4], Wenjiang Luo[4] and Lei Gao[4]

[1] School of Infrastructure Engineering, Dalian University of Technology, Dalian 116024, China
[2] China Railway 19th Bureau Group Co. Ltd., Beijing 10855, China
[3] China Railway 19th Bureau Group 5th Engineering Co. Ltd., Dalian 116199, China
[4] China Railway 19th Bureau Group Rail Transit Engineering Co. Ltd., Beijing 101300, China
[*] Corresponding Author: Hui Qin (hqin@dlut.edu.cn)

Abstract. The ground penetrating radar (GPR) is one of the most recommended methods for tunnel lining inspection, yet the interpreting the GPR data requires significant time and expertise. Recent attempts to automate this procedure using deep learning techniques have made significant progress, yet is still hindered by the need of extensive and accurately annotated data, which is challenging to gather in real scenarios. This paper introduces a domain-adaptive YOLO network, evolving from YOLOv7, which trains on FDTD data for source domain detection and real-world data for target domain adaption. The proposed network enhances the network's ability to generate generalized features across domains and hence improves the detection performance on real GPR data.

1. Introduction

Ground penetrating radar (GPR) is a preferred method for non-destructive tunnel lining defect detection. But interpreting GPR data is labor-intensive and complex. Thus, deep learning has been increasingly used for automated interpretation of GPR data. Wang et al. [1] developed a CNN for detecting tunnel defects of various orientations. Liu et al. [2] introduced a multi-task deep neural network for simultaneous defect identification, lining profiling, and hyperbola shape detection in GPR images. Additionally, Liu et al. [3] employed a YOLOv3 model for automated, fast, and reliable underground pipeline recognition and localization in GPR images.

Deep learning has made progress in GPR data interpretation, but challenges with data acquisition and quality hinder its broader application. Obtaining ample, accurately annotated real GPR data is resource-heavy, and datasets often have a "long-tail" problem, where certain types of tunnel defects are relatively rare in real GPR data, making it difficult for models to learn the typical features of these defects. To overcome these challenges, researchers use the finite-difference time-domain (FDTD) method for training data generation, simulating various defect types in a controlled setting. For instance, Qin et al. [4] used a CNN with FDTD-simulated data for data augmentation. However, simulated data's lack of real-world complexity and noise might limit model performance on actual GPR data, potentially affecting its accuracy and reliability when applying it to the real-world tasks.

Based on the challenges of tunnel lining defects detection in real-world applications using GPR, this paper proposes an object detection network. The proposed network achieves domain adaption between different domain datasets by utilizing domain adaption techniques to the original YOLOv7 object detection network. The network was trained on the annotated FDTD forward simulation GPR dataset and uses unlabeled real GPR data as the target for domain adaption. The training results of the network show that the proposed network exhibits higher precision and recall rate compared to the original YOLOv7 network.

2. Methodology

2.1. Domain adaptive YOLO

The proposed domain adaptive YOLO network, modified from the high-accuracy and fast YOLOv7 [5], incorporates domain classifiers into its body for feature adaption. This adaption helps the network extract generalized features from varied data sources, like FDTD modeling and real GPR data. As shown in Figure 1, the YOLO body captures multi-resolution features suitable for different object sizes, with feature maps at three stages: 80×80×256, 40×40×512, and 20×20×1024. These features are used for domain adaption, enhancing the network's ability to detect multi-scale objects accurately across different domains.

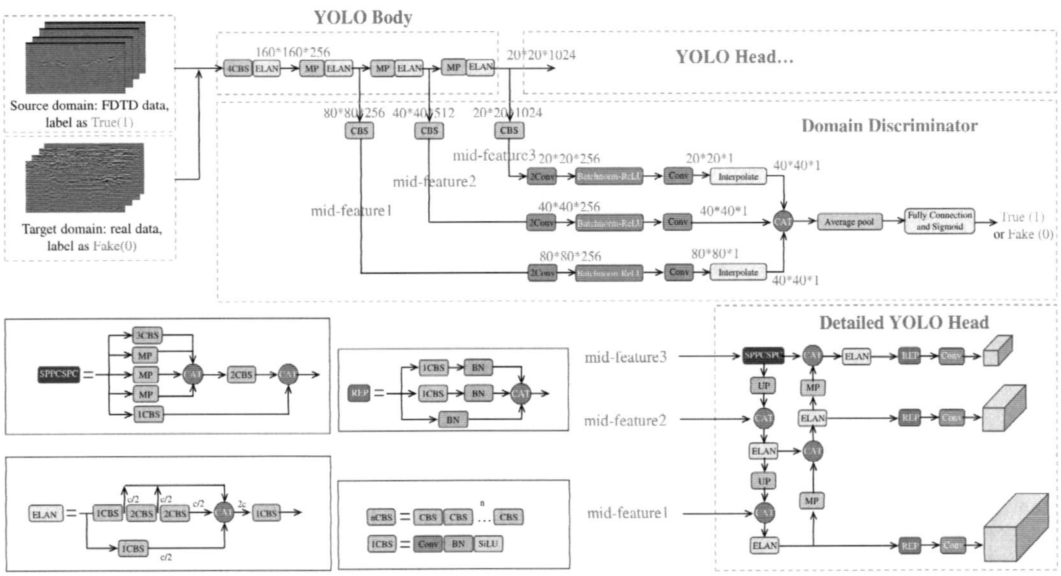

Figure 1. Architecture of the proposed network.

Domain adaptation in the network is achieved through an adversarial learning process, in which GPR data are split into source (labeled True) and target (labeled False) domains. During training, data from both domains are fed into the YOLO network, producing three mid-features that go into a domain discriminator. This discriminator, with multiple convolutional layers, predicts the domain class. The loss for domain classification is calculated using binary cross-entropy.

$$L_d = -\frac{1}{N}\sum_i y_i \ln(p_i) + (1-y_i)\ln(1-p_i) \tag{1}$$

where y_i is the ground truth domain label for the i-th training image, while the p_i is the predicted domain class label for i-th image. The objective of the domain discriminator is to accurately determine whether the images originate from the source or the target domain, that is, minimize this function. Concurrently, the YOLO body is trying to generate the indistinguishable mid-features for domain discriminator, that is, minimizing the following function:

$$L_g = -\frac{1}{N}\sum_i y_i \ln(1-p_i) + (1-y_i)\ln(p_i) \tag{2}$$

In addition to the domain adaption process in adversarial training, the network also needs to accurately detect object. Therefore, the detection loss L_{det} of the YOLO network is also considered. Only the source domain data were used to compute the detection loss because we assume the FDTD simulated GPR data are easier to be accurately labeled. As a result, the loss function of the adversarial loss L_g and the YOLO detection loss L_{det} together contribute the network's loss function.

$$L_{yolo} = \lambda L_g + L_{\text{det}} \tag{3}$$

where λ represents the proportion of the adversarial loss in the overall loss function.

2.2. Training data

As shown in Figure 2, the source domain data is from FDTD simulations of a three-layer model: undulating rock, tunnel lining, and an air layer for GPR operation. We simulated 400 GPR images for training, and collected real data using a MALA 800 MHz antenna over 1150 meters in a railway tunnel, extracting 17 images with tunnel defects.

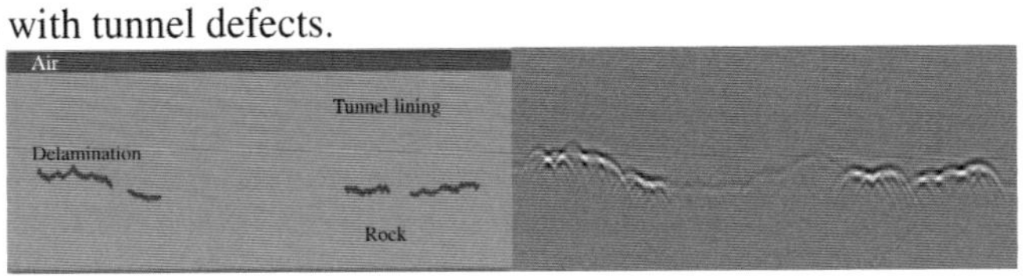

(a) FDTD model (b) Simulated GPR image

Figure 2. An example of FDTD simulated GPR image in source domain dataset.

3. Result

Our network utilizes the Adam optimizer with an initial learning rate of 0.001, starting from pretrained COCO dataset weights. It's trained on manually annotated GPR data from both domains, with the target domain data unannotated. While conducting domain training, we also trained the original YOLOv7 model using GPR data from the source domain and compared the training results with the proposed model to assess the domain adaption performance of the proposed network. The loss and the AP curves of the proposed network during the network training are shown in Figure 3.

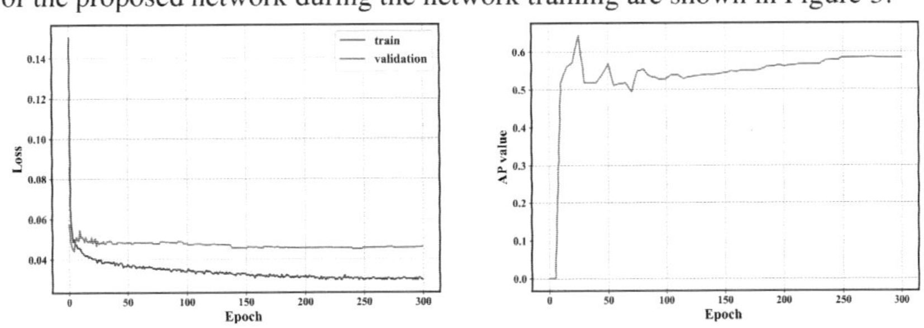

(a) (b)

Figure 3. The training process of the network. (a) loss curve, and (b) AP curve.

Among the 17 actual GPR images in the testing set, there are a total of 22 instances of tunnel lining delamination. The original YOLOv7 network and the proposed domain adaption network performed detection on these 17 images. The confidence score of detected targets higher than 0.2 are considered as positive samples. The samples count for true positive (TP), false positive (FP), false negative (FN), and the calculated precision, recall, and F1 score for both networks are listed in Table 1. The detection examples of the both networks are shown in Figure 4.

Table 1. detection results of both networks

model	TP	FP	FN	precision	recall	F1
Original YOLOv7	14	5	8	73.68%	63.63%	0.68
Domain adaptive YOLOv7	19	3	3	86.36%	86.36%	0.86

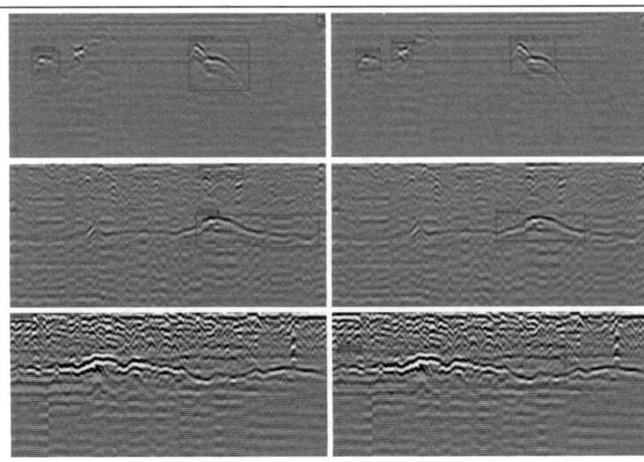

Figure 4. Detection example for (a) original YOLOv7, and (b) domain adaption YOLOv7.

As shown in Table 1 and Figure 4, the proposed network has shown an improvement of 22.73% in detection recall and a 12.68% increase in precision compared to the original YOLOv7 network. The proposed network accurately detects tunnel lining delamination while also having fewer false positive samples.

4. Conclusion

In this paper, to address the issue of poor detection performance caused by the differences between training data and testing data, we modified the original YOLOv7 detection network by applying the domain adaption techniques. We established an FDTD forward simulation dataset as the source domain data, and the field-collected GPR data as the target domain data. During training, the domain discriminator and the YOLO body together form an adversarial, where the domain discriminator encourages the YOLO body to generate domain-generalized features for both the source and target domains. The training results of the network show that the proposed network exhibits higher precision and recall rate compared to the original YOLOv7 network.

5. References

[1] Wang J *et al.* 2022 Arbitrarily-oriented tunnel lining defects detection from Ground Penetrating Radar images using deep Convolutional Neural networks *Automation in Construction* vol 133 p 104044

[2] Liu B, Zhang J, Lei M, Yang S and Wang Z 2023 Simultaneous tunnel defects and lining thickness identification based on multi-tasks deep neural network from ground penetrating radar images *Automation in Construction* vol 145 p 104633

[3] Liu H, Yue Y, Liu C, Spencer B F and Cui J 2023 Automatic recognition and localization of underground pipelines in GPR B-scans using a deep learning model *Tunnelling and Underground Space Technology* vol 134 p 104861

[4] Xiong H., Li J., Li Z., & Zhang Z., (2024). GPR-GAN: A ground-penetrating radar data generative adversarial network. *IEEE Transactions on Geoscience and Remote Sensing* 62, 1-14.

[5] Wang C Y, Bochkovskiy A and Liao H Y M 2023 YOLOv7: Trainable bag-of-freebies sets new state-of-the-art for real-time object detectors 2023 *Proc. of the IEEE/CVF Conf. on Computer Vision and Pattern Recognition (Vancouver)* vol 1 (The united states IEEE) p 7464

Three-dimensional random medium modelling for GPR based on soil fractals

Shilong Guo, Qi Lu*, Sixin Liu, Wenjing Liang and Kexin Liu

College of Geoexploration Science and Technology, Jilin University, Changchun 130062, China

*Email: luqi@jlu.edu.cn

Abstract. Ground penetrating radar (GPR) has a wide range of applications in soil detection. In this article, we model the three-dimensional (3-D) random medium in GPR based on soil fractal characteristics. We use a 3-D Fractal Brownian Motion (FBM) spectral density function to simulate non-uniform media and discuss the effects of Hurst exponent and scale coefficients on the modeling results. The Hurst exponent is an important indicator in FBM, which reflects the degree of disturbance in the medium. Soil water content (SWC) is a key factor affecting the heterogeneity of soil media, and the distribution of soil dielectric properties also depends on SWC. Therefore, we use SWC data to obtain the Hurst exponent through Rescaled-range (R/S) analysis. Finally, a 3-D random medium model is established based on actual SWC data, and GPR forward simulation is performed on the established model by the finite difference time domain (FDTD) method. The simulation results show that the random medium modeling method based on soil fractal characteristics can provide an effective modeling tool for GPR soil detection.

Keywords: Ground penetrating radar (GPR); soil fractals; three-dimensional random medium; Hurst exponent; Rescaled-range (R/S) analysis

1. Introduction

Ground Penetrating Radar (GPR) is a fast, non-destructive geophysical method to locate objects or interfaces in the soil through the electromagnetic properties of the soil[1], enabling the detection of soil water content, structure etc. Scholars often introduce various autocorrelation functions to build models when researching the theory of stochastic media, which are classified as exponential[2], Gaussian [3] , Von Karman[4], Gaussian elliptical autocorrelation function[5], exponential elliptical autocorrelation function[6] and mixed autocorrelation function[7].

In order to describe the fractal characteristics of soil, most of the previous research content is described by the

Journal of Physics: Conference Series **2887** (2024) 012005 doi:10.1088/1742-6596/2887/1/012005

parameters of soil particles such as particle size, mass and volume. However, the particle size of soil particles is too small for GPR resolution, and it is not accurate to apply the fractal dimension of soil particles to the modelling of GPR 3-D random medium model based on soil fractals. Here we choose the characteristic parameter of soil water content (SWC) as the basis of fractal, and apply the method of R/S analysis to statistically process the SWC data by introducing the Hurst exponent H for description. This is substituted into the statistically significant 3-D fractal Brownian motion spectral density function to simulate the non-uniformity of the actual medium, so as to establish a 3-D random medium model and carry out the GPR forward simulation. Finally, it is verified that the simulation of 3-D random medium in GPR based on soil fractals is reliable.

2. Principles of 3D random medium modelling

A non-uniform medium, i.e. a random medium, usually consists of two scales of non-uniformity, large and small. The large-scale part of the non-uniformity describes the average properties of the medium, i.e. the geological model in the traditional sense; the small-scale part of the non-uniformity is a random perturbation added to the basis of the geological model mentioned above.

In order to carry out the orthogonal simulation of the GPR stochastic medium model, in the stochastic medium, the dielectric constant of the medium at the spatial coordinate point (x, y, z) can be expressed by the following equation as

$$\varepsilon(x,y,z) = \varepsilon_0 + \delta\varepsilon(x,y,z) = \varepsilon_0(1 + \delta) \tag{1}$$

where ε_0 is the inhomogeneous medium parameter for the large-scale part; and $\delta\varepsilon$ is the inhomogeneous perturbation quantity for the small-scale part.

In this paper, we use a statistically significant 3-D fractal Brownian motion spectral density function to model the inhomogeneity of real media, where the anisotropic spectral density function is usually expressed as

$$s(k_x,k_y,k_z) = \frac{s_0}{(a_x^2k_x^2 + a_y^2k_y^2 + a_z^2k_z^2)^{H+1}} \tag{2}$$

where, k_x, k_y, k_z, corresponds to the angular frequency in the x, y and z directions, respectively; s_0 is a normalisation constant; a_x、 a_y、 a_z , corresponds to the scale factor in the x, y and z directions, respectively; and H is the Hurst exponent. A random perturbation of size $N \times N \times N$ is established, which can be computed and found by the 3-D stochastic discrete Fourier inverse transform formula:

$$\delta(x,y,z) = \sum_{l=0}^{N-1}\sum_{m=0}^{N-1}\sum_{n=0}^{N-1} \sigma_{l,m,n} e^{2\pi i(lx+my+nz)} \tag{3}$$

where, $x,y,z = 0,\frac{1}{N},\frac{2}{N}\ldots\ldots\frac{N-1}{N}$, and $\sigma_{l,m,n}$ is satisfied

$$\sigma_{l,m,n} = r_{l,m,n}e^{i\theta_{l,m,n}} \tag{4}$$

$$r_{l,m,n}^2 = \frac{1}{(a_x^2l^2 + a_y^2m^2 + a_z^2n^2)^{H+1}} \tag{5}$$

Combining equations (1) and (3) gives the dielectric constant function of the random medium model, which leads to the random medium model as

$$\varepsilon(x,y,z) = \varepsilon_0(1 + \delta(x,y,z)) \tag{6}$$

where ε_0 is the background dielectric constant.

3. Determination of the Hurst exponent

In order to characterise the fractal features of the soil, we apply R/S analysis of SWC in this article.The SWC data $\{X_i = 1,2,3,4, \ldots, N\}$ is divided into m sub-sequences of length n (m=N/n is an integer), and each sub-sequence is now calculated:

(1)Calculate the polar deviation of the subsequence R_m

(2)Calculate the standard deviation of the subsequence S_m

(3)The rescaled polar deviation of each subsequence is defined as

$$(R/S)_m = R_m/S_m \tag{7}$$

(4)Cycling through step (1) to step (3) for each subsequence yields a new rescaled extreme deviation sequence $(R/S)_m$ for which the mean is computed $(R/S)_n$

(5)Increase the length n of the subsequence and continue the loop of operations step (1) to step (3) until n=N/2, which gives another new sequence $(R/S)_n$

Derive the relationship between the polar deviation of the cumulative deviation of the series and the power of the measured data

$$(R/S)_n = C\tau^H \tag{8}$$

Curve fitting by taking the logarithm of both sides of the above equation, the absolute value of H is the Hurst exponent.

$$\log(R/S)_n = H\log\tau + \log C \tag{9}$$

The SWC data quoted in this paper were collected from a saline soil area, and the soil texture at the two locations was loamy sandy soil and silt loamy soil, respectively, and the R/S analyses of the two areas were carried out,

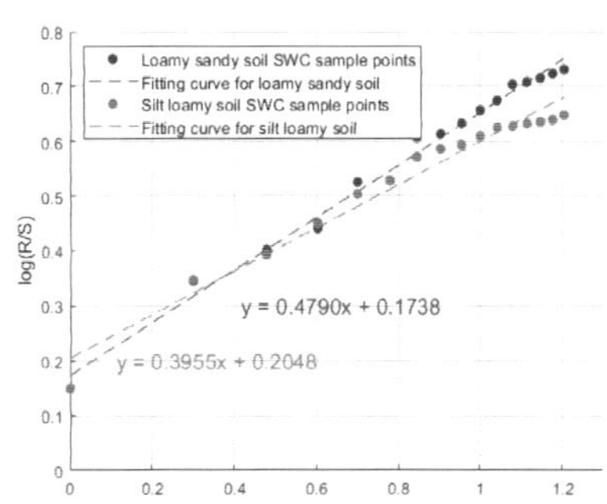

(a)

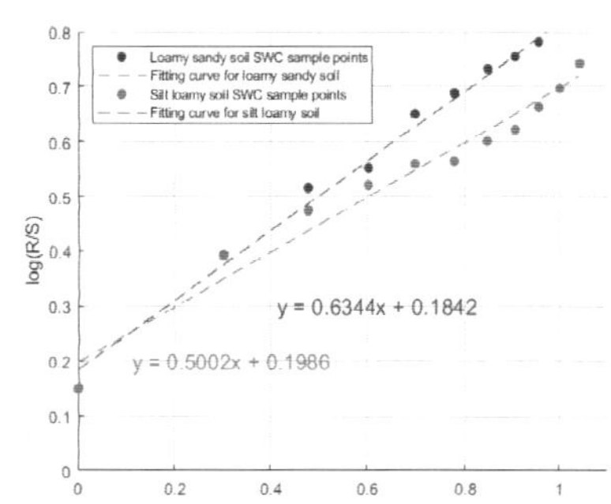

(b)

Figure 1. At sampling points N = 700 (a) and N = 1050 (b) , the results of R/S analyses of SWC in loamy sand and silt loam.

Due to the different number of sampling points, we obtained different numbers of fitted points when rescaling the range. However, when comparing the SWC Hurst exponent of two types of soil textures, we found that they still conform to the same magnitude relationship.

4. Establishment of 3-D random medium model and GPR forward simulation

Here we discuss the different characteristics of the parameters in the principle when modelling a 3-D random medium. Here it is assumed that the model is built with a size of 201 × 201 × 201 (total length of 3 m), and that a 3-D cone function is introduced to suppress the error. For the purpose of the discussion, we change only the value of H and the value of a_x、 a_y、 a_z, respectively, and the rest of the background conditions are the same.

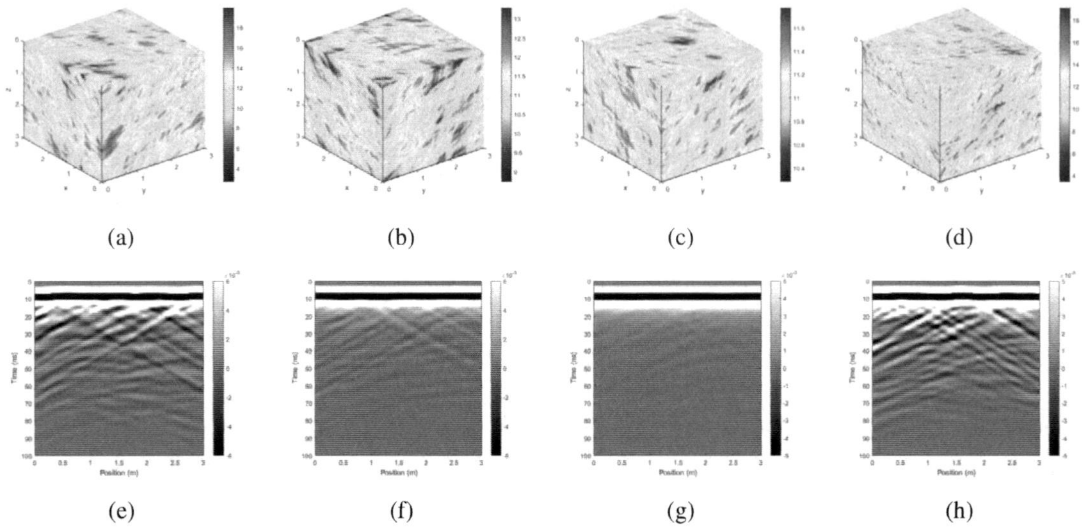

Figure 2. 3-D random medium model with a_x=a_y=a_z=10 (a) , 20 (b) , 30 (c) at H=0.9 and a_x=a_y=a_z=30 at H=0.5 (d) , and its forward simulation results of 100MHz GPR for the profile where the y-axis is located.

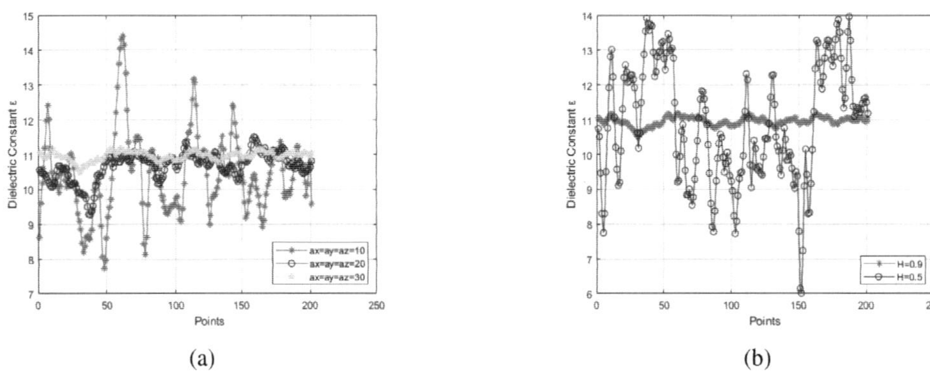

Figure 3. The vertical data distribution at the same location with different scale factors in Fig. 2(a)(b)(c), and the the vertical data distribution at the same location with different Hurst exponent in Fig. 2(c)(d).

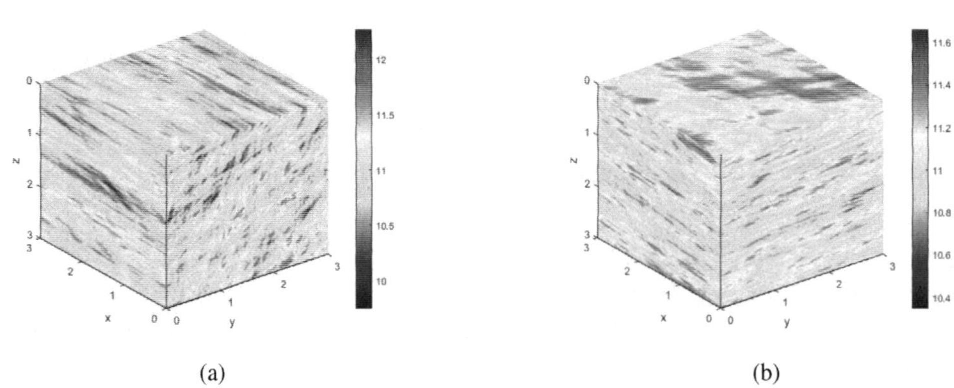

Figure 4. 3-D random medium model at a_x=50, a_y=a_z = 10 (a) and a_x=a_y=50, a_z = 10 (b).

According to Figures 2(a)(b)(c) and Figure 3(a), it can be observed that as the scale factor increases, the range of the model's dielectric constant gradually decreases, and the level of disturbance decreases. According to Figures 2(c)(d) and Figure 3(b), as the Hurst exponent decreases, the smoothness of the model decreases, becoming rougher. From Fig. 4, it can be seen that delamination occurs in the x direction as a_x increases, while delamination occurs in both the x and y directions as a_x and a_y increase simultaneously.

In the FDTD forward simulation calculations, the center frequency of the transmitting antenna is set to 100 MHz and 500 MHz, the model size is 201 × 201×201, the size of the mesh is 0.03 m, the time window is 100 ns, and collect 131 traces of data. The Hurst exponent of the model is 0.9 and the scale factor is a_x=a_y = 50 and a_z =10, after interpolation at different grid sizes the following model is obtained. Figure 5 shows the results of the forward simulation of the model .

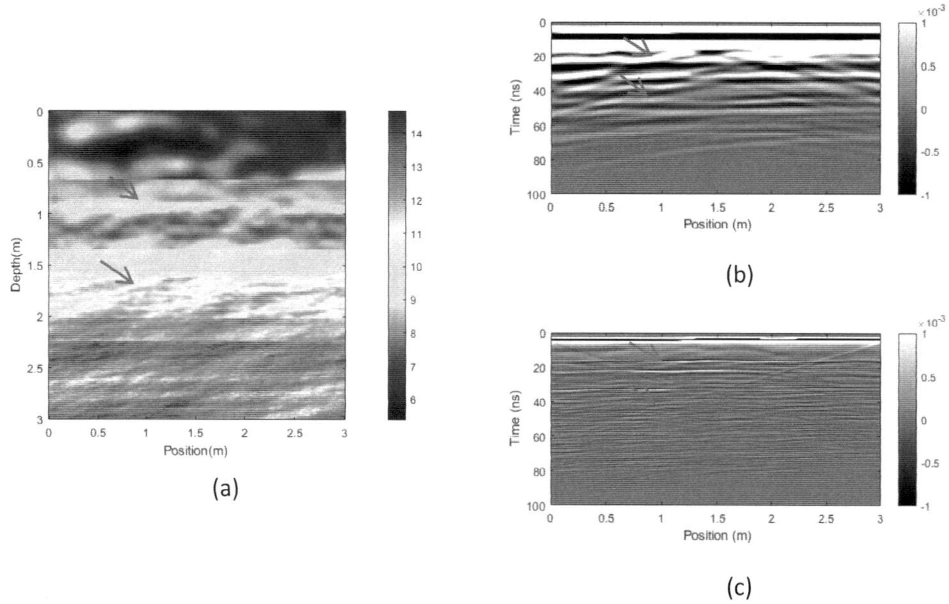

Figure 5. Forward simulation results of the random medium model at 100 MHz and 500 MHz GPR frequencies.

5. Conclusions

In this paper, we simulate a 3-D random medium in GPR based on soil fractals and give the modelling procedure of the random medium model. In constructing the random perturbation, we used the spectral density function of fractal Brownian motion, the parameters used are the scale coefficients in different directions and the Hurst exponent H, and calculated the SWC as its own attribute by using the method of R/S analysis, and got the value of the Hurst exponent H, so as to construct the 3-D random medium model based on the soil fractal. Finally, the GPR forward simulation was carried out, and the results of the forward simulation showed that the application of soil fractal to establish the stochastic medium model has certain practical significance for the application of GPR in agriculture.

Reference

[1] Zeng Z, Liu S, Feng X et al 2010 Ground Penetrating Radar Theroy and Application Beijing: Publishing

House of Electronics Industry

[2] Ikelle, L. T., Yung, S. K., & Daube, F. 1993 2-D random media with ellipsoidal autocorrelation functions. *Geophysics*, **58**(9), 1359-1372.

[3] Ergintav, S., & Canitez, N. 1997 Modeling of multi-scale media in discrete form.

[4] Korn, M. 1993 Seismic waves in random media. *Journal of applied geophysics*, **29**(3-4), 247-269.

[5] GUO ShiLi, YUE GuangHua, CAI WenCai, TIAN PengFei, LIU LanBo. Study on transmission characteristics of GPR in multiphase discrete random medium[J]. *Progress in Geophysics*, 2021, 36(1): 351-360.

[6] GAO Lijun,YANG Jianli,LI Jun,BAO Xiangsheng. Building an equivalent fracture model and performing forward. *Geophysical Prospecting for Petroleum*. 2020, 59(3): 404-408

[7] WANG Baoli, LIN Ying, ZHANG Guangzhi, YIN Xingyao. Study on seismic stochastic inversion method based on characteristic parameters of inhomogeneous media[J]. *Oil Geophysical Prospecting*, 2021, 56(6): 1301-1310.

Acknowledgement

This work was funded by the Natural Science Foundation of China (Grant 42274190).

Effects of Measuring Angles on Detection of Millimeter-Scale Fractures with Fluid Flow using Full-Polarimetric GPR

Zejun Dong[1,2], Xuan Feng[1,2,*], Haoqiu Zhou[1,2], Cai Liu[1,2]

[1] College of Geo-Exploration Science and Technology, Jilin University, Changchun 130026, China.
[2] Key Laboratory of Applied Geophysics, Ministry of Natural Resources, Jilin University, Changchun 130026, China
*Correspondence: fengxuan@jlu.edu.cn

Abstract. The fractures with fluid flow will lead to the damage of rock cave artifacts and the detection of millimeter-scale fractures is a challenge in artifact conservation. The fluid flow in fractures and the infiltration in surrounding rocks based on hydrological modeling was simulated first. Based on the results, electric permittivity and conductivity models at different time lapse were established, and a series of models with different rotation angles were generated through coordinate rotation for full-polarimetric ground penetrating radar (FP-GPR) simulations. Subsequently, the relationships between H, α, P_s, P_d, and P_v parameters and measurement angles and infiltration time are analyzed. The results indicate that α, P_s and P_d have good variation consistency with infiltration time, while H and P_v do not show a significant relationship with the infiltration time. In addition, the analysis results of the five parameters all indicate that the measurement angle has a significant impact on the parameters at the beginning of infiltration, which is related to the frequency of the adopted electromagnetic waves. The results provide important theoretical support and technical guidance for the use of time-lapse FP-GPR to detect fractures with fluid flow in field cases.

1. Introduction

Millimeter-scale fractures are often difficult to detect due to their small aperture, which exceeds the resolution limit of most feasible detection schemes using ground penetrating radar (GPR). In recent years, full-polarimetric ground penetrating radar (FP-GPR) was applied to characterize underground structures[1], providing potential possibilities for identifying microcracks. FP-GPR uses both of horizontal (H) and vertical (V) polarized antennas to obtain the four polarization data (HH, HV, VH, and VV) [2],[3]. Now, FP-GPR technology can identify features that have strong signals on GPR profiles, which, however, requires that the aperture of the features to be within the resolution range of the used FP-GPR and the features with sizes out of this range cannot be detected. However, when the features are filled with fluid like water, the fluid will also infiltrate into the surrounding rocks and form a seepage zone [4],[5]. As a result, the received FP-GPR responses are no longer comes only from individual fractures, but mostly from the volume effect of the seepage zones, which provides new way to identify small fractures using FP-GPR[6].

In order to confirm whether the signals on the GPR profiles are caused by water flow and seepage zones, this paper adopts the method of time-lapse monitoring to achieve this goal[7]. Since the change during the measurements are only the fluid flow, by comparing GPR profiles measured at different time, the place where the GPR profiles change can be determined as fractures. Time-lapse technology has been applied in lots of geophysical technologies and time-lapse GPR has been reported to be feasible in monitoring groundwater flow [8]. However, the time-lapse FP-GPR method has not yet been applied to monitoring fluid flow.

This study simulated the flow of water in fractures and the infiltration process in surrounding rock by solving the Richards' equation[9]. A series of models of electric permittivity (EP) and electrical conductivity (EC) are subsequently established according to the water content models. In addition, EP and EC models were generated at different rotation angles through coordinate rotation equations. These models were used for FP-GPR forward simulation. For simulated data, we use Freeman decomposition to analyze the time-lapse FP-GPR data at different measurement angles, and obtained the relationship between different polarization characteristics and water content changes, as well as the influence of measurement angles on the results. This study provides theoretical guidance for using full polarization ground penetrating radar to detect seepage microcracks.

2. Methodology

This section introduces the methods used in this article, including hydrological simulation methods for obtaining changes in water content distribution in fracture and seepage areas, as well as methods for establishing models from different measurement angles.

2.1. Hydrological Modeling

In order to simulate the fluid flow and infiltration in fracture and seepage zones, this paper calculates the flow of water by solving the Richards equation[9]:

$$\frac{\partial \theta_w(h)}{\partial t} = \nabla \cdot \left[K(\theta_w) \nabla h \right] - \frac{\partial K(\theta_w)}{\partial z} \tag{1}$$

where t denotes time [min]. θ_w represents the water content. K refers to the hydraulic conductivity [cm/min]. z is the positive upward spatial coordinate [cm]. h is the pressure head [cm]. Besides, θ_w(h) represents the water retention function and it is usually described by the van Genuchten model as follows[10]:

$$\theta_w(h) = \begin{cases} \theta_r + \dfrac{\theta_s - \theta_r}{(1+|\kappa h|^n)^m} & h<0 \\ \theta_s \quad h \geq 0 \end{cases} \tag{2}$$

where θr and θs represent the residual water content and the saturated water content, respectively. κ refers to the inverse of the air-entry value [cm^{-1}]. n is used to describe the pore-size distribution and m satisfies m=1−1/n.

By solving Richards' equation with the van Genuchten model, the fluid flow and infiltration can be simulated, and the distributions of water content at different time points can be obtained. Subsequently, the commonly applied petrophysical relationship TOPP equation can be used to transform θ_w into EP, as follows[11]:

$$\varepsilon_r = 3.03 + 9.3\theta_w + 146\theta_w^2 - 76.7\theta_w^3 \tag{3}$$

In addition, the water content θ_w can also be correlated with EC through Rhoades et al.'s model[12]:

$$\sigma = \left(c\theta_w^2 + d\theta_w\right)\sigma_w + \sigma_s \tag{4}$$

where c and d are two constants usually set to 1.85 and 3.85×10^{-2}. σ_w and σ_s represent the bulk ECs of the solid media and the water, which are usually set to 0.075 S/m and 5.89×10-4 S/m, respectively.

In order to establish a real dielectric model for FP-GPR simulation, the solution is achieved in the Comsol software according to the properties of the sandstone. The constructed 3D EP and EC models are shown in first row in Fig.1. The EP of the background is 3 and the black curves in the figures are smooth fractures with an aperture of 5 mm.

2.2. Establishment of models for different measurement angles

In order to study the influence of different measurement angles on the analysis results of FP-GPR data, this paper uses a rotation model to generate multiple models with different rotation angles, and conducts forward simulations with the same fixed antenna. The rotation of the model is achieved through the coordinate rotation matrix method[13]. When a coordinate point (x_1, y_1, z_1) is rotated by an angle of φ around the z-axis, its coordinates (x_2, y_2, z_2) can be expressed as follows:.

$$\begin{bmatrix} x_2 \\ y_2 \\ z_2 \end{bmatrix} = \begin{bmatrix} \cos\varphi & \sin\varphi & 0 \\ -\sin\varphi & \cos\varphi & 0 \\ 0 & 0 & 1 \end{bmatrix} \begin{bmatrix} x_1 \\ y_1 \\ z_1 \end{bmatrix} \tag{5}$$

So the coordinates before rotation can be expressed as follows:

$$\begin{bmatrix} x_1 \\ y_1 \\ z_1 \end{bmatrix} = \begin{bmatrix} \cos\varphi & \sin\varphi & 0 \\ -\sin\varphi & \cos\varphi & 0 \\ 0 & 0 & 1 \end{bmatrix}^{-1} \begin{bmatrix} x_2 \\ y_2 \\ z_2 \end{bmatrix} \tag{6}$$

Based on this method, it is possible to generate the model after it has been rotated by an angle of θ , f. First, an original model and an empty model are created as the rotated model. By using (6), each coordinate point (x_2, y_2, z_2) in the rotated model can be calculated to correspond to the coordinate (x_1, y_1, z_1) in the original model, and then the electrical parameters such as the EP at the original model (x_1, y_1, z_1) are assigned to the rotated model at the rotated model (x_2, y_2, z_2) to generate the rotated model. It should be noted that in order to avoid the calculated pre-rotation coordinates (x_1, y_1, z_1) exceeding the coordinate range of the original model, for the square model, the size of the side length of the original model should be at least $2^{1/2}$ times of that of the post-rotation model, i.e., the length of the sides of the original model should be greater than or equal to the lengths of diagonal lines on the corresponding faces of the post-rotation model.

During the model generation process, the rotation angle was set from 0° to 90° with an interval of 5°, so 19 angular models were generated, and for the 6 models in Fig. 1, a total of 6 × 19 = 114 models were generated in this paper. Due to the large number of models, only the 0°, 30°, 60°, and 90° models are shown in Fig. 1.

3. FP-GPR Simulations and Data Analyses

3.1. FP-GPR Simulations

The simulated results are shown in Fig.2, the 18 radargrams are composed of simulated single channel signals obtained from different measurement angles, so the horizontal axis of the images is the measurement angle. (a) - (f) correspond to the model simulation results of the six seepage moments in Fig.1, representing the time-lapse FP-GPR measurement results. From the radargrams, it can be observed that at the first and second time points, the amplitude of the reflected wave caused by seepage will undergo drastic changes with the measurement angle.

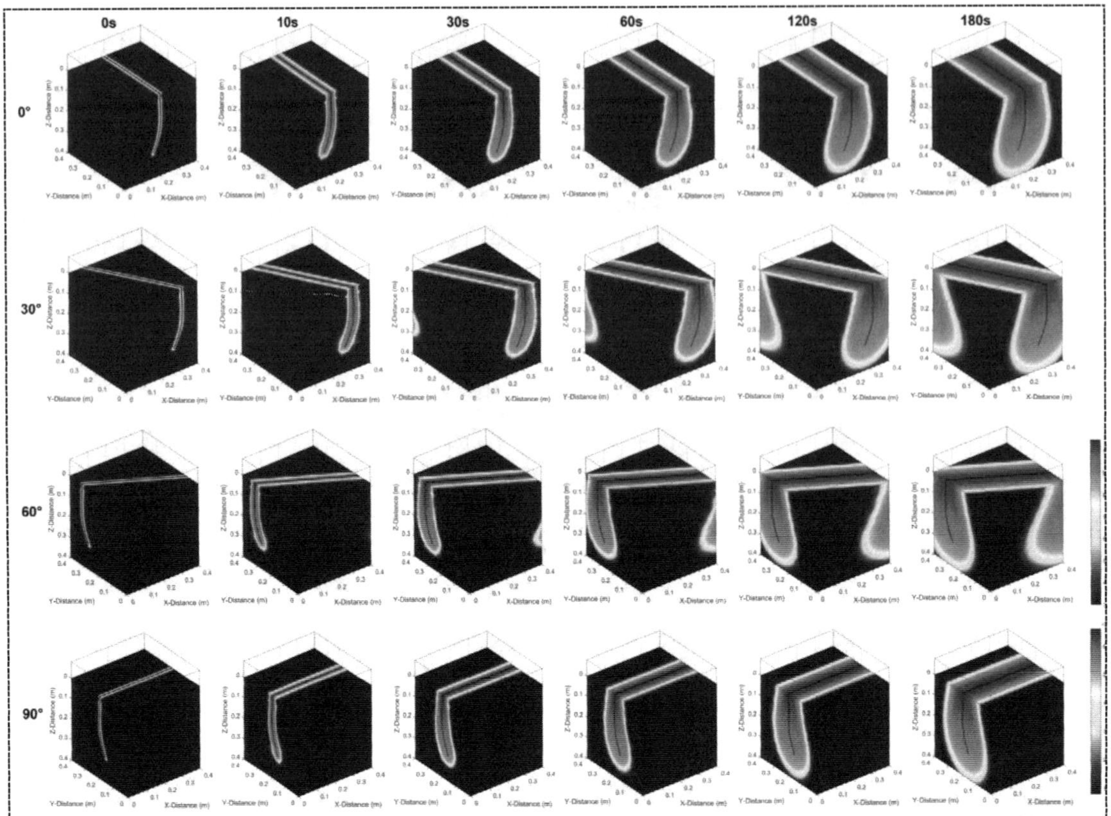

Figure 1. The 3D EP and EC models based on the water content distributions obtained by hydrological modelings and corresponding models at 30°, 60°, 90° measurement angles.

3.2. Data Analyses

In order to analyze the relationship between water content and time-lapse FP-GPR data as well as measurement angles, we performed Freeman decomposition[14],[15] on the reflected wave signal in Figure 2, and obtained P_s, P_d, and P_v characteristics as shown in Fig.3.

Fig.3 (a) - (c) show the variation of the three parameters P_s, P_d, and P_v with the measurement angle, and the curves of different colors represent six different seepage moments; Fig.3 (d) - (f) show the variation of P_s, P_d, and P_v parameters with infiltration time, and the curves of different colors represent the results at different measurement angles. The results indicate that the three parameters are not significantly affected by the measurement angle during 0s of seepage; When the seepage reaches 10s and 30s, the three parameters change dramatically with the change of measurement angle; As the seepage continues, the influence of the measurement angle on the results gradually weakens. In order to analyze the variation of the three parameters with seepage time more clearly, this paper also selected the variation curves of P_s, P_d, and P_v parameters with seepage time at two measurement angles of 0° and 90°, and presented them in Fig.3 (g) - (i). The results show that the curves obtained at different measurement angles vary between the two curves of 0° and 90°. In addition, it can be observed that P_s gradually decreases with increasing infiltration time, P_d gradually increases with increasing infiltration time, and P_v does not have a good linear relationship with infiltration time when the measurement angle is fixed.

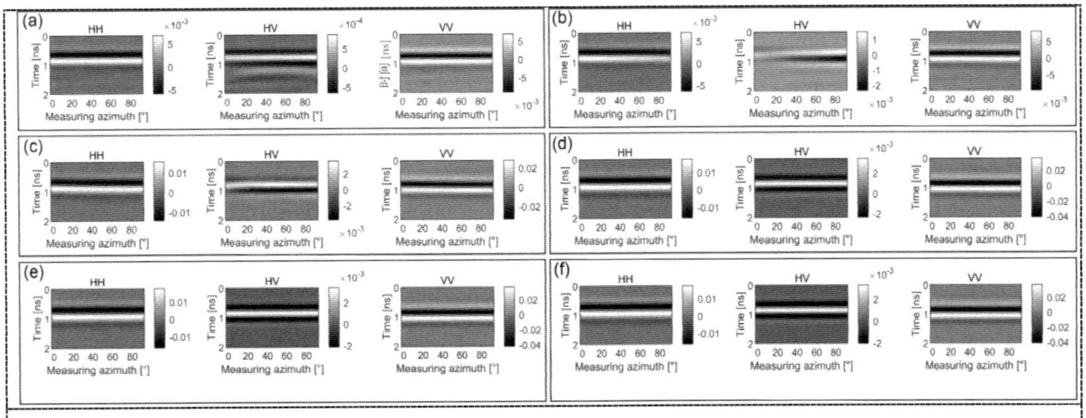

Figure 2. The FP-GPR simulation results based on the models in Fig.1. (a)-(f) are the simulated results for the models at 0 s, 10 s, 30 s, 60s, 120 s, and 180 s, respectively.

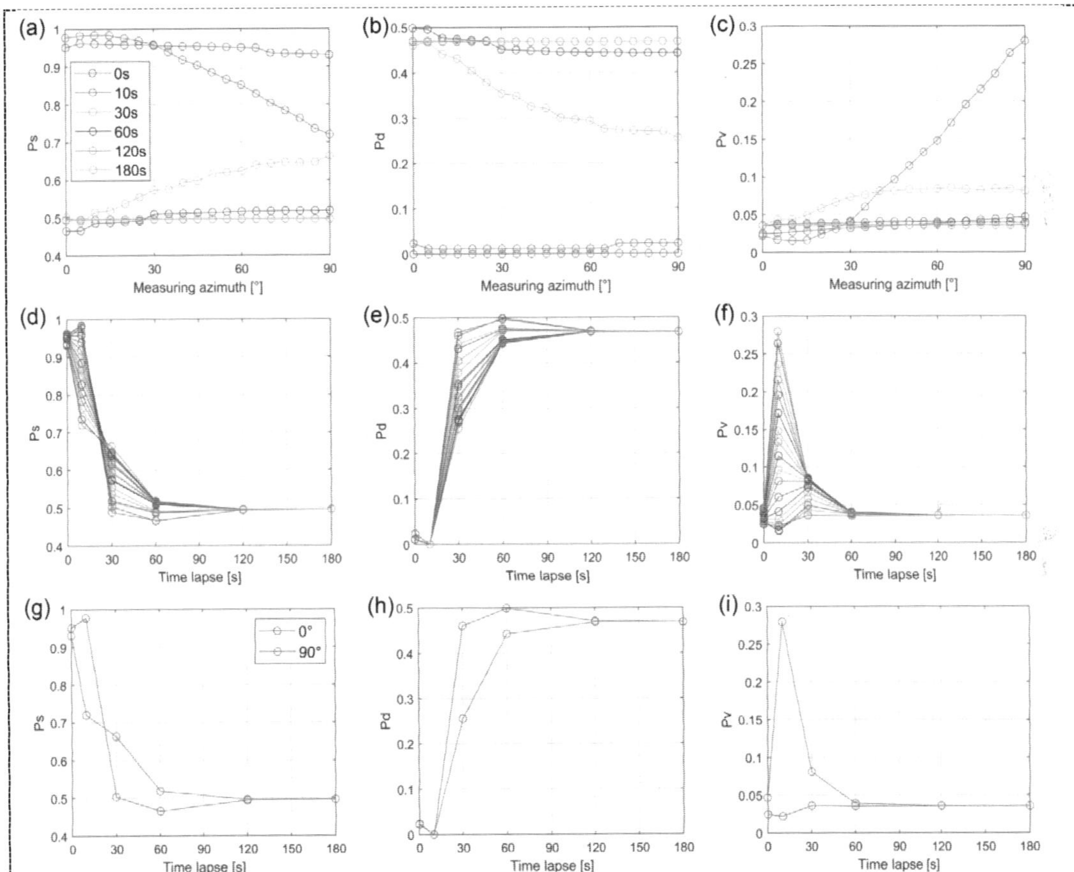

Figure 3. Freeman decomposition results. (a) P_s results varing with measurement angles; (b) P_d results varing with measurement angles; (c) P_v results varing with measurement angles; (d) P_s results varing with infiltration time; (e) P_d results varing with infiltration time; (f) P_v results varing with infiltration time; (g) P_s results of 0° and 90° models varing with infiltration time; (h) P_d results of 0° and 90° models varing with infiltration time; (i) P_v results of 0° and 90° models varing with infiltration time.

The analysis results of the three parameters all indicate that the measurement angle has a significant impact on the parameters during the first three moments of seepage. This result may be related to the frequency of the electromagnetic waves used. In the first three moments, the volume of the seepage zone is very small, only in the vicinity of the fracture. On the surface of the model, the seepage zone and fractures containingcan water can be approximated as a dipole scatterer with infinite length and small width compared to the wavelength of the electromagnetic waves used, namely a cylinder scatterer [16]. This type of scatterer is highly sensitive to measurement angles, and the amplitude of its measurement values varies according to the cosine function of the measurement angle [13]. This method can also be used to detect the direction of linear bodies (such as pipelines). As the seepage area gradually expands, the target body composed of fractures and seepage areas gradually changes from dipole scatterers to plan scatterers with uneven water content. The width of the seepage area perpendicular to the direction of the fracture direction is equivalent to the wavelength of the electromagnetic wave. When the measurement angle changes, the changes in the FP-GPR signals are very small and no longer show a cosine function change with the measurement angle [13]. In summary, in actual measurements, due to the unknown direction of cracks, if low-frequency electromagnetic waves are selected for detection, the polarization analysis results in the early period may be affected by the measurement angle and present abnormal phenomena [17]. In this case, the parameter changes in the later period should be taken for further analyses. In practice, considering the attenuation effect of water on electromagnetic waves, in order to improve the detection depth, lower frequency electromagnetic waves are usually selected for detection. Therefore, this phenomenon may have a significant impact on actual detection [18].

4. Conclusion

This article uses hydrological simulation and time-laps FP-GPR forward simulation to analyze the influence of measurement angles on the parameters P_s, P_d, and P_v obtained by Freeman decomposition of measurement data when fluid flow and infiltration in fracture and seepage zones. The results indicate a good consistency between the parameters P_s and P_d and the infiltration time. As the crack infiltration zone expands, the P_d parameter increases while the P_s parameter decreases. There is no significant relationship between P_v parameters and seepage time. Secondly, the analysis results of the three parameters all indicate that the measurement angle has a significant impact on the parameters during the period before seepage, which is related to the frequency of the electromagnetic waves used. In practice, if low-frequency electromagnetic waves are selected for detection, the polarization analysis results in the previous period may be affected by the measurement angle and exhibit abnormal phenomena. In this case, the parameter changes in the subsequent period should be taken as the standard.

5. References

[1] Wang Y, Feng X, Zhou H et al., 2021. Water ice detection research in Utopia Planitia based on simulation of Mars rover full-polarimetric subsurface penetrating radar. *Remote Sensing*, **13**, 14, 2685.

[2] Dong Z, Feng X, Zhou H, Liu C, Zhang M, Liang W, 2023. Effects of Uniaxial Bianisotropic Media on Full-Polarimetric GPR Signatures. *IEEE Transactions on Geoscience and Remote Sensing*, **61**, 4503120.

[3] Cloude S R, Pottier E, 1996. A review of target decomposition theorems in radar polarimetry. *IEEE Transactions on Geoscience and Remote Sensing*, **34**, 2, 498–518.

[4] Kobayashi T, Ko K, Choi S, Choi J, 2022. Orthogonal Dual Polarization GPR Measurement for Detection of Buried Vertical Fault. *IEEE Geoscience and Remote Sensing Letters*, **19**, 4022805.

[5] Irving J, Knight R, 2003. Saturation-Dependent Velocity Anisotropy in Borehole Radar Data. *Symposium on Applied Geophysics Engineering and Environment Problem*, 1491.

[6] Dong Z, Feng X, Zhou H et al., 2024. Super-Resolution Detection of Millimeter-Scale Fractures With Fluid Flow Using Time-Lapse Full-Polarimetric GPR and Anisotropy Analysis. *IEEE Transactions on Geoscience and Remote Sensing*, **62**, 4500816.

[7] Liu H, Zhong J, Ding F, Meng X, Liu C, Cui J, 2022. Detection of Early-stage Rebar Corrosion Using a Polarimetric Ground Penetrating Radar System. *Construction and Building Materials*, **317**, 24, 1–10.

[8] Tsoflias G P, Halihan T, Sharp J M, 2001. Monitoring pumping test response in a fractured aquifer using

ground-penetrating radar. *Water Resource Research*, **37**, 1221–1229.

[9] Zhang M, Feng X, Bano M, Xing H, Wang T et al, 2022. Review of Ground Penetrating Radar Applications for Water Dynamics Studies in Unsaturated Zone. *Remote Sensing*, **14**, 23, 5993.

[10]Van Genuchten M T, 1980. A Closed-form Equation for Predicting the Hydraulic Conductivity of Unsaturated Soils. *Soil Science Society of America Journal*, **44**, 892–898.

[11]Topp G C, 1980. Electromagnetic Determination of Soil Water Content: Measurements in Coaxial Transmission Lines. *Water Resource Research*, **16**, 3,574–592.

[12]Rhoades J D, Manteghi N A, Shouse P J, Alves W J, 1989. Soil Electrical Conductivity and Soil Salinity: New Formulations and Calibrations. *Soil Science Society of America Journal*, **53**, 433-439.

[13]Zhou H, Feng X, Ding C, Dong Z, Liu C, Liang W, 2022. Heterogeneous Weathering Process of Lunar Regolith Revealed by Polarimetric Attributes Analysis of Chang'E-4 Lunar Penetrating Radar Data Acquired During the Yutu-2 Turnings. *Geophysical Research Letters*, **49**, 13, e2022GL099207.

[14]Zhou H, Feng X, Dong Z, Liu C, Liang W, An Y, 2022. Local freeman decomposition for robust imaging and identification of subsurface anomalies using misaligned full-polarimetric GPR data. *Remote Sensing*, **14**, 3, 804.

[15]Zhou H, Feng X, Zhou B et al., 2022. Polarized Orientation Calibration and Processing Strategies for Tianwen-1 Full-Polarimetric Mars Rover Penetrating Radar Data. *IEEE Transactions on Geoscience and Remote Sensing*, **60**, 5119914.

[16]Zhou H, Feng X, Dong Z, Liu C, Liang W, 2022. Multiparameter Adaptive Target Classification Using Full-Polarimetric GPR: A Novel Approach to Landmine Detection. *IEEE Journal of Selected Topics in Applied Earth Observations and Remote Sensing*, **15**, 2592–2606.

[17]Li J., Zeng Z., Sun J., Liu F., (2012) Through-wall detection of human being's movement by UWB radar. *IEEE Geoscience and Remote Sensing Letter,* 9(6), 1079-1083.

[18]Li J., Liu L., Zeng, Z. , & Liu, F. . (2014). Advanced signal processing for vital sign extraction with applications in UWB radar detection of trapped victims in complex environments. *IEEE Journal of Selected Topics in Applied Earth Observations & Remote Sensing*, 7(3), 783-791.

Acknowledgments

This work was supported by the National Key Research and Development Program of China under grant 2021YFC1523401, the National Natural Science Foundation of China under grant 42304157, the Postdoctoral Fellowship Program of CPSF under Grant GZC20230934, the China Postdoctoral Science Foundation under Grant 2024M751089, the Scientific Research Program of Jilin Provincial Department of Education under Grant JJKH20241253KJ.

CNN based approach for root zone soil moisture prediction by ground penetrating radar

Kexin Liu, Qi Lu*, Shilong Guo and Wenjing Liang

College of Geoexploration Science and Technology, Jilin University, Changchun 130062, China

*E-mail:luqi@jlu.edu.cn

Abstract. Traditional soil moisture prediction methods often face challenges in providing spatially and temporally comprehensive root zone soil moisture (RZSM) with complex soil environment. In this paper, a novel method is proposed for predicting RZSM using images obtained from ground penetrating radar (GPR) data and processed with the convolutional neural network (CNN). The CNN encompasses two major stages: classification and regression. Initially, a CNN is tailored to classify tree root images obtained by GPR into distinct water content categories. Subsequently, through transfer learning, the pre-trained classification network is adapted for regression to predict continuous soil moisture levels. Then, the field experimental results demonstrate the efficacy and robustness of our approach, offering promising avenues for enhancing root zone water management and sustainability.

Keywords: ground penetrating radar (GPR), root zone soil moisture (RZSM), convolutional neural network (CNN), classification, regression.

1. Introduction

The ground penetrating radar (GPR) is an advanced geophysical method that detects and analyzes various soil properties using high frequency electromagnetic waves [1]. This method is highly effective for detecting and mapping variations in soil moisture content (SMC), which is essential for various applications in agriculture, hydrology, and environmental monitoring [2]. The coarse roots exhibit reflection hyperbolas in the GPR profiles, and their shapes are influenced by different SMC. Therefore, the SMC in the root zone can be predicted through reflections from coarse roots [3].

The powerful feature extraction capabilities of convolutional neural network (CNN) make it highly suitable for analyzing and explaining the GPR data obtained from complex environment [4]. Travassos et al. highlight the advancements in GPR brought about by neural network (NN) and machine learning techniques [5]. The significant benefits of deep NNs in classification and regression tasks demonstrate their substantial potential for efficiently determining SMC once the network has been trained. Wang et al. introduce a novel method that combines time-frequency analysis with CNNs to effectively determine the SMC as a classification task, achieving high accuracy in categorizing different moisture levels within the soil [6].

On the basis of previous studies, this paper provides a systematic methodology utilizing CNNs for RZSM prediction. The inputs of the networks are GPR profiles with coarse root reflection. Two stages CNNs are designed: one for classification task and another for regression task. Furthermore, we conducted field experiments in the tree forests to obtain root zone GPR data in Baicheng City, China. The accuracy of the proposed method for predicting RZSM is further explored through this experiment.

2. Method

We develop a novel technique for RZSM classification and prediction using CNNs by GPR images. The inputs of the networks are GPR profiles. Our methodology is structured into two stages: classification and regression. In the classification stage, a CNN architecture is designed to categorize tree root images obtained from GPR based on water content levels. The CNN comprises convolutional layers followed by batch normalization, the rectified linear units (ReLU) activation, and max-pooling layers. Supervised learning with categorical cross-entropy loss is employed for network training. Subsequently, we use the transfer learning to repurpose the pre-trained classification network for regression, enabling the adaptation of already learned features and representations from the classification task to effectively predict continuous variables such as water content. Feature extraction from the classification network is followed by the construction of a new CNN architecture for regression, featuring fully connected layers and a regression output layer. The regression network is trained to predict continuous water content values using supervised learning with mean squared error loss.

2.1. Classification network

Firstly, we use a CNN for the classification task, as shown in Figure 1. The network includes an encoder, which processes input images to extract meaningful features, and a fully connected layer for classification. This architecture is designed to capture discriminative features in GPR images, allowing for accurate classification into distinct water content categories. We classify GPR profiles of root zone into three categories, which are collected using the finite-difference time-domain method under varying water content conditions. Then, we analyze the features learned by the classification network.

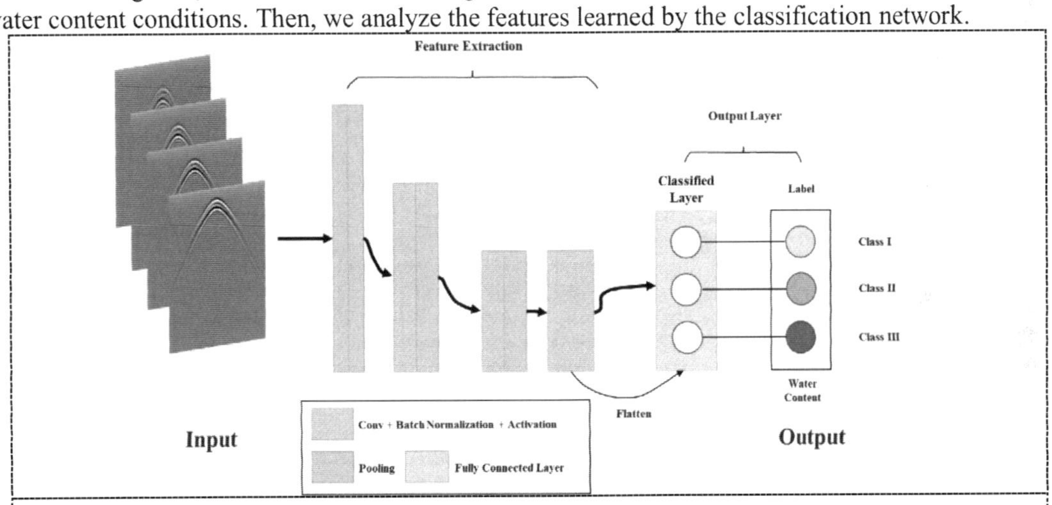

Figure 1. The classification network architecture.

The encoder network for feature extraction follows a typical convolutional network architecture and consists of four layers. Each layer includes a convolutional block, which contains a convolution unit, a batch normalization unit, and an activation function unit. If the input features at the $(k+1)$th layer are Y^k, the output data of the convolution unit for the qth channel, noted as X^{k+1}, can be expressed as:

$$X^{k+1}[q][x] = \mathrm{Conv}\{Y^k\} = \mathrm{B}^{k+1}[q] + \sum_{m=0}^{C-1}\sum_{i=0}^{R-1} Y^k[m][sx+i]W^{k+1}[q][m][i], 0 \le q < Q, 0 \le x < N, \quad (1)$$

where $B^{k+1}[q]$ represents the bias vector for the qth output feature map at the $(k+1)$th layer.

After each convolutional layer, batch normalization is applied to stabilize the training process and enhance the speed of convergence. Then, we use the ReLU as activation function to introduce non-linearity and aid feature learning within the network, can be represented as:

$$y = \mathrm{ReLU}(x) = max(0, x) \quad (2)$$

Journal of Physics: Conference Series **2887** (2024) 012007 doi:10.1088/1742-6596/2887/1/012007

Thus, the output of one convolution block in each layer consists of a combination of convolutional output, batch normalization, and activation units. After each convolution block in every layer, a max-pooling layer with a window size of 3 and a stride of 3 is applied, resulting in output subsampled by a factor of 3. It's important to note that the last convolutional layer is directly connected to the classification layer without a max-pooling layer afterward.

The classes in the label sequences are represented using One-Hot Encoding. The output of the classification layer is a probability value between 0 and 1. Cross-entropy loss is utilized to quantify the loss performance between the labels and the outputs, acting as the optimization objective during training. If we denote the true probability distribution as p and our computed model probability as $\hat{p}$, then the cross-entropy is calculated as:

$$L = -\frac{1}{M}\left[\sum_{j=1}^{M}\sum_{i=1}^{K} p_j(i)\log \hat{P}_j(i)\right] \quad (3)$$

where K is the number of classes, which is 3 in our work.

2.2. Regression network

Transfer learning is used to leverage the knowledge gained from the pre-trained classification network and adapt it to the task of quantitative prediction of water content. The CNN part of the classification network is repurposed for the regression task. In this paper, the original softmax layer designed for classification purposes is replaced with three fully connected layers, as shown in Figure 2. The flattened features obtained from the previous section are then used for the regression process. Due to the application of transfer learning, only the weights in the fully connected layers require training in the regression network. In order to train these weights, we use the mean squared error cost function.

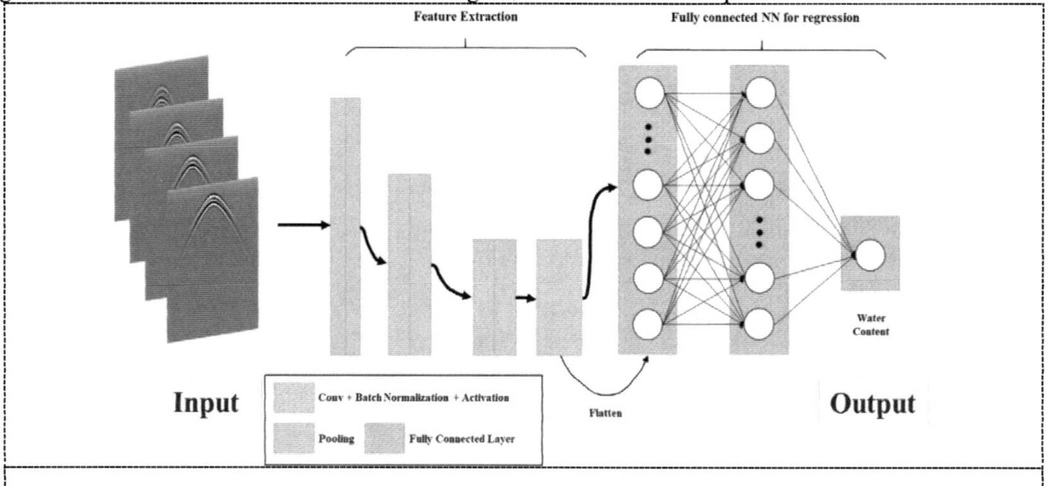

Figure 2. The regression network architecture.

3. Field experiment example

3.1. GPR data acquisition and pre-processing

We conducted field experiment at arbor forest land in Baicheng City, China. Survey lines L1 to L5 with the length of 17 m are arranged parallel to each other. To compare the accuracy of CNN based RZSM prediction method, sampling points were set on lines L1 and L5 and soil samples were collected using the ring knife sampling method. Then, the SMC at each sampling point was obtained by drying and weighing (D&W) method in the laboratory.

In our experiment, the CrossOver GPR system was used, which has dual band GPR antennas with main frequencies of 170 MHz and 600 MHz. The data collection method is common-offset measurement. The trace interval is 0.04 m and the number of sampling points is 600 of each trace. The raw profile of the GPR is shown in Figure 3 (a). The pre-processed GPR profiles after automatic gain, band pass filter, and delete mean trace are shown in Figure 3 (b)-(c), which can see clear reflections from coarse roots.

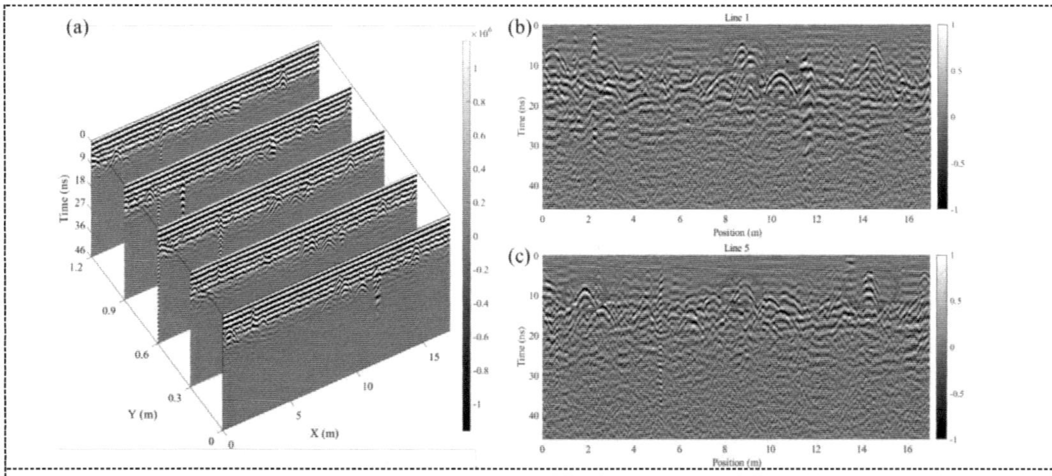

Figure 3. GPR raw profile (a) and pre-processing results (b)-(c).

3.2. Root zone soil moisture prediction

We use the CNN proposed to classify and predict water content of each coarse root and form a scattering dataset for profile soil water storage. Then, we use the inverse distance weighted interpolation method to obtain the 3D visualization of the experimental area, as shown in Figure 4. To test the RZSM measurement accuracy of GPR, the RZSM values determined by the novel method were compared with the results obtained from D&W method. The root mean square error between them is lower than 1.0%, indicating a high level of accuracy in the RZSM predictions provided by the novel method. Therefore, the novel method achieves rapid and accurate prediction of RZSM.

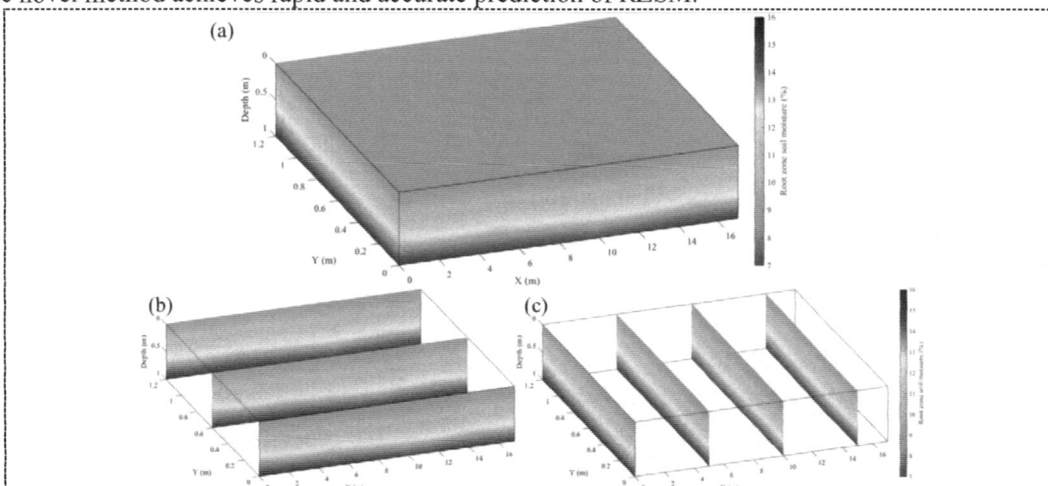

Figure 4. The 3D visualization of RZSM (a)-(c).

4. Conclusion

This paper proposes a new CNN based prediction method for RZSM by GPR, the CNN mainly includes two stages: classification and regression of the water content, demonstrating the potential of deep learning techniques in RZSM prediction. Research has found that the use of transfer learning strategies can quickly obtain RZSM. In field experiment, the error in RZSM measured by CNN method and D&W method is within a reasonable range. Therefore, this novel method is a feasible method for detecting the RZSM. This method of great significance for promoting agricultural water management practices.

References

[1] Zajícová K and Chuman T 2019 Application of ground penetrating radar methods in soil studies: A review *Geoderma* **343** 116-129

[2] Huisman J A, Hubbard S S, Redman J D and Annan A P 2003 Measuring soil water content with ground penetrating radar: A review *Vadose Zone J.* **2** 476-491

[3] Liu X, Chen J, Butnor J R, Qin G, Cui X, Fan B, Lin H and Guo L 2020 Noninvasive 2D and 3D mapping of root zone soil moisture through the detection of coarse roots with ground-penetrating radar *Water Resour. Res.* **56** e2019WR026930

[4] Krizhevsky A, Sutskever I and Hinton G E 2017 ImageNet classification with deep convolutional neural networks *Commun. ACM* **60** 84-90

[5] Travassos X L, Avila S L and Ida N 2021 Articial neural networks and machine learning techniques applied to ground penetrating radar: A review *Appl. Comput. Inform.* **17** 296-308

[6]Xiong H., Li J., Li Z., & Zhang Z., (2024). GPR-GAN: A ground-penetrating radar data generative adversarial network. *IEEE Transactions on Geoscience and Remote Sensing* 62, 1-14.

Acknowledgement

This work is supported by the National Natural Science Foundation of China under Grant 42274190.

Investigating the variation of iron-titanium content in lunar farside regolith along the route of the Yutu-2 rover with lunar penetrating radar.

Huaqing Cao[1], Jing Li[1]*, Chang Zhang[1]

1 College of Geo-exploration Science and Technology, Jilin University, Changchun 130026, China
*Corresponding author's e-mail: inter.lijing@gmail.com

Abstract. The Chang'e-4 lander successfully landed in the Von Kármán crater on the farside of the Moon in 2019, collecting a large amount of scientific data to analyze the surface material and subsurface structures of the Von Kármán crater. In this study, we processed the high-frequency radar data from the first 25 lunar days collected by the Lunar Penetrating Radar along the route of the Yutu-2 rover. Using the peak frequency shift method, we calculated the loss tangent of the shallow regolith layer ranged from 3×10^{-3} to 5.5×10^{-3}. The estimated TiO_2 and FeO content in the regolith is 11.2 wt% - 14.7 wt%, revealing a heterogeneous distribution of iron and titanium content along the CE-4 landing site.

1. Introduction

The lunar regolith, which covers the Moon's surface, contains crucial records of the Moon's geological history. The distribution of FeO and TiO_2 within the regolith is closely related to lunar volcanic activity. Research on the abundance of FeO and TiO_2 in the regolith has primarily been conducted using samples collected during the Apollo missions, as well as remote sensing spectral data and ground-based radar. These researches indicate a significant difference in the distribution of FeO and TiO_2 between the nearside and farside of the Moon. The nearside has a thinner regolith with higher FeO and TiO2, while the farside has a thicker regolith with relatively lower content.

In January 2019, the Chang'e-4 lander successfully landed in the Von Kármán crater within the South Pole-Aitken Basin on the farside of the Moon, conducting in-situ measurements. The Yutu-2 rover, equipped with a dual-frequency Lunar Penetrating Radar (LPR), performed high-resolution imaging of subsurface layers down to tens of meters with centimeter-level precision, making it a crucial tool for studying the physical properties of the lunar regolith. The travel of Yutu-2 rover was limited and data was constrained, in contrast, Ground Penetrating Radar (GPR) has developed several mature methods for estimating signal loss, particularly the frequency shift method. This method, even with an unknown wavelet, can calculate frequency attenuation using a preset Ricker wavelet and is similarly applicable to Chang'e-4 LPR data, ultimately yielding high-resolution titanium and iron content distributions. In this study, the high-frequency data from the first 25 lunar days of the CE-4 lunar radar were processed, and the loss tangent of the regolith was calculated using the peak frequency shift method, allowing for the estimation of FeO and TiO_2.

Journal of Physics: Conference Series **2887** (2024) 012008 doi:10.1088/1742-6596/2887/1/012008

2. Data and method

2.1. CE-4 LPR Channel-2 data

The Yutu-2 rover uses the same LPR system as the original Yutu rover from the Chang'e-3 mission, which explored Mare Imbrium on the Moon's nearside[1][2]. LPR has two channels: Channel-1, which operates at a low frequency (bandwidth of 40–80 MHz), providing a spatial resolution of several meters and deep penetration capabilities, and Channel-2, which operates at a high frequency (250–750 MHz), offering a finer spatial resolution of 30 cm but with shallow penetration depth. Channel-2 records 2,048 echoes per trace, with a sampling interval of 0.3125 ns and a time window of 640 ns. For this study, we utilized the entire 25 lunar days of LPR Channel-2 data, amounting to a total of 15,468 raw valid data tracks.

2.2. Loss tangent and FeO+ TiO2 content

Maximum frequency f_m is defined as the frequency at which the spectrum reaches its highest amplitude. The maximum frequency is determined by setting the derivative of the amplitude spectrum with respect to frequency to zero[3]:

$$\left(\frac{\partial X(f,t)}{\partial f}\right)_{f=f_m}=0$$

$X(f,t)$ is the amplitude spectrum, assuming the wavelet is a Ricker wavelet with a central frequency of 500 MHz, the amplitude spectrum can be represented as $A_0\rho\left(\frac{f}{f_0}\right)^2 e^{-\left(\frac{f}{f_0}\right)^2} e^{-\pi f tan\delta_a t}$, ρ is the Fresnel reflection coefficient, and $e^{-\pi f tan\delta_a t}$ is the attenuation term. Therefore, the f_m can be derived as,

$$f_m(t) \approx f_0-\frac{\pi}{4}f_0^2\ tan\ \delta_a t$$

To calculate the loss tangent, we applied a Short-Time Fourier Transform to obtain the peak frequency at each moment.

The relationship between the loss tangent and the FeO + TiO2 content measured in all regolith samples from the Apollo missions can be expressed as[4]:

$$tan\delta=10^{(0.030\times(\%TiO_2+\%FeO)-2.676)}$$

3. Results and Discussion

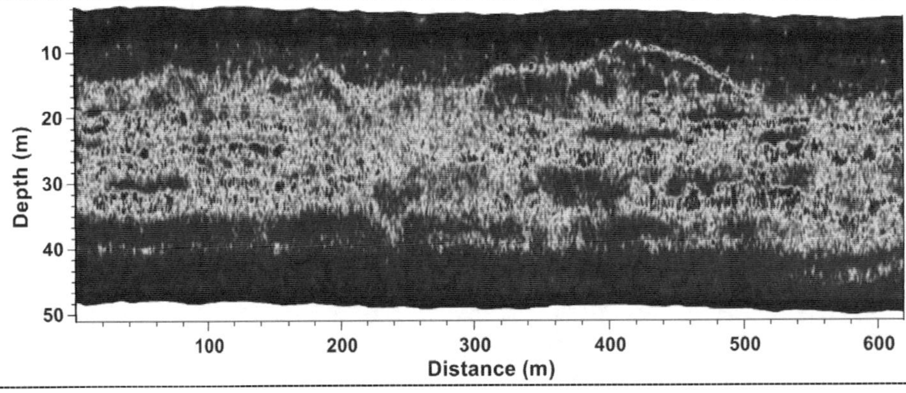

Figure 1. The radar profile of CE-4 channel-2B.

Figure 1 shows the radar profile after data preprocessing, least-squares migration imaging, and time-frequency analysis. During the first 25 lunar days of its traverse, the Yutu-2 rover encountered relatively minor surface elevation changes, and the overall terrain was quite flat. The signal in the top 5 meters of the entire profile is relatively weak and lacks strong reflections, likely indicating a finely weathered regolith layer that has undergone long-term weathering. Current geological interpretations of the high-frequency radar data from the Chang'e-4 landing site suggest that the upper 12 meters exhibit weak internal reflections and uniform material, likely representing a regolith layer that has undergone prolonged space weathering. The multilayered structures found between 12 and 30 meters depth are likely attributed to ejecta from impact craters[5].Beneath a distance of 200-300 meters from the rover's path, a distinct bowl-shaped structure is evident, which has been interpreted as a buried impact crater[5].

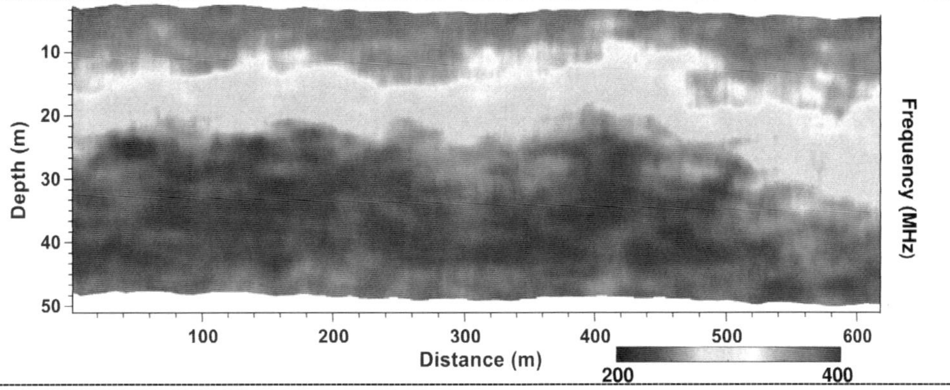

Figure 2. The peak frequency of CE-4 channel-2B.

Figure 2 illustrates the variations in peak frequency within the radar profile, reflecting distinct differences in the dielectric properties across three layers. The peak frequency chart shows noticeable frequency shifts, with a particularly rapid frequency attenuation observed around the 400-meter mark along the rover's path.

Figure 3 shows the average loss tangent for each lunar day, with a noticeable peak at the 400 meters mark where the loss tangent is significantly higher than at other locations. This corresponds well with Figure 2, where the frequency decreases most rapidly at the same location, resulting in the highest calculated loss tangent. Typically, the attenuation in the regolith is influenced by the iron and titanium content. The higher loss tangent in this area corresponds to a higher concentration of these elements. Similarly, Figure 4 indicates that the iron-titanium content at this position is markedly higher than in other regions. Analysis of the radar profile in Figure 1 suggests that this area might contain excavated material from deeper layers, potentially high in iron-titanium content due to the presence of underlying basaltic material.

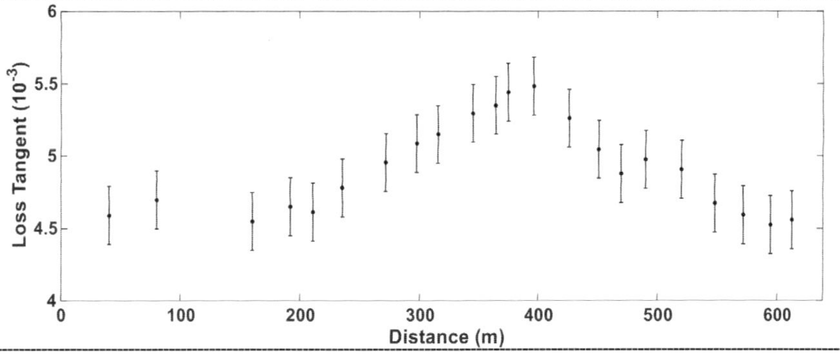

Figure 3. Estimated loss tangent of the top 5 m regolith layer for each lunar day.

As shown in Figure 4, the abundance of FeO and TiO_2 along the entire traversal path ranges between 11.2 and 14.7 wt%. The FeO and TiO_2 is relatively low in the first 200 meters, reaching its peak at 14.7 wt.% around the 400 meters. Overall, the abundance of FeO and TiO_2 along the path is significantly lower compared to the 23 wt.% to 30 wt.% observed at the CE-3 landing site [4]. This difference aligns with the distinct geological backgrounds of the Chang'e-3 and Chang'e-4 landing regions. Chang'e-3 landed in the basaltic unit of Mare Imbrium on the lunar nearside, where the surface is rich in high titanium-iron content materials. In contrast, the Chang'e-4 landing area is believed to be covered with ejecta from the Finsen crater, which is composed of low titanium-iron highland materials.

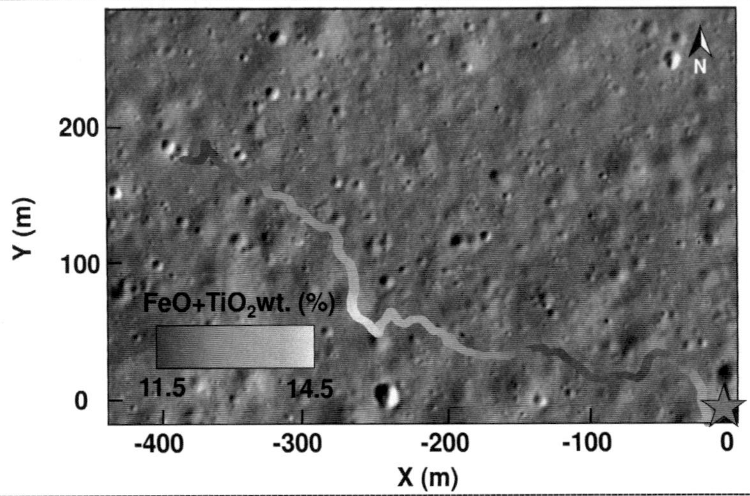

Figure 4. The estimated abundance of TiO_2 and FeO along the route of Yutu-2 rover. The red star is the CE-4 landing site.

4. Conclusion

In this study, the peak frequency shift method was used to estimate the loss tangent and abundance of FeO and TiO_2 in the top 5 meters of the lunar regolith. The loss tangent ranged from 4.3×10^{-3} to 5.5×10^{-3}, while the abundance of FeO and TiO_2 varied between 11.2 wt% and 14.7 wt%. These results reveal a heterogeneous distribution of FeO and TiO_2 in the surface materials at the Chang'e-4 landing site.

5. References

[1] Fang G Y et al., Lunar penetrating radar onboard the Chang’e-3 mission.2014 Res. Astron. Astrophys. 14 1607

[2] Su Y et al., Data processing and initial results of Chang'e-3 lunar penetrating radar. 2014 Res. Astron. Astrophys. 14 1623

[3] Sebastian Emanuel Lauro, Jamaledin Baniamerian, Barbara Cosciotti, Elisabetta Mattei , and Elena Pettinelli, Loss tangent estimation from ground-penetrating radar data using Ricker wavelet centroid-frequency shift analysis. 2022 GEOPHYSICS 87 H1

[4] Ding et al., Compositional variations along the route of Chang’e-3 Yutu rover revealed by the lunar penetrating radar. 2020 Progress in Earth and Planetary Science. 7 32

[5] Zhang L et al., Rock abundance and evolution of the shallow stratum on Chang’e-4 landing site unveiled by lunar penetrating radar data. 2021 Earth and Planetary Science Letters 564 116912

Acknowledgments

These works are supported by the Natural Science Foundation of China (42222407).

Journal of Physics: Conference Series **2887** (2024) 012009 doi:10.1088/1742-6596/2887/1/012009

Study on wavefield characteristics of irregular cavity boundaries based on GPR

Jianyu Ling[1] and Rongyi Qian[1*]

1. School of Geophysics and Information Technology, China University of Geosciences, Beijing 100083, China

Abstract. Cavities pose a serious threat to road safety. Accurately determining the boundary of cavity is crucial for evaluating the risk level, but current research in this area is relatively weak. Therefore, we conducted a study on the reflection characteristics of irregular cavity boundaries based on GPR. Starting from numerical simulations, we designed two irregular cavity models and analyzed their boundary reflection characteristics. We found that diffraction often forms at the boundaries of cavities, since it represents the change of the underground medium along the survey line. Moreover, after performing Kirchhoff migration, the range of anomalies in the profile is smaller than the size of cavities, and the shape of the lateral and bottom boundaries is different from the real. Subsequently, we analyzed the field data validated through drilling. We found that compared with forward data, diffraction is relatively unremarkable in field data but they are always exist, and we can determine the boundaries of the cavity based on them. In addition, to obtain the real shape and size of the cavity, it is necessary to calculate the velocity of electromagnetic waves in the cavity, due to we need to use this velocity to correct the position and shape of the events. This study is of great significance for the localization of cavity boundaries and risk assessment.

1. Introduction

Road collapse is often caused by cavities formed by the construction of urban underground space, long-term high load operation of roads, and soil erosion, posing a serious threat to urban safety. Therefore, it is necessary to carry out regular detection on the road. Compared with traditional road detection methods, GPR has the advantages of fast, non-destructive, high resolution and minimal impact on traffic, making it an ideal choice for road cavity detection (Liu et al., 2021; Lee et al., 2022). At present, researches on GPR road cavity detection mainly focus on cavity identification. These studies are usually based on a large number of numerical simulation experiments combined with deep learning technology, and satisfactory results are obtained in the field data (Yamaguchi et al., 2022; Li et al., 2023). However, research on the reflection characteristics of cavity boundaries is relatively weak, which leads to significant errors in the shape and size of the cavities we obtained. Therefore, studying the wavefield characteristics of cavity boundaries is of great significance for risk assessment and the formulation of road repair plans. This study clarified the boundary reflection characteristics of irregular cavities through numerical simulation experiments and analysis of field data. At the same time, we conducted a thorough analysis of the reasons for the inconsistency between anomalies in the profile and the model, and proposed possible solutions.

2. Numerical simulation experiments

In this study, two cavity models were set up, including the top protrusion (Model 1) and the bottom protrusion (Model 2), as shown in figure 1. Subsequently, we used gprMax to conduct numerical simulation experiments (Giannopoulos, 2005; Warren et al., 2016). The numerical simulation parameters used were shown in table 1, and the obtained GPR profiles were shown in figure 2a and 2b. It can be seen that the reflection from the interface is very clear, and diffraction is well-developed at the boundaries of the cavity. Due to the spherical propagation of electromagnetic waves, although the transmitting and receiving antennas do not above the cavity, the receiving antenna can still record reflected waves from the boundary of the cavity. As the transmitting and receiving antennas gradually approach the cavity, the two-way travel time decreases, resulting in a hyperbolic diffraction in the profile. Therefore, the existence of diffraction can reflect the change of the underground medium along the survey line and the boundary of the cavity will inevitably result in diffraction. Limited by the diffraction wave coverage, it is difficult for us to identify the side and bottom boundaries of the cavity, thus we performed Kirchhoff migration. The processing results are shown in figure 2c and 2d, and the diffraction is well suppressed. But we found that the anomalies corresponding to the lateral and bottom boundaries in the profile have a tendency to shrink towards the center of the cavity, as shown by the red arrows in the figure. In other words, the area of anomalies in the profile is smaller than that cavities in the model, and the shape of the anomaly boundary is different from the real. We believe that this is caused by the significantly higher velocity of electromagnetic waves in the cavity compared to the surrounding medium. That is to say, different two-way travel time may corresponds to the same depth, resulting in differences between the shape of the events and cavities. Therefore, high precision GPR velocity is needed to determine the true depth corresponding to the event, so as to obtain the real shape and size of the cavity.

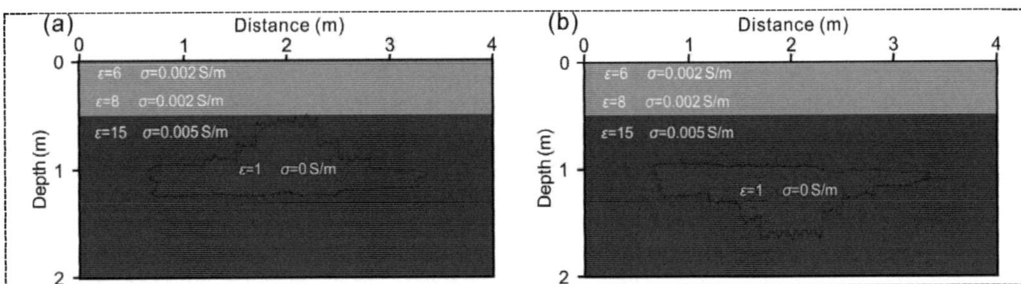

Figure 1. Cavity models: (a) top protrusion; (b) bottom protrusion.

Table 1. Numerical simulation parameters.

Model size (m)	4×2×0.002
Grid spacing (m)	0.002×0.002×0.002
Antenna frequency (MHz)	400
Trace spacing (m)	0.02
Time window (ns)	60

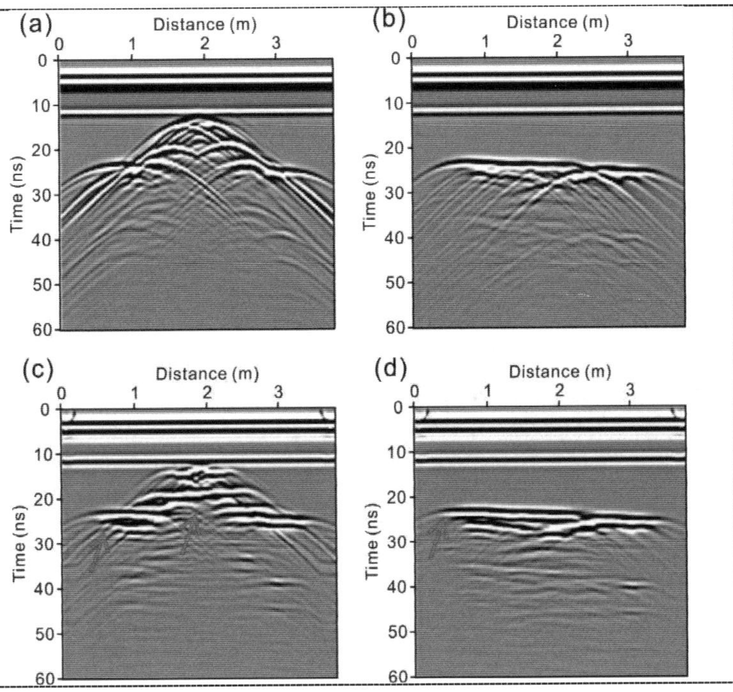

Figure 2. Synthetic GPR data: (a) and (b) correspond to Model 1 and Model 2 ,respectively; (c) and (d) are the results after the processing of Kirchhoff migration.

3. Field data analysis

To clarify the wavefield characteristics of the cavity boundaries in the field data, we selected a segment of field data that has been verified through drilling for analysis, as shown in figure 3, where the red box contains a cavity. We found that the diffraction at the boundaries of cavity is less pronounced than that in the synthetic data, but they always exist. We believe that this is due to the heterogeneity of the underground medium, which causes background reflection to mask diffraction. Subsequently, we delineated the boundaries of the cavity based on the strong amplitude area and diffraction vertices, as shown by the red dashed line in the figure. We obtained the same results as numerical simulation experiments, that is, the range of anomaly is smaller compared to the real size of cavity. Through field data, it has been proven again that the accurate morphology of the cavity requires high precision GPR velocity.

4. Conclusion

We studied the GPR wavefield characteristics of irregular cavity boundaries through numerical simulation experiments and field data analysis. We found that: (1) Diffraction is an important basis for the identification of cavity boundaries; (2) The anomalies corresponding to the lateral and bottom boundaries of the cavity in the GPR profile tend to shrink towards the center of the cavity, that is, the anomaly range is smaller than the real size of the cavity; (3) High precision GPR velocity is essential to restore the real shape and size of the cavity. This study is of great significance for evaluating the risk level of cavities and guiding road restoration.

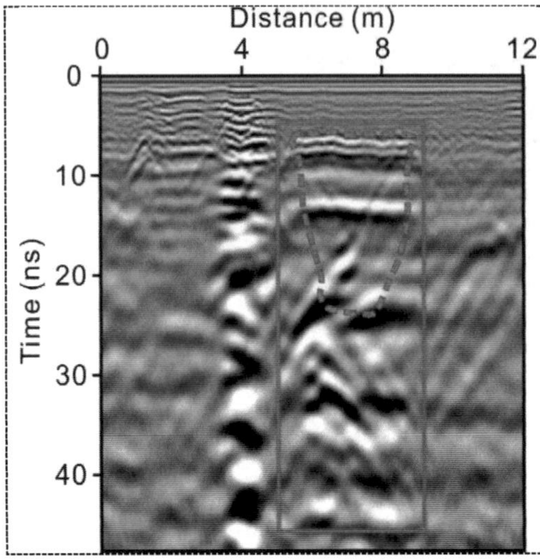

Figure 3. Field GPR data. The red box contains a cavity, and the red dashed line is the boundaries of the cavity.

5. References

[1] Lee S, Park J-J and Cho B H 2022 Management of cavities under flexible pavement road network in metropolitan area: Detection, evaluation, and rehabilitation *Dev. Built Environ.* **12** 100091

[2] Li F, Yang F, Qiao X, Xing W, Zhou C and Xing H 2023 3D ground penetrating radar cavity identification algorithm for urban roads using transfer learning *Meas. Sci. Technol.* **34** 055106

[3] Liu H, Shi Z, Li J, Liu C, Meng X, Du Y and Chen J 2021 Detection of road cavities in urban cities by 3D ground-penetrating radar *Geophysics* **86**(3) WA25-WA33

[4] Yamaguchi T, Mizutani T, Meguro K and Hirano T 2022 Detecting Subsurface Voids From GPR Images by 3-D Convolutional Neural Network Using 2-D Finite Difference Time Domain Method *IEEE J. Sel. Top. Appl. Earth Observ. Remote Sens.* **15** 3061-3073

[5] Giannopoulos A 2005 Modelling ground penetrating radar by GprMax *Constr. Build. Mater.* **19**(10) 755–762

[6] Warren C, Giannopoulos A and Giannakis I 2016 gprMax: Open source software to simulate electromagnetic wave propagation for ground penetrating radar *Comput. Phys. Commun.* **209** 163-170

Acknowledgments

This study is supported by Natural Science Foundation of Beijing Municipality (8212016) and Open Research Fund of Key Laboratory of Engineering Geophysical Prospecting and Detection of Chinese Geophysical Society (CJ2021GC02).

Deep learning-based cross-hole GPR travel time tomography

H Cheng[1], S F Hu[2], and Y H Zhao[1]

[1] School of Ocean and Earth Science, Tongji University, Shanghai, 200092.
[2] School of Mathematics and Computer Sciences, Nanchang University, Nanchang, 330031.

Abstract. As a non-destructive geophysical detection method that is not limited by the target depth, cross-hole radar can accurately identify the positions and properties of underground cavities. Traditional cross-hole radar methods face significant challenges in stability, accuracy, and computational efficiency. In recent years, with the rapid development of deep learning and its remarkable achievements in the field of imaging, it has been applied in various fields. As a deep learning network structure for image segmentation, the Unet model can realize "end-to-end" image output. Therefore, the Unet model of convolutional neural network is introduced into cross-hole radar travel time tomography in this paper. A deep-learning trave time tomography method based on the nonlinear relationship between travel time data and models is proposed. Firstly, the models with cavities of different permittivities are constructed. The 2D traveltime images of transmitting and receiving pairs are calculated based on the GPR travel time fast-picking algorithm. Then, a Unet model is built to learn the nonlinear mapping relationship between the 2D travel time images and models with cavities. Finally, the intelligent travel time tomography method based on the Unet model of underground cavities is realized. Experiment results of models demonstrate the effectiveness of the proposed method.

1. Introduction

In recent years, with the vigorous development of urban construction, the construction of large-scale urban projects has developed towards super-large, ultra-deep, and clustered. The precise detection of underground cavities has always been a challenge in the field of geophysical detection. For underground targets with deep burial depths and small scales, ground detection methods have significant limitations. Cross-hole radar, as a non-destructive geophysical detection method, can accurately identify the positions and properties of cavities (such as shape, filling mediums, etc.) without the limitation of deep depth.

There are three imaging methods in traditional cross-hole radar detection, two of which are tomography methods based on travel time and attenuation, and the last is the inversion method based on full waveform. For the cross-hole radar method, tomography is based on the characteristics of electromagnetic waves. The travel time and amplitude are affected by the dielectric characteristics of the underground medium [1]. The third method, the full waveform inversion algorithm, was originally proposed in the field of seismology, and then is introduced into the field of radar detection [2-3]. However, traditional tomography and inversion methods face significant challenges in terms of stability, accuracy, and computational efficiency.

In recent years, with the rapid development of computational power, artificial intelligence techniques such as deep learning algorithms have achieved remarkable performance in the field of imaging. Deep learning has demonstrated remarkable advantages in describing nonlinear mapping relationships between various data domains, such as images and labels, and different types of images, especially for inverse problems [4]. Therefore, many scholars at home and abroad have applied deep learning to the field of ground-penetrating radar detection [5-6]. In 2018, Dinh et al. [7] proposed an algorithm for automatic rebar location and detection in GPR images based on the combination of traditional image processing technology and deep convolutional neural network. In 2019, Lei et al. [8] proposed to combine the trained deep learning framework Faster R-CNN with the data enhancement strategy to realize the feature recognition of hyperbola. Zhang et al. (2023) studied an inversion framework based on generative adversarial networks, which realized the direct conversion of detection images from cross-hole radar into corresponding 2D underground structure images, and the results were verified on real data [9]. In 2015, Ronneberger et al. [10] proposed a "U" type neural network based on the full convolutional neural network for the first time, which is a kind of neural network that fuses data features of the same level through jumping connections, to realize image feature segmentation more accurately and "end-to-end" learning of images.

Therefore, based on previous studies, an intelligent travel time tomography method of cross-hole radar based on the Unet model is proposed in this paper. Firstly, the models with cavities of different relative permittivity are constructed. Then, the travel time data of cross-hole radar is picked up based on the fast-picking algorithm. The corresponding first arrival time of each transmit-receive pair picked up is constructed into a travel time 2D matrix (image) as the input of the network, with the corresponding model as the label. The training datasets of the Unet model are generated. Then, a classical Unet model is constructed to learn the nonlinear mapping relationship between the travel time images and the models with cavities. Finally, experimental results demonstrate that the proposed method can accurately and efficiently reconstruct the underground cavities.

2. Methodology

2.1. Unet model

The neural network establishes the nonlinear relationship between samples and labels by learning a large number of datasets. As a kind of neural network model for image segmentation, Unet model can extract features of different levels and is widely used as an “end-to-end” network.

Based on the Unet model, the network structure as shown in figure 1 is established in this paper. The network consists of two parts: a contracting encoder on the left and an expanding decoder on the right. The encoder part, structurally similar to convolutional neural networks, is composed of two convolutional layers with a kernel size of 3×3, each layer followed by a nonlinear activation function (ReLU). Then a max pooling layer of step size 2 is used for downsampling, doubling the number of feature channels. The decoder part consists of an upsampling layer (2×2) and two convolutional layers (3×3). The upper sampling layer halves the number of feature channels and the same level of feature maps between the encoder and decoder are connected. The convolutional layers are identical to the encoder part, each followed by a ReLU activation function. In the last two layers, a 1×1 convolutional layer is used to map 64 components to an image with 1 channel. The bilinear interpolation layer is applied to realize the size transformation of images to map the physical models with cavities.

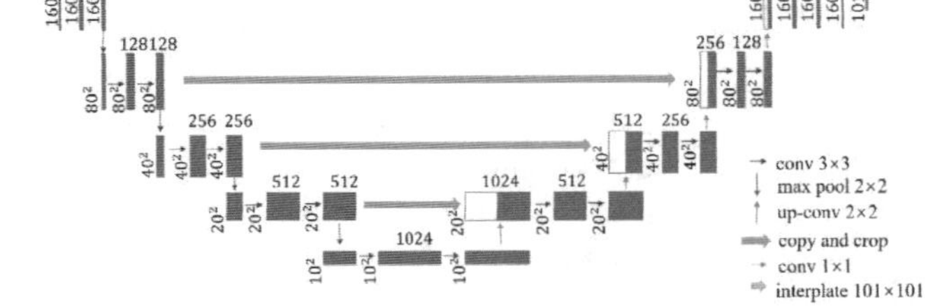

Figure 1. The network structure of Unet model.

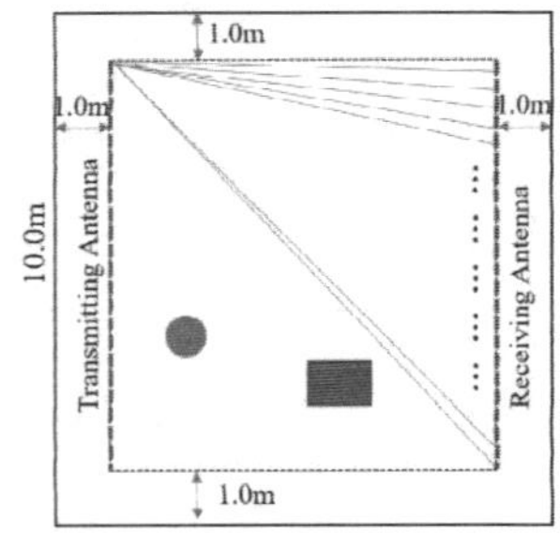

Figure 2. Underground models

Journal of Physics: Conference Series **2887** (2024) 012010 doi:10.1088/1742-6596/2887/1/012010

3. The generation of the datasets and network training

3.1. The generation of the datasets

For the training of the Unet neural network, a large number of datasets are required. In this paper, 1200 underground 10m×10m models (shown in figure 2) with randomly generated cavities are generated. In each model, 1~2 rectangular cavities with different lengths and widths or circular cavities with different radiuses are generated. The size of each cavity is 0.5^2m^2~2.0^2m^2. The cavity mediums in the model are divided into three types: water, air, and sand respectively with the dielectric constants of 81, 1, 5. The relative dielectric constant of the background is 20. As shown in Figure 2, the transmitting antenna is fixed on the left side of the model. The coordinate of the first transmitting point is (1.0m, 1.05m), and the coordinate of the last transmitting point is (1.0m, 9.0m). The spacing of each transmitting point is 0.05m. The receiving antenna is on the right side of the model, and the positions are from (9.0m, 1.05m) to (9.0m, 9.0m) with a spacing of 0.05m. In the simulation process, the fast-scanning method is adopted to calculate the travel time of each transmitting and receiving pair, constructing the 2D travel time images. Figure 3(a)(b) randomly shows two sets of the generated training datasets. The left images are the underground models with cavities as the labels of training. The right images are the corresponding 2D travel time images. It can be seen from figure 3 that the travel time images corresponding to different models have significantly different features.

All datasets are divided into training sets of 90% and test sets of 10%. The test sets are used to verify the performance of the Unet network.

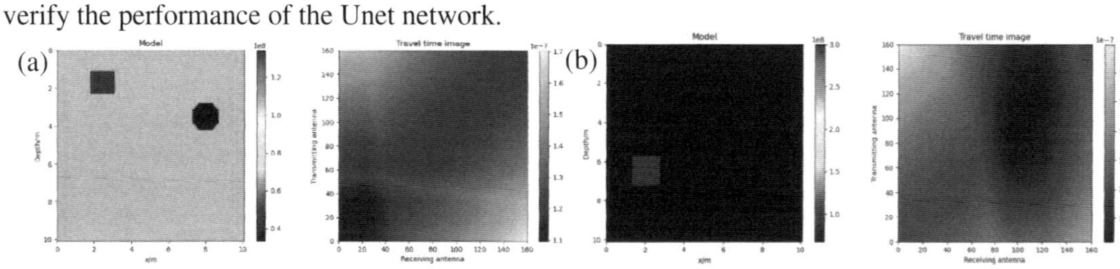

Figure 3. Two datasets in the training datasets (underground models (left) and 2D travel time images(right))

3.2. The network training

Different training parameters will lead to different results. Significant training parameter settings in this paper are: 500 epochs, a batch size of 16, a learning rate of 1.0e-3, and the Adam gradient descent algorithm is adopted. The input size of the structure is 160×160 and the output size is 101×101, with a single channel.

Figure 4 illustrates the mean square error between the real models and the predicted models of the training datasets during the training process. As shown in Figure 4 with the increasing number of epochs, the mean loss on the training datasets decreases steadily. The mean loss stabilizes around 200 epochs and eventually converges to a stable value close to zero by 500 epochs.

4. Experiments

To evaluate the recovery performance of our proposed travel time tomography method based on deep learning, experiments are performed using test sets. The prediction results are shown in figure 5 based on the trained Unet model. The 2D travel time images are on the left; the real models are in the middle; the prediction results based on the right travel time images are shown on the right. As shown in figure 5 (a)(b), the cavities of the models can be reconstructed effectively and accurately.

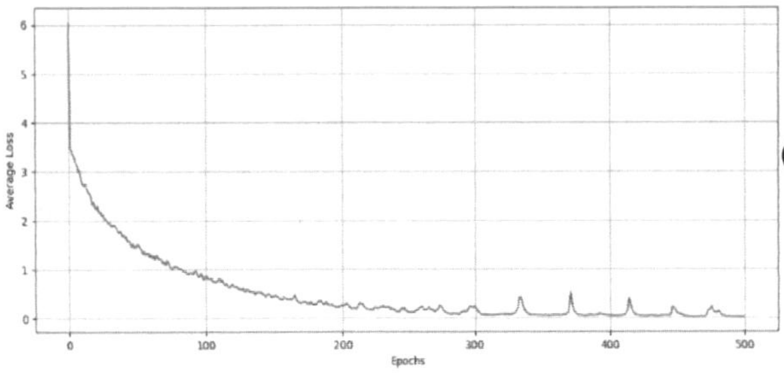

Figure 4. The mean loss of the training datasets during the training process

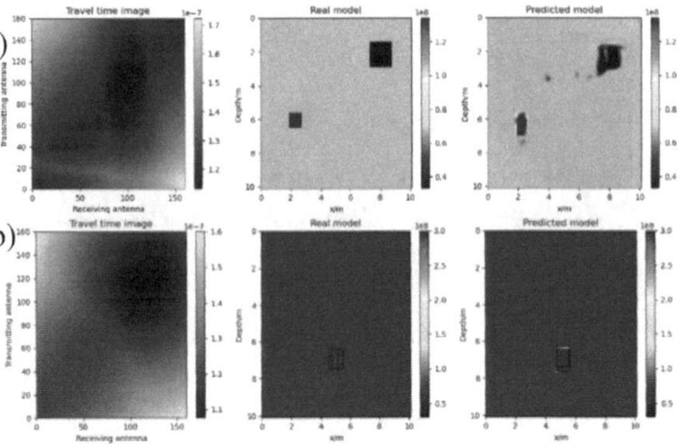

Figure 5. The experiment results. (left: travel time images; middle: real models; right: predicted models)

5. Conclusion

In this paper, the travel time tomography method of underground cavities based on the Unet model is proposed by constructing the 2D travel time images of cross-hole radar and the models with different cavities as labels. The results of experiments show that the trained Unet model can effectively reconstruct the model with cavities. Once the training process of the neural network is completed, the calculation time and the cost of model reconstruction can be negligible. The fast intelligent tomographyima
method of the 2D travel time images of the cross-hole radar is realized.

6. References

[1] Hanafy S and Al Hagrey S A. Ground-penetrating radar tomography for soil-moisture heterogeneity[J]. *Geophysics*, 2006, 71(1): K9–K18.

[2] Ernst J R, Green A G, Maurer H et al. Application of a new 2D time-domain full-waveform inversion scheme to crosshole radar data[J]. *Geophysics*, 2007, 72(5): J53–J64.

[3] Kuroda S, Takeuchi M and Kim H J. Full-waveform inversion algorithm for interpreting crosshole radar data: a theoretical approach[J]. *Geosci J*, 2007, 11(3): 211–217.

[4] Zhu J-Y, Park T, Isola P, et al. Unpaired image-to-image translation using cycle-consistent adversarial networks[A]. *in Proc. IEEE Int. Conf. Comput. Vis. (ICCV),Oct.2017*, pp 2223–2232.

[5] Liu B, Ren Y, Liu H, et al. GPRInvNet: Deep Learning-Based Ground-Penetrating Radar Data Inversion for Tunnel Linings[J]. *IEEE Trans. Geosci. Remote Sens.*, 2021, 59(10): 8305–8325.

[6] Ji Y, Zhang F, Wang J, et al. Deep Neural Network-Based Permittivity Inversions for Ground Penetrating Radar Data[J]. *IEEE Sensors J.*, 2021, 21(6): 8172–8183.

[7] Dinh K, Gucunski N and Duong T H. An algorithm for automatic localization and detection of rebars from GPR data of concrete bridge decks[J]. *Automation in Construction*, 2018, 89: 292–298.

[8] *Xiong H., Li J., Li Z., & Zhang Z., (2024). GPR-GAN: A ground-penetrating radar data generative adversarial network. IEEE Transactions on Geoscience and Remote Sensing 62, 1-14.*

[9] Zhang D, Wang Z, Qin H, et al. GAN-Based Inversion of Crosshole GPR Data to Characterize Subsurface Structures[J]. *Remote Sensing*, 2023, 15(14): 3650.

[10] Ronneberger O, Fischer P and Brox T. U-Net: Convolutional Networks for Biomedical Image Segmentation[J]. *arXiv*, 2015.

Ground penetrating radar urban road underground target classification algorithm using sequential spectral and time-domain features

F R Li[1], W X Shi[1,3], F Yang[1], M X Xu[1], L Fang[1], Y J Fang[1] and Y L Wen[2]

[1]School of Artificial Intelligence, China University of Mining and Technology (Beijing), Beijing, China.
[2]China Railway Guangzhou Group Co., Ltd., Guangzhou, China.

Abstract. Ground-penetrating radar (GPR), a highly efficient non-destructive detection method, finds extensive use in urban road underground target detection. Existing GPR data recognition algorithms often rely on singular time-domain spectrogram features, leading to potential misjudgements. To address this, we propose a novel algorithm based on sequence spectra and time-domain features. Serialized radar data, transformed through wavelets, is combined with time-domain images for input, enabling classification through a multi-scale convolutional neural network. Experiments show improved accuracy in underground target classification, offering a fresh perspective on intelligent GPR data recognition.

1. Introduction

With the continuous advancement of urbanization, the importance of urban road maintenance is increasingly highlighted [1]. Ground-penetrating radar (GPR) is an efficient, non-destructive, and high-precision geophysical detection method [2], widely applied in civil engineering and urban road exploration [3, 4]. GPR emits electromagnetic waves into the ground and receives electromagnetic reflections generated due to variations in the electrical and magnetic properties of the subsurface media [5]. By analysing changes in features such as amplitude, phase, and energy of the received electromagnetic waves, GPR aims to detect the subsurface environment. However, the interpretation of GPR anomalies related to road pavement diseases currently relies heavily on manual analysis, resulting in slow processing speeds and a strong dependence on the expertise of personnel [6], leading to the challenge of data ambiguity.

The GPR intelligent recognition algorithm is a current research focus. Liu, Z. et al. utilized the YOLOv5 model, combined with 2D slices from three perspectives of a 3D GPR, for asphalt layer damage detection on roads [7]. Huang, Z. et al. processed vertically oriented 2D radar slices from adjacent channels through binarization and region connection to extract cavity areas [8]. Practical testing on roads in Guangzhou, China revealed a depth detection of 0.3m-0.4m for cavity identification. Mesecan et al. obtained 2580 training images through cross-validation and forward simulation. After n-line flattening preprocessing and image transformations, they extracted

[3] Corresponding author: W X Shi
The work was supported by the national key research and development program of China, grant number: 2021YFC3090303 and 2021YFC3090304.

underground target feature vectors and employed KNN for recognition [9]. Li, F et al. focused on brightness and directional features, using a multi-scale visual attention pyramid for rapid screening [10]. However, these methods often solely rely on texture and visual features, requiring large datasets and limiting generalization [11]. Li, F et al. proposed a 3D convolution-based algorithm for recognizing depth-time-frequency statistical features of underground targets. Using multi-scale 3D receptive fields, the algorithm extracts detailed variations and overall trends. It constructs features through seven directional slices at different levels, partially addressing deficiencies in applying frequency domain features in intelligent GPR algorithms [12].

To address the aforementioned issues and further explore the application of GPR signal frequency domain features in deep learning algorithms, a novel underground target classification algorithm is proposed. This algorithm combines GPR signal sequence spectra and time-domain features, introducing innovation through:

1. Serializing radar signals based on electromagnetic wave propagation and radar travel direction, generating sequence frequency spectrum data through wavelet transformation for various directions.
2. Construct pseudo three-channel images with radar charts and two directional sequence spectra as input data for the algorithm, integrating frequency domain and texture variation features of signal change sequences.
3. Employing a multi-scale convolutional neural network to extract signal features at different receptive fields, facilitating effective classification of underground targets.

2. Theory of the algorithm

The algorithm utilizes 2D radar images as input data, constructs sequence spectra and texture features, and feeds them into a multi-scale CNN, as depicted in Figure 1.

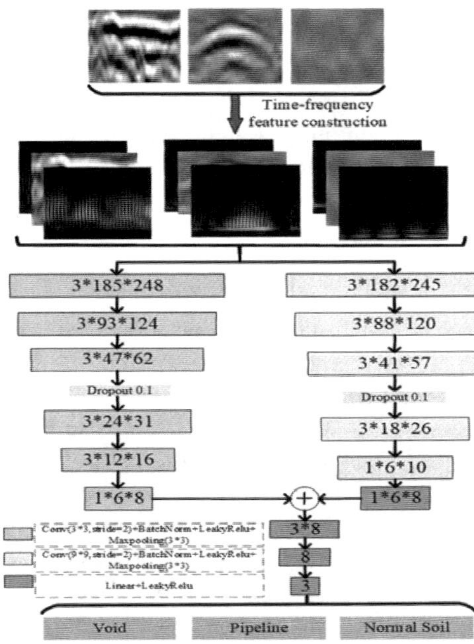

Figure 1. Algorithm processing stream.

Using 2D radar images as algorithm input, the data is adjusted to a standard 32*32 input size through bilinear interpolation [13]. Prior to spectrum conversion, radar data needs to be serialized. In this study, signal serialization is performed separately from the electromagnetic wave propagation direction and the radar traveling direction. Connect the data points indicated by the labels in Figure 2(a) and 2(b) respectively, from small to large, in order to transform the two-dimensional radar signal into a one-dimensional signal sequence. This one-dimensional signal sequence can reflect the changing characteristics of underground targets in different directions in the radar signal.

To extract the spectral data of a sequence signal using the Morlet wavelet, the formula for wavelet transformation is given by Equation (1) [14]. In this equation, $y(p)$ represents the serialized radar signal, a is the scale parameter, and b is the translation parameter.

$$W_{\psi}^{y}(a,b)=\frac{1}{c_{\psi}}\int_{-\infty}^{\infty}y(p)\psi(\frac{p-b}{a})dt \tag{1}$$

The formula for the Morlet wavelet is given as shown in equation (2) [15], where $c_{\psi}=\sqrt{\pi/\beta}$, $\beta=\omega_0^2$ and ω_0 are defined in a particular application so that the admissibility condition is valid.

$$\psi(p)=e^{-\beta(t^2/2)}e^{j\omega_0 t} \tag{2}$$

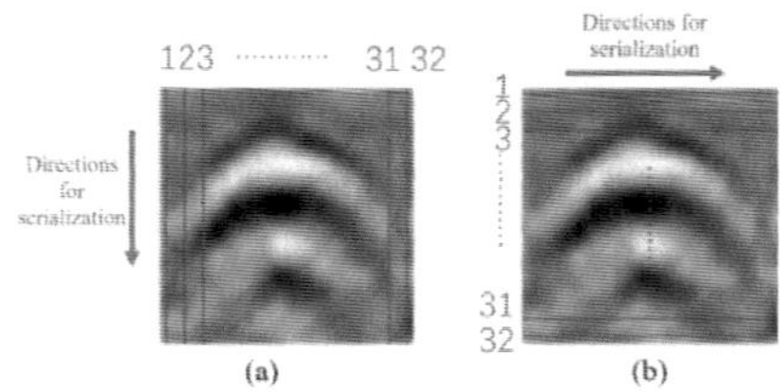

Figure 2. Schematic diagram of 2D radar signal serialization.

The electromagnetic wave propagation and radar movement create two sets of sequential spectrum data. Combining these with radar patterns forms pseudo three-channel data, used as input in the multi-scale convolutional neural network shown in Figure 1. The kernel size in the convolutional layer determines the range of feature extraction, with different sizes emphasizing distinct aspects of features [16]. Small kernels focus on local details [17], while large ones reflect overall trends. Using large convolutional kernels of sizes 3*3 and 9*9 reflects the broad trends of the target and helps extract features with different emphases. Each convolutional layer module comprises Conv2D [18], Batch Normalization (BatchNorm) [19], and Leaky Rectified Linear Unit (LeakyRelu) for non-linear mapping [20]. Features from multiple scales are integrated through a fully connected network to enhance the algorithm's feature representation capability [21].

3. Algorithm Training and Application

3.1. Dataset

To validate the algorithm for underground target classification with ground-penetrating radar, we used pipeline, cavity, and normal soil as classification targets. Data was collected using the GR-400 radar from China University of Mining and Technology (Beijing) with a 400MHz antenna frequency. A total of 168 normal soil samples, 162 pipeline samples, and 170 cavity samples were collected. The data were randomly divided into an 85% training set and a 15% test set. Examples of normal soil, pipelines, and cavities are shown in Figure 3.

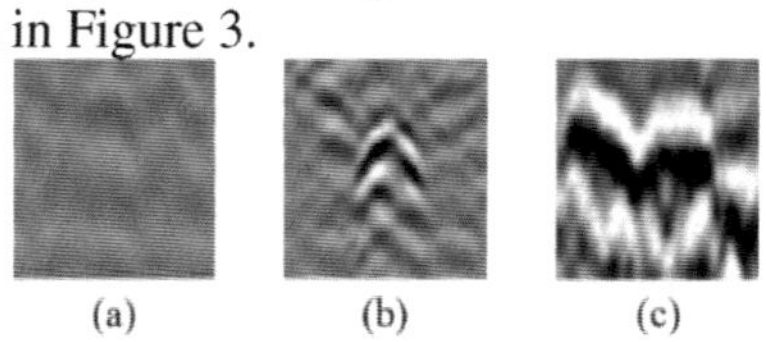

Figure 3. Example of the GPR underground target data. (a) The normal soil. (b) The pipeline. (c) The void.

3.2. Practical application of the algorithm

The algorithm uses Adam [22] as the optimizer and Cross Entropy Loss [23] for the multi-classification task of detecting underground targets in urban road ground-penetrating radar. Experiments were conducted on an Intel(R) Xeon(R) Gold 6133 CPU and NVIDIA GeForce RTX 4080 GPU. The initial learning rate during training was set to 1e-4 with a batch size of 128. Training progress, including training set loss and test set accuracy, is shown in Figure 4.

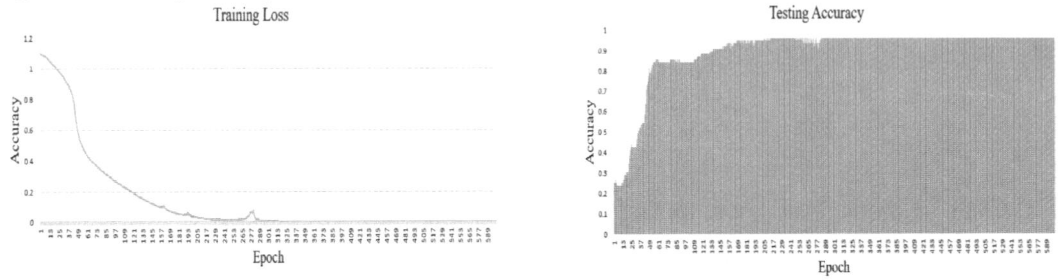

(a) (b)

Figure 4. Training process (a) Training loss. (b) Testing accuracy.

Observing Figure 4, around 600 iterations, the training set's loss plateaus, and the test set accuracy stabilizes at about 96%. With model parameters stable, we stopped the training.

4. Experimental analysis and comparison

To validate the proposed sequence spectrum and the referenced multi-scale convolutional structure, experiments were conducted using only radar spectra and a single-size convolutional kernel to construct a CNN model. Classification experiments for underground targets with ground-penetrating radar were carried out in the same experimental environment and data. The accuracy results are depicted in Figure 5.

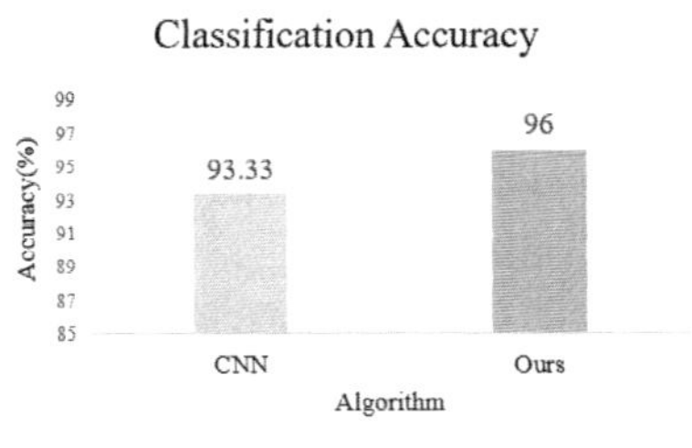

Figure 5. Comparison of classification accuracy.

In Figure 5, the CNN achieves a 93.33% accuracy in classifying underground targets using ground-penetrating radar without the introduced sequence spectrum and multi-scale convolutional structure. In contrast, our proposed algorithm attains a higher accuracy of 96%, showcasing the effectiveness of extracting sequence spectrum from serialized signals in multiple directions. The combination with time-domain texture images and a multi-scale convolutional structure significantly enhances the accuracy of underground target classification.

5. Conclusion

This paper addresses the underutilization of frequency domain features in intelligent recognition algorithms for underground targets in GPR. It proposes a GPR algorithm for urban road underground target identification that combines multi-directional sequence frequency spectra and time-domain texture features. By serializing and applying wavelet transform to 2D radar signals, the algorithm obtains frequency spectrum data from different directions and integrates radar patterns, providing a comprehensive basis for classification. Experiments show improved classification accuracy for urban road underground targets using GPR. Further research should integrate 3D GPR data to explore time-frequency classification methods for underground targets in complex environments.

Acknowledgments
The work was supported by the national key research and development program of China, grant number: 2021YFC3090303 and 2021YFC3090304.

6. References

[1] J. Y. Ling, R. Y. Qian, K. Shang, L. Y. Guo, Y. Zhao, and D. Y. Liu, "Research on the Dynamic Monitoring Technology of Road Subgrades with Time-Lapse Full-Coverage 3D Ground Penetrating Radar (GPR)," *REMOTE SENSING,* vol. 14, no. 7, APR 2022, Art no. 1593, doi: 10.3390/rs14071593.

[2] L. Yang, "The Development History and Its Research Status of the Technology of Ground Penetrating Radar," presented at the PROGRESS IN INDUSTRIAL AND CIVIL ENGINEERING, PTS. 1-5, 2012.

[3] F. Tosti and C. Ferrante, "Using Ground Penetrating Radar Methods to Investigate Reinforced Concrete Structures," *SURVEYS IN GEOPHYSICS,* vol. 41, no. 3, pp. 485-530, MAY 2020, doi: 10.1007/s10712-019-09565-5.

[4] X. W. Liu, X. J. Dong, and D. I. Leskovar, "Ground penetrating radar for underground sensing in agriculture: a review," *INTERNATIONAL AGROPHYSICS,* vol. 30, no. 4, pp. 533-543, OCT 2016, doi: 10.1515/intag-2016-0010.

[5] A. X. Zhang, Y. S. Jiang, and W. B. Wang, "A 3-D imaging method using cross-traverse data for ground-penetrating radar," presented at the 2002 3RD INTERNATIONAL CONFERENCE ON MICROWAVE AND MILLIMETER WAVE TECHNOLOGY PROCEEDINGS, 2002.

[6] E. Skartados *et al.*, "Ground Penetrating Radar Image Processing towards Underground Utilities Detection for Robotic Applications," presented at the 2018 INTERNATIONAL CONFERENCE ON CONTROL, ARTIFICIAL INTELLIGENCE, ROBOTICS & OPTIMIZATION (ICCAIRO), 2018.

[7] Z. Liu, W. X. Wu, X. Y. Gu, S. W. Li, L. T. Wang, and T. J. Zhang, "Application of Combining YOLO Models and 3D GPR Images in Road Detection and Maintenance," *REMOTE SENSING,* vol. 13, no. 6, MAR 2021, Art no. 1081, doi: 10.3390/rs13061081.

[8] Z. Huang, G. Xu, J. Tang, H. Yu, and D. Wang, "Research on Void Signal Recognition Algorithm of 3D Ground-Penetrating Radar Based on the Digital Image," *Frontiers in Materials,* Original Research vol. 9, 2022. [Online]. Available: https://www.frontiersin.org/articles/10.3389/fmats.2022.850694.

[9] I. Mesecan, B. Cico, and I. O. Bucak, "Feature Vector for Underground Object Detection using B-scan Images from GprMax," presented at the 2019 8TH MEDITERRANEAN CONFERENCE ON EMBEDDED COMPUTING (MECO), 2019.

[10] F. R. Li, F. Yang, R. Yan, X. Qiao, H. J. Xing, and Y. J. Li, "Study on Significance Enhancement Algorithm of Abnormal Features of Urban Road Ground Penetrating Radar Images," *REMOTE SENSING,* vol. 14, no. 7, APR 2022, Art no. 1546, doi: 10.3390/rs14071546.

[11] F. R. Li, F. Yang, X. Qiao, W. T. Xing, C. Zhou, and H. J. Xing, "3D ground penetrating radar cavity identification algorithm for urban roads using transfer learning," *MEASUREMENT SCIENCE AND TECHNOLOGY,* vol. 34, no. 5, MAY 1 2023, Art no. 055106, doi: 10.1088/1361-6501/acb6e3.

[12] F. R. Li, F. Yang, X. Qiao, Z. L. Hu, X. X. Wu, and H. J. Xing, "3D ground penetrating radar road underground target identification algorithm using time-frequency statistical features of data," *NDT & E INTERNATIONAL,* vol. 137, 2023 JUL 2023, Art no. 102860, doi: 10.1016/j.ndteint.2023.102860.

[13] C. B. Luo, Y. Zhang, H. Lin, and Ieee, "Efficient Interpolation Scheme for Multilevel Fast

Multipole Algorithm," presented at the 2015 IEEE INTERNATIONAL SYMPOSIUM ON ANTENNAS AND PROPAGATION & USNC/URSI NATIONAL RADIO SCIENCE MEETING, 2015.

[14] R. Büssow, "An algorithm for the continuous Morlet wavelet transform," *Mechanical Systems and Signal Processing,* vol. 21, no. 8, pp. 2970-2979, 2007/11/01/ 2007, doi: https://doi.org/10.1016/j.ymssp.2007.06.001.

[15] E. L. Schukin, R. U. Zamaraev, and L. I. Schukin, "The optimisation of wavelet transform for the impulse analysis in vibration signals," *MECHANICAL SYSTEMS AND SIGNAL PROCESSING,* vol. 18, no. 6, pp. 1315-1333, NOV 2004, doi: 10.1016/j.ymssp.2004.01.008.

[16] L. Sun, W. Shao, Q. Zhu, M. Wang, G. Li, and D. Zhang, "Multi-scale multi-hierarchy attention convolutional neural network for fetal brain extraction," *Pattern Recognition,* vol. 133, p. 109029, 2023/01/01/ 2023, doi: https://doi.org/10.1016/j.patcog.2022.109029.

[17] Y. Tang, L. Zhang, F. Min, and J. He, "Multiscale Deep Feature Learning for Human Activity Recognition Using Wearable Sensors," *IEEE Transactions on Industrial Electronics,* vol. 70, no. 2, pp. 2106-2116, 2023, doi: 10.1109/TIE.2022.3161812.

[18] J. F. Zhao, X. Mao, and L. J. Chen, "Learning deep features to recognise speech emotion using merged deep CNN," *IET SIGNAL PROCESSING,* vol. 12, no. 6, pp. 713-721, AUG 2018, doi: 10.1049/iet-spr.2017.0320.

[19] Xiong H., Li J., Li Z., & Zhang Z., (2024). GPR-GAN: A ground-penetrating radar data generative adversarial network. IEEE Transactions on Geoscience and Remote Sensing 62, 1-14.

[20] X. H. Zhang, Y. X. Zou, W. Shi, and Ieee, "Dilated Convolution Neural Network with LeakyReLU for Environmental Sound Classification," presented at the 2017 22ND INTERNATIONAL CONFERENCE ON DIGITAL SIGNAL PROCESSING (DSP), 2017.

[21] R. Chen, G. Vivone, G. H. Li, C. L. Dai, and J. Chanussot, "Multi-scale feature learning via residual dynamic graph convolutional network for hyperspectral image classification," *INTERNATIONAL JOURNAL OF REMOTE SENSING,* vol. 45, no. 3, pp. 863-888, FEB 1 2024, doi: 10.1080/01431161.2024.2305179.

[22] N. Bohra and V. Bhatnagar, "Group level social media popularity prediction by MRGB and Adam optimization," *JOURNAL OF COMBINATORIAL OPTIMIZATION,* vol. 41, no. 2, pp. 328-347, 2021 FEB 2021, doi: 10.1007/s10878-020-00684-z.

[23] H. S. Yao, D. L. Zhu, B. Jiang, and P. Yu, "Negative Log Likelihood Ratio Loss for Deep Neural Network Classification," presented at the PROCEEDINGS OF THE FUTURE TECHNOLOGIES CONFERENCE (FTC) 2019, VOL 1, 2020.

Journal of Physics: Conference Series **2887** (2024) 012012 doi:10.1088/1742-6596/2887/1/012012

A New 3D Mapping Method of Buried Pipelines from GPR B-scans

Y Li[1], X T Xiao[1] and L Ge[2]

[1] School of Electrical Engineering and Information, Southwest Petroleum University, Chengdu, China (xt_xiao@foxmail.com)
[2] School of Mechanical and Electrical Engineering, Southwest Petroleum University, Chengdu, China

Abstract. Ground penetrating radar (GPR) is a non-destructive testing method originally developed for the investigation of underground space and is now widely used in pipeline detection. To map the underground pipelines using GPR, the main challenges include the estimation of depth and orientation through two-dimensional profiles. In this paper, a new 3D mapping framework of buried pipelines from GPR B-scans is proposed. After obtaining sufficient B-scans through parallel survey lines, the proposed procedure mainly includes four stages: data pre-processing, wave velocity estimation, 2D migration and binarization, and 3D stacked image display. Canny edge detection is used to extract hyperbolic features and estimate the propagation velocity of electromagnetic waves in underground media. Migration and binarization are conducted on pre-processed B-scans before 3D stacking. The results show that the proposed method can accurately and intuitively display the position and direction of pipelines by generating three-dimensional images.

1. Introduction

Pipelines, as a safe and economical means of transporting a large amount of oil, gas, water, etc., have been widely used around the world. After decades of development, pipeline detection technology has achieved a transformation from destructive to non-destructive, and from trenching to trenchless. Ground penetrating radar (GPR) is a non-destructive testing (NDT) method originally developed for the investigation of underground space and has become one of the most popular methods in pipeline detection [1].

In the past decade, thanks to the development of GPR detection technology and the application of machine learning algorithms, the surveying and mapping technology of pipelines based on GPR has developed rapidly. To map the underground pipelines using GPR, the main challenges include the estimation of depth and orientation through two-dimensional profiles. To overcome the above issues, some studies have been done by: developing the antenna array [2], combining GPR and Global Position System (GPS) data [3], fusing multisource data [4], using multi-polarization detection [5], etc. It can be seen that existing research either requires improvements to the detection system or collects complex data. A fast, universal, and easily understandable method is still needed.

In this paper, a new 3D mapping framework of buried pipelines from GPR B-scans is proposed. First, the pipeline positioning accuracy was improved through data pre-processing, such as direct wave removal, gain processing, and time zero correction. Then, Canny edge detection algorithm was used to extract hyperbolic features and estimate the propagation velocity of electromagnetic waves in underground media. The spatial frequency domain interpolation (SFI) algorithm was used for migration.

Finally, the three-dimensional display of the detection results of buried pipelines was achieved through binarization and image stacking. The proposed method was evaluated on two numerical pipeline models with different settings.

2. Methodology

The 3D mapping of buried pipelines from raw GPR data mainly includes four stages: data pre-processing, wave velocity estimation, 2D migration and binarization, and 3D stacked image display. The specific implementation process of each step is introduced below.

2.1. Data Pre-processing

To improve the imaging accuracy, data pre-processing was performed before imaging. First, in order to reduce pipeline positioning error, this paper selected the first negative peak of the first wave to correct the time zero 0. Second, direct wave was removed using the time domain rectangular filtering method. Its basic principle is to truncate the echo signal by setting a rectangular window with the same size as the signal length of the direct wave in the time domain, and only the signal within a specific range is retained. Finally, gain processing was conducted to improve the strength of the reflected echo signal while weakening the clutter signal. Generally, the electromagnetic wave energy follows the law of exponential change, so the exponential gain form was selected as the time gain function, given by:

$$y = a^x - 0.5 \tag{1}$$

where, a is a constant, usually between 1.1~1.5, and x is the time sampling point, and the maximum value of the time gain function should be limited and cannot be increased infinitely.

2.2. Wave Velocity Estimation

According to the detection principle of GPR, when the transmitting and receiving antennas are co-located or the distance between them can be ignored, the buried depth of pipelines can be calculated by:

$$h = \frac{v \times \Delta t}{2} \tag{2}$$

where, v is the propagation velocity of radar wave in underground medium, h is the buried depth of the upper surface of the pipeline, and Δt refers to the two-way travel time between the upper surface of the pipeline and the ground, respectively. A two-step calculation process was used to estimate the wave velocity. Firstly, Canny edge detection algorithm was used to extract the hyperbola shape. Then, the vertex of the hyperbola and multiple edge points were selected to calculate the discrete velocities. The final wave velocity is the average of multiple discrete wave velocities, that can be calculated by 0:

$$v = \frac{1}{n}\sum_{i=1}^{n}\left(2|x_i| / \sqrt{t_i^2 - t_0^2}\right) \tag{3}$$

where n is the number of selected edge points, x_i is the horizontal distance between a random edge point and the vertex, t_i is the two-way travel time at a random edge point, and t_0 is the two-way travel time at the vertex. The more the number of edge points, the more accurate the estimation of wave velocity.

2.3. Two-Dimensional Migration and Binarization

This paper introduces the migration imaging algorithm of synthetic aperture radar (SAR) - spatial frequency domain interpolation (SFI) algorithm to reconstruct the B-scans 0. The purpose of migration was to restore the reflected and diffracted waves received on the ground to the actual position. The proposed wave velocity estimation method improved the imaging depth accuracy.

Then, the target and the background were distinguished by binarization processing. In the binarized image, the white part represents the background information, while the black part is the target. The pixels of the edge detection image were adapted to the detection area [6]. So, the location of the pipe can be estimated by the centroid of the target area, as described in:

$$\begin{cases} X = \frac{1}{m}\sum_{i=1}^{m} x_i \\ Y = \frac{1}{m}\sum_{i=1}^{m} y_i \end{cases} \tag{4}$$

where m is the number of pixels in the target area, x_i and y_i are the horizontal and vertical coordinates of the ith pixel, respectively.

2.4. Three-Dimensional Image Display

The stacked 2D images displayed the 3D mapping of buried pipelines. In order to obtain sufficient GPR data in 3D space, multiple survey lines were set in parallel along the position perpendicular to the pipeline, with equal intervals between the survey lines. Then, B-scan data obtained from various survey lines were processed separately. Finally, according to the detection order, all imaging results are stacked together to realize the three-dimensional image display.

3. Numerical Study

To validate this method, two simulation models of buried non-metallic pipelines were established in this study using gprMax.

3.1. Parallel Pipeline Model

In the parallel pipeline model, three pipelines with different buried depths and diameters are buried in the ground, as shown in Figure 1. The size of the model is 4m*2m*2m, with the discretization spatial step of 0.02m. The relative permittivity of the underground medium is 3. The center frequency of the antenna is 270MHz and the moving step is 0.02m. The time window is 40ns. Six survey lines were set up every 0.3m in the direction of y-axis. The scanning distance is 3.6m long. The pipe diameters are 0.1m, 0.2m and 0.3m, respectively. The horizontal spacing between pipes is 0.8m and the longitudinal spacing is 0.3m.

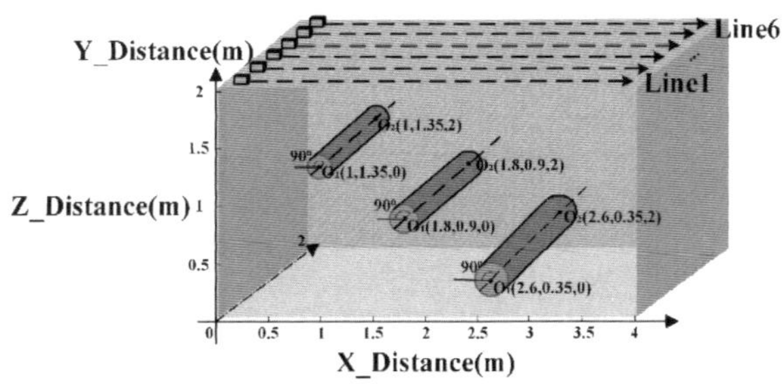

Figure 1. Parallel pipeline simulation model

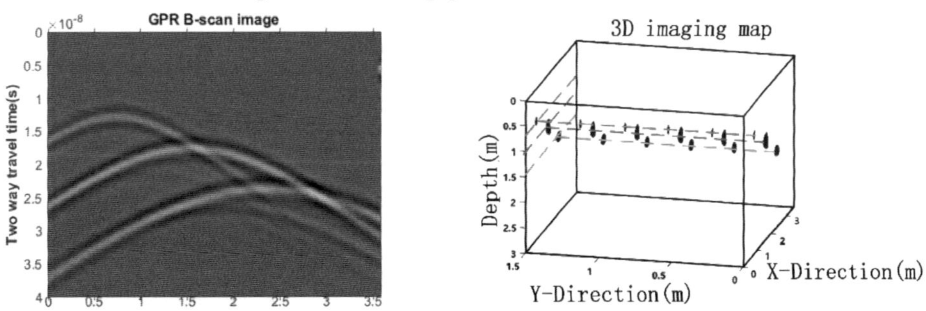

Figure 2. Results of parallel pipeline model: (a) preprocessed B-scan and (b) 3D mapping

After pre-processing and the whole procedure, the corresponding results are shown in Figure 2. It can be seen from the B-scan in Figure 2(a) that due to the close distance between pipelines, the hyperbolas appear overlapping phenomenon, although different reflected signals can still be distinguished. The signal strength of the pipeline gradually increases from top to bottom, because the echo signal strength is not only related to the buried depth of the pipeline, but also related to the size of the pipeline. In Figure 2(b), the time information has been converted into depth information. The hyperbolic effect has been eliminated and focused into one point. By manually connecting the centroids of all survey lines, the direction of each pipeline can be obtained.

3.2. Crossed Pipeline Model

In the crossed pipeline model, two pipelines are buried in the ground with an angle of 45° between the pipelines and a vertical spacing of 0.5m. The burial depths of the pipelines are 1.0m and 1.5m, and the pipe diameters are 0.2m and 0.3m, respectively. The relative permittivity of the underground medium is 4. The remaining media and detection parameters are consistent with the previous model.

The corresponding 3D imaging side view and top view are shown in Figure 3. The angle between the two pipelines and the buried depths can be observed in Figure 3(a). From Figure 3(b), the angle between the two pipelines is approximately 45 °.

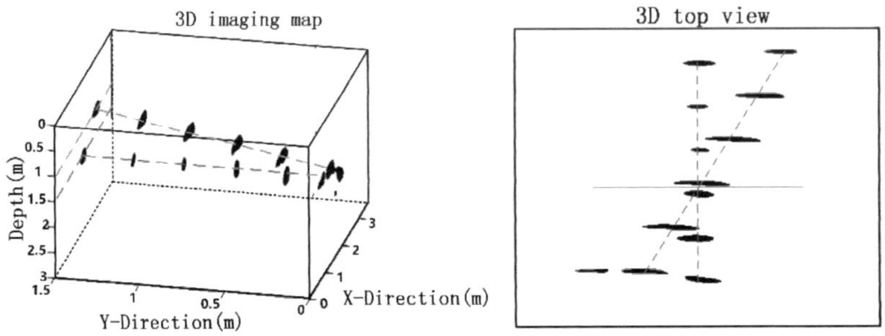

(a) (b)

Figure 3. Results of crossed pipeline model: (a) 3D side view and (b) top view

4. Conclusion

This paper proposed a new 3D mapping method of buried pipelines from GPR B-scans and studied the validity and performance numerically in two pipeline settings: parallel and crossed. The results show that the proposed method can accurately and intuitively display the position and direction of pipelines by generating three-dimensional images, providing convenience for data interpretation in pipeline detection. In future work, experimental validation of real-world scenarios will be conducted. The automatic recognition of pipeline routing based on 3D images will also be studied.

Reference

[1] Ge L, Zhang C, Tian G et al. 2021 *Chinese J. of Mechanical Engineering* **34** 97
[2] Liu H, Huang X, Han F, Cui J, Spencer B F, and Xie X 2019 *IEEE J. Sel. Top. Appl. Earth Observations Remote Sensing* **12(7)** 2080–87
[3] Wu Q, Zhou X, and Chen H 2020 *IEEE Access* **8** 118114–24
[4] Zhou X, Chen Q, Jiang B, and Chen H 2022 *IEEE Trans. Geosci. Remote Sensing* **60** 1–11
[5] Li J, Zhang N, Xu H, Wang B, Liu L, and Zhao M 2022 *IEEE Sensors J.* **22(15)** 15517–25
[6] Li Y, Xu Y, Xiao X T, Guan B R, Ge L 2023 *12th International Workshop on Advanced Ground Penetrating Radar (IWAGPR), Lisbon, Portugal* 1-4
[7] Xie F, Wu C G-W, Lai W W-L and Sham J F-C 2018 *Tunnelling and Underground Space Technology* **76** 76-91
[8] Yigit E, Demirci S, Ozdemir C, et al. 2010 *Microwave and Optical Technology Letters* 2010 **49(10)** 2534-40

Acknowledgments

This work was supported by natural science starting project of SWPU (No. 2023QHZ003), and the Science & Technology Program of Sichuan Province (No. 2023ZHCG0020).

Journal of Physics: Conference Series **2887** (2024) 012013 doi:10.1088/1742-6596/2887/1/012013

Design of dual-channel ground penetrating radar on Tianwen-2 mission

Shuanlao Li[1], Ruidong Liu'[1], Shi Zheng[1], Yinguang He[1], Xiaoqiang Zhang[1],Hongxing Dang[1], Xiaomin Tan[1] and Peng Liu[2]

[1]Xi'an Branch China Academy of Space Technology, Xi'an,710100, China

[2]National Space Science Centre, Beijing, 100190, China

Abstract. The Asteroid Interior Structure Detection Radar(AISDR) is a low-frequency Ground Penetrating Radar(GPR) onboard the Tianwen-2 spacecraft, which aims to detect the internal structure of asteroid. The AISDR has two channels for both deep penetration and high resolution. Field tests have been conducted and the results demonstrate that the performance of AISDR will fulfill the objectives of the Tianwen-2 mission.

1. Introduction

China plans to operate an asteroid exploration mission named TIANWEN-2, which will achieve "close detection, sampling and return" of near-Earth asteroid 2016HO_3 and then detect an active asteroid named 311P closely [1].

The observations of asteroid-like bodies and especially their internal structure are of main interest for science as well as planetary defense [2-4]. One of the most important science objectives of the Tianwen-2 mission is to detect the internal structure of the asteroid. A step-frequency ground-penetrating radar, named Asteroid Interior Structure Detection Radar (AISDR) is equipped on the prober. The radar is used to obtain the echoes from the surface and sub-surface of the asteroid which will be applied to study the internal structure.

2. INSTRUMENT DESCRIPTION

In order to satisfy the requirements of deep penetration and high resolution, two channels are used: the low frequency channel with lower frequency that can penetrate the asteroid with metre-scale resolution, and the high frequency channel with ultra-wide bandwidth that bring out decimeter-scale resolution. The dual-channel can detect the asteroid division of time synchronously.

A stepped-frequency radar does not operate in a pulse mode, but rather transmits through the transmitter antenna a number of continuous-wave (CW) signals, each at a single frequency for a time step duration τ [5]. The system bandwidth W used by the instrument depends on the number of transmitted frequencies N and their spacing Δf, such that

$$W = N \cdot \Delta f$$

AISDR's functions are divided among the main units shown in Fig. 1. The various modules described here are integrated in the electronic box inside the spacecraft,while the HF antenna and LF antenna (used for the transmission and reception signal) is mounted at the bottom of the spacecraft。

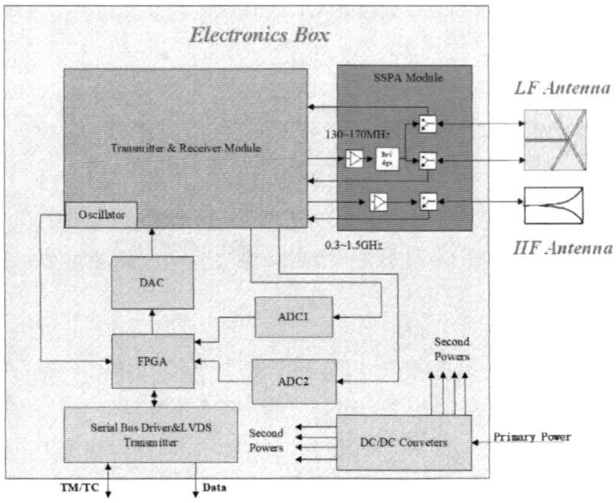

Figure 1. The AISDR's function diagram

2.1. The Electronic Box

The AISDR electronic box acts as the radar controller and generates power and transmits the signal through the antennas. The control of the instrument is operated by the Payload Management Unit(PMU) using the serial bus. The electronic box consists of digital&power unit, transmitter&receiver module and solid-state power amplifier (SSPA) module. The instrument contains the following functions.

- Stepped-frequency generation: delivers a set of N frequencies from 130MHz to 170MHz for the LF channel and from 300MHz to 1.5GHz for the HF channel. N and the frequency step can be set on the fly.
- Transmitter: for the low-frequency channel, amplifies the signal and polarization selection; for the high-frequency channel, up-conversion frequency and amplifies the signal.
- Receiver: amplifies the echoes from the asteroid sensed by antenna, for the low-frequency channel, includes low noise amplifier, filter, automatic gain control; for the high-frequency channel, low-noise amplifiers, filters, mixers, automatic gain control are included.
- Clock management: distributes the clock to satisfy the requirements of analog-to-digital converter(ADC), digital-to-analog converter(DAC) and field-programmable gate array(FPGA).
- Data pre-processing: analog-to-digital converts the signal from receiver and send data to FPGA, which deal with digital down conversion(DDC), finite impulse response(FIR) filter, extraction. The processed data is delivered to LVDS transmitter for transmission to the PMU.

- Radar controller: includes system control and operational parameters control. Radar receives a command from PMU, responses and entries work mode corresponding to the command. The operational parameters have default values, and can be modified by injection instructions during in-orbit operation if necessary.

2.2. The Antenna System

The low-frequency antenna of the AISDR takes the form of a symmetric oscillator antenna. Two pairs of oscillators placed orthogonally to form a bi-linear polarization. The dimension of the antenna is limited by the spacecraft envelope.To meet the envelope requirements, the low-frequency antenna is designed as a deployable antenna. To simplify the design and reduce the weight of the deployable mechanism, hinges have been chosen as the driving mechanism during deployment. The structure of the oscillator is also similar to a tape spring. When the oscillator is collapsed, it is wrapped around the oscillator support and locked by a locking string. After the antenna is unfolded, the oscillator is expanded by its own elastic force. The biggest advantage of the fuse-type lock-and-release device is low cost, tiny impact, light and reusable, which meets the requirements of ultralight design.

For achieving ultra-wideband (UWB) performance of the high frequency antenna of AISDR, the Vivaldi slot-plane antenna is adopted. The Vivaldi antenna is a planar structure similar to a horn opening, and its structure is usually progressively opened by slotted radiation port. There are many kinds of slotted lines, the exponential curve is used. The fan-shaped opening formed by the exponential curve is relatively far away from each other, which is the main working region. Depending on the frequency band, the receiver and transmitter regions of the electromagnetic wave are shifted in the whole slot of the radiation port to achieve its ultra-wideband performance.

a. Electronic box **b.** Low-frequency antenna **c.** High-frequency antenna

Figure 2. The photos of AISDR products

3. EXPERIMENTAL VALIDATION

In order to verify the performance of AISDR, the ground verification test is carried out. Due to the large beam footprint coverage of the antennas, it is necessary to perform the test in flat terrain to avoid the effects of surface clutter. As a result, Chagan Lake in northeast China was chosen as the test site. In winter, the ice is up to 0.5 m thick, with relatively clearly surface and sub-surface layers. In addition, the area of Chagan Lake is large enough, and the surface is flat, which can reduce the effects of surface clutter as much as possible.

In the test, the UAV was mounted with AISDR and flown to a certain altitude to simulate the attitude of the on-orbit station. To validate the radar performance, simultaneous measurements of the ice were also performed using a ground commercial radar. Surface and subsurface targets can be clearly distinguished from the echoes of the high-frequency channel of AISDR. With an ice dielectric of 3.15, the thickness of the ice layer is about 0.64m, which is consistent with the test results of the ground commercial radar and the field sampling tests. For the low-frequency channel of AISDR, the surface and sub-surface of the ice lake can't be distinguished due to lower resolution.

Figure 3. AISDR detecting on frozen lake

Figure 4. Commercial GPR test

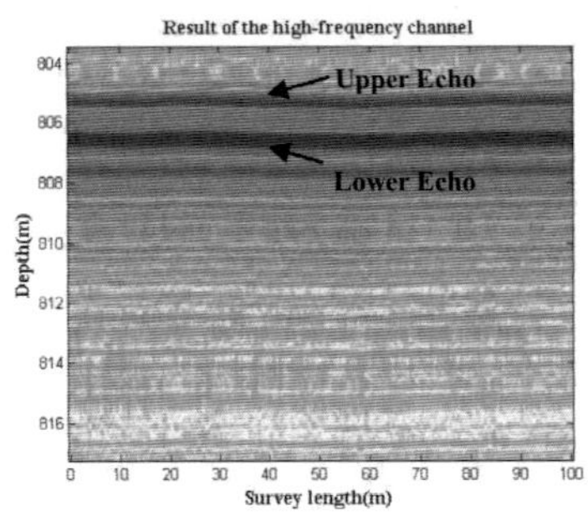

Figure 5. AISDR detecting result

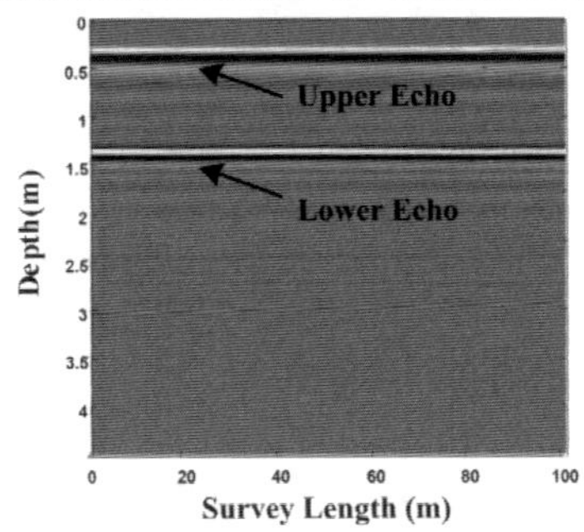

Figure 6. Commercial GPR test result

4. CONCLUSION

The design and field test of AISDR have been described. The test results demonstrate that the performance of AISDR will fulfill the objectives of theTianwen-2 mission.

5. References

[1] Announcement Opportunities for Scientific payloads and Piggyback Projects of Asteroid Exploration Mission [EB/OL]. [2019-04-19].

[2] Yang M F, et al. Progress and prospect in planetary exploration of solar system, Chinese Space Science and Technology, 2023, 43(05): 1-12

[3] Alain H, et al. A radar package for asteroid subsurface investigations: Implications of implementing and integration into the MASCOT nanoscale landing platform from science requirements to baseline design，Acta Astronautica 156(2019)317-329

[4] A. Herique, et al, Direct observations of asteroid interior and regolith structure: science measurement requirements. Advances in Space Research, 62(8), 2141–2162

[5] Zhang R Q， Huang J C， He R W， et al. The development overview of asteroid exploration[J]. Journal of Deep Space Exploration，2019，6（5）：417-423，455

[6] Ciarletti V, et al, WISDOM GPR Designed for Shallow and High-Resolution Sounding of the Martian Subsurface, Proceedings of the IEEE, Volume 99, Issue 5, Pages 824-836

Study on detection and estimation method of PCCP filament breaking

Ya Li Ren[1] Jian Le Zhao[1] Shan Shan Huang[1, 3] Xue Chao Sun[1] and Yi Liu [1, 2]

[1]The 22nd Research Institute of China Electronics Technology Group Corporation,No.36,Xianshan East Road,Chengyang District,Qingdao,Shandong Province.

[2]Corresponding author,email:liuyi3821@163.com.

[3]XiDian University,School of Science and Technology,266 Xinglong Section xifeng Road,XI'an City,Shanxi Province.

Abstract.PCCP pipeline traditional comparison method needs to consume a lot of manpower and time, in order to save a lot of manpower and time, improve the detection efficiency, this project for the problems existing in the PCCP pipeline detection, through the study of different pipeline types, analysis of different pipeline waveform, put forward a joint actual wire detection and wire quantity estimation method, put forward the PCCP detection information processing method, solve the problem of interpretation without reference waveform contrast, interpretation frequency. The data of each section of pipeline collected is assembled with equal length, and the selection rules of reference waveform are formulated, so as to realize the difference comparison between interpretation waveform and reference waveform, and assist professional interpreters in rapid interpretation.

Keywords: PCCP pipeline, broken wire breaking detection, reference waveform, difference comparison

1.Introduction

Prestressed steel tube concrete tube (PCCP) is a kind of composite water pipe, composed of high strength steel tube, circumferential prestressed steel wire and concrete[1].It combines the tensile resistance, sealing of steel and concrete compression resistance, corrosion resistance, with high sealing, high strength and high impermeability, has been widely used in the south-north water Diversion project, water resources allocation in northern Hubei[2], water supply in northeast China and Ningxia water supply project[3]. However, the fracture of the circumferential prestressed steel wire often leads to the rupture of the PCCP pipeline, which may cause leakage, pipe explosion and other accidents, bringing huge losses. Therefore, it is very important to study the detection technology of PCCP pipeline fracture, which can find the position and quantity of steel wire fracture in time, which is helpful to pipeline maintenance and early warning work[4]. The wire breaking detection equipment is composed of transmitting coil, receiving coil, PWBD-006 wire breaking detector, battery and detection vehicle, etc. The equipment moves in the pipe to identify the wire fracture position through the excitation magnetic field generated by the transmitting coil and the receiving voltage signal recorded by the receiving coil.

Figure 1. Photograph of PCCP broken filament detection equipment

At present, engineering testing and online monitoring are used, the main methods include percussion echo, far field eddy electromagnetic, ultrasonic and X ray, etc., far field eddy electromagnetic method is the main survey method, and others are used for review and verification[5]. We mainly used the orthogonal electromagnetic method to collect signals through the PCCP filament detection equipment and determine the location and number of filaments.

2. PCCP detection, analysis problem analysis

2.1. PCCP data analysis and interpretation technique

To accurately estimate the number of PCCP wire breaks, the calibration and alignment method is often used. Cut a different number of wire at a specific location as a standard sample; then, the measured data is compared with the standard sample. But directly analyzing the raw data remains challenging. Therefore, we used MATLAB to turn the electromagnetic signal into images, and manually read the energy and phase of the curve, to realize the filament identification, positioning and quantity estimation. At present, the data collected by our PCCP fracture detection equipment has the problem of poor consistency. Influenced by pipeline type, environment, equipment parameters and other factors, the fracture position, number of broken wire and normal conditions are irregular, and the sample library cannot be established. Traditional detection methods rely on manual interpretation, which consumes a lot of manpower and time, and lacks statistical tools to support it[6]. Therefore, it is urgent to develop efficient information processing means.

3.Algorithm design

PCCP pipeline detection algorithm on the basis of the traditional detection method, the introduction of benchmark curve concept, through the splicing data and benchmark curve difference calculation processing, combined with manual interpretation, pipeline detection data semi-automatic processing, solve the traditional data processing process manual workload, data length, high error rate, interpretation difficulties.

3.1. Select the amount of interpretation features

PCCP pipeline detection data by Matlab four images, respectively is orthogonal component image, same component image, amplitude image, phase image (as shown in the figure below), by calculating the slope of the waveform of the four images, found that the same component image, amplitude image, phase image, waveform feature fluctuations, no regularity, not easy to analyze. The waveforms of orthogonal component images have small characteristic fluctuations and obvious rules, which can analyze the detection data of PCCP pipeline by using orthogonal component images.

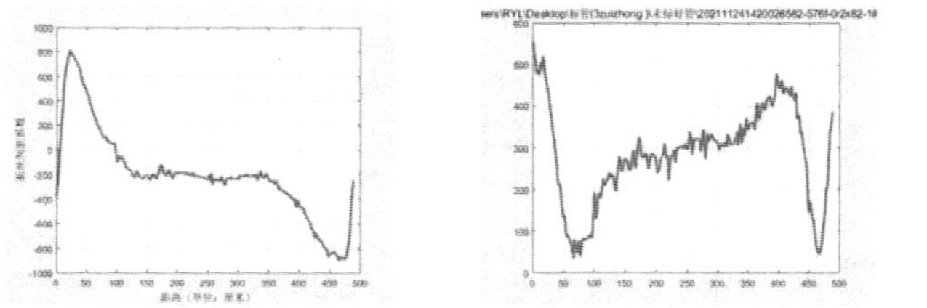

Orthogonal component image Codirectional component image

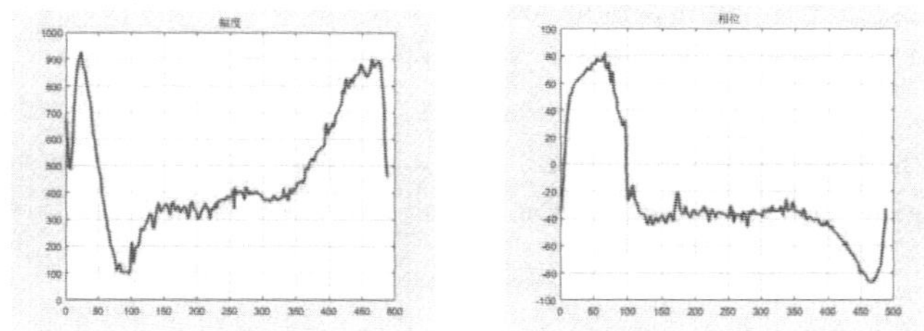

Amplitude image Phase image

Figure 2. Different image types

3.2. Cell data length consistency

Through the analysis of the detection data files, it is found that the length of each PCCP cell is inconsistent. In order to solve this problem, the benchmark cell data (as shown in the figure below) is selected, and the waveform alignment is conducted by manual intervention or partial waveform interpolation, so as to ensure that the waveform length of each pipe is consistent.

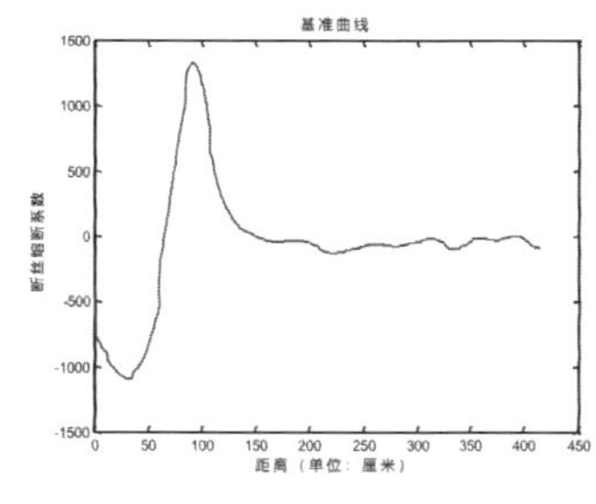

Figure 3. Reference curve

3.3. Read the curve stitching and the reference curve selection

Traditional interpretation way for each section of the pipeline waveform alone interpretation, this way efficiency is very low, and the lack of reference curve contrast, cause high misjudgment rate, therefore, the first the same length cell curve detection curve (length, can be customized by the reader), at the same time make a length of the benchmark curve for its reference comparison.

3.4. Difference value comparison

In the detection data, most of them are normal data. In order to highlight the abnormal data as much as possible, the detection, curve and benchmark curve of equal length and length are differentially processed to eliminate the consistent parts of the two curves, and focus on the comparative analysis of the inconsistent parts. In the process of interpretation, suspected points can be amplified and confirmed manually interpreted, or can be modified and marked. After all the interpretation, statistical anomaly points are carried out to form the evaluation report of the whole pipeline.

4. Silk number estimation rules

4.1. Estimation of calibration data

Peel the mortar protective layer of the selected DN2600 PCCP calibration pipe to expose the prestressed steel wire according to the exposed prestressed steel wire in the order of 1,3,5,10..., and collect the data in the case of various broken wires. For the original phase spectrum data by normalization, background signal suppression, broken wire signal enhancement algorithm processing, can get phase spectrum broken wire discrimination curve, as shown in Figure 5~ figure 8, the abscissa is distance (direction is from the socket to the socket, less than 600cm because the calibration only one pipe and the detector itself is about 1 meter length, so only collect about 500cm distance data), ordinate is broken wire judgment coefficient.

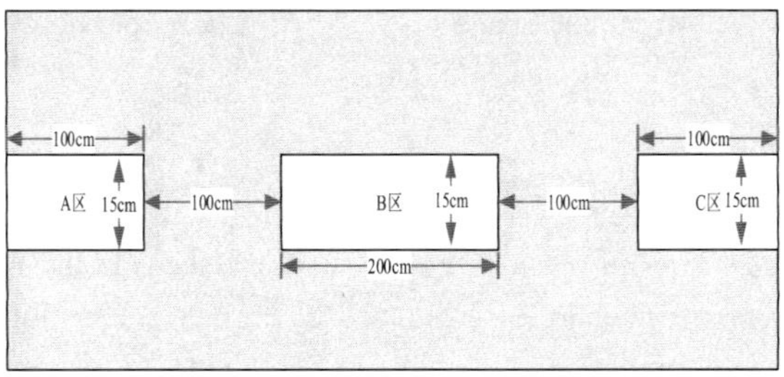

Figure 4. Schematic diagram of the fenestration of the protective layer of the calibration pipe

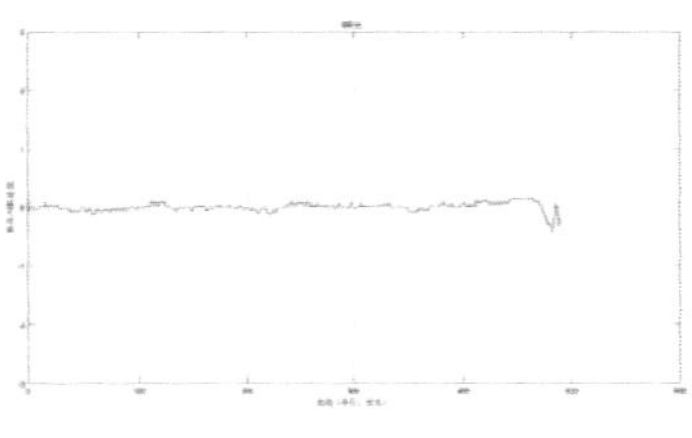

Figure 5. Nonbreak curve

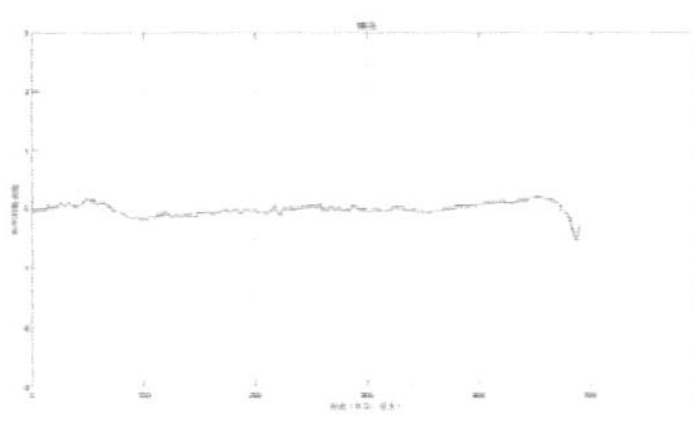

Figure 6. 1 broken wire curve

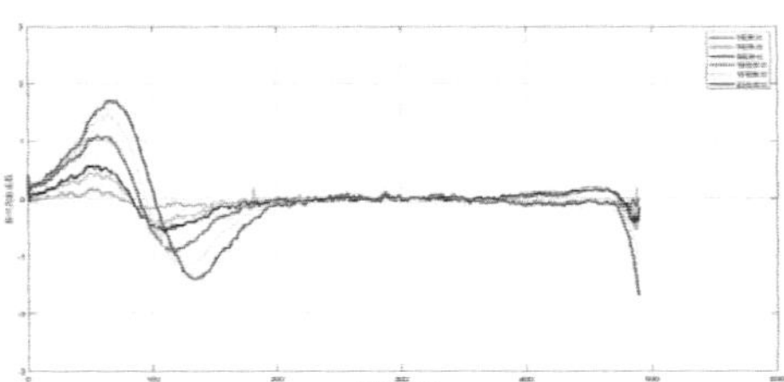

Figure 7. 1~20 broken filaments

In Figure 5, the curve trend in the whole tube section is smooth, although there are fluctuations, but not much, which is a typical unbroken filament feature curve diagram. In Figure 6, the curve first sinks in the range of 0 to 120 cm, and passes through the zero point at about 80cm. Its shape is similar to a sinusoidal curve. This is the typical characteristic of single filament curve, and also the method to judge a small amount of local filament:

- For the "sinusoidal" curve with large peak value, the number of broken wires can be estimated according to the size of peak value;
- The position of the curve over the zero point is used to determine the position of the filament break.

In Figure7 puts the curve of broken filaments in a diagram, and the change trend of the curve with the number of broken filaments can be seen more intuitively.

4.2.The energy integration method of calibration data establishes the standard characteristic energy spectrum of the broken wire

Since the number of broken wire is related to the amplitude and width of the broken wire signal, it is actually related to the energy of the broken wire signal, the energy integration method is adopted to transform the characteristic signal of the broken wire into the corresponding energy, which avoids the uncertainty caused by the comprehensive judgment of multiple parameters. As shown in Figure 8, the standard peak area method of filament characteristic energy spectrum is as follows:

$$Y_i^E = \int_{n1}^{n2} x_{i,n}^2 d_n = x_{i,n1}^2 + x_{i,n1+1}^2 + \dots + x_{i,n2}^2 \qquad (1)$$

$i = 0,1,3,5,10,15,20,25$ is the number of broken roots; $n = 1 \sim 1250$ are the length of data, the point interval is 0.4cm, covering PCCP length 5m. Y_i^E is the standard characteristic energy spectrum of the root, and n_1 and n_2 are the distance coordinates of the positive and negative extreme points of the characteristic curve of the broken filament standard sample.

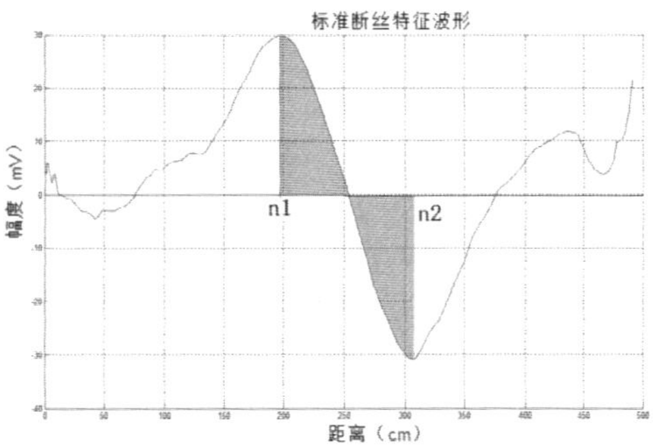

Figure 8. Methods for calculating energy spectral peak integration

In the actual detection process, section PCCP may have multiple filament defects, and when the interval is small, its abnormal signal is likely to not present the complete positive peak-negative peak characteristic waveform due to mutual superposition. For the practicality of the standard characteristic energy spectrum, considering the basic symmetry of positive and negative peaks, the standard characteristic energy spectrum of the number of broken filaments will not affect the estimation result of the number of broken filaments. The characteristic energy spectrum of broken filament established according to the energy area method is shown in Figure 9, and its quantified values are shown in Table 1.

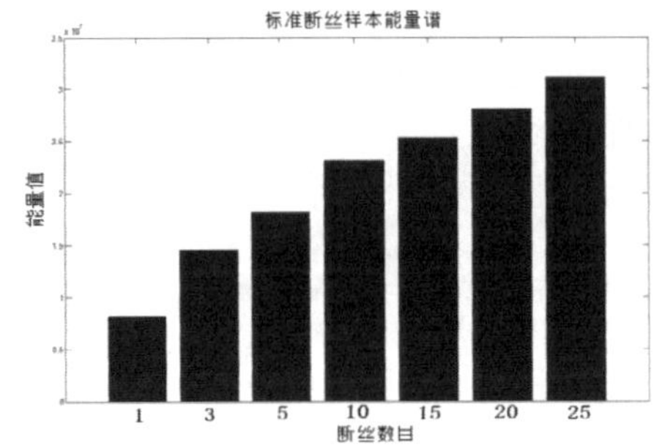

Figure 9. Broken filament standard spectra

Table 1. The relationship between the amplitude of the characteristic energy spectrum and the number of broken aments

Number of broken wires (root)	1	3	5	10	15	20	25
The amplitude of the energy(10^4)	2.72	3.97	5.27	5.93	6.39	6.67	7.13
proportion (%)	7.1	10.4	13.9	15.6	16.8	17.5	18.7

5. Conclusion

5.1. Evaluation results

Then we analyzed the 2 PCCP pipeline on the right line of the deep ditch bridge and the upstream of the left line.

The comparison of the results obtained by this method and the actual test results on the right line of the deep ditch bridge 2 PCCP pipeline is shown in Figure 10.

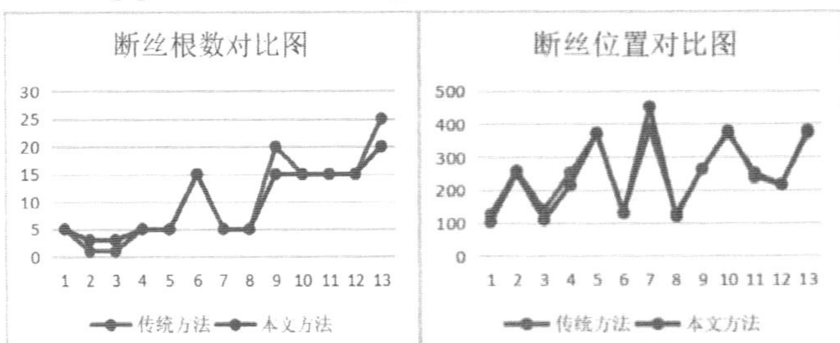

Figure 10. comparing the traditional method and the current method

The comparison of the results obtained by the method of the 2PCCP pipeline upstream of the left line of the deep trench bridge with the actual test results is shown in Figure 11:

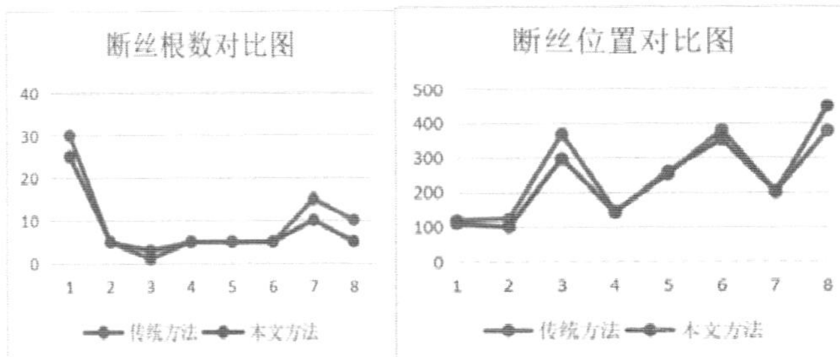

Figure 11. Comparing the traditional method and the current method

5.2. Feasibility

Through the analysis of the broken wire number, position and root number of 2 PCCP pipe upstream of the deep ditch bridge, comparing with the traditional results, found: 14 traditional detection, 18,4 difference; the maximum difference between position and root number is 23m and 5 roots. The traditional workload is 4 people per day, the actual 1 person per day, the results are similar, but the time and manpower are halved to improve efficiency.

The number, position and number of roots on the right line of the deep ditch bridge analyze the analysis with the traditional results: 11 places and 3 differences; the maximum difference between position and roots is 23m and 5 roots. The traditional workload is 4 people per day, the actual 1 person per day, the results are similar, but the time and manpower are halved to improve efficiency.

By comparing the traditional methods and the methods of this paper, the judgment gap between the two methods is not very big, but the detection efficiency of this method is twice that of the traditional method. It shows the feasibility of this method.

References

[1] Shen Y ,Phys,Markus Träuble,et al.This paper is published as part of a PCCP Themed Issue on: Electrocatalysis: theory and experiment at the interface[J].[2024-02-28].

[2] Chai J D , Head-Gordon M .Long-range corrected hybrid density functionals with damped at om-atom dispersion corrections.[J].Physical Chemistry Chemical Physics Pccp, 2008, 10(44):6615-6620.DOI:10.1039/b810189b.

[3] Lee D H , Xu J , Meng Y S .An advanced cathode for Na-ion batteries with high rate and excellent structural stability[J].Physical Chemistry Chemical Physics Pccp, 2013, 15(9).DOI:10.103 9/c2cp44467d.

[4] Hawtin R W , Quigley D , Rodger P M .Gas hydrate nucleation and cage formation at a wa ter/methane interface[J].Physical Chemistry Chemical Physics Pccp, 2008, 10(32):4853-4864.DOI: 10.1039/b807455k.

[5] S.E,Brandt,and,et al.Percutaneous compression plating (PCCP) versus the dynamic hip screw for pertrochanteric hip fractures: preliminary results[J].Injury, 2002.DOI:10.1016/S0020-1383(02)0 0080-3.

[6] Chai J D , Head-Gordon M .Long-range corrected hybrid density functionals with damped at om-atom dispersion corrections.[J].Physical Chemistry Chemical Physics Pccp, 2008, 10(44):6615-6620.DOI:10.1039/b810189b.

Journal of Physics: Conference Series **2887** (2024) 012015 doi:10.1088/1742-6596/2887/1/012015

Detection Method for Asteroid Interior Structure Verified by Scale Model

Zheng Shi[1,*], Chen Gang[1], Li Shuanlao[1], Liu Tao[2], Dai Shun[3], Liu Ruidong[1], Dang Hongxing[1]

[1]Xi'an Branch, China Academy of Space Technology, Xi'an,710100, China,
[2]College of Electronic Science and Technology, National University of Defense Technology, Changsha 410073, China
[3]Key Laboratory of Lunar and Deep Space Exploration, National Astronomical Observatories, Chinese Academy of Sciences, Beijing 100012, China
*shizheng_xjtu@163.com

Abstract. We have developed a radar scheme for detecting the asteroid interior structure. Using the auto rotation of an asteroid, radar system can hover at a fix distance from the asteroid, and continuously measure the internal structure echoes at different spin angles, which can be processed to obtain the internal structure profile image. Through simulation and scale model test, we have verified the internal structure imaging results and conformed the effectiveness of the hovering radar method. The proposed radar scheme was selected as the science payload for internal structure in China's asteroid exploration program aiming at Asteroid 2016HO3.

1. Introduction

Asteroids retain materials from the early solar system, thus many deep space missions including Dawn, Hayabusa, Rosetta, have been performed to explore asteroids. The interior structure plays an important role in understanding the origin and evolution of the asteroids and the solar system. Radar Technique has been approved to be an important method for interior structure detection of a celestial body and widely applied in deep space missions.

The types of deep space object internal structure detection radar can be divided into bi-static radar, mono-static orbiter radar and mono-static rover radar. Bi-static radar measures the transmission waves through the internal structure, such as the CONSERT[1] onboard the Rosetta mission, which obtained the dielectric and material composition of the comet 67P. Mono-static orbiter radar for mars or lunar can move along with the spacecraft, such as the SHARAD[2] onboard the MRO mission and the LRS on KAGUYA mission, and can obtain the subsurface structure under the flight line. Mono-static rover radar, such as the RoSPR for the Tianwen-1 mission, obtains the internal structure below the rover movement route [3]. For the asteroid interior structure, we developed a new detection method, mono-static orbiter hovering radar. The radar hovers at a fixed distance with the spacecraft, and forms an observation line using the auto-rotation of the asteroid. The detection method can obtain asteroid internal structural profile by the motion of the target instead of the spacecraft. By this reason, the

Journal of Physics: Conference Series **2887** (2024) 012015 doi:10.1088/1742-6596/2887/1/012015

mono-static hovering radar has low hardware cost and considerable scientific output, and is being applied to China's asteroid exploration program aiming at Asteroid 2016HO3[4].
In this paper, the detection principle of the proposed mono-static orbiter hovering radar is presented in chapter 2, and chapter 3 presents the simulation results and scale model verification, and chapter 4 concludes the asteroid interior structure detection method.

2. Asteroid interior structure detection method

2.1. Interior structure classification
Some asteroids are monolithic pieces of rock, while most asteroids are aggregated from pieces of their parent asteroid by gravity. In some case, an asteroid forms an onion-like structure with a core, a mantle, and a crust like the Earth. In the other case, asteroids retain rubble-pile structure. Thus, the asteroid interior structure can be divided into three types: monolith, layered onion and rubble pile.

2.2. Detection principle
The mono-static orbiter hovering radar will hover at a fixed distance from the asteroid while the asteroid continued to spin for interior structure detection. The radar will complete one measurement by transmitting electromagnetic waves to the asteroid and receiving the reflected echoes from the discontinuity of the internal medium of the asteroid. During one cycle of the asteroid rotation, the radar implements continuous measurements, which can receive reflected echoes at different rotation angles.
The schematic diagram of the mono-static orbiter hovering radar is as following. The receive signals include surface echoes and subsurface echoes, which contain the information of asteroid interior structure. Through processing and interpreting the radar signals in one rotation period, we can estimated the asteroid interior structure.

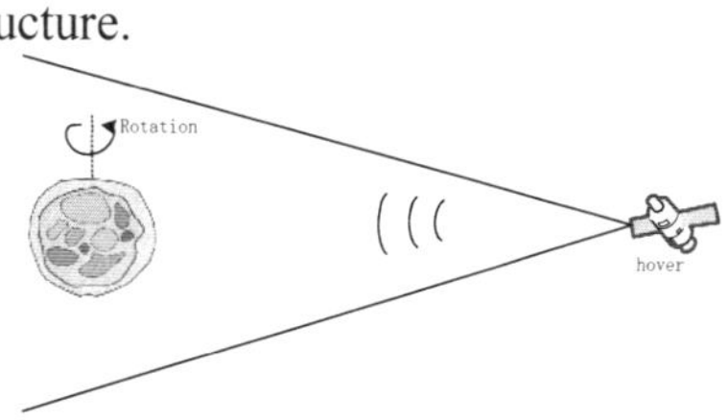

Fig. 1. Schematic diagram of the mono-static hovering radar.

2.3. Transmitted signal
To ensure the safety of the detector, the hovering radar needs to stay at a relatively far distance from the small object target, which leads to a large space attenuation. For the radar system, high average transmitting power is required, so the frequency domain radar signal is suitable. Considering the spin line speed of an asteroid is generally not too fast, the stepping frequency signal can be selected to synthesize large bandwidth through long time summation.
At one observing position, an A-Scan signal is obtained by pulse compression technique with high range resolution, which is decided by the system bandwidth B as $\Delta R = v/2B$, where v represents the propagation velocity inside the asteroid medium.

2.4. Imaging method
For interior structure profile imaging, a large amount of A-Scans at different observing angles are required. Considering the target asteroid usually has a long rotation period, the same scatters can be observed from multiple angles while the probe hovering. It forms an Inverse Synthetic Aperture (ISAR) imaging model. The multiple received signals corresponding to one scatter can be coherently accumulated for obtaining high resolution along the rotation direction and imaging the interior structure profile.

Back Projection (BP) is a suitable imaging algorithm for interior structure in the reason that it is easy to compensate velocity changes in the medium.

For an interior scatter A in Fig. 2, transmitted waves propagate through vacuum with length ΔR_{i1} and interior medium with length ΔR_{i2} for arriving at scatter A, and the received waves would play the same wave.

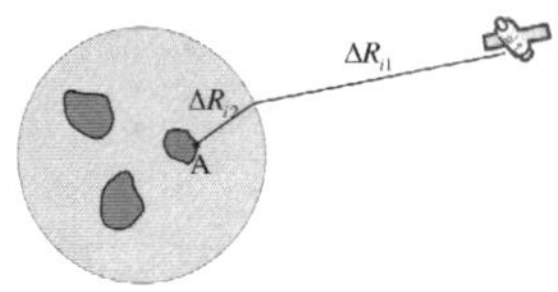

Fig. 2. Imaging of the interior structure profile.

Then the total propagating time corresponding to scatter A would be

$$\Delta \tau_i = \frac{2\Delta R_{i1}}{c} + 2\int_{\Delta R_{i2}} \frac{\sqrt{\varepsilon(\vec{r})}}{c} d\vec{r} \tag{1}$$

where $\varepsilon(\vec{r})$ represents the relative permittivity at position $\vec{r}$. BP algorithm calculates the propagating time of every interior points and accumulates the coherent echoes.

3. Result and discuss

3.1. Simulation result

Assume a rubble pile asteroid with the interior permittivity contribution as Fig. 3 (a). Size of this asteroid is about 40 m. The probe hovers at about 50 m from asteroid's center. The asteroid's rotation forms the equivalent radar observations at the white points in Fig. 3 (a). The received signals after pulse compression are shown in Fig. 3 (b). By applying BP algorithm, we get the asteroids interior imaging result as illustrated in Fig. 3 (c).

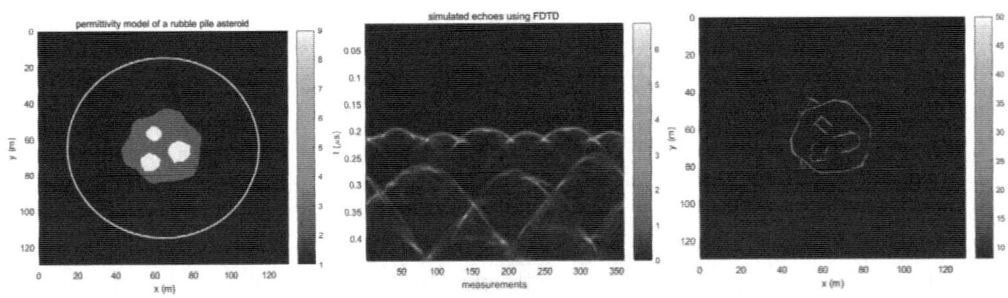

(a) (b) (c)

Fig. 3. Simulation of asteroid interior structure detection. (a) Asteroid model and radar positions(white points); (b) B-Scan results after pulse compression; (c)Imaging result through BP algorithm

In Fig.3(c), three interior targets are clearly recognized, especially the surface near the asteroid surface. Comparing Fig.3(c) and Fig.3(a), the asteroid interior structure imaging result is highly consistent with the real interior structure.

3.2. Scale model result

In order to verify the internal structure detection results of mono-static hovering radar, a scale model anechoic chamber test was built, with the target reduced to about 30cm, and the auto rotation was imitated by a turntable, and the working frequency was increased to 1~18GHz. Radar signals were transmitted and received by vector network analyzer and standard antennas. The scale model with irregular shape and complex internal structure was produced by 3D printing technology.

To build a rubble-pile asteroid model, the scale model was filled with sand, among where two balls were buried. One ball was empty air ball, and the other one was a water ball, as shown in the Fig. 5 (a). The B-scan data within a rotation period is shown Fig. 5 (b), and the internal structural profile after BP imaging is shown Fig. 5 (c).

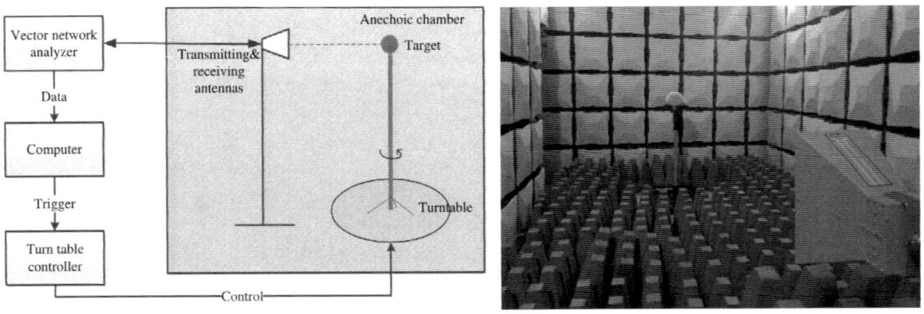

(a) (b)

Fig. 4. Scale model test. (a) box diagram; (b) real photo

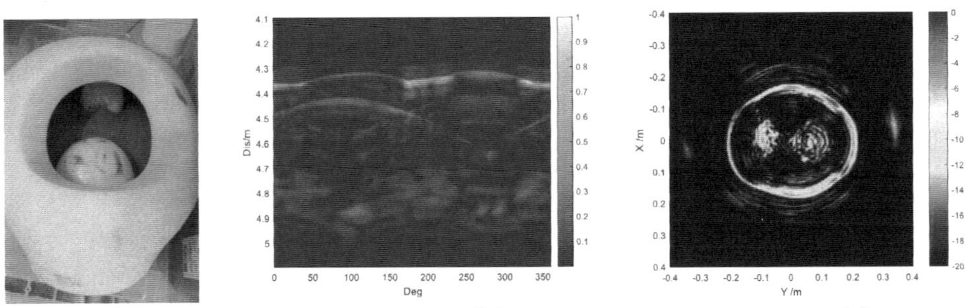

(a) (b) (c)

Fig. 5. Interior structure imaging results (a)rubble-pile analogy; (b) B-scan; (c) imaging result

It can be seen that in Fig. 5(b), the echoes of two internal spheres can be collected at multiple rotation angles. In the results of Fig. 5(c), after multi-angle echo synthesis, the two internal spheres are accumulated to the correct position, and the internal structure profile is correct.

4. Conclusion

For asteroid internal structure detection, we proposed a mono-static orbit hover radar to detect asteroid internal structure, using the asteroid auto rotation to obtain enough measurements to form a profile result. The simulation test and scale model test show that the internal structure profile can be clearly imaged and the effectiveness of the proposed detection method is validated.

5. References

[1] WLODEK KOFMAN, ALAIN HERIQUE, YVES BARBIN, JEAN-PIERRE BARRIOT, 2015, Properties of the 67P/Churyumov-Gerasimenko interior revealed by CONSERT radar, *Science*, Vol. 349

[2] Renato Croci, Roberto Seu, and Enrico Flamini, The SHAllow RADar (SHARAD) Onboard the NASA MRO Mission, *Proceedings of the IEEE*, Vol.99,No.5,May2011

[3] Bin Zhou, S. S. 2020. "The Mars rover subsurface penetrating radar onboard China's Mars 2020 mission." *Earth and Planetary Physics.*

[4] Ruidong Liu 2023, Design and Ground Experiment of Asteroid Internal Structure Detection Radar (AISDR) onboard Tianwen-2 Mission, *IET International Radar Conference 2023*

Journal of Physics: Conference Series **2887** (2024) 012016 doi:10.1088/1742-6596/2887/1/012016

GPR Image Retrieval and Localization Based on Spatial Prior Constraints

Beizhen Bi[1], Yuwei Chen[1], Liang Shen[2*], Pengyu Zhang[1], Xiaotao Huang[1] and Tian Jin[1]

[1]College of Electronic Science and Technology, National University of Defense Technology, Changsha, Hunan 410073,China (bibeizhen@nudt.edu.cn)
[2]Test Center, National University of Defense Technology, Xi'an 710106, China

Abstract. Localizing Ground Penetrating Radar (LGPR) offers the distinct advantage of being unaffected by weather and light conditions changes. As a novel auxiliary driving localizing system, LGPR enhances the robustness of the auto-drive system. Traditional ground penetrating radar localizing technologies rely heavily on high-quality map data, exhibiting optimal performance in environments with complete, high-precision maps. However, deviations from the mapped trajectory can compromise localization accuracy. We introduce a novel approach: a ground penetrating radar image retrieval and localization method based on spatial prior constraints of array antennas to address this challenge. This method aims to mitigate mismatches in ground penetrating radar navigation and localization encountered under incomplete map coverage. We performed a comparative analysis with the LGPR baseline and then validated them experimentally on our dataset. The results show that the proposed method achieved significant enhancement in accuracy. In particular, the superiority of the proposed method is demonstrated in low overlap GPR image matching and localisation.

1. Introduction

With the continuous development of autonomous driving, ground penetrating radar systems have provided a new research direction for existing navigation and localization methods. MIT first proposed the concept of LGPR [1], which realizes navigation and localization by utilizing relatively stable subsurface features and aligning subsurface data. Compared to camera and LiDAR assisted driving localization schemes, LGPR relies less on aboveground information and is less susceptible to weather effects and changes in lighting conditions. The LGPR system, as a new assisted driving localization system, complements the field of autonomous driving by enhancing the resistance of existing localization technologies to common failure modes and robustness to environmental changes.

The existing GPR navigation and localization technology mainly focuses on map-based matching. Literature [2] uses A-scan to achieve GPR antenna localization on known underground maps. Literature [3-6] uses B-scan images and 3D maps to achieve registration and localization. The literature [3] completed the design of a fully autonomous navigation and localization system based on LGPR under adverse weather conditions. Literature [5] [6] proposed an end-to-end network-based regression method to solve GPR localization under different weather conditions. Existing studies have shown better localization results with complete maps. Still, due to the limited width of the GPR, it is difficult to get exclusive map data from a single measurement, and the problem of splicing and mapping multiple sets of trajectories is still tricky, so incomplete map data often occurs in LGPR. The localization effect will deteriorate when the driving route deviates from the map data. Therefore, it is necessary to solve the problem of stable localization in the case of an incomplete map.

Journal of Physics: Conference Series **2887** (2024) 012016 doi:10.1088/1742-6596/2887/1/012016

In this paper, we propose a GPR image retrieval and localization method that can improve localization accuracy by utilizing the GPR array antenna as a prior constraint to address the issue of mismatches in LGPR when the map is incomplete.

2. Method

2.1. LGPR based on image retrieval

LGPR based on image retrieval mainly includes: map construction and real-time image retrieval localization. As shown in the left of Fig.1, the complete image corresponding to the real-time image can be extracted from the map when the map is complete so that the real-time image can be directly utilized for retrieval and localization. As shown in the middle of figure1, when the map is incomplete, the map cannot obtain the complete image corresponding to the real-time image, and the corresponding location only partially overlaps with the real-time image. As the overlap area between the real-time image and the map decreases, the invariant information that can be used for matching decreases, and the redundant information in the non-overlapping area increases accordingly, which significantly reduces the match ability between the images and seriously affects the retrieval results.

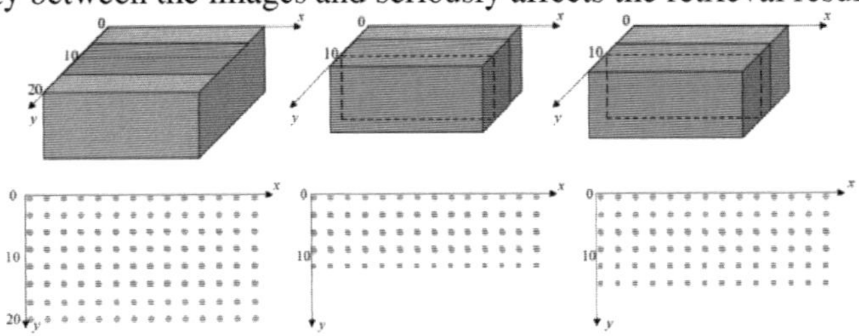

Figure 1. Maps and real-time trajectories under different offsets.

To reduce the adverse effects of redundant information, we use A-scan data to retrieve the positions of each channel of the GPR antenna array. Localization can be achieved by determining the position of each channel and fitting it to obtain the position of the antenna array. However, due to the limited amount of information in the A-scan, many erroneous results may occur during retrieval. To improve accuracy, we perform robust feature extraction and similarity measurement design on A-scan while utilizing spatial prior information of array antennas for constraints in order to obtain precise localization results. The basic framework is shown in Fig.2:

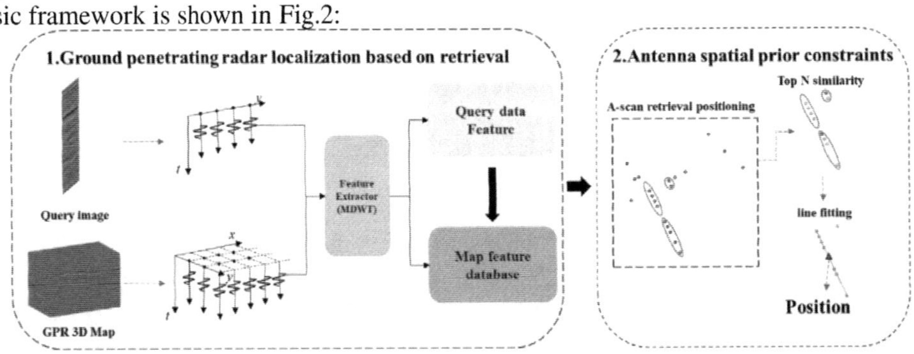

Figure 2. Overview of the GPR image query based spatial prior constraints framework.

2.2. Similarity measurement based on multi-level discrete wavelet transform

The literature [5] proves that low-frequency information is more advantageous for localization results in LGPR. Therefore, we decompose the A-scan from the frequency domain and use the wavelet transform to obtain more stable low-frequency features. We are using dB wavelet for multi-level decomposition of A-scan data. On multi-level wavelet decomposition, approximate coefficients are input for the next generation. Repeat the above decomposition process to construct A-scan data features for all approximation coefficients: $x=[x_1^l, x_2^l, x_3^l ...]$.Using Euclidean distance as a similarity measure for

A-scan features: $M = \|\hat{x} - x\|_2$ .The smaller the numerical value, the smaller the distance between features, and the higher the feature similarity.

2.3. Retrieval and localization based on spatial prior constraints

Map feature database is constructed using the multi-level discrete wavelet transform to extract features from the A-scan in the map. The A-scan data in real-time B-scan are transformed to obtain a single-channel data feature set F . The data features of m channels are searched for similarity in W , and the position corresponding to the minimum value is taken as the position of the feature to be queried. The real-time image channel position set P is obtained.

Using the spatial prior information of array channels, construct a spatial prior constraint matrix R , and calculate the distance D between channels based on channel position information:

$$R = \begin{bmatrix} 0 & d_c & 2d_c & \cdots & (m-1)d_c \\ d_c & 0 & d_c & \cdots & (m-2)d_c \\ 2d_c & d_c & 0 & \cdots & (m-3)d_c \\ \vdots & \vdots & \vdots & \ddots & \vdots \\ (m-1)d_c & (m-2)d_c & (m-3)d_c & \cdots & 0 \end{bmatrix} \tag{1}$$

$$D = \begin{bmatrix} 0 & \|p_1 - p_2\|_2^2 & \|p_1 - p_3\|_2^2 & \cdots & \|p_1 - p_m\|_2^2 \\ \|p_1 - p_2\|_2^2 & 0 & \|p_2 - p_3\|_2^2 & \cdots & \|p_2 - p_m\|_2^2 \\ \|p_1 - p_3\|_2^2 & \|p_2 - p_3\|_2^2 & 0 & \cdots & \|p_3 - p_m\|_2^2 \\ \vdots & \vdots & \vdots & \ddots & \vdots \\ \|p_1 - p_m\|_2^2 & \|p_2 - p_m\|_2^2 & \|p_3 - p_m\|_2^2 & \cdots & 0 \end{bmatrix} \tag{2}$$

Among them, d_c represents channel interval and p represents channel positions. The distance between effective search pairs should satisfy the formula:

$$abs(R_{ij} - D_{ij}) \le \Lambda \tag{3}$$

Among them, Λ is the acceptable distance error threshold, $\Lambda = (\sqrt{2} - 1)d_c$.

Corresponding search channels can be obtained based on valid searches. The first N channels with smaller feature distances are taken as valid channels. The channel layout should conform to a linear distribution, so a straight line is fitted to the channel positions to obtain the slope matrix A:

$$A = \begin{bmatrix} NaN & & & & \\ a_{12} & NaN & & & \\ a_{13} & a_{23} & NaN & & \\ \vdots & \vdots & \vdots & \ddots & \\ a_{1N} & a_{2N} & a_{3N} & \cdots & NaN \end{bmatrix} \tag{4}$$

The slope value with the highest frequency of occurrence is the slope of the fitted array line, and the intercept b of the line l is determined based on the corresponding coordinates. Calculate the position of the array center based on the line and the corresponding channel number as shown in Equation (6), where $\theta = \text{atan}(a)$. The position of the center of the array (px, py) is the result of the LGPR.

$$(px, py) = (x_i + (10 - C)\cos(\theta) \cdot l, y_i + (10 - C)\sin(\theta) \cdot l) \tag{5}$$

3. Experimental and Result

3.1. Dataset

The system platform used for data collection mainly includes 3D-radar, DMI, RTK and vehicle platform. To reduce the impact of multi-sensor errors, data collection is strictly collected according to the preset track. The GPR uses fixed interval sampling with a sampling interval of 0.75m and collects three sets of data: non-offset, offset 6 channels, and offset 12 channels.

3.2. Result

Non-offset data is used as the map data, while offset 6 channels and offset 12 channels are used as two real-time data for localization. As shown in Figure 3, the image retrieval localization result is considered correct when the retrieval result is within two red boundary lines and deviates from the actual value when it is opposite.

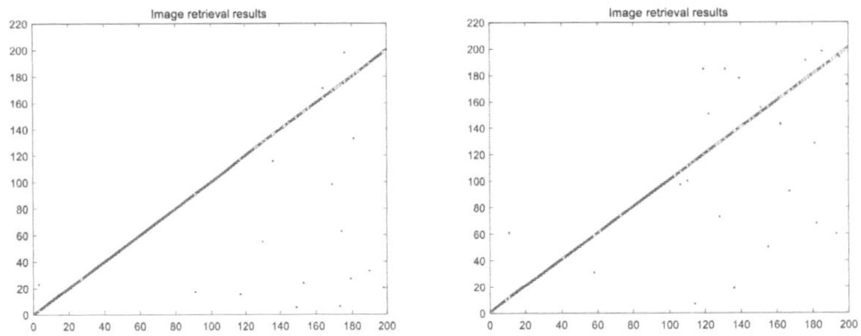

(a) Offset 6 channels (b) Offset 12 channels **Figure 3.** Image retrieval results

When the localization error is within 0.16m, we consider the localization result to be accurate. Comparing it with the baseline proposed by MIT, we can obtain the localization accuracy as shown in Table 1. The results show that the proposed method has higher localization accuracy compared to the baseline method, and its advantages are more detailed when the offset increases, which can ensure the robustness of LGPR.

Table 1. Comparison of localization accuracy

Offset	Baseline	Proposed
6 channels	0.4900	0.8550
12 channels	0.1300	0.6100

4. Conclusion

This paper proposes a retrieval and localization method that utilizes A-scan for similarity measurement and combines spatial prior constraints of array antennas to address the issue of high mismatch rate in low overlap image registration under incomplete map data. Compared with existing methods, it improves the accuracy of LGPR in the case of incomplete maps, providing important guarantees for achieving safe and reliable GPR localizing.

5. References

[1] M. Cornick, J. Koechling, B. Stanley, and B. Zhang, "Localizing ground penetrating radar: A step toward robust autonomous ground vehicle localization," Journal of field robotics, vol. 33, no. 1, pp. 82–102, 2016.

[2] E. Skartados, A. Kargakos, E. Tsiogas, I. Kostavelis, D. Giakoumis,and D. Tzovaras, "GPR antenna localization based on a-scans," in 2019 27th European Signal Processing Conference (EUSIPCO). IEEE, 2019, pp. 1–5.

[3] T. Ort, I. Gilitschenski, and D. Rus, "Autonomous navigation in inclement weather based on a localizing ground penetrating radar," IEEE Robotics and Automation Letters, vol. 5, no. 2, pp. 3267–3274, 2020.

[4] Z. NI, S. YE, C. SHI, J. PAN, Z. ZHENG, and G. FANG, "A deep learning assisted ground penetrating radar localization method," Journal of Electronics and Information Technology, vol. 44, no. 4, pp. 1265– 1273, 2022.

[5] T. Ort, I. Gilitschenski, and D. Rus, "Grounded: A localizing ground penetrating radar evaluation dataset for learning to localize in inclement weather," The International Journal of Robotics Research, vol. 42, no. 10, pp. 901–916, 2023.

[6] B. Bi, L. Shen, P. Zhang, X. Huang, Q. Xin, and T. Jin, "TSVR-net: An end-to-end ground-penetrating radar images registration and location network," Remote Sensing, vol. 15, no. 13, p. 3428, 2023.

Acknowledgments

The authors would like to thank the State Key Laboratory, Changsha University of Science and Technology, Changsha, China for providing equipment support.

Journal of Physics: Conference Series **2887** (2024) 012017 doi:10.1088/1742-6596/2887/1/012017

Enhanced Scoring Instance Segmentation Network for Subsurface Targets Recognition from GPR B-Scans

Feifei Hou[1], Xue Yin[1], Hai Liu[2,*], and Boxuan Qiao[1]

[1]School of Automation, Central South University, Changsha 410083, PR China
[2]School of Civil Engineering, Guangzhou University, Guangzhou 510006, PR China
[*]Corresponding author: hliu@gzhu.edu.cn

Abstract. Ground Penetrating Radar (GPR) is a non-destructive technique used for detecting subsurface structure and space. However, the interpretation of field GPR data is quite labour-intensive and time-consuming. This paper develops a deep learning-based approach for the detection and segmentation of subsurface targets based on GPR data. Since existing deep learning methods do not always accurately reflect the true quality of segmentation masks and exhibit locality when handling details in complex scenes, the proposed method employs the Mask Scoring R-CNN (MS R-CNN) framework as its primary framework and makes additional modifications to its backbone and neck architecture. Experimental results on GPR bridge dataset show notable improvements in loss metrics, segmentation accuracy, and bounding box precision. These enhancements have effectively reduced overall losses and enhanced the performance of Average Precision index of the enhanced MS R-CNN, demonstrating the effectiveness of our approach in the accurate identification of underground objects under complex environments.

1. Introduction

Ground Penetrating Radar (GPR), well-established in the geophysical field, is extensively utilized in the realm of civil engineering to detect unseen elements such as embedded rebar, subterranean installations, or voids beneath roads. Traditional methods for underground target detection in GPR applications, such as Otsu's thresholding and Full Waveform Inversion, have significantly evolved since the late 20th century to enhance segmentation accuracy and advance subsurface exploration technology. Although they have advanced target segmentation, they are limited by their reliance on manual feature extraction. After that, numerous deep learning methods, with technologies such as CNNs[1], U-Net[2], GANs[3, 4], and GCNs[5], further revolutionized accuracy and efficiency, particularly with Mask R-CNN's advanced segmentation. However, using classification confidence as a mask quality score has limitations, as it may not accurately reflect mask quality due to occlusions or algorithmic constraints, especially in complex environments [6, 7].

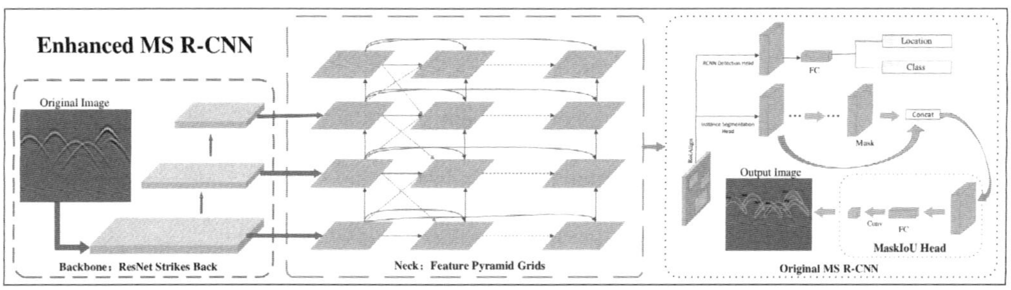

Figure 1. Illustration of GPR target detection and segmentation framework.

This research introduces a deep learning-enhanced methodology that builds upon the Mask Scoring R-CNN (MS R-CNN).[8] framework, emphasizing enhancements in the backbone and neck regions. Our approach integrates the 'ResNet Strikes Back' (ResNet SB) architecture for more efficient feature extraction and employs 'Feature Pyramid Grids' (FPG) for the consolidation of multi-level features. Additionally, the incorporation of a Region Proposal Network (RPN) alongside a mask IoU head significantly boosts the precision of target detection and mask quality assessment. Figure 1 shows the proposed framework.

2. Enhanced MS R-CNN framework

This section initially previews the fundamental structure of the original MS R-CNN, followed by details of the Enhanced MS R-CNN model's structural improvements. The framework of the Enhanced MS R-CNN is depicted in Figure 1.

The MS R-CNN builds upon the Mask R-CNN by introducing a MaskIoU head for Enhanced mask quality evaluation, improving the accuracy of instance segmentation by ensuring that mask quality is considered alongside classification scores. This refinement addresses the discrepancies between classification confidence and mask accuracy in prior models. To improve the performance of the previous framework, the MS R-CNN model is enhanced by incorporating:

(1) The ResNet SB serves as the backbone network, integrating FPG within the feature pyramid to bolster feature fusion and address information loss at higher levels. The AdamW optimizer is employed for network optimization (see Equation 1) and a loss function with weight penalties is used to counter overfitting (see Equation 2), ensuring both accuracy and robustness.

$$\theta_{t+1} = \theta_t - \frac{\eta}{\sqrt{\hat{v}_t} + \grave{o}}\hat{m}_t \tag{1}$$

where θ_{t+1} represents the updated parameter values for the next iteration of the model process, θ_t is the current parameters, η denotes a fraction of the learning rate, $\hat{v}_t$ represents the square root of the bias-corrected second raw moment estimate, $\hat{m}_t$ is the bias(corrected first moment estimate) and $\grave{o}$ is a small constant.

$$L = L_{\text{original}} + \lambda \sum \theta^2 \tag{2}$$

where L represents the total loss function, L_{original} is the original loss function, λ is the parameter and θ denotes the sum of the squares of the model parameters.

(2) Integrating FPG into the feature pyramid for Enhanced feature fusion, and minimizing information loss in higher-level features. FPG merges multi-scale and multi-level features from the ResNet SB into a comprehensive representation, improving the detection of small targets and internal features within GPR images. (3) The head and loss function retain the original MS R-CNN's MaskIoU head learn the quality of predicted instance masks and L2 loss to accurately predict the IoU score. These structures collectively elevate the model's performance in detecting and segmenting 'pec' metal targets in GPR images, offering an accurate solution for single-category object recognition and segmentation tasks.

3. Experimental Results

3.1. Experimental Settings and Datasets

To validate the proposed method, two datasets were used: a simulated dataset containing 1000 images generated by gprMax. The simulated dataset is divided into three groups of images, each containing 1-9 targets, 10-15 targets and 16-20 targets, respectively, for each image. The ratio of the number of images in each group is 5:3:2. Another on-site dataset consists of 80 images collected from Neyland Greenway bridge at the University of Tennessee, Knoxville, USA, with GPR equipment mounted with an antenna of 2000 MHz. The experimental setup included an NVIDIA GeForce RTX 3090 GPU and utilized PyTorch 1.13.1 with CUDA 11.7 for GPU acceleration. The experimental setup for the model evaluation was configured with a batch size of 2, a minimum positive Intersection over Union (IoU) of 0.3, and

Journal of Physics: Conference Series **2887** (2024) 012017 doi:10.1088/1742-6596/2887/1/012017

IoU thresholds for negatives and positives set at 0.3 and 0.7, respectively. The model was trained over 40 epochs with a specified learning rate of 2×10^{-4}. The entire dataset is split into a training-test ratio of 9:1, combining simulated and field data.

3.2. Performance Evaluation

In this section, four experimental subjects are considered: the original MS R-CNN, Enhanced MS R-CNN I with only the backbone updated, Enhanced MS R-CNN II with only the neck updated, and Enhanced MS R-CNN III, which updates both the backbone and neck (excluding the Enhanced MS R-CNN mentioned in Section 3). Figure 2 shows the effectiveness of the Enhanced MS R-CNN III model, demonstrated by reduced losses in classification, bounding box(bbox), and mask predictions. The model maintains high accuracy in object identification and precision in localization and segmentation. In scenarios with minimal target overlap, it achieves detection rates with over 80% confidence in both simulated test data and field data. Even in complex environments with scattered targets, it identifies over 95% of objects, maintaining high confidence. Notably, it distinguishes true targets from diffraction waves in field datasets, avoiding most false positives. Furthermore, Figure 2 (c1-c2) shows the model's capability to accurately identify underground PVC targets, which have a radar response similar to metal pipes, with average confidence exceeding 80%, underscoring its generalizability and practicality.

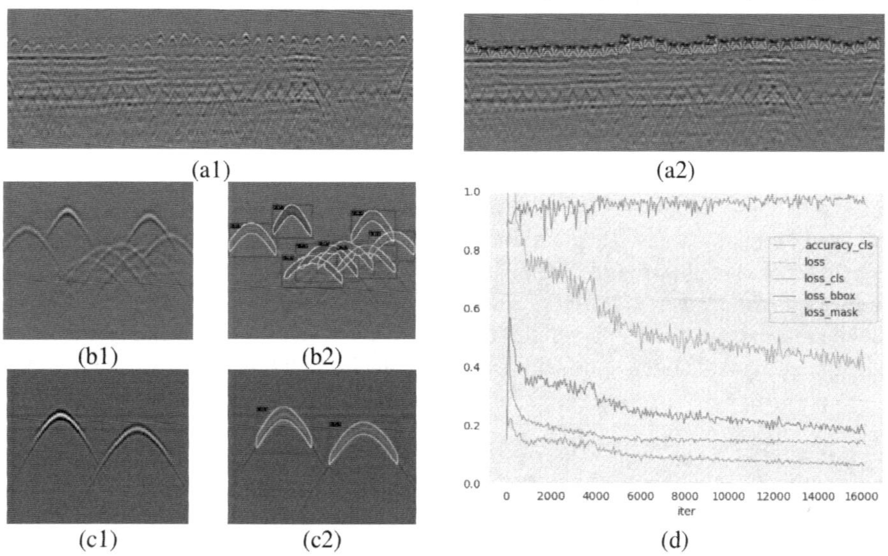

Figure 2. (a1-a2) illustrates results on field data, (b1-b2) illustrates results on simulated test data, (c1-c2) shows the results on simulated data of 'PVC' and (d) shows the loss distribution of Enhanced MS R-CNN III.

3.3. Ablation Experiments

The various performances of the Enhanced model are shown in this section.

Table 1. Loss performance comparison of different frameworks.

	Framework details		**Loss performance**		
Method	**Backbone**	**Neck**	**total loss**	**rpn_bbox**	**mask_iou**
Original MS R-CNN	ResNet-50	FPN	0.598	0.0027	0.027
Enhanced MS R-CNN I	ResNet SB	FPN	0.601	0.0035	0.026
Enhanced MS R-CNN II	ResNet-50	FPG	0.597	0.0037	0.027
Enhanced MS R-CNN III	ResNet SB	FPG	**0.525↓**	**0.0022↓**	**0.020↓**

Table 1 displays a loss performance comparison of different frameworks. The Enhanced MS R-CNN III outperformed other models in the GPR echo image dataset, achieving the lowest overall loss of 0.525, which is approximately 7% lower than its predecessor. It demonstrated higher accuracy in target localization and classification, as evidenced by the lowest losses in RPN classification and bbox. The model also excelled in AP metrics, particularly at lower IoU thresholds. Table 2 presents a segmentation and bbox performance comparison of different frameworks. It illustrates the Enhanced MS R-CNN III's superior detection accuracy with a bbox mAP of 0.576 and significant improvements in bbox mAP_50 and bbox mAP_75, suggesting enhanced capabilities in detecting and segmenting underground metal objects with GPR images.

Table 2. Segmentation and bbox performance comparison of different framework.

	segmentation performance			**bbox performance**		
Method	**mAP**	**mAP_50**	**mAP_75**	**mAP**	**mAP_50**	**mAP_75**
Original MS R-CNN	0.509	0.904	0.545	0.574	0.938	0.625
Enhanced MS R-CNN I	0.508	0.901	0.543	0.577	0.943	0.638
Enhanced MS R-CNN II	0.504	0.911	0.528	0.575	0.940	0.620
Enhanced MS R-CNN III	**0.512↑**	**0.912↑**	**0.545-**	**0.576↑**	**0.946↑**	**0.639↑**

4. Conclusion and Future Work

The performance of the experiments indicates that the proposed enhancements to the MS R-CNN have been instrumental in advancing the detection of bridge steel bars. When benchmarked against the original MS R-CNN and its subsequent iterations, the Enhanced MS R-CNN demonstrates a superior capability to accurately detecting and segmenting targets across various environments. The results reveal outstanding performance on field data, simulated data, and other targets exhibiting radar responses similar to metal pipes, characterized by a low rate of false detections and a high degree of accuracy in segmentation tasks. This underscores the generalization ability and effectiveness of our deep learning approach in interpreting complex GPR data for subsurface explorations. However, there are still some limitations to the effectiveness of current methods. For example, accurately obtaining the target shape against various backgrounds is still far from sufficient. We also need to perform localization detection and structural state analysis on underground steel targets. Furthermore, the field dataset used in our experiment is limited, and we need to collect more data on site to improve the accuracy of the method.

Reference

[1] Zhang J, Yang X, Li W, Zhang S, Jia Y. *Automatic detection of moisture damages in asphalt pavements from GPR data with deep CNN and IRS method.* Automation in Construction, 2020. **113**: p. 103119.

[2] Dai Q, He Y, Lei Y, Lei J, Wang X, Zhang B. *GPR Data Reconstruction Using Residual Feature Distillation Block U-Net.* IEEE Journal of Selected Topics in Applied Earth Observations and Remote Sensing, 2023.

[3] Xiong H, Li J, Li Z, Zhang Z. *GPR-GAN: A Ground-Penetrating Radar Data Generative Adversarial Network.* IEEE Transactions on Geoscience and Remote Sensing,2023.**62**: p.1-14.

[4] Fazeel A, Rottmayer J, Mehta R, Bajcinca N, editors. *GPR-GANs: Generation of Synthetic Ground Penetrating Radargrams Using Generative Adversarial Networks.* in *2021 11th International Workshop on Advanced Ground Penetrating Radar (IWAGPR)*. 2021. IEEE.

[5] Shao Z, Shi D, Han A, Vasnev A, Guo Y, Gao J. *Enhancing framelet GCNs with generalized p-Laplacian regularization.* International Journal of Machine Learning and Cybernetics, 2023: p.1-21.

[6] Liu Z, Yeoh JK, Gu X, Dong Q, Chen Y, Wu W, et al. *Automatic pixel-level detection of vertical cracks in asphalt pavement based on GPR investigation and Enhanced mask R-CNN.* Automation in Construction, 2023. **146**: p. 104689.

[7] Hou F, Lei W, Li S, Xi J, Xu M, Luo J. *Enhanced Mask R-CNN with distance guided intersection over union for GPR signature detection and segmentation.* Automation in Construction, 2021. **121**: p. 103414.

[8] Huang Z, Huang L, Gong Y, Huang C, Wang X. *Mask Scoring R-CNN*, in *2019 IEEE/CVF Conference on Computer Vision and Pattern Recognition (CVPR)*. 2019. p. 6402-6411.

Journal of Physics: Conference Series **2887** (2024) 012018 doi:10.1088/1742-6596/2887/1/012018

Simulation analysis of ground-penetrating radar in seasonal freezing soil area with temperature-varying media

Guojiang Lu, Qinghua Chen and Xianwu Zhang*

Key Laboratory of Signal Detection and Processing, College of Information Science and Engineering, Xinjiang University, Urumqi 830000, China
* Corresponding author's e-mail: zxw@xju.edu.cn

Abstract. Affected by the seasonal freezing area climate, the shallow engineering geological electromagnetic environment exhibits seasonal variations, affecting the propagation process of electromagnetic waves in shallow geological soil. This brings excellent difficulties in accurately identifying the detection target. In this paper, Kaba-he, the seasonal freezing area of Xinjiang, is taken as an example to build a model of the dielectric constant of the temperature variable soil medium in different seasons, and utilizing the high-order O(2,4) FDTD algorithm implemented with MATLAB software. This experiment provides the propagation velocity of electromagnetic waves in shallow subgrade soil of Kaba-he during different seasons and, based on numerical simulation results, gives the calculated depth of the underground targets buried in other seasons. This research can provide theoretical guidance for the application of GPR in Kaba-he.

Key words: GPR, Seasonal freezing soil area, FDTD, Forward modeling

1. Introduction

According to relevant research, temperature significantly impacts the dielectric constant of soil[1, 2], and some researchers have established complex soil dielectric constant models that include many factors, such as temperature[3-5]. Due to the climate in the seasonal freezing area, the shallow engineering geological electromagnetic environment exhibits seasonal variations, impacting the propagation of electromagnetic waves in shallow geological soil. This poses significant challenges for accurately identifying the detection target. In reality, the underground soil medium is complex, making it challenging to establish geological dielectric parameter models under complex environmental conditions, and it isn't easy to conduct field verification. Numerical simulation techniques have emerged as a practical solution, and as a standard electromagnetic calculation method, FDTD is widely used in the numerical simulation of GPR[6].

In this paper, taking the Kaba-he in the seasonal freezing area of Xinjiang, the relative dielectric constant model of soil in different seasons of this seasonal freezing area has been constructed. Then, numerical simulations using the advanced O(2,4) FDTD algorithm on the MATLAB platform were conducted to investigate the seasonal variations of relative dielectric constants in the temperature-changing soil of this seasonal freezing area. These simulation experiments aim to elucidate the propagation process of electromagnetic waves within the soil of the roadbed during different seasons in the seasonal freezing area. The calculated depth of GPR detection in other seasons is analyzed for the same buried target.

2. Forward simulation based on FDTD

2.1. Data presentation

Based on relevant research, the relative dielectric constant of soil and the data of subgrade soil temperature and moisture content in Kaba-he can be obtained(Figure 1a to Figure 1c)[7,8]. According to these data, the average moisture content of the subgrade soil of Kaba-he in March and December is 12.36%, and the average moisture content of the subgrade soil of Kaba-he in June and October is 19.97%.

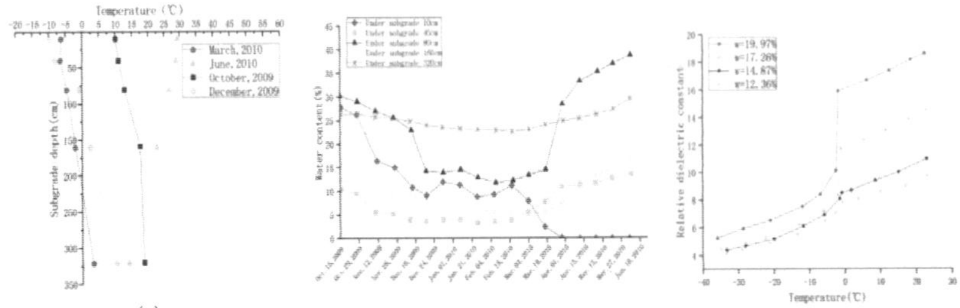

(a) (b) (c)

Figure 1 (a) Subgrade soil temperature in different months, (b) Subgrade soil moisture in different months, (c) Relative dielectric constant of soil under different temperature and water content.

2.2. Modeling and simulation of temperature-varying medium

Based on the temperature distribution data of each layer of subgrade soil in Kaba-he and the soil relative dielectric constant data under different moisture content and temperature conditions, the soil relative dielectric constant data at depths of 0.1m, 0.8m, 1.6m, and 3.0m within the subgrade in Kaba-he can be obtained. The soil relative dielectric constants between layers were assumed to vary linearly; the variations of soil relative dielectric constants with depth for March, June, October, and December in the Kaba-he can be shown in Figure 2a to Figure 2b.

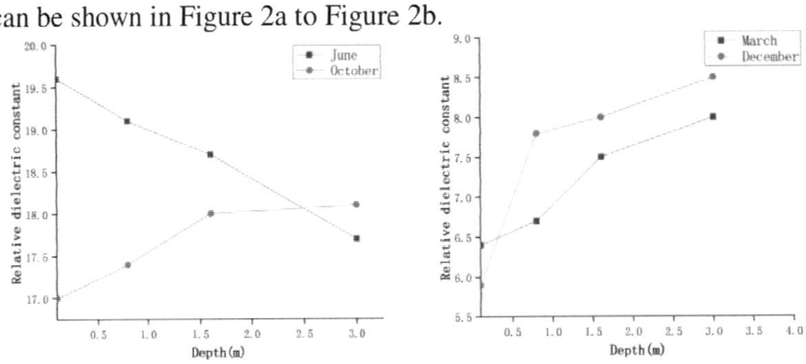

(a) (b)

Figure 2 (a) Soil relative dielectric constant distribution data in-depth direction in June and October, (b Soil relative dielectric constant distribution data in depth direction in March and December.

The study primarily considers the influence of soil relative dielectric constant variations on the electromagnetic wave propagation process, with the soil conductivity set to zero and the relative magnetic permeability of soil set to one. Based on the above data(Figure 2a to Figure 2b), two-dimensional temperature-dependent soil relative dielectric constant models can be constructed. In these models(Figure 3a to Figure 3d), the relative dielectric constants of soil vary according to the curves in the depth direction while remaining constant in the horizontal direction. The model's dimensions are 3m in length and depth, with spatial steps of 0.01m in both horizontal and depth directions. CPML boundary conditions were applied at the boundaries with a thickness of 10 grids. To analyze electromagnetic wave propagation, a steel rebar with a radius of 0.05m was placed at a horizontal distance of 1.5m and a depth of 2.5m, with a relative dielectric constant of 300 and a conductivity of $1.0 \times 10^8 \, S/m$.

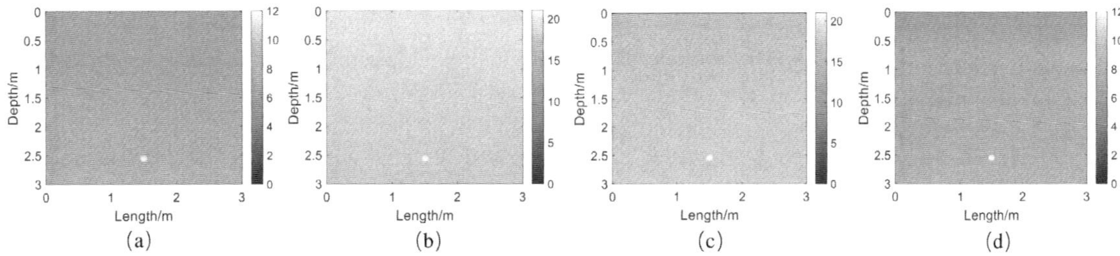

Figure 3 Models of the relative dielectric constant of soil with temperature variation in different seasons.
(a)Model in March, (b) Model in June, (c) Model in October, (d) Model in December.

In MATLAB software, numerical simulations can be conducted using the high-order O(2,4) FDTD algorithm for the temperature-dependent soil relative dielectric constants models during different seasons in Kaba-he. The simulations employed the profile method for GPR numerical simulations, with a 200MHz Ricker wavelet used as the transmitted antenna signal. The time step was 0.02ns, and the time window was 100ns. The spacing between the transmitting and receiving antennas was 0.01m, with a stepping interval of 0.01m. The simulation results, as depicted in Figure 4a to Figure 4d, indicate that the travel times of electromagnetic waves are 44.20ns in March, 72.12ns in June, 69.80ns in October, and 45.80ns in December for a target located at a depth of 2.5m underground.

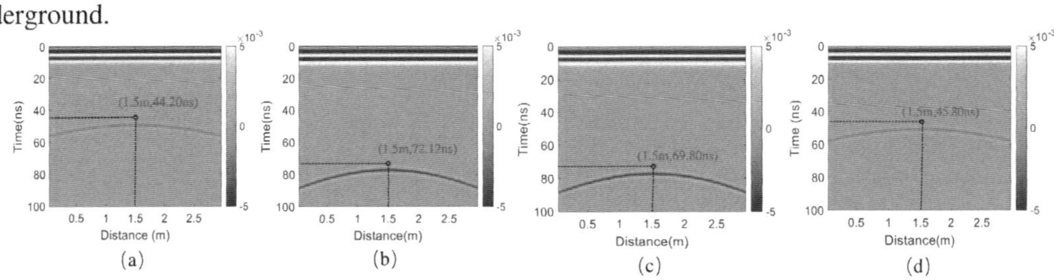

Figure 4 Simulation results of soil medium models in different seasons.
(a)Result in March. (b)Result in June. (c)Result in October. (d)Result in December

The average propagation velocity of electromagnetic waves in the subgrade soil during different seasons can be calculated based on the travel times of electromagnetic waves obtained from the GPR profile, considering that the underground targets are located at a depth of 2.5m. The average propagation velocities of electromagnetic waves in the subgrade soil during different seasons in Kaba-he can be presented in Table 1.

Table 1 Average propagation rate of electromagnetic waves

	March	June	October	December
Mean wave velocity m/s	$1.127\times10^{8}m/s$	$6.912\times10^{7}m/s$	$7.147\times10^{7}m/s$	$1.085\times10^{8}m/s$

Continuing this analysis, disregarding the influence of temperature and moisture content on soil relative dielectric constant, the computed detection depths of GPR for underground targets buried in different months are illustrated in Figures 5a to 5d. Taking Figure 5a as an example, the underground target in this scenario was buried in March, and the propagation velocity of electromagnetic waves remains constant across different months' soil conditions. Consequently, the computed detection depths of GPR for underground targets buried in various months can be derived based on the travel time of electromagnetic waves in the subgrade soil in different months, aiming for the underground target at a depth of 2.5m. Thus, for an underground target buried in March, the computed detection depth of GPR in March is 2.50m, matching the actual depth of the underground target. For June, the computed detection depth of GPR is 4.077m, representing an increase of 1.577m compared to the exact depth. For October, the computed detection depth of GPR is 3.942m, indicating a rise of 1.442m compared to the actual depth. For December, the computed detection depth of GPR is 2.596m, representing an increase of 0.096m compared to the actual depth. And the information represented in Figure 5b to Figure 5d holds a similar significance to that of Figure 5a.

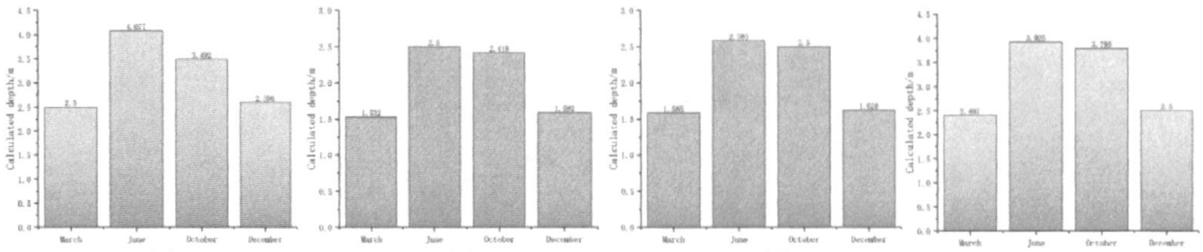

(a) (b) (c) (d)

Figure 5 Underground targets buried in different months, calculation depth for detection in other seasons. (a)Buried in March, (b)Buried in June, (c)Buried in October, (d)Buried in December

3. Results and discussion

Overall experiments, based on the subgrade's temperature characteristics and the roadbed's moisture content characteristics in Kaba-he different months, models of relative dielectric constants under various temperature and moisture content conditions were constructed. Through FDTD simulation, it was determined that for a target located 2.5m underground, the average propagation velocities of electromagnetic waves in the underground medium are approximately in spring (March), $6.912\times10^7 m/s$ in summer (June), $7.147\times10^7 m/s$ in autumn (October), and in winter (December) in Kaba-he. These findings can provide valuable references for estimating wave velocities during GPR detection in different seasons in Kaba-he.

This paper investigated the influence of temperature and moisture content on the soil's relative dielectric constant and analyzed its impact on GPR detection results. However, temperature and moisture content affect the soil's relative dielectric constant and influence other properties, such as soil conductivity. Moreover, factors beyond temperature and moisture content, such as soil salinity, also affect the electromagnetic properties of soil. A comprehensive analysis of these influences on GPR detection in seasonal freezing areas is still under further investigation.

4. References

[1] Szyplowska A Lewandowski A, Kafarski M, et al 2023 Influence of Temperature on Soil Dielectric Spectra in the 20 MHz–3 GHz Frequency Range *J. IEEE Transactions on Geoscience and Remote Sensing,* 61: 1-10.

[2] Xu X, Zhang W and Wang Y 2022 Measuring and modeling the dielectric constant of soil during freezing and thawing processes: An application on silty clay *J. Acta Geotechnica,* 1-20.

[3] Mironov V, Karavayskiy A, Lukin Y, et al 2020 A dielectric model of thawed and frozen Arctic soils considering frequency, temperature, texture and dry density *J. International journal of remote sensing,* 41(10): 3845-3865.

[4] Kabir H, Khan M, Brodie G, et al 2020 Measurement and modelling of soil dielectric properties as a function of soil class and moisture content *J. The Journal of microwave power and electromagnetic energy,* 54(1): 3-18.

[5] Xuxia D, Feng M, ShuPing Z and et al 2022 Investigation on the influence of temperature on soil dielectric constant *J. Journal of Glaciology and Geocryology,*44(02):634-642.

[6] Luo T, Lai W and Giannopoulos A 2020 Forward modelling on GPR responses of subsurface air voids *J. Tunnelling and Underground Space Technology,* 103: 103521.

[7] Qian L 2012 The Study on the highway subgrade temperature and humidity variation characteristics and mechanical properties in Xinjiang *D.Chang'an University,* 51-53.

[8] Weidong Z 2020 Experimenal study on the thermophysical properties of Genhe silty clay under freezing_thawing condition *D. Inner Mongolia University,* 36-37.

Probing multilayer road structures using the GPR "Terrazond"

Ilya Tseplyaev[1], Sergey Shipilov[1], Ivan Fedyanin[1], Vitaly Khmelev[1], Rail Satarov[2], Dmitry Romanov[3,5] and Roman Eremin[4]

[1] Radiophysics Faculty, Tomsk State University, 634050 Tomsk, Russia
[2] Limited Liability Company "Radiovision", Russia
[3] Research and Production Association "Terrazond", Moscow, Russia
[4] Limited Liability Company "TIM", Moscow, Russia
[5] Author to whom any correspondence should be addressed.

E-mail: info@terrazond.ru

Abstract
The paper describes a method for obtaining information about the thickness of road pavement layers using the "Terrazond" multichannel ground penetrating radar. The method includes the use of signal pre-processing to eliminate common-mode interference on radar data of the road surface. The results of applying the methodology to data obtained during sounding of a road test site are presented. Using the example of a radar data of a test section, it is shown that three iterations of the proposed filtering contrastively highlight three layers of the road structure. The results of automated construction of the boundaries of road layers in the GeoReader program using machine learning technology are shown.

1. Introduction

The diverse characteristics of materials, variations in the thickness of pavement layers, and the existence of imperfections and damage on the road surface and in layers present difficulties in precisely identifying the boundaries between layers. The existing signal processing methods used to determine the boundaries of layers have disadvantages that limit their application in practical conditions.

Previously, authors developed a signal processing method for filtering common-mode interference [1], which largely eliminates the problem of accurately determining layer boundaries. The use of data processing allows for improving the image quality obtained using GPR and making it more accurate and reliable for studying road structures to determine layer boundaries and thickness [2]. It also allows for detecting other underground objects, making GPR very useful in engineering geology, construction, and other industries. At the same time, the definition of the boundaries between layers of the road structure can be done using machine learning on large data sets in an automated manner, which would

Journal of Physics: Conference Series **2887** (2024) 012019 doi:10.1088/1742-6596/2887/1/012019

significantly reduce time and resource costs, decrease reliance on operator expertise, and increase the overall efficiency of the data interpretation process. It is known that, by using computer vision algorithms and neural networks, it is possible to implement the functionality of recognizing various defects on the road surface, such as cracks, potholes and voids. This allows quickly and accurately determining their location and size, which helps prevent accidents and improves road safety [3].

The article [4] notes that the accuracy of measuring of the characteristics of the underground environment using neural networks depends on the quality of data obtained using GPRS. For example, the accuracy of methods may decrease if there is strong noises in the data. Therefore, using processed data such as based on common mode interference filtering will allow more widespread use of neural networks for processing big data. This data is then used to create digital models of roads, recording their actual condition and increasing the efficiency of planning repairs.

This paper demonstrates the main characteristics of the multi-channel ground penetrating radar “Terrazond” and the data obtained from it when probing various sections of highways. The results of processing this data using common-mode interference filtering and further processing with machine learning methods are also presented.

2. Hardware of Terrazond GPR

As the main instrument for collecting GPR data, we used the wide-range multi-channel GPR Terrazond (figure 1). This instrument has several advantages over common mono-pulse ground-penetrating radar systems. Let's list some important feature of this radar. Firstly, this device is a distributed antenna array with high-speed electronic switching, allowing for simultaneous sensing over 32 channels and a capture of up to 2.5 m in pavement width (figure 2). The radiation pattern of one element retains its characteristics within the operating frequency range (figure 3). This allows us to evaluate the characteristics of road surfaces and inhomogeneities across the entire scanning plane of a road (figure 4), rather than just along a single route. Secondly, the radar is equipped with a high-speed signal generator operating on the principle of LFM location, which provides sensing at speeds up to 130 km/h. The GPR consists of a UWB antenna unit of receiving and transmitting antennas operating in the frequency range from 0.71 to 2.6 GHz, a sensitive high–speed LFM generator provides the operation of the GPR at speeds up to 130 km/h (with a recording density of GPR tracks - every 10 cm), a high-speed electronic relay, a programmable logic integrated circuit (FPGA) for receiving, processing and storing geodata (figure 5). The ground-penetrating radar antenna array is fixed in front or behind the car at a height of 30-40 cm from the road surface (figure 6). Currently, the ground-penetrating radar Terrazond is serial-produced. It use for the road laboratories by organizations such as RosdorNII and Rosavtodor.

The advantages of using multi-channel georadar systems over single-channel ones are:

1. Measurement of the radar response from the roadway in both transverse and longitudinal directions during the movement of a vehicle equipped with a radar.
2. Data processing, and obtaining a three-dimensional array, as well as amplitude maps linked to geolocation.
3. The possibility of using the common depth point method, which allows the procedure for calculating electrophysical properties of roadway layers. This opens up opportunities to conduct a more detailed analysis of these layers [2, 5, 6].
4. The software implements the functions of primary signal processing in real time with all channels. Such as converting the received LFM signal in frequency representation into a time sequence of data, filtering common-mode interference, linking each received radar trace to interpolated GPS coordinates [7].

Figure 1. Multi-array ground radio imaging system «Terrazond».

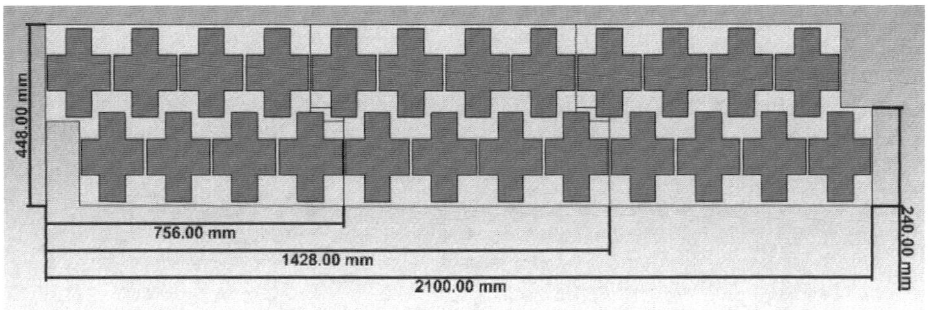

Figure 2. 8-32 elements antenna array.

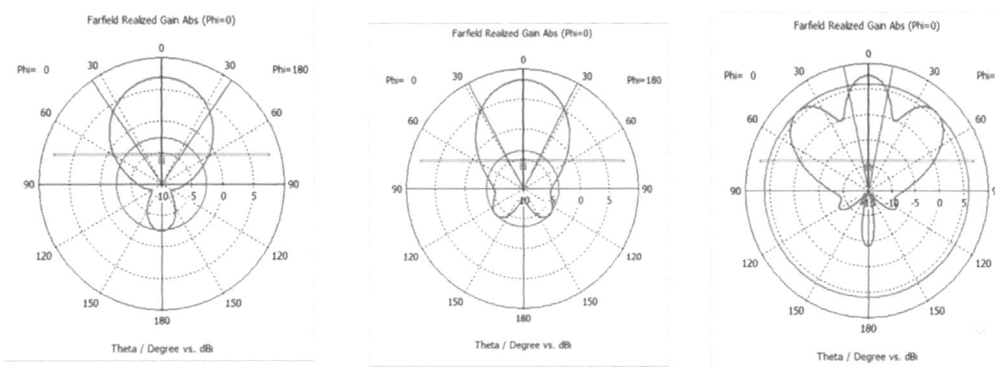

Figure 3. Radiation patterns of antennas element.

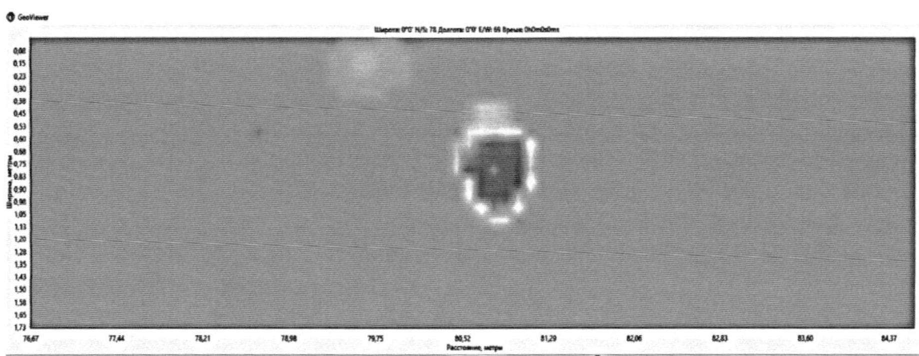

Figure 4. An amplitude map with heterogeneity obtained in multichannel mode.

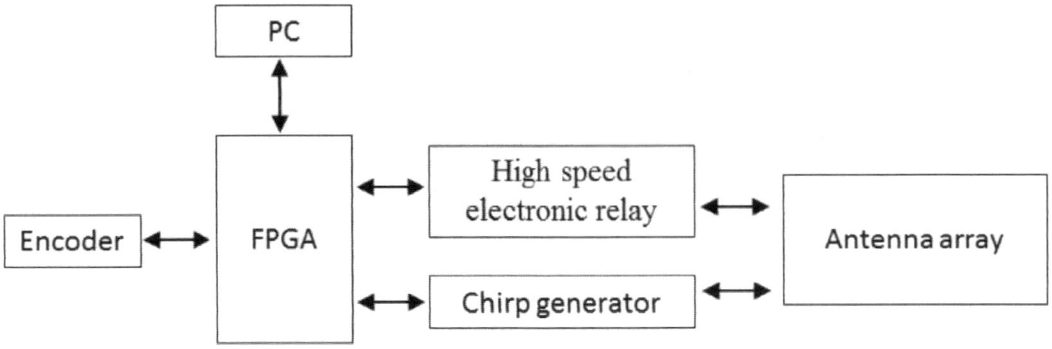

Figure 5. Structural diagram of the device.

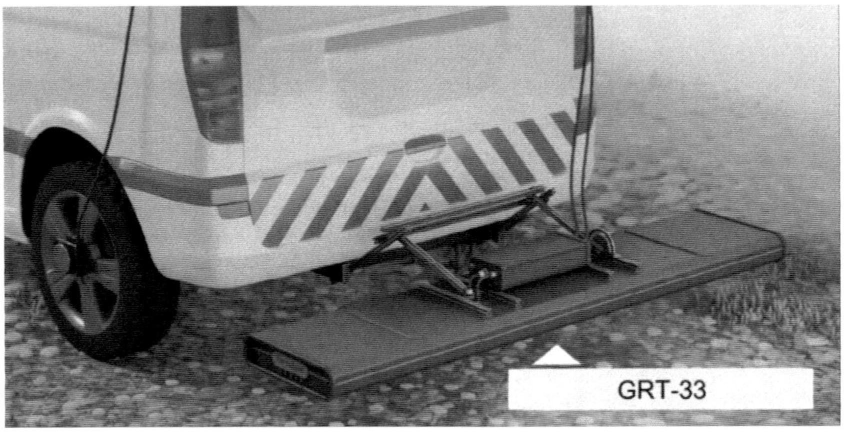

Figure 6. GPR mounting on a road car of a laboratory.

3. Software of Terrazond GPR

Common-mode interference in the radar image is caused by multiple interference between the antenna and the air-environment boundary. These disturbances appear as multiple bands, which prevent the determination of the boundaries between layers of pavement. Its occurrence is mainly due to the repetitive reflection of the probing signal from the road surface and other reflectors associated with the design of the antenna array housing and its mounting on the car. Its occurrence is mainly due to the repetitive reflection of the probing signal from the road surface and other reflectors associated with the design of the antenna array housing and its mounting on the car. The authors proposed [1] an algorithm for eliminating common-mode interference using an iterative procedure to sequentially eliminate common-mode from the upper layer down to the lower layer. Figure 7 shows the initial radar image of the 160-meter test section of the M-12 road polygon, Moscow, obtained in 2022. Figure 8 shows the results obtained after three iterations of radar signal filtering.

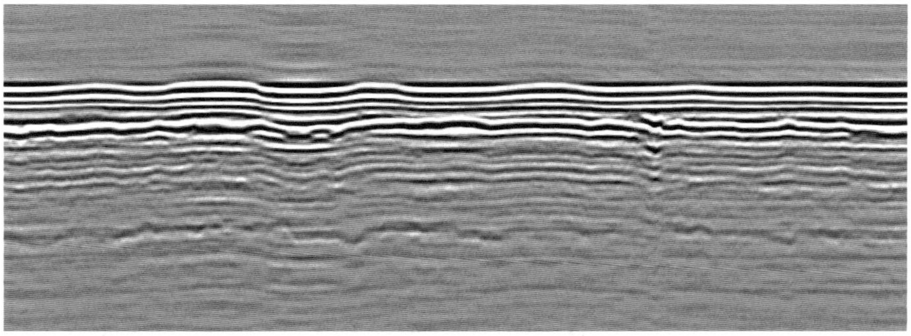

Figure 7. A set of initial radar images of a 160-meter-long road section.

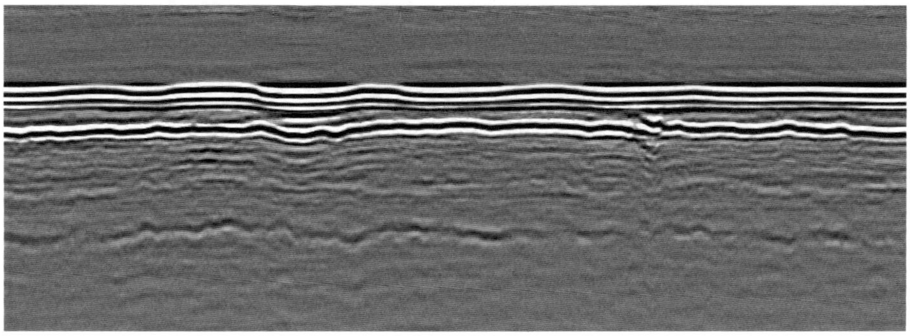

Figure 8. Radar image of a road section after three iterations of interference filtering.

The figure shows that the localization of the amplitude maxima of the reflected signal corresponding to 3 layers significantly increases after filtering. This allows us not only to talk about increasing the accuracy of determining the location of the boundary, but in some cases also about the possibility of separating layers in general. This feature gives hope for the application of this technology as pre-filtration in the task of constructing layer boundaries using neural networks.

The methodology for calculating the thickness of pavement layers based on georadar sounding data using Terrazond equipment provides for data processing using machine learning methods in GeoReader software [5]. The results of such processing are shown in Figure 9.

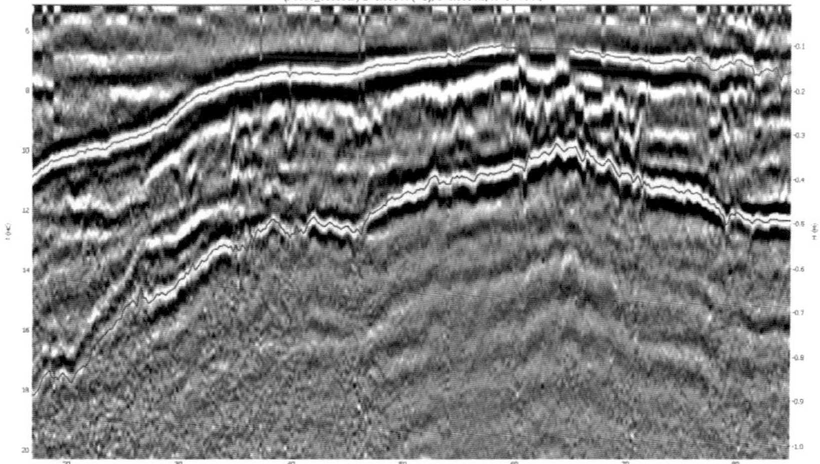

Figure 9. The result of processing the received data using machine learning.

As a result of the survey of highways using the method described above, it is possible to create an amplitude map to monitor compliance of pavement thickness with design requirements and present results in CAD and BIM solutions along with laser scanning data (figure 10).

Figure 10. Display asphalt concrete thickness information from GPR with point cloud from 3d laser scanning inside "Credo 3D Scan" environment.

4. Conclusion

A method for obtaining information on the thickness of pavement layers has been developed for the Terrazond multichannel ground-penetrating radar. This method involves the use of signal preprocessing to eliminate common-mode interference on roadway radars. The method has been tested on the results of probing the test site of the M-12 road. The example of the radar image of the test area shows that three iterations of the proposed filtering contrast three layers of road structure. For further processing of the results, we used the technique of automated boundary construction for road layers in the GeoReader program, which includes machine learning methods. The results can be presented as an amplitude map showing changes in the thickness of the road layers, superimposed onto a three-dimensional city map in the form of a cloud of laser scanning points.

Acknowledgments

This study was supported by the Tomsk State University Development Programme («Priority-2030»).

References

[1] Gorst A, Tseplyaev I, Eremeev A, Shipilov S, Romanov D and Eremin R 2023 *Remote Sensing* **15** 2751
[2] Karim, Hussein & Al-Qaissi, Ammar 2014 *Eng. & Technology Jour.* **32** 1240-1250
[3] Yang X; Guan J; Ding L; You Z; Lee V C; Hasan M R M ; Cheng X 2021 *J Traffic Transp. Eng.* **8** 1000–1021
[4] Liu Z, Gu X, Chen J, Wang D, Chen Y and Wang L 2023 *Autom. Constr.* **146** 104698
[5] Sukhobok Y A and Kurbatov M S 2019 *CAD GIS Roads* **2** 26–31
[6] Romanov D, Zykov A, Fedyanin I, Sukhobok Y 2020 *Russian Physics J.* **63** 214-220
[7] Li J., Liu C, Zeng, Z., & Chen, L (2015). GPR Signal Denoising and Target Extraction with the CEEMD Method. *IEEE Geoscience and Remote Sensing Letter,* 12 (8), 1615-1619.

Journal of Physics: Conference Series **2887** (2024) 012020 doi:10.1088/1742-6596/2887/1/012020

A holographic subsurface imaging radar with high plane resolution: principle and application

He Zhihua[1], Liu Tao[1], Song Xiaoji[1], Chen Cheng[2], Jin Guanghu[1], Huang Chunlin[1] and Su Yi[1]
1 Collage of Electronic Science and Technology, National University of Defense Technology, Deya road 109, Changsha 410073, China (skynismile@163.com)
2 Test Center, National University of Defense Technology, Guangming road 5, Xi'an 710106, China

Abstract. Holographic subsurface imaging radar uses the penetration of electromagnetic waves into non-metallic media to detect and image the interior of the media, which can intuitively reflect the structure and abnormal distribution of objects in the media in the form of images. By adopting the design methods of integrated radar RF front-end, fast high-density spatial sampling and autofocus imaging processing, a holographic subsurface imaging radar with high plane resolution is realized. The system weighs only 2.5 kg and achieves a plane resolution of 2.5mm, which has high detection efficiency and high portability. The application tests of imaging resolution, penetration of different materials, penetration of biological tissues and non-destructive detection of materials were carried out, which verified the feasibility and effectiveness of the design scheme.

1. Introduction

Holographic subsurface imaging radar (HSIR) is a kind of penetrating imaging radar system based on coherent continuous wave and distance plane projection imaging theory. It can visually display the internal structure and anomalies of the medium in the form of images, and its high-resolution imaging ability provides a new technical means for non-destructive penetrating detection[1-4]. Compared with non-linear junction detection, metal detector, wireless signal detection, infrared signal detection and other detection methods, HSIR has intuitive results and good resolution characteristics, and can detect metal and non-metal objects at the same time. Compared with X-ray backscattering, it has better penetrability and safety, and can be used in the fields of hidden objects detection, non-destructive testing of non-metallic materials and so on.

According to different application requirements, a series of research work on the development and application of HSIR system has carried out. Bormann Technical University in Moscow has successfully developed the RASCAN series holographic surface penetrating radar system, and systematically carried out practical applications in mine detection, non-destructive detection of buildings and materials, archaeological detection, safety detection and so on. RASCAN series radar adopts continuous wave system, and the frequency range is 1.6 GHz~6.8 GHz. The echo signal is collected by manually moving the antenna to form a two-dimensional aperture, and finally the target image is reconstructed by processing software[5]. StructureScan Mini XT radar introduced by American GSSI company[6]. Conquest series products launched by Sensors&Software Company of Canada[7]. It is a special penetrating imaging radar for the detection of steel mesh in the wall. This kind of system adopts the pulse waveform system with the time width of nanosecond/sub-nanosecond to obtain a large bandwidth to achieve high resolution in the range. The frequency range of the transmitted signal is generally 1~2 GHz, and the detection of steel mesh in the wall can be realized by manual scanning. The working frequency band of the above system is low, and the resolution of typical imaging plane is in the order of cm. Although an automatic scanning device is also designed, but the scanning speed is slow and the position accuracy is difficult to control[8]. Two-dimensional MIMO radar can realize fast spatial sampling by using the electric scanning method of transceiver antenna[9-11]. However, in the imaging scene close to the medium, there are some application limitations, such as insufficient effective space sampling density and insufficient antenna beam coverage. Also, multiple RF transceiver components are adopted, which leads to high system power consumption and cost.

Considering the application requirements of millimeter-level high-resolution penetrating imaging and portability, this paper introduces a high-resolution HSIR system. In the first section, the working principle and key technologies of the system are introduced. In the second section, the typical practical test application of the system is given, and finally the conclusion is given.

2. Working principle and design method

2.1. Working principle

The schematic diagram of detection principle of HSIR is shown in Figure 1. The radar probe actively emits and receives electromagnetic wave signals, and obtains the backscattered echo of the scanning area through spatial scanning. By imaging the backscattered echo and other data processing, a high-resolution two-dimensional image can be obtained, and the spatial distribution of scattering coefficients of different targets in the medium can be restored, thus intuitively reflecting the outline of abnormal areas or targets and the electromagnetic characteristics difference with the surrounding medium, providing image basis for anomaly identification and confirmation.

To realize light weight, portability and high plane resolution, this system is designed to work in K-band, using step-frequency waveform, and its working bandwidth is 1GHz. It is necessary to solve the problems of integrated radar RF front-end design, fast high-density space sampling design and autofocus imaging processing to achieve high resolution and lightweight design.

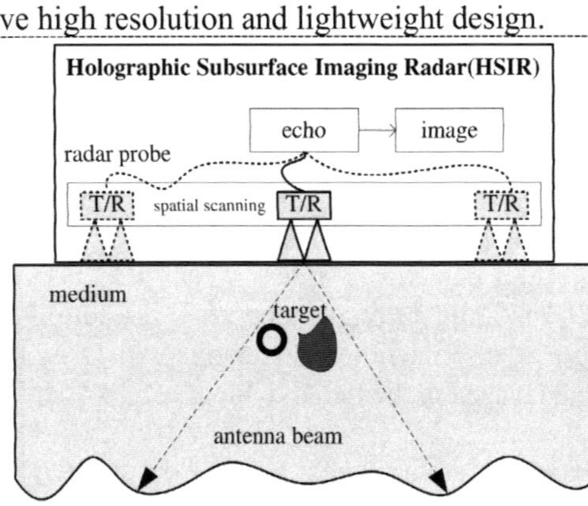

Figure 1. Schematic diagram of detection principle of HSIR.

2.2. RF front-end design of integrated radar

In this paper, the integrated design of radar RF front-end is adopted, and the high performance, low power consumption and light weight of radar RF front-end are ensured by optimizing the component structure and adopting the lightweight and integrated design idea under the condition of ensuring the instantaneous high dynamic range required for imaging detection of weak targets in the medium. Figure 2 is the block diagram of the integrated radar RF front end. In layout design, the laminated design scheme of multi-layer microstrip board and signal conditioning and acquisition board is adopted to ensure the miniaturization of the system and reduce the complexity of the single board. The microstrip transceiver antenna and RF transceiver circuit are respectively arranged on the front and back of the multilayer microstrip board, which adopts the form of microstrip patch antenna, which has the advantages of relatively high efficiency, good processing consistency, small volume and weight, etc. The optimized design ensures that the antenna isolation reaches 30 dB, and reduces the influence of direct waves on the detection of deep buried weak targets to an acceptable level. The RF transceiver circuit includes step frequency source, power divider, power amplifier, low noise amplifier, mixer, attenuator and filter, etc., and the interference influence is reduced by shielding shell. The signal conditioning and acquisition board integrates functional modules such as intermediate frequency circuit, analog-to-digital converter (ADC), power supply and connector, and realizes signal filtering, variable gain amplification and 24-bit ADC sampling. The integrated design of analog and digital is adopted, which lays the foundation for lightweight RF front-end. After optimization design, the length, width and height of the integrated radar RF module are only 35 mm×28 mm×10 mm, and the weight is less than 20g.

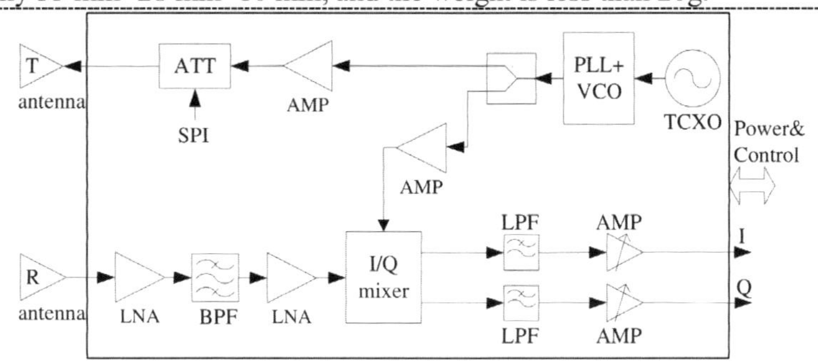

Figure 2. Integrated radar RF front end block diagram.

2.3. *Design of Fast and High-Density Spatial Sampling*

Penetrating imaging radar needs to reconstruct the scanned area image through spatial scanning. The common scanning methods include hand-held probe scanning, mechanical scanning and electrical scanning. The hand-held probe scanning equipment is simple, but the scanning speed is slow, the position accuracy is low and the repeatability is poor[12]; The electric scanning mode has the fastest scanning speed, but the spatial sampling density is low, the isolation between channels is not high, and the cost is relatively high[13]; Mechanical scanning has the advantages of both, which can realize automatic scanning with high repeatability. Using a single transceiver probe also ensures the advantages of consistency and low cost, and the scanning speed can also meet the requirements of rapid penetration imaging.

The existing mechanical scanning method is usually carried out in rectangular coordinates and generally consists of two linkage moving mechanisms in X direction and Y direction[14]. When the RF front-end of the radar moves, there are acceleration and deceleration links, so it is difficult to improve the speed. A complete scan is at the minute level, as shown in Figure 3(a), where Δx and Δy are the sampling intervals representing the X direction and the Y direction respectively, which is need to be smaller than the size of the highest resolution unit to meet the sampling theorem $\delta_r \approx \lambda/4$, and λ is the wavelength of electromagnetic waves in the medium is determined by the highest working frequency of the system.

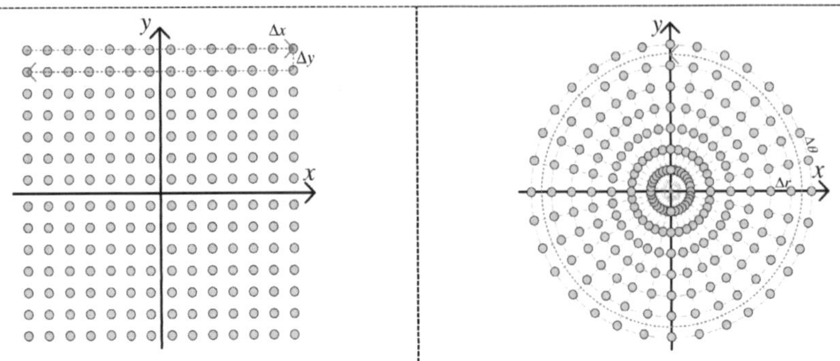

(a) Spatial sampling with cartesian grid (b) Spatial sampling with polar grid

Figure 3. Schematic diagram of two-dimensional spatial scanning comparison.

In this paper, the polar coordinate spatial sampling scheme is adopted, and a small and compact scanning servo system is constructed through reasonable scanning structure design. Through precise timing control of uniform circular motion along the polar axis and linear stepping along the radial direction, the integrated radar RF front-end is driven to run continuously and at high speed in the scanning plane, and with the synchronous control of the RF front-end, rapid polar coordinate spatial scanning is realized, and repeated start-stop time is also avoided by adopting the form of continuous uniform circular motion. The detection speed can be increased to less than 10 seconds, and the grid position control accuracy of the scanning device is better than 0.1 mm, which ensures the rapid detection ability, high plane resolution imaging accuracy and image positioning accuracy of penetrating imaging. Figure 3(b) shows a schematic diagram of spatial sampling in polar coordinates, in which Δr is the radial sampling interval and $\Delta\theta$ is the angular sampling interval. Assuming that the maximum scanning radius is $R_{\max}$, in order to satisfy the sampling theorem, the following requirements should be met

$$\Delta r < \delta_r, R_{\max} \times \Delta\theta < \delta_r \tag{1}$$

2.4. *Autofocusing imaging processing algorithm design*

Let the radar probe be located at the grid point (x, y) at a certain moment, where x represents the horizontal coordinate and y represents the vertical coordinate, and the backscattered echo data is $s(x, y, \omega)$. Imaging processing includes spatial two-dimensional Fast Fourier Transform (FFT), wavefield migration and spatial two-dimensional Inverse Fast Fourier Transform (IFFT). Firstly, the spatial FFT transform of $s(x, y, \omega)$ is carried out to obtain a two-dimensional wavenumber spectrum $S(k_x, k_y, k)$, in which k_x and k_y are the transverse and longitudinal wavenumber variables of the field of view are respectively represented, and $k = \omega/v$ is the total wavenumber, where ω represents the angular frequency of electromagnetic waves and v is the propagation speed of electromagnetic waves in the medium (hereinafter referred to as wave velocity), and then the wave field migration is carried out by multiplying the spatial spectrum by the following formula

$$H(k_x, k_y, k, z) = \frac{j2\pi(2k)^3 z_0^3}{k_z^4} \exp(jk_z z) \tag{2}$$

Where z_0 is the depth of the imaging plane to be focused, z indicating the distance between the interior of the medium and the surface of the medium, and $k_z = \sqrt{k^2 - k_x^2 - k_y^2}$ represents the wave number variable in the depth direction, and j is the imaginary unit. Stolt interpolation is performed on the spatial spectrum after wave field migration, so that the data are arranged at equal intervals in the wave number domain. Finally, the imaging results can be obtained by spatial IFFT. At this time, the obtained two-dimensional image is the spatial projection of the target echo on the radar scanning plane, which can be expressed as

$$p(x, y; z, v) = \mathrm{IFFT}_{2D}\left\{S\left(k_x, k_y\right) \times H\left(k_x, k_y, z_0\right)\right\} \tag{3}$$

The key step of the penetrating imaging algorithm is to reconstruct the matched filter in the wavenumber domain shown in Formula (2), in which z_t is the depth of the target and v_t is the wave velocity of the medium are two essential imaging parameters, and the parameter mismatch will cause the imaging defocus. Manual adjustment of parameters is also highly dependent on the professional knowledge level and experience of operators, which takes a long time, thus affecting the detection efficiency and effect.

Autofocusing imaging is used to solve the imaging problem with unknown parameters. According to the image characteristics of radar penetrating imaging, a focusing function operator (i.e. cost function) is constructed, and the focusing effect of imaging results is quantitatively evaluated. According to the evaluation results, the deviation degree of imaging parameter depth z_t and electromagnetic wave velocity v_t in the medium relative to the true value is estimated, and the optimal imaging parameters are obtained by iterative processing of numerical optimization, and finally the self-focusing fast, high plane resolution and fully automatic imaging of penetrating imaging radar system is realized. The optimization process adopted can be expressed as follows

$$z, v = \arg\max_{z,v}\left\{\frac{std\left(\left|p(x, y; z, v)\right|\right)}{mean\left(\left|p(x, y; z, v)\right|\right)}\right\} \tag{4}$$

Where $std(\cdot)$ represents the standard deviation and $mean(\cdot)$ represents the mean value of the image respectively, and when the contrast of the image is maximum, the optimal focusing parameters z_t and v_t are obtained

2.5. System composition

Through the above design, the composition diagram of the radar system finally designed in this paper is shown in Figure 4, which is mainly composed of target scattering field acquisition module, echo acquisition module, data processing module, control module, communication module and image display terminal PAD. The target scattering field acquisition module is driven by a microminiature 2D spatial scanning device, and the integrated radar RF front-end has the function of transmitting and receiving radar signals, and completes high-speed and high-density scattering echo sampling in two-dimensional space under the driving of the scanning device. The microminiature 2D spatial scanning device includes a precision micro-drive rod and a subminiature precision ceramic gear mechanism, which greatly reduces the structure and weight of the target scattering field acquisition module while ensuring the high-precision scanning of the two-dimensional space by the RF front end of the integrated radar. The echo acquisition module receives the acquired echo signal, and realizes the filtering, amplification and analog-to-digital conversion of the echo signal to obtain echo data. The data processing module completes echo data preprocessing, autofocus imaging processing, image stray noise suppression and other processing steps to obtain a two-dimensional high-resolution image. The communication module transmits the image results to the image display terminal in a wireless way, which can further complete the functions of selecting the best focused image, displaying and saving the image. The control module provides control instructions for the target scattering field acquisition module, the echo acquisition module and the communication module, and coordinates the target scattering field acquisition, echo acquisition and communication transmission functions.

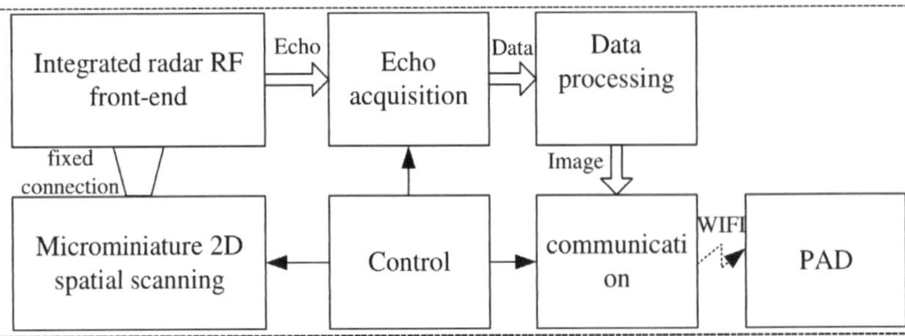

Figure 4. Diagram of the proposed HSIR.

The physical diagram of the proposed HSIR system is shown in Figure 5. The system consists of a penetrating imaging radar host and a portable display terminal PAD, which is suitable for single-person holding operation. Compared with the existing system, it is greatly improved in portability, detection speed and automation. Its imaging plane resolution is better than 2.5 mm, detection speed can reach 10 seconds/frame, each detection area is 200 cm^2, and the weight of the host is only 2.5 kg.

Figure 5. Physical drawing of the proposed HSIR system.

3. Typical application

3.1. Resolution test

Figure 6 shows the experimental results of imaging resolution. The gypsum board with a thickness of 1 cm and the coin copper foil are used as the target. The width of the copper foil is 5 mm, and the shortest distance from the edge of the coin is 2.5 mm. As shown in Figure 6 (a), Figure 6 (b) is the penetrating imaging result, which can clearly distinguish the interval between the coin and the copper foil, and the number "1" in the middle of the coin is also clearly visible, indicating that the imaging resolution of this system can reach 2.5 mm, which can reflect the subtle differences of the target through the medium.

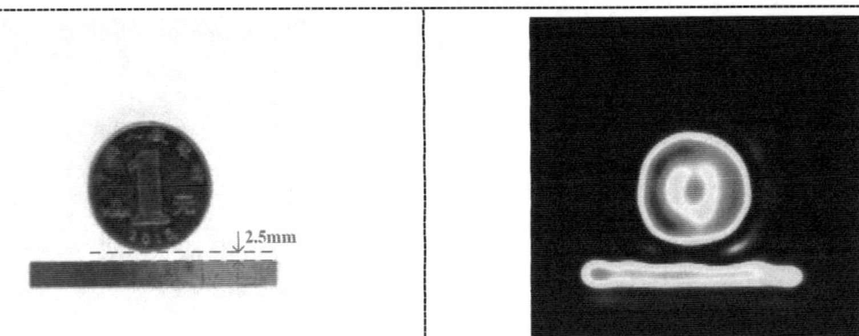

(a) Coin and copper foil with small gap

(b) Imaging result

Figure 6. Experimental results of imaging resolution.

Journal of Physics: Conference Series **2887** (2024) 012020 doi:10.1088/1742-6596/2887/1/012020

3.2. Penetration ability of different materials

The penetrating imaging radar system can realize the penetrating imaging ability of non-metallic targets under different media conditions. Figure 7 shows a non-metallic cross-shaped target with a width of 6.6mm and a length of 5cm. Figure 8 shows the common dielectric materials and their penetrating imaging results, including 0.8cm thick gypsum board, 0.6cm thick multi-layer wooden board, 1.6cm thick density board, 1.5cm thick foam wall and 1cm thick concrete. From the corresponding imaging results, it can be seen that different media have different electromagnetic and geometric properties, showing different clutter characteristics, and the more uniform the materials, the clearer the target.

Figure 7. Non-metallic cross-shaped target.

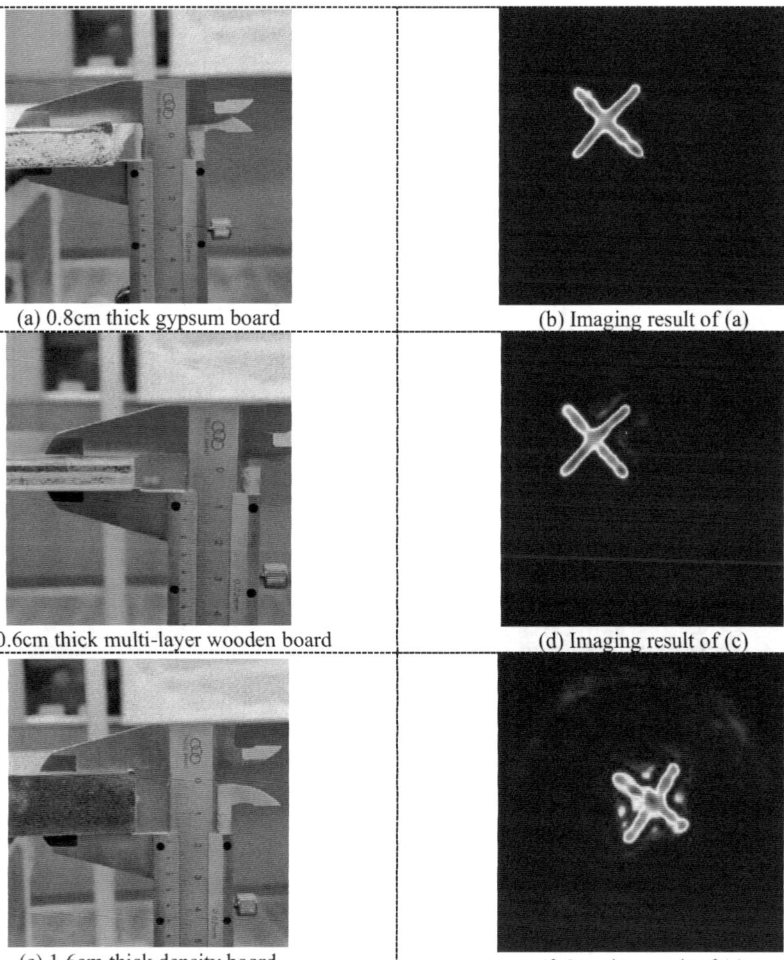

(a) 0.8cm thick gypsum board

(b) Imaging result of (a)

(c) 0.6cm thick multi-layer wooden board

(d) Imaging result of (c)

(e) 1.6cm thick density board

(f) Imaging result of (e)

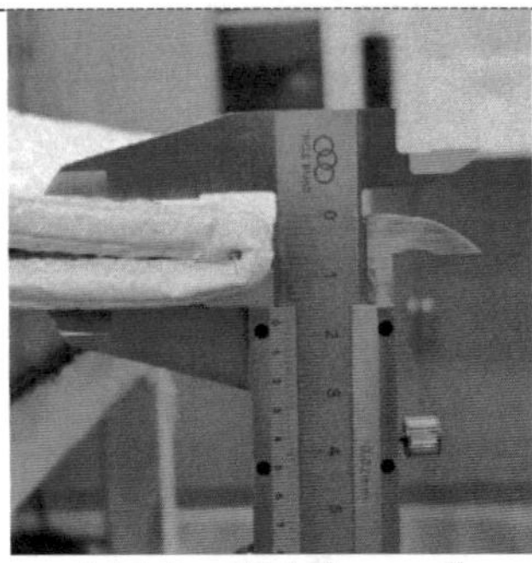

(g) 1.5cm thick foam wall

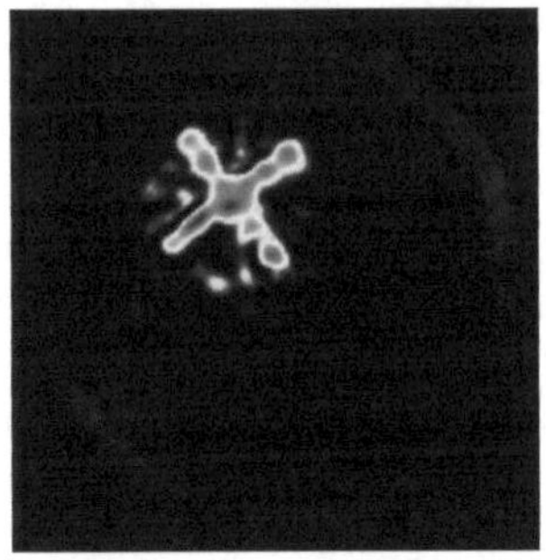

(h) Imaging result of (g)

(i) 1cm thick concrete

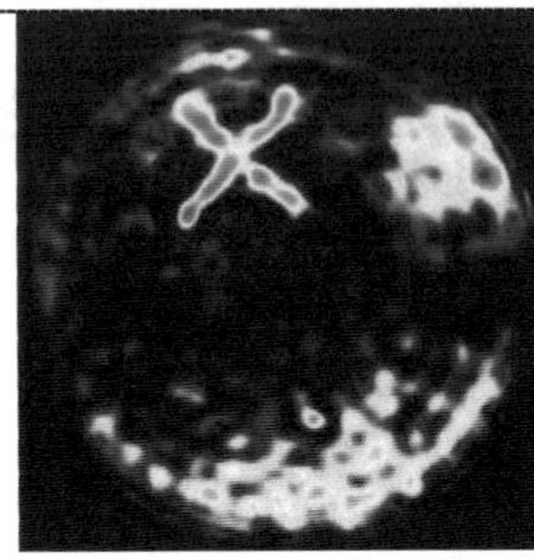

(j) Imaging result of (i)

Figure 8. The common dielectric materials and their penetrating imaging results

3.3. Biological tissue medical detection

Figure 9 shows that the system can penetrate the ABS plastic medium with a thickness of 5cm to realize the imaging of palm targets, and can clearly see the structural characteristics of palm bones, indicating the potential of the system in rapid medical imaging of biological tissues.

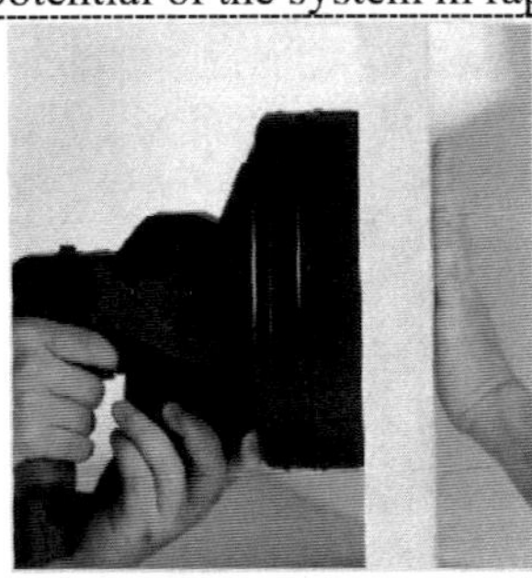

(a) Detection scenario

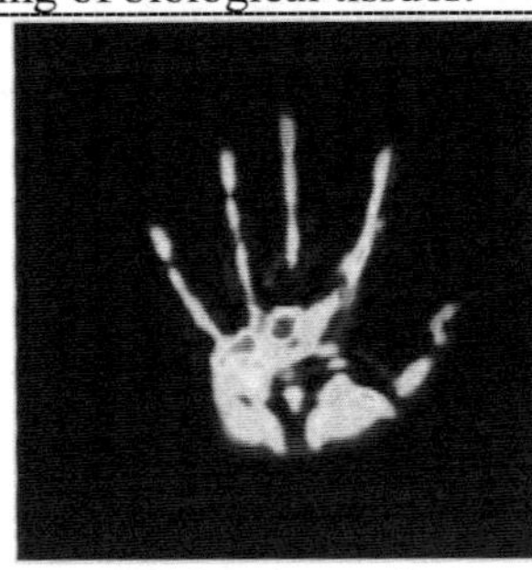

(b) Imaging result

Figure 9. Experimental results of palm bones.

3.4. Non-destructive testing of materials

Figure 10 shows the effect of the system in non-destructive testing of materials. Figure 10 (a) shows a composite material sample with a defect area, with a thickness of about 2cm, a defect area width of 2.5cm and a height of 3 mm. Figure 10 (b) shows the penetrating imaging result, and the image reflection characteristics caused by material defects can be clearly seen. The target area is marked by a dashed white line, and the strong reflection pattern outside the dashed white line is caused by the strong contrast difference between the sample and the surrounding air environment.

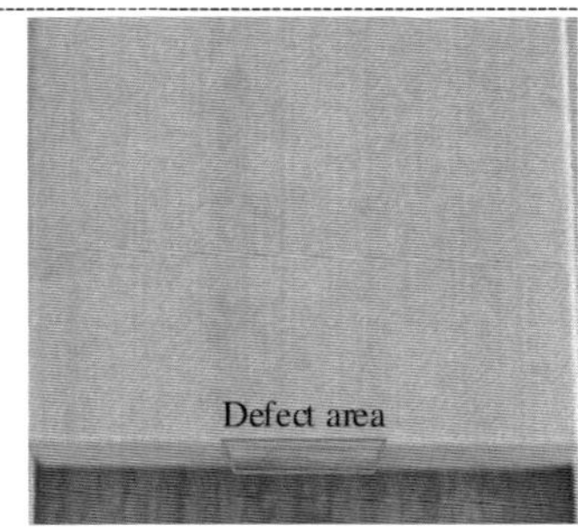

(a) Composite material sample with a defect area

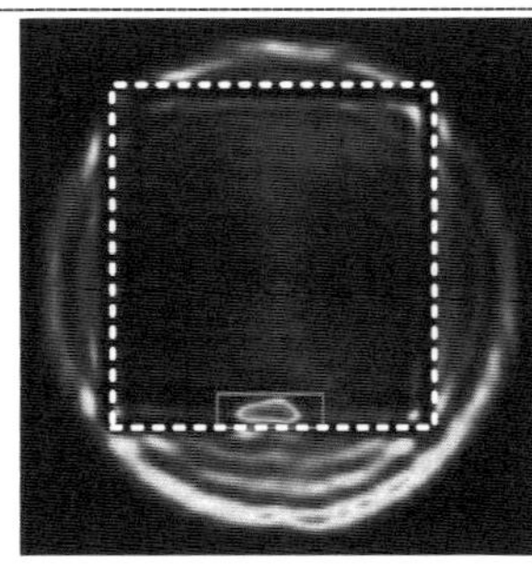
(b) Imaging result

Figure 10. Experimental results of composite materials with a defect area.

4. Conclusion

In this paper, a millimeter-scale high-plane resolution penetrating imaging radar system is realized. The system can penetrate marble, ceramic tile, concrete, wood board, gypsum board, composite board and other materials, and has good detection effect on both metal and nonmetal materials in the medium. Its main machine weight is only 2.5 kg, and the imaging plane resolution can reach 2.5 mm It has the characteristics of millimeter-scale high-plane imaging resolution, rapid detection ability and portability, and can be applied to high-resolution wall penetrating imaging detection and hidden objects detection. The next step is to adopt automatic image interpretation method to realize intelligent image detection and recognition.

References

[1] Gaikovich, K. P. , Gaikovich, P. K. , Maksimovitch, Y. S. , & Badeev, V. A. . (2017). Subsurface near-field microwave holography. *IEEE Journal of Selected Topics in Applied Earth Observations & Remote Sensing*, 9(1), 74-82.

[2] Zhuravlev A., Razevig V., Rogozin A., & Chizh M.. (2023) Microwave Imaging of Concealed Objects With Linear Antenna Array and Optical Tracking of the Target for High-Performance Security Screening Systems. *IEEE Transactions on Microwave Theory and Techniques*, 71(3), 1326-1336.

[3] Chen, C. , He, Z. , Song, X. , Liu, T. , & Su, Y. . (2022). A subspace projection approach for clutter mitigation in holographic subsurface imaging. *IEEE geoscience and remote sensing letters*, 19,1-5.

[4] Amineh, R. K. , Mccombe, J. J. , Khalatpour, A. , & Nikolova, N. K. . (2014). Microwave holography using point-spread functions measured with calibration objects. *IEEE Transactions on Instrumentation & Measurement*, 64(2), 403-417.

[5] Ivashov S. I., Razevig V. V., Vasiliev I. A., Zhuravlev A. V., Bechtel T. D., & Capineri L.. (2011). Holographic Subsurface Radar of RASCAN Type: Development and Applications. IEEE *Journal of Selected Topics in Applied Earth Observations and Remote Sensing*, 4(4),763 -778.

[6] GSSI. StructureScan mini XT. (2023). https://www.geophysical.com/products/structurescan-mini-xt

[7] CONQUEST® 100 Concrete Scanner. (2023). https://www.sensoft.ca/products/conquest100/overview/

[8] Vivoli E., Bossi L., Bertini M.. (2023). Error assessment of microwave holography inversion for shallow buried objects. *2023 12th International Workshop on Advanced Ground Penetrating Radar (IWAGPR).*

[9] Ahmed S. S., Schiessl A., Schmidt L. P.. (2011). A Novel Fully Electronic Active Real-Time Imager Based on a Planar Multistatic Sparse Array. *IEEE Transactions on Microwave Theory and Techniques*, 59(12),3567-3576.

[10] Spreng, T. , Prechtel, U. , Schonlinner, B. , Ziegler, V. , & Siart, U. . (2013). UWB near-field MIMO radar: Calibration, measurements and image reconstruction. *Radar Conference*.

[11] Ghasr, Mohammad Tayeb, Horst, Matthew J., Dvorsky, & Matthew R. et. al. Wideband Microwave Camera for Real-Time 3-D Imaging.

[12] Razevig, V. V. , Zhuravlev, A. V. , Chizh, M. A. , Ivashov, S. I. , & Bugaev, A. S. . (2017). Hand-held Radar with Video Positioning System for Microwave Imaging. *Progress in Electromagnetics Research Symposium-spring* (3052-3055).

[13] Khushaba R. N., Hill A. J..(2022). Radar-Based Materials Classification Using Deep Wavelet Scattering Transform: A Comparison of Centimeter vs. Millimeter Wave Units[J]. IEEE Robotics and Automation Letters, 7(2), 2016 - 2022.

[14] Zhuravlev, A. , Ivashov, S. , Razevig, V. , Vasiliev, I. , & Bugaev, A. . (2014). Automated data acquisition system for holographic subsurface radar. *2014 15th International Conference on Ground Penetrating Radar (GPR)*. IEEE.

Journal of Physics: Conference Series **2887** (2024) 012021 doi:10.1088/1742-6596/2887/1/012021

Unmanned unexploded ordnance detection system based on magnetic method

Y L Xu[1], Y J Zhou[1], K Ding[1,3], R Q Lv[1], M W Wu[2] and Z Wang[2]

[1]Science and Technology on Near-Surface Detection Laboratory, Wuxi 214035, China
[2]The 50th Research Institute of China Electronics Technology Group Corporation, Shanghai 200331, China
[3]Corresponding Author's E-mail: winfast113@sina.com

Abstract. Magnetic method is one of the main methods of UXO detection. This paper uses cesium optical pump magnetic detection module combined with UAV to build an unmanned UXO detection system, and introduces the detection principle and composition of the system. At the same time, the system is used to detect four kinds of UXO targets. The test results show that the system can detect ferromagnetic UXO targets, and can locate the position of UXO targets through magnetic anomaly images. Finally, the limit detection range of four kinds of targets is tested, and the limit detection range of 155 projectiles can reach 4.5m, which meets the actual detection requirements

1. Introduction

Unexploded ordnance (UXO) mainly includes mines, bombs and other weapons left over during war and other military activities. AS a form of pollution that threatens human life and the living environment, UXO is now distributed in 83 countries and regions around the world[1]. Many countries in the world have set up special projects to study this problem. At present, the commonly used UXO detection methods mainly include magnetic method, electromagnetic induction, ground penetrating radar and radioactive detection. Among them, magnetic method has become an important means of UXO detection due to its advantages of convenient total field measurement and large detection depth[2]. The CS Series magnetometers of Canada's Scintrex company are suitable for measuring large-scale magnetic changes, and have been widely used in high-precision magnetic measurement systems on land and on the sea[3]. The U.S. Army Environmental Center (AEC) led a series of research on the detection of UXO based on magnetometer[4], and used the G858 magnetometer of geometry company to detect, which greatly improved the resolution of magnetic measurement. In this paper, an unmanned UXO detection system is built based on the cesium optically pumped magnetometer, and the detection tests of four kinds of UXO targets are carried out to verify the detection performance of the system.

2. Detection principle and system composition

After being magnetized by geomagnetic field, UXO targets such as aerobombs will superimpose secondary magnetic field on the basis of geomagnetic field, resulting in magnetic anomaly. The magnetic susceptibility of the magnetic object, the strength and direction of the residual magnetization, the size, geographical location and buried depth of the magnetic object are the main factors affecting the distribution characteristics and intensity of the magnetic field. For example, the magnetic anomaly curve of the vertical component Z_a of an ellipsoid is shown in Figure 1. When the cylinder extends downward and is finite in length, when it is magnetized along its axis, if the upper end is S-pole and the lower end is N-pole, the magnetic field of the cylinder on the surface is the superposition of two monopolar magnetic fields with different polarities, which is very similar to the anomaly of UXO invading soil, and its magnetic field can be expressed as:

$$Z_a = \frac{\mu_0 m}{4\pi}\left[\frac{h}{\left(x^2+h^2\right)^{3/2}} - \frac{h+2l\sin\beta}{\left[\left(x-2l\cos\beta\right)^2+\left(h+2l\sin\beta^2\right)\right]^{3/2}}\right] \tag{1}$$

Where Z_a is the magnetic anomaly value, x is the horizontal axis coordinate, μ_0 is the permeability, h is the buried depth of the top, m is the strength of the magnetic pole, and l is the length of the cylinder. Due to the regular shape of UXO its magnetic anomaly signal has regularity. According to the static magnetic field theory, the position, attitude, depth and other information of the target can be obtained from the distribution of the known magnetic field.

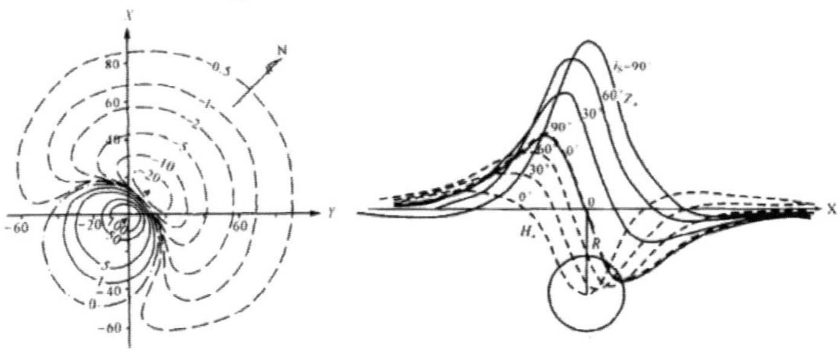

Figure 1. Spatial distribution model of magnetic anomalies in ellipsoid.

The unmanned UXO detection system is mainly composed of cesium optical pump magnetic detection module, UAV platform and ground station platform. The composition block diagram is shown in Figure 2. Cesium optical pump magnetic detection module is used to measure the total geomagnetic field. It is mainly composed of cesium optical pump probe, signal processing unit, magnetic compensation unit and data processing unit. Cesium optical pump probe is the optical unit of the module. The signal processing unit includes an analog part and a digital part. The analog part mainly completes the generation of magnetic field signal together with the optical unit; The digital part converts the magnetic field signal into digital signal, and simultaneously collects the data of the magnetic compensation unit and sends it to the data processing unit. The magnetic compensation unit is composed of a three-component fluxgate magnetometer, which is used to calculate the compensation coefficient and obtain the value of the interference field according to the compensation coefficient, so as to obtain the data close to the real total magnetic field. The data processing unit integrated the received inertial navigation information with the data of two cesium optical pump magnetometers, and sent it to the ground information management syste.

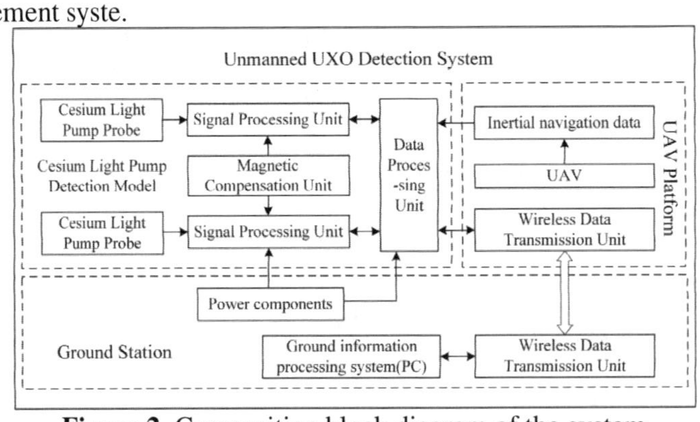

Figure 2. Composition block diagram of the system.

3. Experimental verfication

Figure 3 shows the physical picture of the system. The UAV is equipped with a cesium optical pump magnetic detection module, a differential satellite positioning device, and a wireless data transmission module. It executes the corresponding flight path according to the instructions of the ground station to detect unexploded ordnance, and transmits the measured data to the ground station in real time through the wireless data transmission module. The system was used to detect 60 mortar shells, 82 mortar shells, 122 aerial bombs and 155 aerial bombs. The buried depths of the bombs were 0.6m, 0.8m, 1.5m and 2m respectively. The 16m*20m square area is selected as the test area, and the projectiles of different diameters are buried in several times, and the position coordinates of the projectiles are recorded. The UAV detection route is edited and designed according to the coordinate information of the selected area. The UAV flies back and forth in the north-south direction, with a flight altitude of 0.5m, a flight speed of 1m/s, and an interval of 1m for each route. A total of 16 routes are planned to be flown, as shown in Figure 4. After the flight, according to the magnetic anomaly data collected by the detection module in real time, the magnetic anomaly image of each target can be drawn, as shown in Figure 5.

Figure 3. Physical drawing of the system.

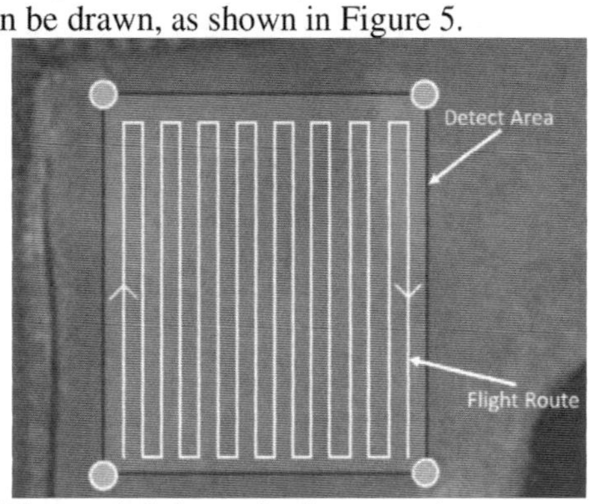

Figure 4. Schematic diagram of UAV route.

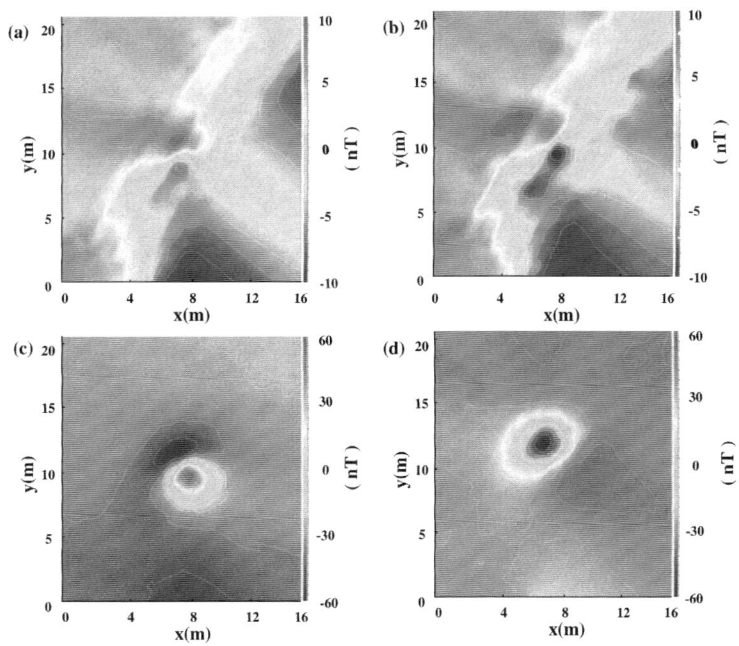

Figure 5. Magnetic anomaly image (a) 60 mortar shell (b) 82 mortar shell (c) 122 aerial bomb (d) 155 aerial bomb.

It can be seen from the magnetic anomaly image that the magnetic anomaly shape of each target is complete, and the position of the target can be determined according to the extreme point of the magnetic anomaly. According to the measured data, the variation range of magnetic anomalies of the four targets are -6~9 nT, -9~10 nT,-55~60 nT,and -55~50 nT, respectively. The shape and intensity of magnetic anomaly have reached the credible standard of magnetic anomaly, and the four UXO targets can be effectively detected. The magnetic anomalies of bullets 60 and 82 are positive in the north and negative in the south, while bullets 122 and 155 are negative in the north and positive in the south. The magnetic anomalies of the 60 projectile and the 82 projectile are different from the distribution characteristics of ferromagnetic materials in the northern hemisphere, which are negative in the north and positive in the south. It is mainly caused by the different directions of the remanence of the bomb model. Its remanence effect exceeds the effect of magnetic induction on the field value and plays a major role in the distribution of the magnetic field, which should be paid attention to in the actual detection. On this basis, the limit detection range test of four targets was carried out, and the test results are shown in Table 1. It can be seen that the limit detection distance of small targets is smaller than that of large targets, and the limit detection distance of 155 projectiles is 4.5m. When the UAV flying height is 1m, it can effectively detect 155 projectiles within 3.5m buried depth.

Table 2. Limit detection range and magnetic anomaly value of each detection target.

Target object	Extreme detection distance (m)	Magnetic anomaly value (nT)
60 mortar shells	1.3	2.3
82 mortar shells	1.6	2.3
122 bomb	3.5	2.3
155 bomb	4.5	2.4

4. Conclusion

Aiming at the demand of UXO detection, this paper proposes an unmanned UXO detection system based on magnetic method, and introduces the detection principle and system composition. The system is used to detect four kinds of UXO targets, and the magnetic anomaly images of each target are drawn. The magnetic anomaly image reflects the spatial distribution of the magnetic anomaly of the target, and can provide the location information of the target. The abnormal magnetic anomaly phenomenon found in the experiment increases the understanding of the observed anomaly morphology. In the actual measurement, in order not to omit the possible UXO, excavation verification should be carried out for all credible anomalies. Finally, the limit detection range of four targets is tested, and the limit detection range of 155 projectile is 4.5m, which verifies the performance of the system.

References

[1] McGinnity C and Kolster M E 2024 Towards Automated Target Picking in Scalar Magnetic Unexploded Ordnance Surveys: An Unsupervised Machine Learning Approach for Defining Inversion Priors. *Remote Sensing* ***16*(3)** p 507

[2] Zheng X Tian Y and Wang B 2023 A Magnetic Gradient Tensor Based Method for UXO Detection on Movable Platform *IEEE Transactions on Geoscience and Remote Sensing* ***61*** pp 1-9

[3] Stanley J M 2001 UXO and Hazardous Metallic Waste Detection Technologies for Use in Litigation-Conscious Environments. In *63rd EAGE Conference & Exhibition* (European Association of Geoscientists & Engineers) p15

[4] Li J., Zeng Z., Sun J., Liu F., (2012) Through-wall detection of human being's movement by UWB radar. *IEEE Geoscience and Remote Sensing Letter,* 9(6), 1079-1083.

Advancements and applications of GMS3: a surface and subsurface unified database system

J M S Rodrigues Neto[1], J Shinohara[1,2], Y Kato[1,2], K Tanida[1], A Okino[1], T Sato[3]

[1] Canaan Geo Research, 2-1-4 Imazaike, Matsuyama, Japan (rodrigues@canaan-geo.jp)
[2] Center for Disaster Management Informatics, Ehime University, 10-13 Dogohimata, Matsuyama, Japan
[3] Iwane Laboratories, 7-8-3 Maruyamanishi, Sapporo, Japan

Abstract. GPR Mobile Mapping System 3D (GMS3) combines three-dimensional ground-penetrating radar (GPR) subsurface data with omnidirectional camera surface mobile mapping system (MMS) imagery to create a high-precision unified surface-subsurface database where underground features are easily localizable from the surface. In the context of Japan, the system is generally able to detect cavities as deep as 1.5 m (depending on soil water content) with a maximum horizontal positioning error of a few cm. The main application of GMS3 in Japan is road subsurface cavity investigations aimed at sinkhole prevention, detecting an average of 0.53 cavities per surveyed km in Japan's national roads, and 4.76 cavities per km in municipal roads in the case of Matsuyama City. We also evaluate the feasibility of 3D mapping of buried pipes (necessary before undergrounding works) at the Johoku Campus of Ehime University. Since both services show increasing demand in Japan due to the country's deteriorating infrastructure and its susceptibility to natural disasters, GMS3 is continuously being developed to stand out as a vital tool in subsurface surveying both in the Japanese and the international context.

1. Introduction

Development of the GPR Mobile Mapping System 3D (GMS3) [1] started in 2015 by Canaan Geo Research, Ltd., and since 2017 counts with collaboration by Japan's Ehime University. The system utilizes ground-penetrating radar (GPR) and an omnidirectional camera-based mobile mapping system (MMS) to generate an integrated surface-subsurface spatial information database, allowing for comprehensive and precise visualization of subsurface features (Figure 1). The technology was introduced in the 17th International Conference on GPR (GPR 2018) [2], and experiments with the system conducted in a dedicated test field were presented in the 18th International Conference on GPR (GPR 2020) [3]. The current paper aims to further demonstrate the developments of GMS3 applied to road subsurface cavity detection and undergrounding-aimed buried pipe 3D modelling.

2. GMS3 technology

2.1. Ground-penetrating radar (GPR)

Multichannel arrays enable 3D visualization of features' geometry, allowing for differentiation between, for example, buried pipes and cavities (Figure 2). Although performance varies with the utilized array's antenna interval, wavelength, sampling and trigger settings, the multichannel GPR array commonly utilized in GMS3 surveys in Japan allows for detection of features as deep as 1.5 m. Japan's generally high soil moisture content is the main limiting factor in achieving greater penetration depths, but since most of the buried pipes and risky cavities in the country are located within that interval, deeper surveys are not usually demanded.

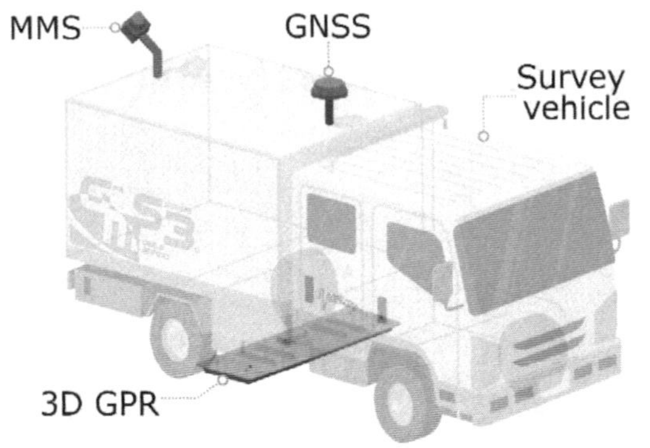

Figure 1. Illustration depicting the fundamental components of GMS3. Carried by a survey platform (in this case, a truck vehicle), the 3D GPR and the MMS compose the system's main sensors, while GNSS provides initial geopositioning basis.

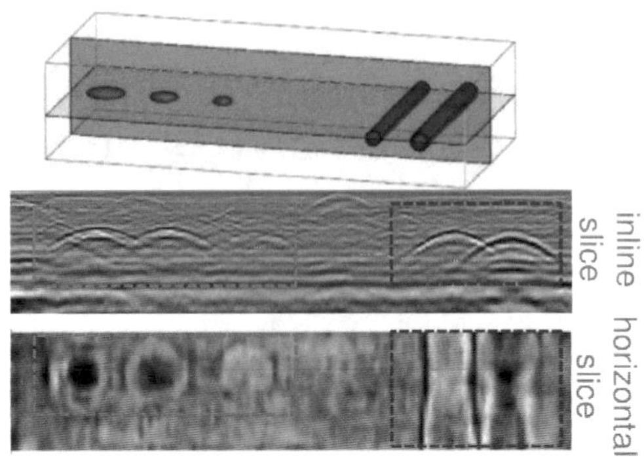

Figure 2. Illustration of subsurface survey of cavities (red) and buried pipes (blue) with 3D multichannel GPR (top), and respective inline and horizontal slice radargrams enabling feature detection, localization and differentiation (bottom panels).

2.2. Mobile mapping system (MMS)

The MMS utilized in GMS3 was developed in cooperation with Iwane Laboratories, Ltd., and is based on camera vector (CV) processing of omnidirectional camera video. CV processing consists of 1) tracking distinguishable reference points in each frame, 2) calculation of the camera's (and thus, the GPR's) position and posture in three-dimensions for each frame based on tracked reference points, and 3) correction of GNSS data with the calculated CV trajectory. CV processing and correction allows for high-precision positioning of the collected GPR data even in situations where GNSS signal reception is insufficient or unstable. After processing, the geopositioned omnidirectional imagery is also utilized to produce high-resolution georeferenced orthoimage maps covering the GPR-surveyed swaths.

2.3. Subsurface data and surface data unification

During GMS3 surveying, GPR subsurface data and omnidirectional camera surface data are concurrently collected in separate hardware and software platforms. GNSS data are also logged, allowing for geopositioning basis and data synchronization through accurate GPS time. After processing, both camera imagery and GPR data are linked to CV-corrected navigation data for positioning, and the final results are simultaneously visualized in the dedicated software GMS3 Viewer (Figure 3).

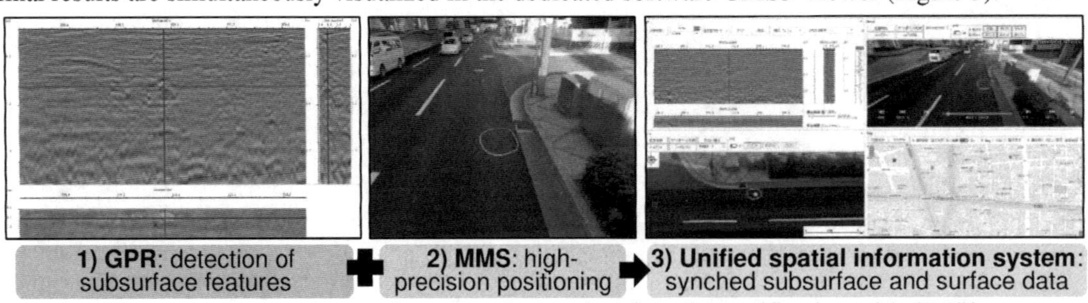

Figure 3. Diagram illustrating subsurface and surface data unification with GMS3.

3. GMS3 applications

Since its development in 2015, GMS3 has been used in various applications where surface and subsurface data integration is advantageous. Such include archaeological investigations, bridge slab

Journal of Physics: Conference Series **2887** (2024) 012022 doi:10.1088/1742-6596/2887/1/012022

diagnosis, and comparative surveying for construction works. The two most substantially employed GMS3 applications, however, are road subsurface cavity detection for sinkhole prevention, and buried pipe 3D modelling for undergrounding works. These will be described in the following sections.

3.1. Road subsurface cavity detection

According to data released in the Japan's Ministry of Land, Infrastructure, Transport and Tourism (MLIT) homepage, between 2020 and 2022, Japan has approximately 10,000 subsurface cavity-induced road sinkhole occurrences per year, averaging to about 27 every day (Table 1). Although many causative factors lead to subsurface cavity formation (soluble geology, groundwater flow, earthquakes), MLIT reports show that the most common cause for sinkholes in the Japanese context is cracking and/or damage of buried pipes. Japan has been facing many infrastructure maintenance issues in recent years attributed to deterioration of constructions (including underground) from the country's economic bubble period in the mid 80's. Japan's high earthquake activity is another substantial aggravating factor for subsurface cavity formation. The Japanese government is expected to respond to the increasing cavity issue with more cavity detection survey orders in the following years, as well as expansion of investigations to prefectural and municipal roads as opposed to solely national roads.

GMS3 covered up to 16,944 km in cavity surveys between the fiscal years of 2016 and 2023 (average of 2,118 km per year). Between 2020 and 2022, GMS3 cavity detection in national roads averaged 0.53 possible cavities per surveyed km (with a peak of 0.93 in 2022). However, in one case of survey of municipal roads (characterized by a more intense buried pipe network) in Matsuyama City, southwestern Japan, cavities per surveyed km averaged 4.76 during the same 3-year period, with a peak of 8.25 in 2022 (Table 1). The disparity of cavity detection rate between national and municipal roads evidences the major role of buried pipe infrastructure in subsurface cavity formation.

Table 1. Records of road sinkhole occurrences in Japan between 2020 and 2022, and GMS3 survey results in national roads and one case of municipal roads (Matsuyama City) during the same period.

Year	Sinkholes		Our survey (national roads)			Our survey (Matsuyama City)		
	Total	Daily	Surveyed length	Possible cavities	Possible cavities/km	Surveyed length	Possible cavities	Possible cavities/km
2020	9,124	25	3,529 km	948	0.28	6.0 km	19	3.17
2021	9,967	27	1,455 km	1,352	0.93	5.7 km	47	8.25
2022	10,548	29	1,434 km	548	0.38	4.9 km	14	2.86

3.2. Buried pipe 3D modelling

Undergrounding is a term utilized to refer to the replacement of otherwise overhead communication and electricity cables into the subsurface. Benefits of undergrounding in urban infrastructure include reduction in the risk of power outages and accidents during extreme weather conditions, reduction of operating costs, and aesthetic improvement. As pointed out by Japan's MLIT in 2017, while most major cities in developed countries of Europe and Asia have reached 100% undergrounding rates, Japan stands out as an odd exception: only 8% of Tokyo's, and 6% of Osaka's electricity and communication cables are underground. The situation poses a particular threat due to the country's susceptibility to natural disasters such as earthquakes and typhoons, reason why the Japanese government is planning for massive undergrounding works in most of the country's major cities. Before any undergrounding work, a subsurface survey for underground feature mapping is mandatory to prevent accidents or damage to existing utilities, such as buried pipes. 3D GPR is particularly efficient in buried pipe mapping as visualization of features' geometry in map view easily differentiates long-shaped pipes and individually-shaped objects such as rocks or cavities. We evaluated the feasibility of GMS3 in buried pipe mapping and 3D modelling in an experiment at Ehime University's Johoku Campus (Matsuyama City). Using

Journal of Physics: Conference Series **2887** (2024) 012022 doi:10.1088/1742-6596/2887/1/012022

GPR and CV processing assisted with reliable existing maps and direct confirmation, a 3D model of the buried pipes in the study area was created with high location accuracy. The subsurface model was then implemented into the MMS as to make the underground features visible from the surface imagery, constructing an integrated database that allows for practical visualization of subsurface data (Figure 4).

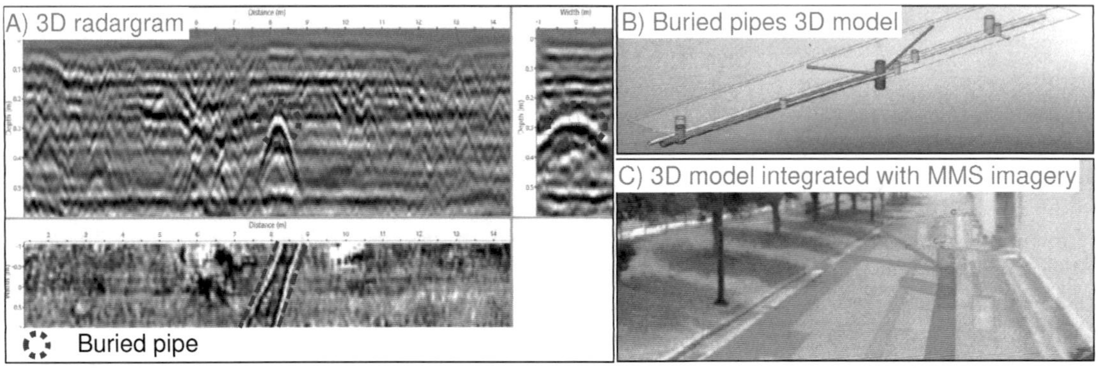

Figure 4. GMS3 buried pipe mapping and 3D modelling experiment at Ehime University: radar data is used to detect and locate buried pipes (A), which are then modelled in 3D (B), and finally integrated into the CV imagery for practical visualization and referencing of subsurface data.

4. Discussions and conclusions

Japan's current era of public infrastructure deterioration and latency in overhead cable undergrounding, coupled with the country's predisposition to natural disasters, create a scenario where subsurface surveying is in high (and likely increasing) demand. Although 3D GPR is the most effective tool in most subsurface utility locating services, the more orthodox methods solely utilize GNSS measurements to localize GPR information, leading not only to inaccurate data in situations where satellite signal reception is unfavourable, but also to difficulty in visualizing features in the field without clear beaconing references. GMS3 has a substantial advantage in subsurface mapping as it allows for high-precision and comprehensible feature localization. Concerning data interpretation difficulty, though many developments have been discussed concerning the use of machine learning techniques for cavity and utility detection in GPR [4], manual diagnosis by experts is still crucial for most cases. Some GMS3 road subsurface cavity investigations also require secondary boring surveying for confirmation purposes. This study presented the latest developments of GMS3, as well as its current and prospecting applications in the context of Japan. The technology utilizes integrated GPR and MMS to provide highly accurate and comprehensible localizing of subsurface features with a unified surface and subsurface database. While further adjustments for higher accuracy and efficiency are being continuously developed, it is expected that GMS3 will be a useful tool not only in response to Japan's increase in demand for subsurface survey business, but also in the international context.

5. References

[1] Shinohara J, Baba T 2018 Underground exploration apparatus and underground exploration vehicle Japan patent JP6431573B

[2] Shinohara J, Kato Y, Okino A, Shukla E A, Baba T. GMS3 a unified system of ground penetrating radar and camera vector for efficient road infrastructure maintenance *2018 17th International Conference on Ground Penetrating Radar (GPR)* pp 1-4

[3] Kato Y, Niizaki T, Baba T, and Shukla E A 2020 Advancement of GMS3 a unified system of ground-penetrating radar and camera vector technology *18th International Conference on Ground Penetrating Radar (GPR)* pp 152-155

[4] Chun P J, Suzuki M, Kato Y 2023 Iterative application of generative adversarial networks for improved buried pipe detection from images obtained by ground-penetrating radar *Computer‐Aided Civil and Infrastructure Engineering* **38**(17) pp 2472-2490.

Research on the application of 3D GPR in the detection of road collapse diseases in loess region of Ningxia

Zhaotao Yan[1], Zhaofa Zeng[1]*, Baizhou An[2], Jiayu Lu[2] and Boyang Sun[2]

1 College of Geo-exploration Science and Technology, Jilin University, Changchun 130026, China
2 Geophysical and Geochemical Exploration Institute of Ningxia Hui Autonomous Region (Autonomous Regional Deep Earth Exploration Center), Yinchuan 750001, China
*Corresponding author's e-mail:zengzf@jlu.edu.cn

Abstract. In recent years, the loess region of Ningxia is faced with severe challenges of road collapse diseases. Such diseases have the characteristics of strong suddenness, high frequency and difficult prediction. Coupled with complex causes, traditional detection methods are often difficult to achieve effective, rapid detection and accurate response, posing a major threat to the local social economy and public safety. This paper focuses on the application of three-dimensional radar technology in road collapse diseases detection in the loess region of Ningxia, aiming to explore an efficient, non-destructive and rapid diseases detection and verification method, so as to achieve timely and accurate detection and verification of collapse diseases. In this study, the Guyuan area with complex geological conditions was selected as a demonstration, and advanced 3D ground penetrating radar equipment was used to conduct fine scanning of the structures under the roads in the selected area. Through comparative analysis of radar images and actual geological structures, the 3D radar could accurately identify and locate potential underground voidage, cavitation and porosity and other hidden hazards. It provides the possibility for the early detection and prevention of collapse diseases. The research results of this paper verify the applicability and reliability of 3D ground penetrating radar in road collapse diseases in loess area of Ningxia, providing a reference example for road safety detection under similar geological conditions, and also providing strong support for road safety management and emergency response.
Keywords: road collapse diseases, three-dimensional ground penetrating radar, Ningxia loess region, non-destructive rapid detection, road safety

1. Introduction

In the world, the frequent occurrence of urban road collapse diseases has become a serious public safety problem, especially in the loess region of Ningxia in western China. In recent years, such diseases have caused more and more serious loss of life and property and social impact. The collapse accident in Xining in 2020, which killed 29 people and injured 25 others, once again warned us of the destructive power and unpredictability of road collapse diseases. The occasional and unpredictable nature of road collapse diseases, as well as the uncertainty of its occurrence place, time and frequency, bring great difficulties to diseases prevention and control, and make the specific attribution extremely complicated. Ningxia Hui Autonomous Region, especially Guyuan City, is prone to frequent road collapse diseases due to its special loess geological structure. In the face of this urgent situation, it is particularly important to improve the detection system of road collapse hidden dangers and develop efficient hidden dangers identification technology.

This paper is based on the loess area of Ningxia, with Guyuan city as the research focus, and is committed to forming a complete and accurate understanding of road collapse hazards. Based on the scientific analysis of recent road collapse cases, the geophysical response rules of road collapse hazards are explored, and the advantages of 3D radar non-destructive rapid detection are utilized to conduct high-precision scanning of underground structures and quickly identify potential collapse risks[1]. Through the application of this technology, its application effect under the complex geological conditions in the loess area is verified, which provides scientific and technological support for the early warning and prevention of road collapse diseases, and contributes to the protection of road safety and public welfare.

2. 3D GPR method

3D GPR technology shows great potential in the field of non-destructive testing with its advantages of non-contact, high resolution and all-weather operation[2][4]. Compared with the traditional 2D GPR, the 3D GPR system adopts array antenna technology, sets up multiple channels, and carries out high-density and fast full coverage scanning, which greatly shorens the detection cycle and improves the work efficiency. Moreover, data are obtained simultaneously in both horizontal and vertical directions to obtain the underground 3D data body, providing more intuitive and comprehensive underground information.[5][6]

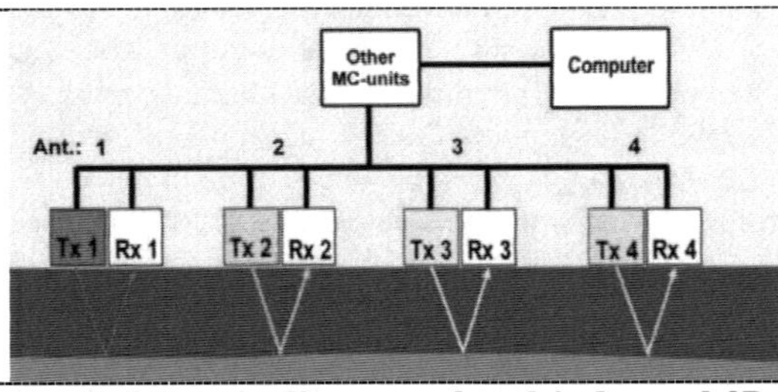

Figure 1. Schematic diagram of multi-channel 3D GPR

3. Detection of road collapse diseases in loess region of Ningxia

This radar detection selected some provincial and national roads in Guyuan City for typical demonstration, full coverage scanning by truck-mounted three-dimensional geological radar, the abnormal area delimated by three-dimensional radar was remeasured by two-dimensional radar in the shape of a well, the hole anomaly identified by the retest was located by RTK measuring equipment, and drilling verification.

3.1. Inspection of work area

The road collapse detection area is located in Yuanzhou District, Guyuan City, Ningxia Hui Autonomous Region, and the work scope covers part of S202 provincial Highway and G341 National Highway in Guyuan City. The construction pavement is made of asphalt, and the surrounding and subgrade are all yellow soil. A total of 5 3D ground penetrating radar survey lines are arranged, numbered LD2 ~ LD5, with a total length of about 50km, as shown in Figure 2 below.

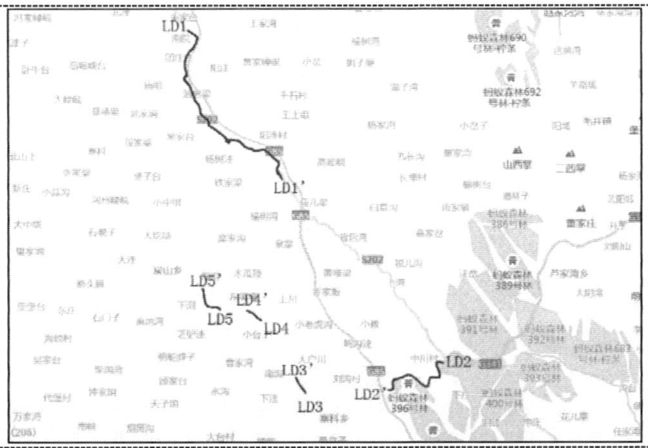

Figure 2. Layout of 3D GPR survey line

3.2. Analysis of characteristics of road collapse diseases and evaluation suggestions

The detection work found out the abnormal development of road in the radar detection area of Guyuan City, and delineated three types of anomalies: cavity, loose and subsidence anomaly. The following are the characteristics analysis, evaluation and treatment suggestions of three typical diseases.

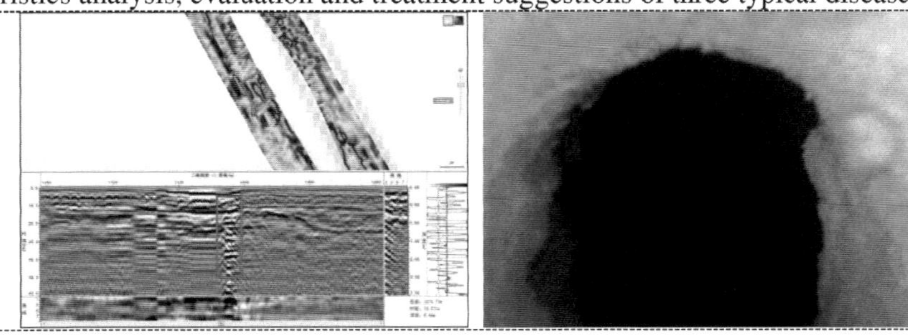

Figure 3. 3D GPR image of the cavity anomaly and borehole verification diagram.

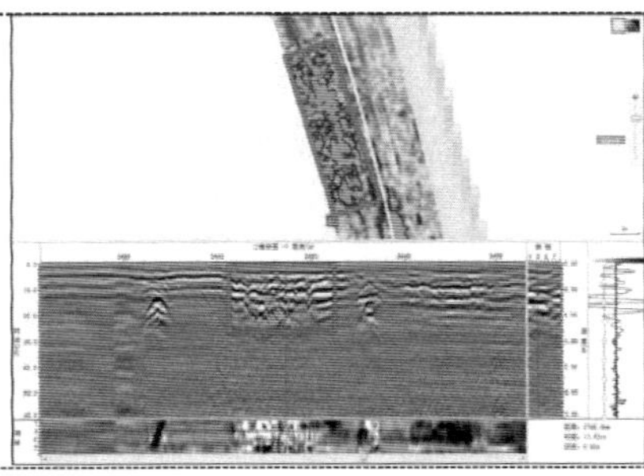

Figure 4. 3D GPR image of the loose anomaly.

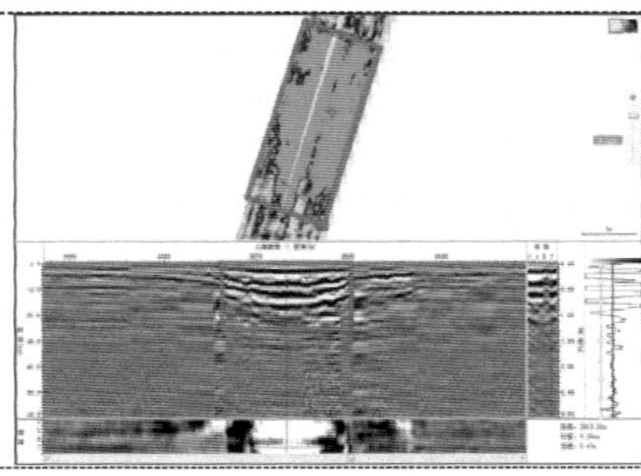

Figure 5. 3D GPR image of the subsidence anomaly.

The cavity anomaly is shown in Figure 3. On the radar profile, the top of the anomaly is irregularly hyperbolic, there are obvious diffraction wave and bottom reflection wave on both sides, and the multiple wave oscillation is obvious and extends to the bottom. This is because the long-term rainfall since mid-July in this area has caused the backpouring of rainwater under the road surface, resulting in soft subgrade, loose interior and containing water. Therefore, this anomaly presents a characteristic structure different from that of a typical cavity in the profile. Combined with the drilling verification and field investigation, this anomaly belongs to an atypical cavity.

The loose anomaly is shown in Figure 4. On the radar profile, there are continuous reflection wave in phase axis interrupted or curved distribution at the boundary of the non-uniform body in the abnormal region, its wavelength becomes longer, its amplitude changes obviously, and the characteristics of the wave group also change obviously. There are multiple reflections below, and multiple waves are obvious. Long-term rainfall causes the originally buried loess layer to become soft, which is more water-rich than the surrounding strata.

The subsidence anomaly is shown in Figure 5. The abnormal area on the radar profile has an obvious downward trend compared with the surrounding normal road section. The radar reflection signal is strong, and the reflected signal with strong amplitude appears and the frequency increases. It can be seen from the slices that the amplitude of the abnormal region is uniform and stronger than that of the surrounding strata. Long-term rainfall causes the subgrade to be eroded by rain and become soft, and then subsidence occurs.

The area of the hole is less than $3m^2$, the height is less than 1m, and the overlap ratio is greater than 3. The abnormal area of the loose body is greater than $15m^2$, the height is less than 1m, and the overlay ratio is greater than 1 and less than 3. The abnormal subsidence area is greater than $15m^2$, the height is less than 1m, and the overlay ratio is greater than 0.5 and less than 1. In these three typical disease locations, the road itself has slight cracks, the rock and soil around the anomaly is collapsible loess, belonging to viscous soil, frequent vehicles, heavy vehicles passing, and general drainage conditions. There is no pipeline, underground structure under the anomaly, no nearby facilities, low personnel density, low property density, so the social impact is low, the risk is general. It is suggested that the cavity should be filled with grouting, and regular inspection and detection should be carried out after treatment. It is recommended that the loose and settling areas should be inspected regularly, and the inspection frequency should not be less than 3 times/month.

4. Conclusion

This survey is mainly aimed at road collapse disasters in Ningxia loess area. Based on 3D ground penetrating radar detection technology, it accurately identifies and locates hidden dangers such as underground voids, voids and looseness, and accurately and timely completes road collapse disaster detection in Ningxia loess area. Meanwhile, based on the detection results, the causes and mechanisms of collapse disasters are analyzed. The risk assessment of the disease body was carried out, and suggestions for post-treatment were put forward. Road subgrade in Guyuan City is mostly collapsible loess, which is prone to collapse under the erosion of hydrostatic water. The road collapse situation in Ningxia loess region can be dealt with through the full coverage detection of key areas, the risk assessment of detected road diseases, and the methods such as grouting filling, soil replacement subgrade and regular inspection can be carried out to ensure the stability and safety of subsequent road subgrade. In addition, the detection depth of road collapse has certain limitations. For the detection of road collapse disaster in loess area, under certain conditions, a more perfect road non-destructive testing system with different depths can be established by integrating multiple geophysical means.

In summary, combined with the field test data in loess area of Ningxia, this paper verifies the applicability and reliability of 3D ground penetrating radar in road collapse disaster in loess area of Ningxia, and also provides a reference example for road safety detection under similar geological conditions, providing possibilities for early detection and prevention of collapse disaster, and providing strong support for road safety management and emergency response.

5. References

[1] LIU Lan-Bo, QIAN Rong-Yi. Ground Penetrating Radar:A critical tool in near-surface geophysics[J]. Chinese Journal of Geophysics (in Chinese), 2015, 58(8): 2606-2617.

[2] Saarenketo T, Scullion T. Ground penetrating radar applications on roads and highways [J]. Detection & Identification, 1994.

[3] Jeng Y, Chen C S. Subsurface GPR imaging of a potentialcollapse area in urban environments[J]. Engineering Geology, 2012, 147: 57-67

[4] Fenning P J, Jack A G, Veness J K. Ground probing radar applications in site investigation; proceedings of the conference proceedings of the british institute of non-destructive testing

international conference, Ndt in civil engineerinG 14-16 april 1993, Liverpool university volume 1, F, 2003 [C].
[5] SyaifuddiN F, Syaifuddin F. Cavities detection with ground penetrating radar in limestone dominated rock formation [J]. 2014.
[6] Song I, Larkin A. Use of ground penetrating radar at the faa's national airport pavement test facility (NAPTF) [M]. Springer International Publishing, 2016.

Acknowledgments
The research was funded by Science and Technology Benefit the people special project of Ningxia Hui Autonomous Region: Ningxia urban road collapse hidden danger comprehensive detection technology promotion and application demonstration (grant number : 2022CMG03006).

doi:10.1088/1742-6596/2887/1/012024

Three-dimensional Reverse Time Migration for Detecting Tree Trunk Defects Using Ground Penetrating Radar

Y Liu [1,2], D Feng [1,2], X Wang [1,2]

[1] School of Geosciences and Info-Physics, Central South University, Changsha 410083, China
[2] Key Laboratory of Metallogenic Prediction of Nonferrous Metals, Ministry of Education, Changsha 410083, China (yx_liu@csu.edu.cn)

Abstract. Due to environmental degradation and external intrusions, trees may develop various defects that diminish their stability and survival rates, leading to significant losses. Ground penetrating radar (GPR), characterized by its rapid and non-destructive advantages, has been regarded as a potentially effective method for assessing tree health conditions. However, existing GPR-based tree imaging techniques are predominantly two-dimensional (2D), making it difficult to diagnose three-dimensional (3D) structural features of defects. Therefore, this paper introduces, for the first time, 3D reverse time migration (RTM) into trunk defect detection. Compared to 2D RTM, 3D RTM can simultaneously back-propagate in-line and cross-line signals, accurately locate reflection waves, and focus diffracted waves, reconstructing the true interfaces and defect positions within the tree trunk. Experimental results on irregular trunk models containing a cavity demonstrate the feasibility of 3D RTM in trunk defect detection, providing specific and comprehensive guidance for formulating tree protection and restoration measures.

1. Introduction

In recent years, the incidence and outbreak area of diseases has been increasing annually due to factors such as climate change [1], international timber trade [2], and international tourism [3]. To promptly detect tree trunk internal defects and monitor their health condition, regular inspection and assessment of living trees have become critical. Ground Penetrating Radar (GPR), due to its non-destructive and simple operation, has been widely used in shallow surface exploration [4][5] and has gradually garnered high attention from forestry workers. The basic principle of GPR is to emit high-frequency electromagnetic waves underground to understand the geological structure. When applied to tree detection, it can obtain data and images of the internal structure and health conditions of tree trunks.

However, the internal media of tree trunks are complex, and electromagnetic waves encountering diseased or defective areas may produce multiple waves and diffraction waves, making it difficult for manual interpretation. Therefore, some scholars have processed radar data to achieve internal imaging and reconstruction of tree trunks [6]-[8]. To further clarify the types of internal trunk defects and implement precise repairs, [9] and [10] introduced full waveform inversion (FWI) and wavefield reconstruction inversion (WRI) into trunk detection, respectively. Both theoretical and field experiments have demonstrated the feasibility of FWI and WRI for the two-dimensional (2D) reconstruction of trees.

In fact, 2D modeling can capture GPR signatures relatively well but it cannot mimic the behavior of real GPR antennas adequately, especially for near-surface fields. The resulting 2D techniques remain utopian, making it difficult to distinguish the true 3D structural features of target bodies. Furthermore, internal defects within tree trunks mainly develop in 3D forms. Considering the demand for subsequent maintenance, we intend to further investigate 3D imaging of internal defects within tree trunks. However, the expensive computational costs indicate that 3D FWI is currently impractical for field testing. In this regard, migration techniques provide solutions [11][12]. Compared to migration techniques such as diffraction stacking migration [13], Kirchhoff migration [12], and frequency-wavenumber migration [14], reverse time migration (RTM) does not require approximations of the wave equation. It can image electromagnetic waves propagating in any direction, making it suitable for complex media.

In the following sections, we briefly introduce the basic principles of 3D RTM based GPR data in trunk detection. Then, we demonstrate the validity of the algorithm using a numerical model of a 3D irregular trunk with a cavity.

2. Methodology

GPR is a shallow subsurface geophysical exploration method that utilizes high-frequency electromagnetic waves for detection [15]. When trees develop defects, changes occur in their dielectric constant and conductivity, forming property boundaries, and leading to different radar echoes between the healthy trunk and the defective trunk (as shown in Figure 1). Maxwell's equations in 3D TM mode can be expressed by:

$$\begin{cases} \nabla \times \mathbf{E} = -\mu \dfrac{\partial \mathbf{H}}{\partial t} \\ \nabla \times \mathbf{H} = \varepsilon \dfrac{\partial \mathbf{E}}{\partial t} + \sigma \mathbf{E} + J \end{cases} \tag{1}$$

where $\mathbf{E}$ is the electric field intensity, $\mathbf{H}$ is the magnetic field intensity, J is the current source. The parameters ε, σ, and μ are the dielectric constant, conductivity, and permeability of the tree trunk, respectively. In this paper, we assume that the permeability is constant. The finite-difference time-domain (FDTD) method is used for forward simulations.

RTM relies on the principle of time consistency. The electromagnetic wavefield recorded by the receiving antenna is regarded as an adjoint source, and it propagates backward in time. Upon extrapolation to the zero moment, all energy from the diffraction and reflection wave returns to the spatial interface where the diffraction and reflection occurred. The migration profile is then obtained by applying appropriate imaging conditions [16]. The workflow of RTM based on GPR data is shown in Figure 2.

In this paper, we adopt the zero-lag normalized cross-correlation imaging condition [17]. However, this imaging condition may result in instability during computation, as the electromagnetic wave field near or far from the source point may approach zero. Therefore, in the computational process, a small real number δ is added to the denominator to mitigate this instability. Its expression is:number δ is added to the denominator to mitigate this instability. Its expression is:

$$\begin{cases} I(x,y,z) = \sum_m \sum_n \dfrac{S_m(x,y,z,t_n) R_m(x,y,z,t_n)}{S_m(x,y,z,t_n) S_m(x,y,z,t_n) + \delta} \\ \delta = \alpha * \max[\sum_m \sum_n S_m(x,z,t_n) S_m(x,z,t_n)] \end{cases} \tag{2}$$

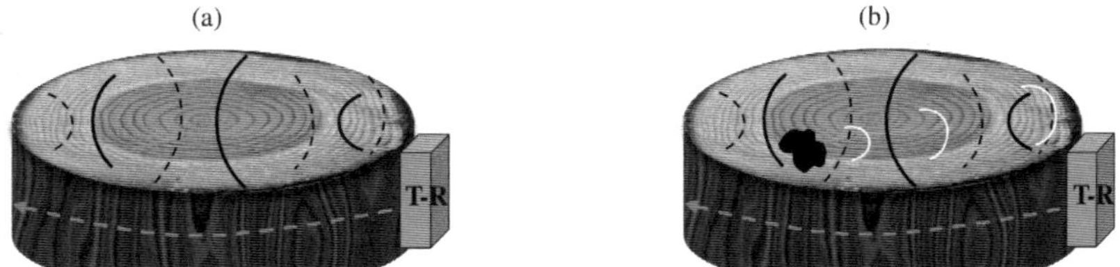

Figure 1. Diagram of detection within (a) a healthy treetrunk and (b) a defective tree trunk.

where $I(x, y, z)$ is the zero-lag cross-correlation RTM result, $S_m(x, y, z, t_n)$ and $R_m(x, y, z, t_n)$ represent the forward and backward wavefields of the mth source at time t_n, respectively. And α is set to 1×10^{-3}.

3. Numerical case

To better fit real tree conditions, an irregular tree model with a cavity, as depicted in Figure 3(a), was constructed in this study. The internal structure consists of three parts: bark, cambium, and sapwood. Based on regression equations, their respective dielectric constants are 5, 8, and 4. The outer bark layer has an irregular shape with a trunk diameter of 0.88m, and a cavity with a diameter of 0.12m is placed inside the trunk. Typically, cavities grow upwards from the bottom of the trunk. To simulate the 3D structure of the cavity simply, it extends along the z-axis for 0.5m.

Forward modeling was performed using the FDTD algorithm in a computational region of 50×50×50 grids. A 900MHz Ricker wavelet was used as the excitation source. The common-offset acquisition was adopted. Two measurement lines were placed around the trunk, at distances of 0.3m and 0.7m, each with 50 traces collected clockwise. The green 'o' in Figure 3(a) represents the combination of the transmitting antenna and receiving antenna.

Figure 3(b) presents the result of 3D RTM, clearly displaying the interfaces and positions of the cavity at 0.0m and 0.3m along the trunk. And, at 0.7m and 1m, the migration result indicates no obvious defect inside the trunk, consistent with the true model. Figure 3(c) shows the 2D RTM results obtained from the two measurement lines. The images reveal a significant cavity at 0.3m and 0.7m along the trunk, contrary to reality. This discrepancy arises because, in actual 3D scenarios, GPR data can be affected by side effects, which are not the interference signals generated by the abnormal body directly below the radar. Compared to 2D RTM, by introducing 3D RTM into trunk detection, this study accurately describes the 3D structural characteristics of internal trunk defects.

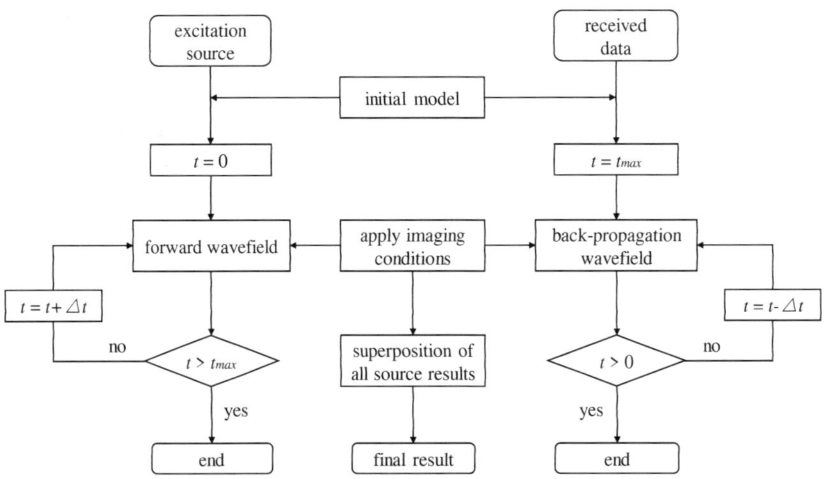

Figure 2. Workflow of 3D RTM based on GPR data.

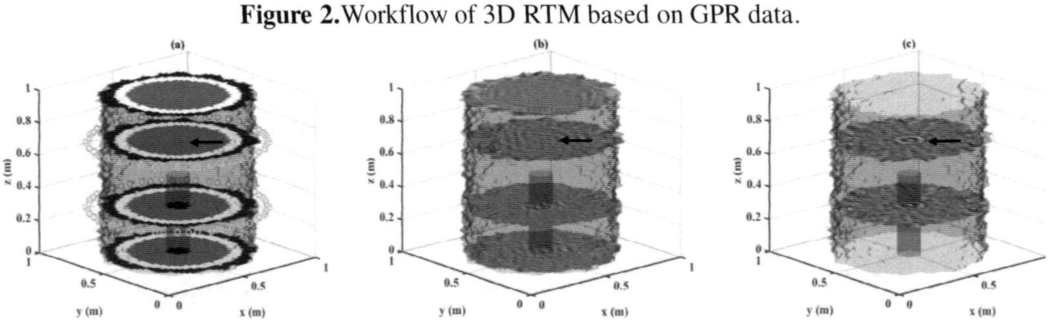

Figure 3. (a) The true tree trunk model containing a cavity, (b) 3D RTM result, (c) 2D RTM result.

4. Conclusion

For utilizing surface-coupled commercial GPR common-offset data, this paper introduces 3D RTM into trunk detection for the first time, achieving 3D localization and imaging of internal trunk defects. Compared to 2D RTM, 3D RTM can alleviate side effects of GPR in subsurface exploration, accurately describe the 3D structure of defects, and provide more specific and comprehensive guidance for formulating tree protection and restoration measures.

Subsequent work needs to focus on three aspects: suppressing interference noise, improving resolution, and enhancing efficiency. Additionally, concerning trunk detection, how to extract the surface of the trunk and how to compensate for attenuation are also worthy of discussion.

5. References

[1] Tubby K V and Webber J F 2010 Pests and diseases threatening urban trees under a changing climate. *Forestry*, **83**(4): 451-459.

[2] Guo Q, Rejmanek M and Wen J 2012. Geographical, socioeconomic, and ecological determinants of exotic plant naturalization in the United States: insights and updates from improved data. *Neobiota*, **12**(2): 262-70.

[3] Santini A, et al. 2012 Biogeographical patterns and determinants of invasion by forest pathogens in Europe. *NP*, **197**(1):238–50.

[4] Peng S, Yang F and Xu X 2015 Application and prospects of the GPR technology on road disease detection in China. *Near-Surface Asia Pacific Conf.*, 476-479

[5] Li S and Liu B, et al. 2017 An overview of ahead geological prospecting in tunneling. *TUST*, **63**: 69-94.

[6] Tosti F, Gennarelli G, Lantini L, Catapano I, Soldovieri F, Giannakis I, et al. 2022 The use of GPR and microwave tomography for the assessment of the internal structure of hollow trees. *IEEE Trans Geosci Remote Sens*, **60**: 1-14.

[7] Fu L and Liu L 2021 Imaging the internal structure of trunks via multiscale phase inversion of ground-penetrating radar data. *Geophys.*, **9**(3): 869-880.

[8] Giannakis I, Tosti F, Lantini L and Alani A M 2019 Health Monitoring of Tree Trunks Using Ground Penetrating Radar. *IEEE Trans Geosci Remote Sens*, **57**(10): 8317-26.

[9] Feng D and Liu Y, Wang X, Zhang B, Ding S, Yu T, et al. 2023 Inspection and Imaging of Tree Trunk Defects Using GPR Multifrequency Full-Waveform Dual-Parameter Inversion. *IEEE Trans Geosci Remote Sens*, **61**:1-15.

[10] Feng D, Liu Y, Wang X, Ding S, Xu D, and Yang J 2023 Quantitative Evaluation for the Internal Defects of Tree Trunks Based on the Wavefield Reconstruction Inversion Using Ground Penetrating Radar Data. *Forests*, **14**(5).

[11] Fisher E, McMechan G A, and Annan A P 1992 Acquisition and processing of wide-aperture ground-penetrating radar data. *Geophys.*, **57**(3): 495-504.

[12] Moran M L, Greenfield R J, Arcone S A and Delaney A J 2000 Multidimensional GPR array processing using Kirchhoff migration. *J. Appl. Geophys*, **43**(2-4): 281-295.

[13] Sattlegger J W and Stiller P K 1974 Section migration, before stack, after stack, or inbetween. *Geophysical Prospecting*, **22**(2): 297-314.

[14] Bitri A and Grandjean G 2008 Frequency–wavenumber modelling and migration of 2D GPR data in moderately heterogeneous dispersive media. *Geophys. Prospect.*, **46**(3): 287-301.

[15] Daniels D J 2004 *Ground Penetrating Radar* vol 1 (London: IEE)

[16] Nguyen B D and McMechan G A 2013 Excitation amplitude imaging condition for prestack reverse-time migration. *Geophys.*, **78**(1): 37-46.

[17] Kaelin B and Guitton A 2006 Imaging condition for reverse time migration. *SEG Technical Program Expanded Abstracts 2006*, 2594-8.

A GPR layered media parameters inversion method under rough surface

S Zhao[1], X Yang[1], L Liao[2] and T Lan[1*]

[1] School of Information and Electronics, Beijing Institute of Technology, Beijing, China
[2] School of Materials Science and Engineering, Chongqing Jiaotong University, Chongqing, China (1030978096@qq.com)

Abstract. Ground Penetrating Radar (GPR) plays an important role in pavement detection and road exploration due to its efficiency. Traditional media parameter inversion algorithms only consider ideal smooth surfaces and neglect the roughness of interfaces, leading to significant errors in practical applications. Therefore, a GPR rough surface-layered medium parameter inversion method based on genetic algorithm is proposed in this paper. The method uses the generalized reflection coefficient and considers the influence of the surface roughness of the medium. Then, a radar echo signal time-domain model is established. Finally, the cost function is optimized by genetic algorithm. The effectiveness and accuracy are validated and analysed by utilizing simulated data through random rough surfaces models.
Keywords. Ground Penetrating Radar (GPR), parameter inversion, rough surface, layered medium

1. Introduction

Ground Penetrating Radar (GPR) has been widely applied in various engineering fields, including highway quality inspection, building damage detection [1], and military exploration [2], etc. GPR layered parameter inversion can provide valuable information such as layer thickness and relative permittivity and has extensive application in the pavements where the medium interfaces can be considered as random rough surfaces whose influence cannot be ignored.

Layered media parameters inversion methods can mainly categorized into time-domain and frequency-domain. Nevertheless, these methods primarily focus on smooth surface media.

Within the operating frequency range of GPR, the influence of interface roughness cannot be ignored. The height distribution of many engineering surfaces follows a Gaussian distribution, and their surface profiles exhibit an exponential autocorrelation function relationship [3]. Relevant studies indicate that the roughness of medium interfaces has minimal impact on the phase of the reflected wave, but it can cause a decrease in the amplitude of the reflected wave. Under narrowband conditions, the frequency characteristics of the scattered echo can be approximated by an exponential function [4]. An improved MUSIC algorithm has estimated time delay and roughness of rough surface media, but did not perform further inversion of the media parameters [5].

In this paper, a GPR layered media parameters inversion method under rough surface based on signal time-domain reconstruction is proposed. This method utilizes generalized reflection coefficients which takes into account the frequency behavior of scattered echoes from rough interfaces to establish a GPR reflection signal model, and then employs a genetic algorithm for global optimization inversion of parameters, which realizes accurate parameter inversion of multilayered media with rough surface.

2. Methodology

The proposed method consists of three parts: defining the generalized reflection coefficient, reconstructing the echo signal in time domain, and optimizing by genetic algorithm.

2.1. Generalized reflection coefficient

When considering non-magnetic lossless media, the reflection coefficient $r_{i,i+1}$ at the interface between the ith layer and the $i+1$th layer can be expressed as:

$$r_{i,i+1} = (\sqrt{\varepsilon_i} - \sqrt{\varepsilon_{i+1}})/(\sqrt{\varepsilon_i} + \sqrt{\varepsilon_{i+1}}) \tag{1}$$

where ε_i denotes the relative permittivity of the ith layer medium.

The generalized reflection coefficient is defined as the ratio of the amplitude of the reflected wave at each layer interface to the incident wave, denoted as $R_{i,i+1}$. Through derivation, the ith generalized reflection coefficient $R_{i,i+1,\text{smooth}}$can be expressed as:

$$R_{i,i+1,\text{smooth}} = r_{i,i+1} \prod_{k=0}^{i-1} (1 - r_{k,k+1}^2) \prod_{k=1}^{i} \mathrm{e}^{-2\alpha_k d_k} \tag{2}$$

where α_k、d_k respectively represent the attenuation coefficient and thickness of the kth layer medium. And the expression for α_k is:

$$\alpha_k = \omega \left\{ \frac{\mu\varepsilon}{2} \left[\left(1 + \left(\frac{\sigma}{\omega\varepsilon} \right)^2 \right)^{1/2} - 1 \right] \right\}^{1/2} \tag{3}$$

where ω、σ、μ represent angular frequency, conductivity and magnetic permeability, respectively.

When the interfaces between multiple layers of media are rough, the rough surfaces can cause changes in the amplitude and phase of the scattered echo. In reference [6], experiments were conducted using the electromagnetic asymptotic method and the PILE method, which showed that the phase of the scattered echo from a rough surface does not undergo significant changes. However, there is a noticeable decrease in amplitude. Additionally, under narrowband conditions, the frequency behavior of the scattered echo approximates an exponential function. In this case, the ith generalized reflection coefficient, denoted as $R_{i,i+1}$, can be expressed as:

$$R_{i,i+1} = \mathrm{e}^{-b_i f} r_{i,i+1} \prod_{k=0}^{i-1} (1 - r_{k,k+1}^2) \prod_{k=1}^{i} \mathrm{e}^{-2\alpha_k d_k} \tag{4}$$

where f is the frequency of the scattered echo, and b_i is the roughness parameter of the ith interface.

2.2. Time-domain reconstruction of received echo signals

When the road is considered as a non-dispersive and non-magnetic medium, some studies simplify the road reflection system as a linear time invariant system [7]. The generalized reflection coefficient sequence $r(t)$ can be expressed as:

$$r(t) = \sum_{i=0}^{N-1} R_{i,i+1} \delta(t - t_i) \tag{5}$$

Let $x(t)$ represent the radar transmission signal, then the received signal can be obtained by inputting it into the LTI system:

$$y_r(t) = \sum_{i=0}^{N-1} R_{i,i+1} x(t - t_i) + n(t) \tag{6}$$

where N is the number of interfaces, including the air-road surface interface. $R_{i,i+1}$ is the defined generalized reflection coefficient, t_i is the signal offset time representing the two-way propagation time from the surface to the bottom of the ith layer, which corresponds to the temporal thickness of the layer medium. The expression of the received signal includes the time parameters and the electrical characteristic parameters of the medium. The reconstructed signal model is shown in figure 1(a).

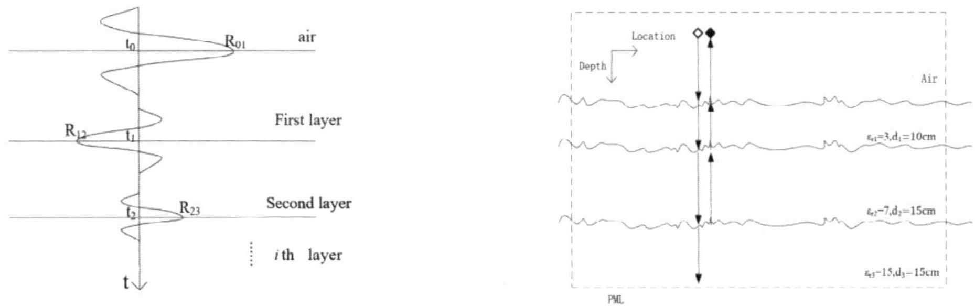

(a) **(b)**

Figure 1. (a) Reconstructed signal model. (b) Three-layer simulation model.

2.3. Genetic algorithm inversion

Based on the actual reflected signal and the reconstructed reflection signal model, the inversion problem can be transformed into:

$$l = \arg\min_{l}\{O(l)\} \tag{7}$$

where $O(l)$ represents the objective function and is expressed as:

$$O(l) = \|y_r(t) - y(t)\|^2 \tag{8}$$

where $y_r(t)$ is the reconstructed signal and $y(t)$ is the actual received echo signal. In this paper, genetic algorithm is employed with $O(l)$ as the optimization objective, and the double travel time of the layer medium t_i, relative permittivity ε_{ri}, and roughness parameter b_i as the optimization objects, where i denotes the layer number.

3. Numerical simulation

In this section, we employ the Monte Carlo method [8] to model random rough surfaces. By controlling two parameters, namely the RMS height σ_h and the correlation length L_c, we can control the roughness of the surface. We then use gprMax [9] to simulate the rough layered media. The parameters are shown in table 1.

Table 1. Simulation parameter.

	σ_{h1}(mm)	L_{c1}(mm)	σ_{h2}(mm)	L_{c2}(mm)	σ_{h3}(mm)	L_{c3}(mm)
Case1	1	7	2	10	2	7
Case2	0.5	7	1	10	1	7
Case3	1	5	2	7	2	5

We use a Ricker wavelet with a center frequency of 1.5 GHz as the GPR transmit waveform. The GPR is simplified as a transmitting and receiving antenna located 20 cm above the surface of the road/metal plate to simulate an air-coupled antenna. As the statistical mean frequency of the Ricker wavelet is extremely close to its theoretical center frequency, and its energy is concentrated around the center frequency [10], we use the center frequency of the Ricker wavelet to represent the frequency characteristic of scattered echoes from rough surfaces. We establish a three-layer simulation model as shown in figure 1(b).

We performed inversion on all three rough models, and the inversion results are shown in table 2. The results indicate that the inversion relative errors for relative permittivity and layer thickness are both within 4%.

Furthermore, we compared our method with the method proposed in reference [11] for the inversion of the Case 1 model. The comparison results are shown in table 2. The results show that our method has a significant advantage in the inversion of relative permittivity and layer thickness.

Table 2. Relative error of inversion results.

Method adopted	Rough model	d_1	d_2	ε_{r1}	ε_{r2}
	Case1	0.42%	1.95%	1.02%	2.97%
Proposed method	Case2	1.02%	0.71%	1.52%	1.49%
	Case3	0.49%	2.36%	1.67%	3.47%
Method in reference [11]	Case1	9.13%	14.07%	13.04%	18.29%

4. Conclusion

In this paper, we studied the propagation of electromagnetic waves in rough layered media and defined the generalized reflection coefficient for rough surfaces to reconstruct the backscattered time-domain data. The proposed layered media parameters inversion method under rough surface shows a high inversion accuracy in simulation models.

Acknowledgment

This work was supported in part by the National Natural Science Foundation of China under Grant 62101042, and Grant 52278146; and in part by the Natural Science Foundation of Chongqing under Grant CSTB2023NSCQ-LZX0062.

References

[1] Lai W W, Dérobert X and Annan P 2018 *NDT & E International* **96** 58-78 https://doi.org/10.1016/j.ndteint.2017.04.002
[2] Torrione P, Morton K D, Sakaguchi R and Collins L M 2014 *IEEE Trans Geosci Remote Sens* **52** 1539-50 https://doi.org/10.1109/TGRS.2013.2252016
[3] Beckman P and Spizzichino A 1963 *The Scattering of Electromagnetic Waves from Rough Surfaces* (London: Oxford: Pergamon)
[4] Pinel N, Bastard C L, Baltazart V, Bourlier C and Wang Y 2011 *IET Radar, Sonar Navig.* **5** 650-6 https://doi.org/10.1049/iet-rsn.2010.0197
[5] Sun M, Bastard C L, Pinel N, Wang Y, Li J, Pan J and Yu Z 2017 *Signal Process.* **132** 272-83 http://dx.doi.org/10.1016/j.sigpro.2016.05.029
[6] Pinel N, Liu L, Bourlier C, Wang Y and Bastard C L 2009 *IEEE Int. Geosc. and Remote Sensing Symposium* **5** 276-9 https://doi.org/10.1109/IGARSS.2009.5417677
[7] Al-Qadi I L and Lahouar S 2005 *Constr Build Mater.* **19** 763-72 https://doi.org/10.1016/j.conbuildmat.2005.06.005
[8] Kuo C H and Moghaddam M 2006 *IEEE Trans. Antennas Propag.* **54** 2917-29 https://doi.org/10.1109/TAP.2006.882160
[9] Warren C, Giannopoulos A and Giannakis I 2016 *Comput Phys Commun* **209** 263-70 https://doi.org/10.1016/j.cpc.2016.08.020
[10] Gholamy A and Kreinovich V 2014 *IEEE SSCI 2014* 11-16 https://doi.org/10.1109/CIES.2014.7011824
[11] Liu R, Yang X, Lan T, Li Y and Qu X 2022 *IEEE Geosci. Remote. Sens. Lett.* **19** https://doi.org/10.1109/LGRS.2022.3172440

3D GPR Detection and Case Analysis of Collapse Hazards for an International Airport Runway

Q F Hu[1,2,4], Z H Zhang[1,4], F Wang[1,2], H Liu[3] and Z Su[2]
[1] College of Civil Engineering, Tongji University, Shanghai, 200092, China
[2] Shanghai Institute of Disaster Prevention and Relief, Shanghai 200092, China
[3] College of Civil Engineering, Guangzhou University, Guangzhou 510006, China
[4] Correspondence Author Q F Hu and Z H Zhang, 2310526@tongji.edu.cn

Abstract. The safety of airport runways is crucial for ensuring the operation of airports and the safety of aircraft flights during taking off and landing. All current methods for assessing airport runways primarily focus on surface routine inspections or surveys, lacking both the capability and available technologies for comprehensive detection of runway and its underground conditions. In this paper, considering a specific airport runway in China that experienced collapse hazards due to cavities, three-dimensional ground-penetrating radar (3D GPR) is explored for detecting potential hazards on the runway and its surroundings. By analyzing the detection data, the formation and distribution of collapses were examined clearly based on B-scan and C-Scan, and then assessed obviously the safety conditions of the airport runway. The research work and its findings provide a new valuable insight of NDT method for guiding comprehensive inspections and maintenance of large airport runways.

1. Introduction

The safety operation and maintenance of runways in airports is of great concern. From 2018 to 2022, the global total number of flights reaches 34.4 million per year, with accident rate of 1.26 accidents per million flights and an annual average of 231 fatalities [1], Regulatory Oversight and Safety Management (ROSM) deficiencies account for 46 percent of aircraft accidents [2]. Currently, the inspection tools for airport runways are mainly focused on surface detection, such as runway friction and profile [3] and the detection of foreign objects or airborne obstacles (e.g., debris, birds) on the runway [4,5]. Existing studies have only focused on single collapse hazard (e.g., voids), such as work by C M Duan [6] and have not considered the negative impact of buried pipe leakage-induced roadway collapses. There have been accidents caused by roadway collapse, settlement, and pipeline leakage in airports, such as the Brazilian "TAM Air Flight 3054" disaster, inundation and collapse in runway areas contributed to the tragedy. K Muthukkumaran used GPR to estimate the damage to the pavement and roadbed soils at Cochin International Airport Limited (CIAL), Kerala, India [7], but subjective factors of this case study

Journal of Physics: Conference Series **2887** (2024) 012026 doi:10.1088/1742-6596/2887/1/012026

are obvious, and the radar frequency is only 400 MHz which precision is less than 3 meters. Most of the current studies are limited to single-point or single-line localization of measurement, and there is fewer related research applied to airport runways.

This paper proposes a new multi-measurement lines detection technology for airport runway inspection and accurately detects a variety of hazards such as runway collapses, voids, sparsity, and pipe leakage by the Ground Penetrating Radar (GPR) technologies of B-scan and C-scan. This research and results provide technical solutions and decision support for comprehensive collapse hazards detection and maintenance of large international airports.

2. 3D GPR Technology and Data Analysis Methods

2.1. 3D GPR Technology

Ground Penetrating Radar (GPR) system consists of host, transmit/receive antenna, display and supporting software, etc., the transmitting antenna (Tx) transmits high-frequency electromagnetic waves to the measured medium, waves are reflected when encounter different geologic bodies during propagation. Then, the receiving antenna (Rx) converts the reflected wave into electrical signal and transmits it to the host computer for analysis and processing [8]. Therefore, the buried depth *h* of the geologic bodies can be calculated by:

$$h = \frac{c}{\sqrt{\varepsilon_r}} \frac{\Delta t}{2} \tag{1}$$

Where c is the propagation speed of electromagnetic waves in the atmosphere, ε_r is the relative permittivity, Δt is wave propagation time. The soil permittivity in the detection field area can be measured on site using Time Domain Reflectometry (TDR).

2.2. C-Scan and Kirchhoff offset processing

Setting the distance between adjacent measurement lines is less than a quarter of the minimum wavelength, and the results of the B-Scan slices are combined to obtain the C-Scan results. Offset processing is utilized to refocus the suppressed diffracted waves using Kirchhoff 3D offset, which significantly improves the accuracy of the estimated surface topography. The fluctuation equation at point (x_0, y_0, z_0) can be expressed as:

$$P_{out}(x_0, y_0, z_0) = \frac{\Delta x \Delta y}{4\pi} \sum_A \frac{\cos\theta}{vr} \frac{\partial}{\partial t} p_{in}(x, y, t) \tag{2}$$

where Δx and Δy are adjacent distances along and perpendicular to the survey line, A is the aperture diameter of the observation, θ is the angle between the direction of wave propagation and the vertical axis, P_{in} is the amplitude of the electromagnetic wave at z=0, and t is the time taken for the wave to propagate to the T/R position.

3. On-site Case Study

3.1. Research background

The test site is located at a large international airport in China, and the geomorphologic unit is a pre-mountain inclined alluvial and floodplain gravelly plain area. Before the test started, collapse pits had already appeared in the subway tunnel crossing under this airport. During the construction of the subway, several subsidence pits appeared near the airport runway, threatening the safe operation of the airport, and the photographs of the subsidence pits are shown in Figure 1.

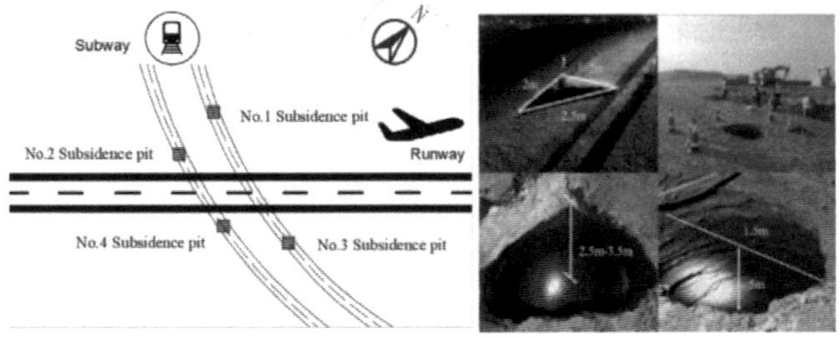

Figure 1. Subsidence pits found in runway of the airport.

3.2. Detection equipment

The DS2000 dual-frequency radar is equipped with 250 MHz and 700 MHz radar antennas, and the dual-frequency radar antennas can be operated synchronously. Main types of detection equipment are listed in Table 1.

Table 1. List of detection equipment for comprehensive airport on-site inspections.

Detection Equipment	Model	Number	Function
Vehicle-mounted 3D GPR	ZRY-TD5200	1	Large scale detection
Hand-powered GPR	Leica DS2000	2	Road disease and pipeline detection
Positioner	GS18	4	GPS data collection
Borescope	/	2	Borehole image data acquisition
3D Laser Scanner	/	1	3D point cloud
Time Domain Reflectometer	/	1	Dielectric constant measurements

3.3. 3D ground-penetrating radar detection method

Airport inspections process can be categorized into three steps:

Step 1. INITIAL SCREENING. Based on the regional function of the airport, different inspection areas are divided according to their importance, Vehicle-mounted 3D GPR is used to do a preliminary screening of the entire airport area to initially determine the extent of the suspected disease locations.

Step 2. PRECISION DETECTION. Based on preliminary results, lay out inspection routes and identify priority inspection areas. Hand-powered GPR was used to do precision detection of potential disease areas to distinguish the type, size, and location of the disease. Processing inspection data by special software to get precise disease information.

Step 3. SPOT DRILLING. On-site excavation to verify potential disease areas and prepare a detection report to guide subsequent roadway disease repair and maintenance decisions.

4. Discussion

Repeated scanning measurements of multiple survey lines were carried out to compare and check the detection results, and the number of measurement lines accumulated 1082, with a total length of 233.7Km, and the surveyed area was more than 23,468 square meters. A total of 14 subsurface road diseases were identified in the airport's runways, taxiways, aprons, and

Journal of Physics: Conference Series **2887** (2024) 012026 doi:10.1088/1742-6596/2887/1/012026

underpasses, of which there are 8 sparse anomalies, 2 voids, and 4 sparse cavities. To ensure the correctness and effectiveness, B-Scan and C-Scan were applied in key areas Taxiway and Runway and take Taxiway A as an example. The scanned results are shown in Figure 2.

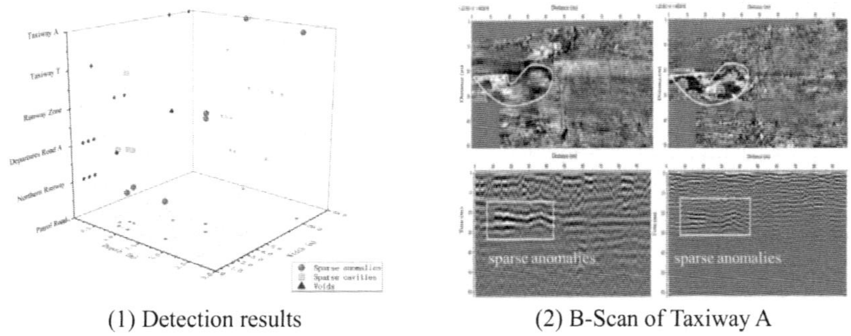

(1) Detection results (2) B-Scan of Taxiway A

Figure 2. Detection results and B-Scan of Taxiway A

On-site Drilling verification based on above results and combined with industrial endoscopy to observe the internal situation, the accuracy of the detection results reaches more than 95 percent. Take the suspected diseases at Taxiway A as an example, the photo of on-site drilling is shown in Figure 3, and it can be clearly inspected there are inner sparse anomalies here.

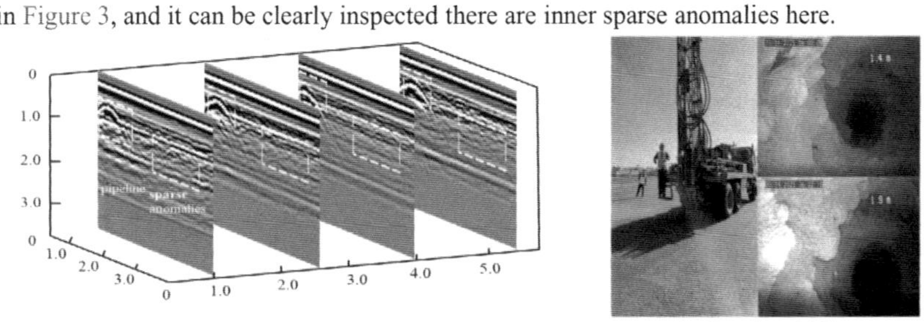

(1) C-Scan of Taxiway A (2) On-site drilling

Figure 3. C-Scan and On-site drilling at Taxiway A

5. Conclusion

This paper proposes 3D GPR technology to detect potential sub-surface diseases and the technologies of B-scan and C-scan applies to the on-site detection data analysis at an international airport in China, which summarizes a set of GPR localization and diagnostic scheme of multi-measurement lines group and forms a more efficient detection technology. On-site spot drilling is confirmed that the accuracy rate reaches more than 95% based on the detection results. This non-destructive testing (NDT) method provides technical and decision-making support for the Regulatory Oversight and Safety Management (ROSM) of large airports.

Acknowledgments

This research was supported by the National Key Research and Development Program of China (Grant No. 2022YFC3801000), Shanghai Science and Technology Innovation Action Plan Project (No. 22dz1201200 and No. 22dz1200402).

References

[1] Said H 2023 *IATA Releases 2022 Airline Safety Performance*. p 1

[2] IATA 2015 *Runway Safety Accident Analysis Report 2010-2014 1st Edition*. p 34
[3] Aytekin Ö, Zöngür U, Halici U 2013 Texture-Based Airport Runway Detection. *IEEE Geosci. Remote Sens. Lett.* **10(3)**: 471-5.
[4] Qian S, Chen Z, Lin M and Zhang C 2015 Significance detection based on conditional random fields and image segmentation. *J. Automat.* **41(04)**: 711-24. (in Chinese)
[5] Liu D, Cao X, Xue B and Li H 2013 Characterization and Detection of Foreign Objects in the Complex Background of Airport Roadway. *Electron. Des. Eng.* **21(02)**: 12-5. (in Chinese)
[6] Duan C, Wu R, Liu J 2011 Thickness Determination Method for Thin Void-layer of Lossy Medium Using Ground Penetrating Radar. *Signal Process.* **27(10)**: 1581-5.
[7] Muthukkumaran K, Raja N, Umaiyan U 2020 Evaluation of Pavement Stripping Using Ground-Penetrating Radar: A Case Study. *Indian Geotech. J.* Vol 51, p197–204
[8] Hu Q, Zheng Z, Liu H and Chen Z 2020 Application of three-dimensional ground-penetrating radar in leakage detection of urban municipal pipelines. *J. Tj U.* **48(07)**: 972-81. (in Chinese)

Research on ground penetrating radar and Muon detection method in Xixia Mausoleum

Jiayu Lu[1], Baizhou An[1]*, Yadong Bai[1], Zhaofa Zeng[2], Xiufen Ren[3], Shuai Zhou[2], Boyang Sun[1] and Zhaotao Yan[2]

1 Geophysical and Geochemical Exploration Institute of Ningxia Hui Autonomous Region (Autonomous Regional Deep Earth Exploration Center), Yinchuan 750001, China
2 College of Geo-exploration Science and Technology, Jilin University, Changchun 130026, China
3 Yinchuan Xixia Mausoleum Management Office, Yinchuan 750001, China
(845941307@qq.com)

Abstract. Xixia Mausoleum is an important historical site in China, which has important research value and archaeological significance. Non-destructive detection technology is of great significance for the detection of Xixia Mausoleum and its surrounding architectural sites. In this paper, GPR and other methods were used in comprehensively archaeological exploration of Mausoleum No. 2. The mausoleum pagoda, central platform and underground in the mausoleum area showed obvious GPR abnormal characteristics which reflects the underground structure. For the mausoleum pagoda and the central platform, GPR and cosmic ray probe were used to image the density structure inside the mausoleum, and the range-ranging GPR was used to conduct regional detection in the mausoleum area and the external wall area to study the characteristics of the underground structure in this area. The solid loess body structure of the mausoleum pagoda and the central platform was found in this detection, and there were certain differences in underground structure between the inner and outer walls, which was inferred to be an unknown ancient building buried. The detection results will be combined with other geophysical methods and technologies in the future to complete the imaging of the internal structure characteristics of Xixia Mausoleum site, and provide innovative and reliable basis for the archaeological exploration of Xixia Mausoleum.

Keywords: Ground penetrating radar; Muon transmission imaging; Xixia Mausoleum; Structural imaging; Anomaly features

1. Introduction

Xixia Mausoleum is the largest and most complete site of Xixia culture in China. It was built from the early 11th century to the early 13th century. Influenced by Buddhist architecture, Xixia Mausoleum organically combines Han culture, Buddhist culture and Dangxiang culture, forming a unique form of Chinese mausoleum architecture. It carries on the history of Tuoba family of Xianbei from Pingcheng of Northern Wei Dynasty to Dangxiang Xixia Dynasty. Xixia Mausoleum region is distributed with 9 imperial tombs, 271 accompanying tombs, 32 flood control engineering sites and 1 large architectural site. The scale is magnificent and well-arranged. Each imperial tomb is an independent building group with longitudinal square facing north to south the scale is similar to the Ming Tombs, absorbing the strengths of Tang and Song tombs since the Qin and Han Dynasties, and influenced by Buddhist architecture. Constitute a unique form of Chinese mausoleum architecture, known as the "Oriental pyramid" reputation, is an important carrier of Xixia culture[1].

At present, among the nine imperial tombs, only Mausoleum No. 3 and No. 6 have been intensively excavated and studied, while the protection of other Xixia Mausoleum sites is limited to anti-weathering and other rescue reinforcement treatment. In the site protection process from archaeological investigation, exploration, excavation to protection planning, there is no corresponding geophysical exploration basis, so it is impossible to carry out further excavation and research[2]. In order to better elucidate the connotation and influence of the Yellow River culture, it is of great significance to carry out systematic archaeological work of Xixia Mausoleum. In the past work, large-scale exploration was mainly carried out on ancient sites, but there is still a big gap in determining the internal structural characteristics of mausoleums, central platforms, underground palaces, tomb tunnels, etc. Among all the nine imperial tombs, Only the underground site of the Mausoleum No. 6 has been excavated. The shape and specific location of the other underground sites have not been clarified, and further exploration is needed. This paper carried out experiments and researches on the Xixia Mausoleum area, taking the mausoleum pagoda, central platform and underground remains in the mausoleum area as the main target geological body with anomaly characteristics such as permittivity and density, and aiming at the archaeological requirements of Xixia Mausoleum, it realized effective collection of high-resolution data of the remains based on ground-penetrating radar detection combined with cosmic ray detector detection. In this way, imaging of underground sites and target identification can be realized.

2. Geophysical experimental study of Xixia Mausoleum

Based on ground-penetrating radar detection [3]-[7], combined with cosmic ray transmission imaging method [8], comprehensive geophysical detection was carried out for Mausoleum No. 2, imaging and target identification were carried out for the mausoleum pagoda, central platform and underground structure at the edge of the mausoleum area.

2.1. Imaging of the inner structure features of the mausoleum pagoda and the central platform

Two kinds of mausoleum structures on the ground were detected by means of circumferential scanning with the integrated ground penetrating radar (GPR). At the same time, Muon transmission imaging method was used to detect the mausoleum pagoda without contact. By combining these two detection methods, the image of the interior structure of the mausoleum is realized.

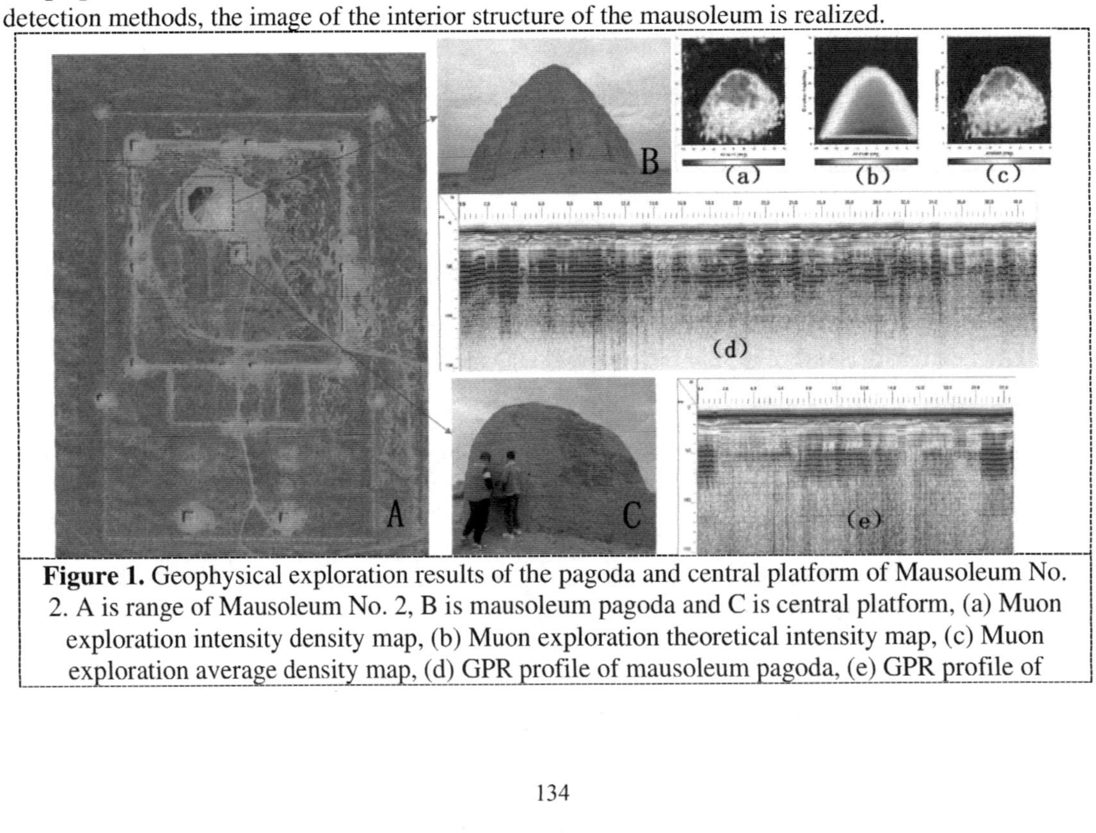

Figure 1. Geophysical exploration results of the pagoda and central platform of Mausoleum No. 2. A is range of Mausoleum No. 2, B is mausoleum pagoda and C is central platform, (a) Muon exploration intensity density map, (b) Muon exploration theoretical intensity map, (c) Muon exploration average density map, (d) GPR profile of mausoleum pagoda, (e) GPR profile of

central platform

As shown in figure 1, according to the average density map of the mausoleum pagoda detected by the cosmic ray muon, there are local high-density bodies in the upper part of the mausoleum pagoda, and the high-density bodies are distributed in pieces in the upper part of the mausoleum pagoda, which is obviously distinguished from the middle and lower parts of the mausoleum pagoda, while the density difference between the middle and lower parts of the mausoleum pagoda is small, and no obvious high-density bodies are found.

(d) the radar wave profile of the cemetery pagoda and (e) the central platform, the first wave of both is clear. Through the relevant data processing such as de-noising, gain and filtering, the electromagnetic wave energy decayed with the depth of detection, and no obvious abnormal wave group features such as diffraction and reflection were found. It can be considered that there is no obvious electrical difference interface between the cemetery pagoda and the central platform. It shows that there are no obvious voids or other anomalies in the internal structure, and the overall structure of the medium is relatively uniform.

Through the analysis and research of the above data, the geophysical exploration results show that no multi-layer structure features are found in Mausoleum No. 2 and the central platform, and it can be inferred that the mausoleum pagoda and the central platform are solid loess structures, but the local density is greatly different.

2.2. *Detection of underground ancient sites in the mausoleum area and between the external walls*

According to the structure of the underground ancient ruins in the cemetery area and between the external walls, the regional detection is carried out by using the transceiver integrated ground penetrating radar, and the characteristics and distribution of the underground structure of the ancient ruins in this area are studied.

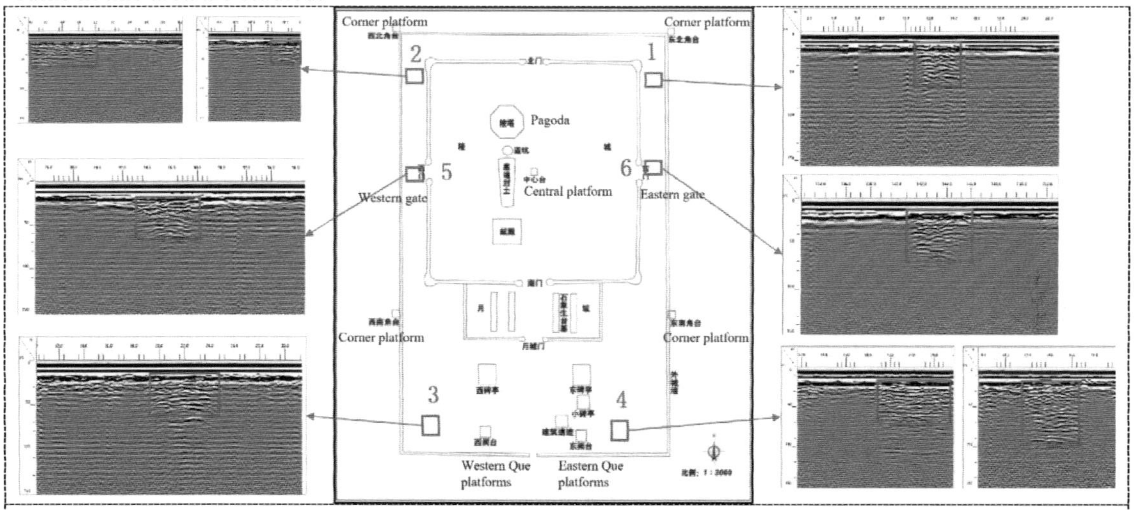

Figure 2. Plane diagram of Mausoleum No. 2 and results of underground anomaly detection.

As shown in figure 2, the plane diagram of Mausoleum No. 2 has delimited six abnormal detection areas. By comparing the radar result data of these six regions, the typical abnormal result profiles of each region are selected for display. These radar profiles have relatively similar wave group characteristics, and there are obvious abnormal wave group interfaces. The abnormal wave group shows the characteristics of strong amplitude and strong energy. The abnormal wave group is downward at both ends, the shape is hyperbolic, local diffraction exists, and multiple waves are chaotic and discontinuous, which is suspected to be an underground abnormal structure.

Through the analysis and research of the above data, the geophysical exploration results show that there is a total of 6 anomaly structure areas in Mausoleum No. 2 area, mainly distributed in the vicinity of the four corner platforms, the eastern and western gate, the eastern and western Que platforms in figure 2, which may be the underground structure of the ancient site.

3. Conclusion

This exploration is aimed at the archaeological needs of the Xixia Mausoleum, carrying out tests and research for the Xixia Mausoleum area, and realizing the imaging of the internal structure of the mausoleum, the central platform and the underground ancient architecture site. According to the results of this experiment, there are two main understandings. First, it is considered that the Xixia Mausoleum and the central platform are solid loess structures, which answers the predecessors' understanding of whether there are multi-layer structures in the mausoleum and the central platform; Second, it is found that there are some underground anomaly traces in the symmetrical positions around the four corner platforms, the eastern and western gate, the eastern and western Que platforms in Mausoleum No. 2 area, and it is speculated that there are ancient buildings or ancient drainage facilities and other relics in the interior of this area, which can be further verified.

To sum up, the results of this experiment have to a certain extent filled the vacancy in the past that only extensive investigation of ancient sites failed to identify the internal structural characteristics of the sites, and provided innovative and reliable basis for the archaeological exploration of Xixia Mausoleum. In the subsequent archaeological exploration work, it can be further carried out a wide range and accurate comprehensive geophysical method detection and imaging, to provide technical support for the archaeological exploration of Xixia Mausoleum.

4. References

[1] Li Yu 2022 The Inheritance and development of the Zhongyuan culture in Xixia reflected in archaeological findings——Centered on the Mausoleum No.3 and No.6 Yinchuan: North Minzu University

[2] Lin J X, Tian G and Shi Z J 2014 Geophysical prospecting study of the satellite tomb in Xixia mausoleum site *Geotechnical Investigation & Surveying* **42** 87-93.

[3] Liu D W, Xu G Y and Huang R D 2002 Application of geophysical prospecting radar in prospecting integrality of ancient tombs *Progress in Geophyiscs* **01** 96-101.

[4] Goodman D 1994 Ground - penetrating radar simulation in engineering and archaeology *GEOPHYSICS* **59** 224-32.

[5] Shokri N H, Amin Z M, Hisham N A A N and Yusup A 2018 Processing Strategies for Archaeology Prospection Using Ground Penetrating Radar (GPR) *Adv Sci Lett* **24** 4074-9.

[6] Sanchez G M, Grone M A, Apodaca A J, Byram R S, Lopez V and Jewett R A 2021 Sensing the Past: Perspectives on Collaborative Archaeology and Ground Penetrating Radar Techniques from Coastal California *Remote Sens-Basel* **13**.

[7] Ristic A, Govedarica M, Pajewski L, Vrtunski M and Bugarinovic Z 2020 Using Ground Penetrating Radar to Reveal Hidden Archaeology: The Case Study of the Wurttemberg-Stambol Gate in Belgrade (Serbia) *Sensors-Basel* **20**.

[8] Liu G R, Ding D T, Jin L, Wang Y X, Wu C, Gao H, Liu J T and Liu Z Y 2023 On the feasibility of applying muography to research on large-scale historical sites Research on Conservation of Cave Temples and Earthen sites 2 76-88.

Acknowledgments

The research was funded by Key Research and Development Plan of Ningxia Hui Autonomous Region: Research and development of key technologies for comprehensive geophysical exploration of Xixia Mausoleum and demonstration of archaeological application of large sites (grant number: 2023BEG02066).

Deviation Aware Mask for Localizing Ground Penetrating Radar Data Alignment

Pengyu Zhang[1], Yuwei Chen[1], Liang Shen[2], Beizhen Bi[1], Tian Jin[1], Xiaotao Huang[1]

[1]College of Electronic Science and Engineering, National University of Defense Technology, Changsha, Hunan 410073, China. (zhangpengyu970524@163.com)
[2]Test Center, National University of Defense Technology, Xi'an 710106, China.

Abstract. This paper proposes a data align pipeline for LGPR (Localization Ground Penetrating Radar). LGPR achieves robot localization by matching prior underground data map, therefore high-precision maps are essential. Utilizing GPR in a time acquisition mode for robot localization enables the technology to generate maps with resolutions far superior to those produced by camera or lidar mapping methods. However, it also introduces several challenges. On one hand, the process of map generation heavily relies on velocity control of the moving platform, and non-uniform motion in the localization scenario can result in feature deformation of GPR images. On the other hand, in the absence of relative a priori position conditions, there exists a certain degree of spatial dimension offset between the multiple data acquisitions from array GPR systems, rendering existing time-series deformation correction schemes ineffective. To address these challenges, by designing a deviation-aware distance, we aim to enhance existing time-series regularization approaches without relying on additional sensor position information, thus achieving synchronized alignment of multiple rounds of ground-penetrating radar data for the generation of high-quality underground map data used to evaluate LGPR performance. In this work, we compare existing synchronization strategies of LGPR with those of other sensors and alignment schemes, and validate the effectiveness of our proposed approach.

1. Introduction

In recent years, LGPR as a novel application of GPR has attracted widespread attention[1], [2], [3]. LGPR achieves vehicle localization by matching prior underground maps, thereby relying on the accuracy of these maps for positioning performance. Unlike other GPR applications, which typically rely on wheel encoder integration for triggering ground-penetrating radar data, LGPR primarily utilizes a time acquisition mode[1], [4], [5]. Due to the simplicity of time acquisition mode and its ease of synchronization with other sensors, coupled with its high-speed frame rate, GPR can generate maps with resolutions surpassing those of existing optical/Lidar methods. Therefore, in LGPR applications, the time acquisition mode is commonly employed, followed by distance normalization to calculate ground-penetrating radar mile positions in standard channel spacing. However, the detection results of this method heavily depend on the control of radar antenna movement speed. Uneven motion platform speed control can lead not only to deviations in the size of radar images but also to distance errors.

In the publicly available dataset GROUNDED[4], [6], Savitzky-Golay filter is employed to synchronize timestamps of multiple time-acquired GPR sequence data with data from other sensors, effectively aligning data collected from wide-ranging sensor setups such as Cameras/Lidar. However, for GPR data with excessively high acquisition frame rates and limited detection ranges, there are still certain deviations, resulting in feature misalignment between the two datasets, which cannot be used for subsequent accurate positioning evaluations. In [2], GPR data calibration by using total stations and wheel encoders, resulting in a high-quality dataset, but at a higher cost, limiting its application in LGPR in extreme scenarios such as underground parking lots or mines. Currently, the production of LGPR maps are all relies on high-precision RTK GPS or other location tags to stitch together multiple data segments. This not only greatly limits the application of LGPR in scenarios with GPS denial conditions such as underground mines or parking lots but also makes it impossible to effectively evaluate existing positioning algorithms.

To address the deformation of time-series data caused by rates, time regularization techniques have been typically employed in previous studies and have achieved performance improvements in tasks such as speech recognition and action recognition[7], [8]. In GPR applications, most of researcher attempt to align two sequences of GPR data using dynamic time warp to improve the detection performance of underground targets, ensuring alignment of GPR lines. However, existing DTW-based solutions still focus on deforming alignment of two-dimensional sequences and cannot address the spatial-temporal dimension offset of GPR three-dimensional sequences.

This paper proposes a dynamic time regularization method that can perceive spatial dimension offset, to achieve regularization and deformation of two incomplete matching datasets, enabling adaptive calibration and alignment without prior location knowledge.

2. Method

In this section, we designed a spatial-aware distance to effectively measure the cross-track direction deviation of GPR sequence data, which was then utilized to generate a lane deviation mask for calibrating two rounds of data. Combined with the exist dynamic time warp techniques, this approach achieved alignment of both spatial and temporal dimensions of the two sequences of GPR data.

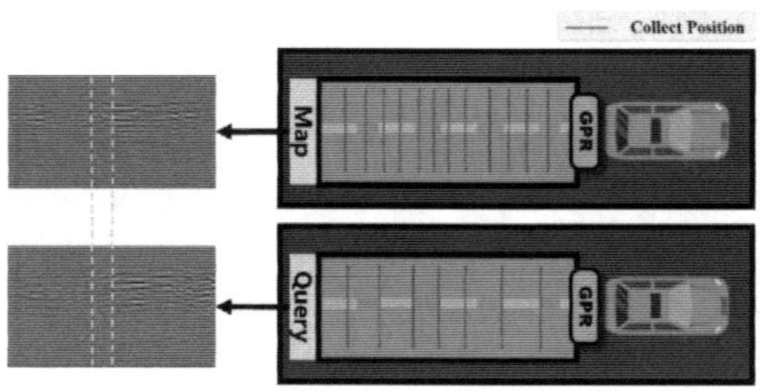

Figure 1. Feature misalignment caused by inconsistent collection rates.

Assuming two sequences of GPR data, seq $X = [x_1, x_2, ..., x_n]$ and seq $Y = [y_1, y_2, ..., y_n]$. As shown in Figure 1, due to certain channel deviation and uneven velocity during data acquisition, the data between the two sequences undergo deformation in both spatial (vertical track) and temporal (along the track) dimensions. Without accurate priors, it is challenging to calibrate them.

To address this issue, as illustrated in Fig. X, we first consider dividing each sequence into N segments of local sub-sequences. We employ the simple Euclidean distance as the basic similarity measure to assess the similarity of different local regions. This is defined as:

$$d(X,Y) = \sqrt{\sum_{i=1}^{n}(x_i - y_i)^2}, \tag{1}$$

where, The smaller the distance, the higher the similarity of two sequence. Then, a distance matrix D can be obtained through the one-to-one correspondence between local sub-sequences. By partitioning the volumetric data sequences collected from multiple GPR acquisitions into multiple local sequences, an $N \times N$ matrix D can be established to describe the correlation between local sub-sequences (i, j) from X and Y. Adaptive Dynamic Time Warping (DTW) is employed to find an optimal deformation path in matrix D, reflecting the deformation and correspondence between the two sequences in the vertical track dimension. Meanwhile, the alignment between sequences X and Y is also indicated by points in this deformation path R.

$$R = \{r_1, r_2, ..., r_k\}, \tag{2}$$

where, K is the length of the optimal path, and we calculate the accumulated lane deviation aware matrix L of the optimal path using $L_{i,j}$. The (i, j) of L represents the minimum cost of the optimal path from $(1,1)$ to (i, j), and $L_{i,j}$ are calculated as follows

$$L_{i,j} = \begin{cases} d_{i,j} & i=1, j=1 \\ d_{i,j} + L_{i,j-1} & i=1, j \in \{2,..,N\} \\ d_{i,j} + L_{i-1,j} & i \in \{2,..,N\}, j=1 \\ \min\{a \cdot d_{i,j} + L_{i-1,j}, d_{i,j} + L_{i-1,j}\} & i, j \in \{2,..,N\} \end{cases} \tag{3}$$

Here, in contrast to the traditional DTW computation process, an adaptive alpha parameter is defined to address certain degrees of feature deformation.

By applying the argmin operation to each row and column, a Lane Deviation-aware Mask (LDMask) can be generated. The LDMask can then be used to achieve feature alignment in the vertical track direction through matrix left and right multiplication. In theory, when there is no cross-track direction deviation in the sequence pairs, D becomes an identity matrix, meaning no left or right shifting operations are applied to the original sequences.

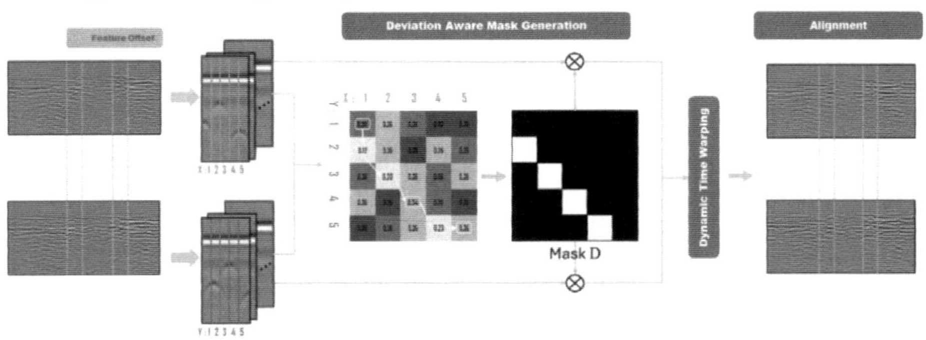

Figure 2. The proposed framework for two GPR sequence data alignment.

As illustrated in the flowchart in Figure 2, the aligned paths of the two rounds can achieve synchronization of sliced sequences along the track dimension using existing DTW or Fast-DTW methods. Cases of lane variation can be handled by segment-wise processing to generate masks for calibration.

3. Experiment and Results

The experimental results were validated on the public dataset GROUNDED and self-collected dataset. The GROUNDED employs a 11-channel array GPR with 126Hz. The self-collected data were obtained using a 15-channel array GPR, employing a stepped-frequency signal. Alignment of the two segments of GPR data ensured consistency between the start and end positions, while not guaranteeing alignment in the channel direction, reflecting the real-world settings of data collection. We selected MIT's Method[4], ADTW [10] and the classical Fast DTW[11] as comparative algorithms for experimental validation.

As depicted in Figure 3, our proposed method still achieves excellent correspondence calibration between the data even in the presence of lane deviation.

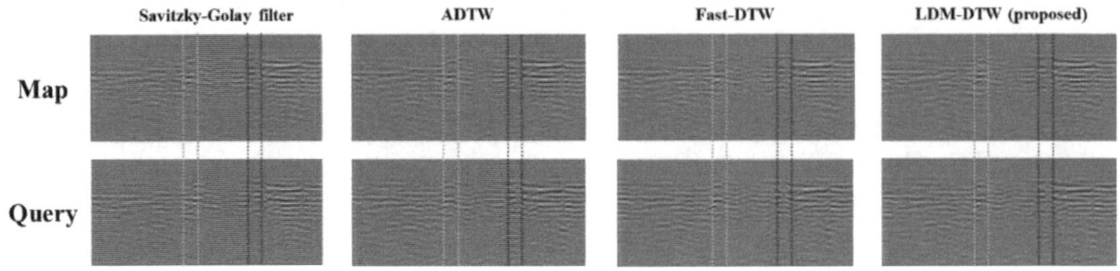

Figure 3. Experiment Results.

4. Conclusion

The proposed method only aligns local features in the vertical orbit dimension of the ground penetrating radar measurement direction. This is based on the movement characteristics of array GPR mapping, and the fact that feature migration mainly changes in the horizontal direction. For special cases that violate this rule (such as differences in yaw angles between multiple radar data acquisitions), future research will further consider and design solutions for synchronous calibration in large-scale underground scenes.

Acknowledgments
Thank you to Changsha University of Science & Technology for providing equipment and technical support.

5. References

[1] Cornick, Koechling, Stanley, Zhang, and BJ, 'Localizing Ground Penetrating RADAR: A Step Toward Robust Autonomous Ground Vehicle Localization', *J FIELD ROBOT*, vol. 2016,33(1), no., pp. 82–102, 2016.

[2] A. Baikovitz, P. Sodhi, M. Dille, and M. Kaess, 'Ground encoding: Learned factor graph-based models for localizing ground penetrating radar', presented at the 2021 IEEE/RSJ International Conference on Intelligent Robots and Systems (IROS), IEEE, 2021, pp. 5476–5483.

[3] T. Ort, I. Gilitschenski, and D. Rus, 'Autonomous navigation in inclement weather based on a localizing ground penetrating radar', *IEEE Robot. Autom. Lett.*, vol. 5, no. 2, pp. 3267–3274, 2020.

[4] T. Ort, I. Gilitschenski, and D. Rus, 'GROUNDED: The Localizing Ground Penetrating Radar Evaluation Dataset.', in *Robotics: Science and Systems*, 2021.

[5] I. Stasewitsch, J. Schattenberg, and L. Frerichs, 'Robust Monte Carlo Localisation Using a Ground Penetrating Radar', in *ROBOT2022: Fifth Iberian Robotics Conference*, D. Tardioli, V. Matellán, G. Heredia, M. F. Silva, and L. Marques, Eds., Cham: Springer International Publishing, 2023, pp. 247–258.

[6] T. Ort, I. Gilitschenski, and D. Rus, 'GROUNDED: A localizing ground penetrating radar evaluation dataset for learning to localize in inclement weather', *Int. J. Robot. Res.*, p. 02783649231183460, 2023.

[7] L. Wang and P. Koniusz, 'Uncertainty-DTW for time series and sequences', in *European Conference on Computer Vision*, Springer, 2022, pp. 176–195.

[8] A. S. Haq, M. Nasrun, C. Setianingsih, and M. A. Murti, 'Speech recognition implementation using MFCC and DTW algorithm for home automation', *Proceeding Electr. Eng. Comput. Sci. Inform.*, vol. 7, no. 2, pp. 78–85, 2020.

[9] S. Jazayeri, A. Saghafi, S. Esmaeili, and C. P. Tsokos, 'Automatic object detection using dynamic time warping on ground penetrating radar signals', *Expert Syst. Appl.*, vol. 122, pp. 102–107, 2019.

[10] S. Salvador and P. Chan, 'Toward accurate dynamic time warping in linear time and space', *Intell. Data Anal.*, vol. 11, no. 5, pp. 561–580, 2007.

Modeling and calculation of ground penetrating radar detection on rough interface of soil media

Jia Yi Zhu, Ying Liu, Yao Yao Liu, Chen Wang, Ya Jing Li

School of Electronic Engineering, Tianjin University of Technology and Education, Tianjin, China (1924400562@qq.com)

Abstract. Ground penetrating radar technology is an effective underground physical detection method. In order to better study the ground penetrating radar echo characteristics of complex soil media rough interfaces, a soil rough surface electromagnetic scattering model based on fractal rough surface geometry modeling and soil media dielectric properties modeling is established. The mixed soil characteristics including sand content, water content, clay content, and sand particle density are analyzed, and the radar echo image characteristic of soil rough media interfaces under different fractal dimensions is discussed.

Keywords: ground penetrating radar, rough surface, soil dielectric model, echo characteristic

1. Introduction

The ability of ground penetrating radar(GPR) to detect soil media makes it important in fields such as civil engineering, geological exploration, archaeology, etc.. However, the complexity of soil media and the influence of rough interfaces on radar signal propagation have a significant impact on the accuracy and reliability of radar imaging [1].

In previous studies, some scholars have focused on exploring the rough interfaces of soil media. However, most of these studies consider soil as a homogeneous non-dispersion medium [2, 3], failing to fully consider the complexity of soil. Additionally, they often treat the soil surface as a smooth interface. In practical engineering, soil is influenced by many factors in radar echo imaging, such as moisture content, sand content, clay content, etc. [4, 5]. Moreover, the soil surface is also a fluctuating random rough surface. Therefore, in order to better study the rough interfaces of soil media with complex structures and analyze their radar echo characteristics, this paper establishes a multi-parameter dielectric model of soil as well as a geometric model of soil random rough surfaces. It investigates and analyzes the radar echo scattering characteristics of soil rough surfaces and conducts time-domain scattering calculations to analyze the radar echo scattering characteristics of rough interface soil media.

2. Modeling of rough interface soil dielectric media

2.1. *Geometric modeling of soil roughness surface*

We employ fractal rough surfaces to simulate soil roughness. Fractals provide a detailed representation of the terrain at various scales, achieved by the convolution of Gaussian noise and the

 doi:10.1088/1742-6596/2887/1/012029

inverse Fourier transform of $1/k_b$, where k represents the wave number and b is a constant related to the fractal dimension. The change in fractal dimension can have a significant impact on roughness. In general, the higher the fractal dimension, the more details and irregularities on the surface, resulting in an increase in roughness. This means that there are more small-scale structures and undulations on the surface, making it rougher. Therefore, this paper adopts fractal rough surfaces to represent soil topography. As illustrated in Figure 1, different fractal dimensions correspond to different rough surface models.

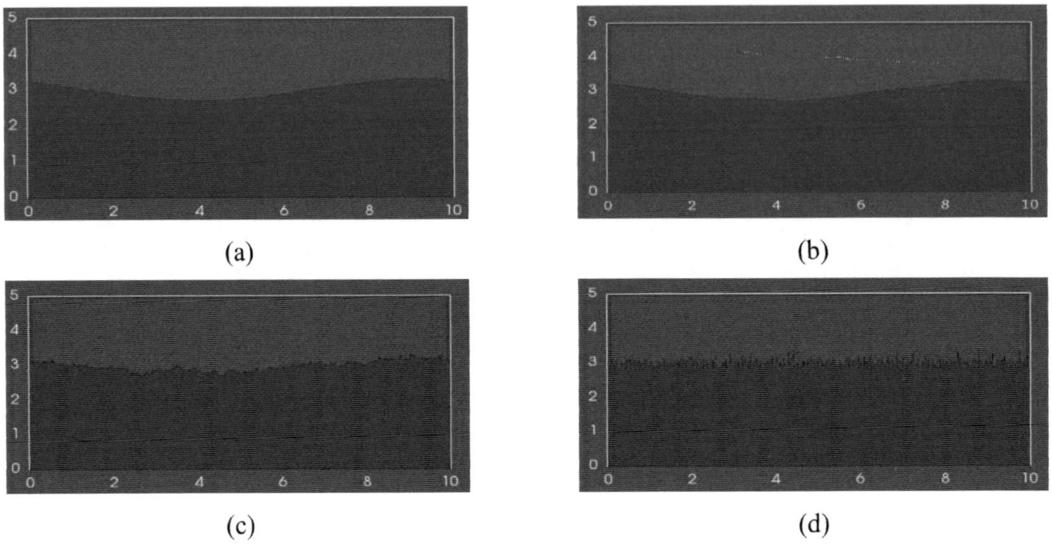

Figure 1. Rough surface models with different fractal dimension of: (a) 0; (b) 1; (c) 2; (d) 3

2.2. *Modeling the dielectric properties of soil media*

We employ the Peplinski's semi-empirical soil model [6], which theorizes that the electromagnetic properties of soil are primarily determined by the proportions of different particle components in the soil and the soil moisture content. Different proportions of soil particles can form different types of soil, each with distinct electromagnetic properties. According to this theory, Peplinski proposed the following expression:

$$\varepsilon'_{(0.3-1.3\mathrm{GHz})} = 1.15[1 + \frac{\rho_b}{\rho_s}(\varepsilon_s^{\alpha} - 1) + m_u^{\beta'} \varepsilon'^{\alpha}_{w} - m_u]^{\frac{1}{\alpha}} - 0.68 \quad (1)$$

$$\varepsilon'' = \varepsilon_0 m_u^{\frac{\beta''}{\alpha}} (\frac{\sigma_f}{w\varepsilon_0} \frac{(\rho_s - \rho_b)}{\rho_s m_u} - \varepsilon''_w) \quad (2)$$

$$\varepsilon_s = (1.01 + 044\rho_s)^2 - 0.062 \quad (3)$$

$$\beta' = 1.2748 - 0.519S - 0.152C \quad (4)$$

$$\beta'' = 1.33797 - 0.603S - 0.166C \quad (5)$$

$$\varepsilon_w = \varepsilon_{w,\infty} + \frac{\varepsilon_{w,s} - \varepsilon_{w,\infty}}{1 + j\omega t_{0,w}} \quad (6)$$

$$\sigma_{f(0.3-1.3\mathrm{GHz})} = 0.0467 + 0.22044\rho_b - 0.411S + 0.6614C \quad (7)$$

where ρ_b is the bulk density of the soil($\mathrm{g} \cdot \mathrm{cm}^{-3}$), ρ_s is the sand particles density, S is the relative dielectric constant of sand grains, ε_s is the dielectric constant of solid materials, typically set to $\varepsilon_s = 4.7$, α is a constant factor, with an optimal value of 0.65 for all soils. m_u represents the volumetric

Journal of Physics: Conference Series **2887** (2024) 012029 doi:10.1088/1742-6596/2887/1/012029

soil moisture content per unit volume, β is an adjustable parameter related to soil type, $\varepsilon_w^{'}$ denotes the real part of the dielectric constant of free water, S and C respectively represent the sand content and clay content, $\varepsilon_{w,s}$ and $\varepsilon_{w,\infty}$ respectively represent the relative dielectric constants of water at frequencies of 0 and infinity. Table 1 shows different soil characteristics parameters.

Table 1. Soil characteristics

Soil types	Sediment content	Clay content	Soil density	Sand grain density	The minimum volumetric moisture fraction of the soil	The maximum volumetric moisture content of the soil
Sandy loam	0.515	0.135	1.4-1.6	2.65	0.05-0.1	0.3-0.4
Silty loam soil	0.172	0.19	1.2-1.6	2.65	0.1-0.15	0.3-0.4
Silty clay	0.05	0.474	1.2-1.5	2.65	0.2-0.3	0.4-0.5

3. Modeling and numerical calculations of time-domain scattering

In this section, a rough interface soil medium model was established using the gprmax software [7]. The model domain was set to 10×5×3m, with discretization step size set to $\Delta x = \Delta y = \Delta z = 0.01m$. The time window was set to 50ns. The GPR radiation antenna uses Ricker pulses with a center frequency of 0.4GHz, with a transmitting and receiving antenna spacing of 0.1m and a sampling interval of 0.05m. The mixed soils in the model are sandy loam, silty loam, and silty clay. According to Table 1, the maximum height of soil roughness fluctuation is set to 0.6m. Figure 2 shows the scattering models of different soil texture with rough interfaces.

(a) A model of sandy loam soil with a fractal dimension of 1

(b) A model of powdery clay with a fractal dimension of 3

Figure 2. Scattering models of different soil texture with rough interface

Figures 3 and Figure 4 respectively show the GPR echo simulation results of the silty loam soil rough surface with a fractal dimension of 2 and the powdery clay rough surface with a fractal dimension of 3. The image results show that the radar echo images of soil rough surfaces with higher fractal dimensions have a more convoluted surface, with more microstructure and detailed features, while the radar echo images of soil rough surfaces with lower fractal dimensions have simpler surface features. In summary, as the fractal dimension of the rough surface increases, the surface roughness increases, and the curve fluctuations of the radar echo image become more severe.

IOP Publishing
doi:10.1088/1742-6596/2887/1/012029

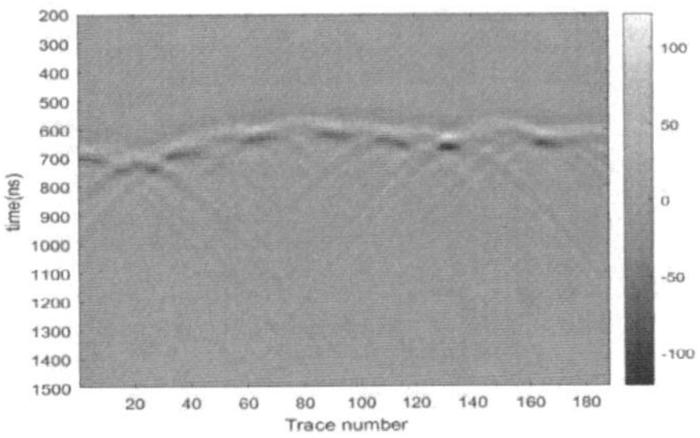

Figure3. Numerical results of Radar echo from silty loam soil rough surface with a fractal dimension of 2

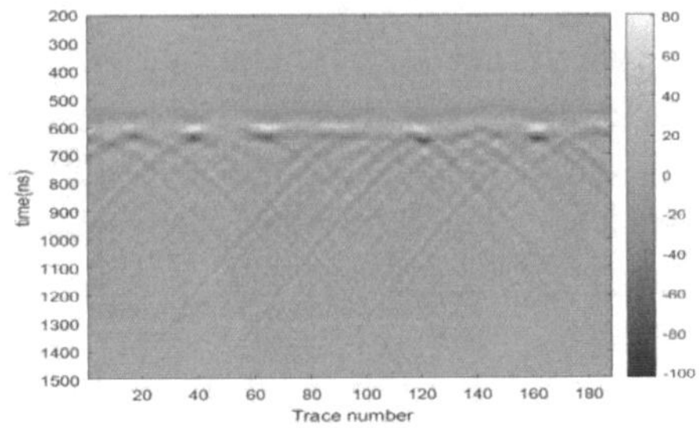

Figure4. Numerical results of Radar echo from powdery clay rough surface with a fractal dimension of 3

4. Conclusion

This paper conducts modeling and simulation calculations of ground penetrating radar detection on rough soil interfaces using the finite difference time domain (FDTD) method. It analyzes the electromagnetic and geometric properties of mixed soils, investigates the influence of fractal dimension on soil roughness, and explores the characteristics of radar echo images on rough soil interfaces. Through this research, a deeper understanding of the micro-structure and macroscopic properties of soil is achieved, providing fundamental data for practical detection operations.

Acknowledgments

This work was supported by the National Natural Science Foundation of China (Grant No.62241109), and Scientific Research Fund of Tianjin Education Commission (Grant No. 2021KJ005).

References

[1] Li J, Guo C C, Wang F M, et al.. The summary of the surface ground penetrating radar applied in subsurface investigation [J]. Progress in Geophys (in Chinese), 2007, 22 (2): 629-637.

[2] Jun Guo, Qinghua Wu, Guozheng Yan, The inverse scattering problem by a crack buried in a piecewise homogeneous medium [J]. IMA Journal of Applied Mathematics, 2015, 80 (4): 1199–1218.

[3] X. Liu, B. Zhang and G. Hu. An uniqueness result for the inverse electromagnetic scattering problem in a piecewise homogeneous meddium [J]. Appl. Anal88 (2009), 1339-1355.

[4] Liang Y, Guo L X, Wu Z S, et al. A study of composite electromagnetic scattering from an object near a rough sea surface using an efficient numerical algorithm [J]. IEEE antennas and wireless propagation letters, 2016, 15: 186-190.

[5] Apostol B F. Scattering of the electromagnetic waves from a rough surface [J]. Journal of Modern Optics, 2012, 59 (18): 1607-1616.

[6] N. R. Peplinski, F. T. Ulaby and M. C. Dobson. Dielectric properties of soils in the 0.3–1.3GHz range [J]. IEEE Trans. Geosci. Remote Sens., 1995, 33 (3): 803–807.

[7] A. Giannopoulos. Modelling ground penetrating radar by GprMax [J]. Construction & Building Materials, 2005, 19 (10): 755-762.

Realistic method for numerical characterization of GPR responses of cracked drainage pipelines in loose-soil areas

Weihuang Shi[1] **and Tess Luo**[1, 2, 3]

[1] Collage of Civil and Transportation Engineering, Shenzhen University, Shenzhen, China
[2] Guangdong Key Laboratory of Urban Spatial Information Engineering, Shenzhen University, Shenzhen, China
[3] National Key Laboratory of Green and Long-Life Road Engineering in Extreme Environment, Shenzhen, China
(2210474170@email.szu.edu.cn)

Abstract. A damaged drainage pipeline poses significant subsurface risks, including the potential for ground collapse. This study addresses the unreliability of previous Ground Penetrating Radar reflection patterns in characterizing drainage pipeline defects due to an incomplete understanding of the leakage mechanism and subsurface degradation. We propose an integrated approach using Computational Fluid Dynamics and Discrete Element Method to construct a realistic pipe-soil spatial distribution model. Converting this spatial model into an electromagnetic model enables numerical simulation of radar wave propagation with the Finite Difference Time Domain method. Our findings reveal that the position of pipeline defects influences soil degradation processes, resulting in distinct GPR reflection patterns. This research contributes to the accurate diagnosis of subsurface hazards related to cracked pipelines by offering a more reliable reference for GPR inspections.

1. Introduction

The persistent seepage of subsurface drainage pipelines can cause an increase in soil water content, soil erosion, formation of underground cavities, and potential geological disasters such as ground collapse. Ground Penetrating Radar (GPR) is widely employed in inspecting underground utilities due to its superior resolution, efficiency, penetrability, and non-destructiveness. Reliable subsurface diagnosis relies on precise characterization of GPR responses. Previous studies have explored GPR patterns associated with seepage through laboratory experiments [1–3]. However laboratory conditions can only replicate limited scenarios, hence current research predominantly relies on numerical simulations to investigate various pipeline defect scenarios. Previous studies have demonstrated the viability of utilizing the Finite Difference Time Domain (FDTD) method to characterize the GPR response of cracked pipelines and got consistent observation with the laboratory research [4,5]. Apart from significant improvements in antenna and excitation signal modelling[6], many methods have been developed to model the complicated subsurface environment, e.g., irregular FDTD grid[7] and stochastic media[4–8]. These studies artificially depicted seepage areas around the pipe using elliptical shapes and increased dielectric constants when establishing the seepage spatial model, without considering the physical mechanism of seepage and cavities formation process. Therefore, the literature is facing a need for a realistic spatial model that can correctly depict the soil degradation caused by cracked pipelines.

Journal of Physics: Conference Series **2887** (2024) 012030 doi:10.1088/1742-6596/2887/1/012030

The coupling of computational fluid dynamics (CFD) and the discrete element method (DEM) has recently become popular for numerically investigating the seepage mechanism and characteristics of soil structure around cracked drainage pipes[9,10]. Therefore, this study couples the soil-water seepage mechanism to characterize the GPR responses of various drainage pipeline defects, by developing a realistic GPR simulation method by integrating CFD-DEM and FDTD to couple the soil-water erosion mechanism and GPR wave propagation.

2. The proposed information communication between CFD-DEM and FDTD

The integration of CFD-DEM and FDTD requires information exchange between models, hence ensuring accurate exchange of relevant parameters and variables at appropriate interfaces is essential.

2.1. Particle-based electromagnetic field

The spatial model established in CFD-DEM described the proportion of soil, water and air in the seepage area, and the water and air are the most critical factors that determine the EM properties of eroded soils. While the EM properties of the medium determine the propagation of radar waves in the medium, this study embeds the degraded soil properties onto the spatial cell in the fluid-particle field based on the proportion of each material.

The four essential properties are relative permittivity ϵ_r, conductivity σ (Siemens/metre), relative permeability μ_r, and magnetic loss σ_r (Ohms/metre). The complex refractive index model (CRIM) was proved reliable in higher frequency[11], and it is applied to estimate the dielectric permittivity based on the proportions of each material in an unresolved cell. The conductivity of unresolved cells is obtained with Archie's law[12], which calculates the conductivity of each unit in the mixed media model through the porosity and saturation of the mixed media.

2.2. Particle-grid transformation

To preserve resolution and mitigate information loss, the information concerning spherical particles is projected onto an initial orthogonal grid, where the grid size corresponds to the average particle size. (Figure 1). When there is an information conflict, the EM property of the grid is defined by the weighted average of the involved volume of the related particles. Subsequently, the grid size is interpolated further until it reaches the convergence criteria of the FDTD simulation Courant–Friedrichs–Lewy (CFL) condition, which imposes a limit on the time-step size based on the grid spacing and the maximum wave propagation velocity in the simulation domain. Finally, radar wave propagation in the spatial model is simulated using FDTD.

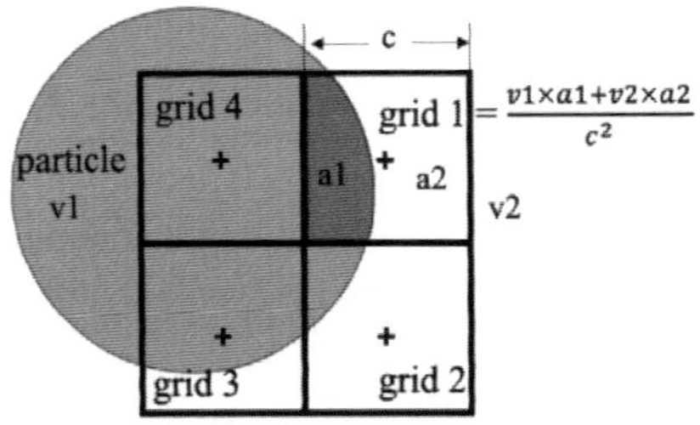

Figure 1. Illustration of information transformation between DEM particle and FDTD grid

3. Experiment and Results

3.1. CFD-DEM simulation of soil erosion caused by a cracked pipeline

By the standardized design of drainage pipelines, which are typically constructed using polyvinyl chloride and buried at a depth of 2 meters, a full-scale model measuring 10 m × 10 m × 8 m (length × width × height) is employed in this study. To mitigate boundary effects during simulation, a pipe with a 1.2-meter outer diameter is centrally buried within the model. The pipe has a wall thickness of 0.01

meters and is inclined at an angle of 0.005. Circular defects with a diameter of 0.5 meters, constituting 40% of the pipe diameter, are positioned at three different locations: the top, side, and bottom of the pipeline. These defects allow the passage of soil particles. The erosion process is influenced by the physical parameters of water and soil. Sandy soil, characterized by over 80% sand and less than 20% clay, is employed in the model. This type of soil is relatively loose and exhibits superior water permeability and air permeability. Generally, due to the low clay content in the sand, the porosity of the sand ranges from 30% to 45%.

Pipeline defects were intentionally designed for the application of CFD-DEM to numerically analyse water seepage and soil erosion resulting from various pipeline defects. More specifically, CFD analysis was carried out using the commercial software Ansys Fluent, while DEM was implemented through PFD3D. In the spatial simulation, water ingress occurs from both the upper end of the gravity drainage pipeline and the ground surface due to rainfall, with outflow taking place from the lower end of the gravity drainage pipeline and the pipeline defect. These dynamic flow conditions result in the displacement of soil particles.

3.2. FDTD simulation of GPR response of cracked drainage pipeline

In this study, FDTD simulations were conducted using GPRMAX, an open-source software for simulating GPR wave propagation. The spatial model was downscaled to improve calculation efficiency, employing 200 MHz Ricker excitation sources. Synchronously, the excitation source and the receiving antenna moved perpendicular to the pipe orientation to simulate a fix-offset survey. The spatial-temporal resolution was determined by the number of iterations and time steps, with a heightened resolution needed to describe the pipe surroundings with varying degrees of infiltration. The step size adhered to the CFL principle, not exceeding the diameter of coarse sand particles, striking a balance between high spatial-temporal resolution and computational efficiency. Additionally, as drainage pipelines in urban areas typically run parallel to the road network, pavement was incorporated above. Table 1 presents the simulation progress and results.

Table 1. Simulation and results of GPR response of cracked drainage pipeline.

Model / Position	CFD	DEM	FDTD
Top	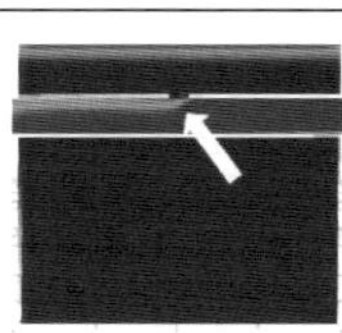	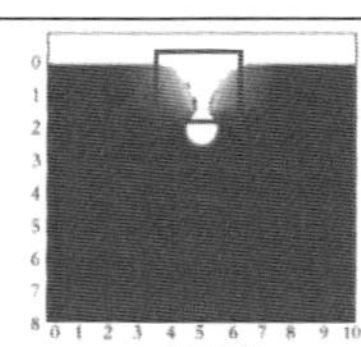	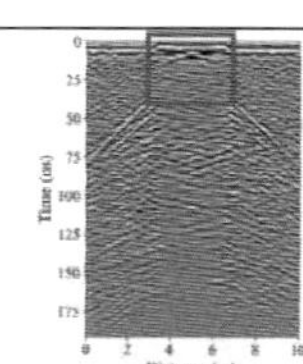
Side	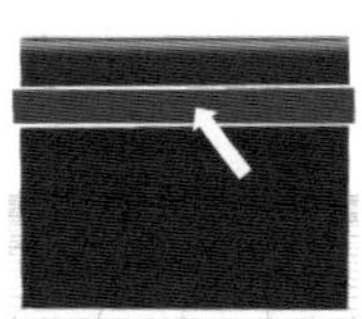	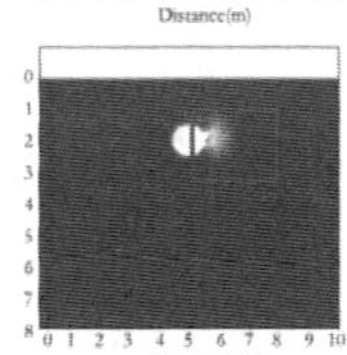	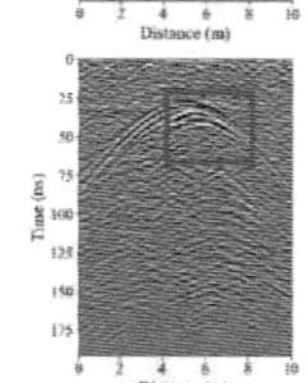
Bottom	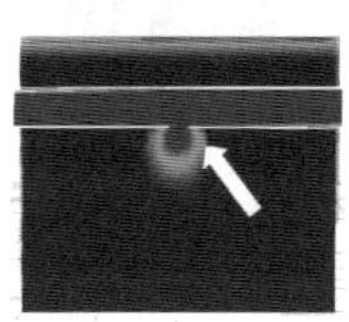	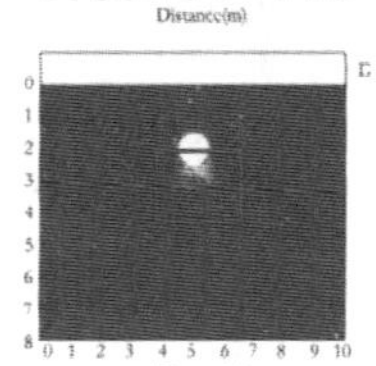	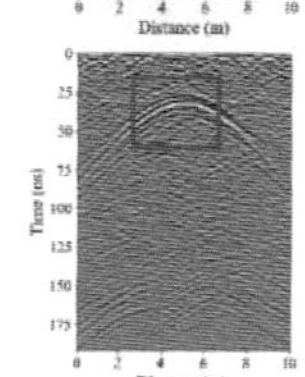

4. Discussion and Conclusion

Table 1 reveals distinct outcomes based on the defect's location on the pipe. When the defect is positioned at the top of the pipe, a gradual soil discharge occurs, leading to surface void formation. The radargram exhibits a bowl-shaped pattern, aligning with observations made by Luo et al.[13]. Conversely, a defect on the side of the pipe induces a small cavity, with water seeping downward into the soil. The radar program displays reverberated reflections, signalling an increase in the water content of the soil surrounding the pipe. In cases where the crack occurs at the bottom of the pipe, no cavity or void is generated, but water seepage downward results in reverberated GPR reflections. The appearance of the pipe bottom at a later time suggests a decrease in wave velocity attributable to an increase in water content. These findings contribute to a nuanced understanding of the varied responses associated with different pipeline defect locations and their implications for subsurface conditions.

In conclusion, this study has addressed the inherent unreliability of previous GPR reflection patterns in characterizing drainage pipeline defects. By integrating CFD and DEM, we successfully constructed a realistic pipe-soil spatial distribution model. The conversion of this spatial model into an electromagnetic model facilitated the numerical simulation of radar wave propagation using the FDTD method. Our research highlights that the positioning of pipeline defects significantly influences soil degradation processes, ultimately leading to distinctive GPR reflection patterns. This integrated approach not only enhances our understanding of subsurface conditions but also provides a more reliable reference for the accurate diagnosis of subsurface hazards associated with cracked pipelines during GPR inspections.

5. Reference

[1] Cheung B and Lai W 2019 Field validation of water-pipe leakage detection through spatial and time-lapse analysis of GPR wave velocity *Near Surface Geophysics* **17** 231–46

[2] Lai W, Chang R, Sham J and Pang K 2016 Perturbation mapping of water leak in buried water pipes via laboratory validation experiments with high-frequency ground penetrating radar (GPR) *Tunnelling and Underground Space Technology* **52** 157–67

[3] Crocco L, Prisco G, Soldovieri F and Cassidy N 2009 Early-stage leaking pipes GPR monitoring via microwave tomographic inversion *Journal of Applied Geophysics* **67** 270–7

[4] Gamal M, Di Q, Zhang J, Fu C, Ebrahim S and El-Raouf A 2023 Utilizing Ground-Penetrating Radar for Water Leak Detection and Pipe Material Characterization in Environmental Studies: A Case Study *Remote Sensing* **15** 4924

[5] Millington T and Cassidy N 2010 Optimising GPR modelling: A practical, multi-threaded approach to 3D FDTD numerical modelling *Computers & Geosciences* **36** 1135–44

[6] Giannakis I, Giannopoulos A and Warren C 2019 Realistic FDTD GPR Antenna Models Optimized Using a Novel Linear/Nonlinear Full-Waveform Inversion *IEEE Transactions on Geoscience and Remote Sensing* **57** 1768–78

[7] Feng D, Wang X and Zhang B 2018 Specific evaluation of tunnel lining multi-defects by all-refined GPR simulation method using hybrid algorithm of FETD and FDTD *Construction and Building Materials* **185** 220–9

[8] Jiang Z, Zeng Z, Li J, Liu F and Li W 2013 Simulation and analysis of GPR signal based on stochastic media model with an ellipsoidal autocorrelation function *Journal of Applied Geophysics* **99** 91–7

[9] Ibrahim A and Meguid M A 2023 CFD-DEM simulation of sand erosion into defective gravity pipes under constant groundwater table *Tunnelling and Underground Space Technology* **131** 104823

[10] Ma Z, Wang Y, Ren N and Shi W 2019 A coupled CFD-DEM simulation of upward seepage flow in coarse sands *Marine Georesources & Geotechnology* **37** 589–98

[11] Zadhoush H, Giannopoulos A and Giannakis I 2021 Optimising the Complex Refractive Index Model for Estimating the Permittivity of Heterogeneous Concrete Models *Remote Sensing* **13** 723

[12] Kennedy W and Herrick D 2012 Conductivity models for Archie rocks *Geophysics* **77** 109

[13] Luo X, Lai W and Giannopoulos A 2020 Forward modelling on GPR responses of subsurface air voids *Tunnelling and Underground Space Technology* **103** 103521

Journal of Physics: Conference Series **2887** (2024) 012031 doi:10.1088/1742-6596/2887/1/012031

Adaptive Variational Mode Decomposition Denoising Scheme for Ground Penetrating Radar Data

Ding Yang[1], Raffaele Persico[2], Cheng Guo[1], Yajie Liu[1], Jiamin Yang[1], Handing Liu[1], Changjin Bai[1], Qing Zhao[1*]

[1] University of Electronic science and Technology of China, Chengdu, Sichuan, China (yuesibi@hotmail.com)
[2] Department of Environmental Engineering DIAM, University of Calabria

Abstract. Addressing the impact of noise on target detection performance in Ground Penetrating Radar (GPR), an adaptive Variational Mode Decomposition (VMD) denoising scheme based on the Whale Optimization Algorithm (WOA) is proposed. An improved adaptive VMD method, based on WOA, is employed in this study for modal decomposition of GPR data, with target modes being retained and noise modes removed to obtain denoised GPR data. It is demonstrated by numerical simulation results that the SNR is increased by 20.846 dB through this scheme, surpassing common denoising methods and achieving the best results in terms of RMSE and MSE. Numerical simulations substantiate the scheme's effectiveness in denoising GPR data.

Key words: *Ground Penetrating Radar, Adaptive Variational Mode Decomposition (VMD), Denoising*

1. Introduction

GPR is extensively utilized as an electromagnetic detection tool in geophysical applications [1-3], including road assessment, subsurface investigations, and oil and gas pipeline detection [4-7]. External electromagnetic interferences, thermal noise, and underground environment complexity can hinder target information observation, degrading detection performance.

Morphological Component Analysis (MCA) efficiently suppresses clutter by decomposing GPR data into targets and clutter [8]. However, it shows a high dependency on the dictionary. Robust Principal Component Analysis (RPCA) surpasses PCA in noise reduction capabilities for clutter suppression within GPR [9]. Robust Non-Negative Matrix Factorization (RNMF) effectively suppresses clutter by decomposing radar data into low-rank clutter and sparse target signals [10].

The Empirical Mode Decomposition (EMD) method effectively separates and emphasizes target signals from noise in GPR data. Enhanced EMD variants, including Ensemble EMD (EEMD), Complete Ensemble EMD (CEEMD), and Improved CEEMD (ICEEMD), address mode mixing issues in EMD, enhancing noise suppression [7]. Variational Mode Decomposition (VMD) represents a novel method, achieving significant noise reduction. However, the core parameters in VMD require manual tuning, limiting its applicability [11]. This study introduces an adaptive VMD-based approach for denoising.

The Whale Optimization Algorithm (WOA) [12] is utilized for a parametric adaptive optimization within the VMD. Optimized parameters are then used for VMD-based modal decomposition of original GPR Bscan data. This process retains target modes and removes noise, yielding denoised GPR Bscan data. The performance of the proposed approach will be compared with other methods, such as mean filtering, RNMF, and EMD, showing better performances.

2. Proposed Scheme

This study presents a noise suppression scheme in GPR data, featuring a workflow depicted in Figure 1. Initially, an average is calculated from all Ascan signals, which is then compressed into a single Ascan, followed by the optimization of VMD parameters using the WOA. Subsequently, modal decomposition of the original GPR Bscan data is performed through VMD using the optimized parameters, with target modes being retained and noise modes removed, ultimately resulting in the acquisition of denoised GPR Bscan data. The concept of this scheme involves enhancing the capability of VMD to extract target signals by the optimization of its core parameters through WOA, thereby effectively suppressing noise.

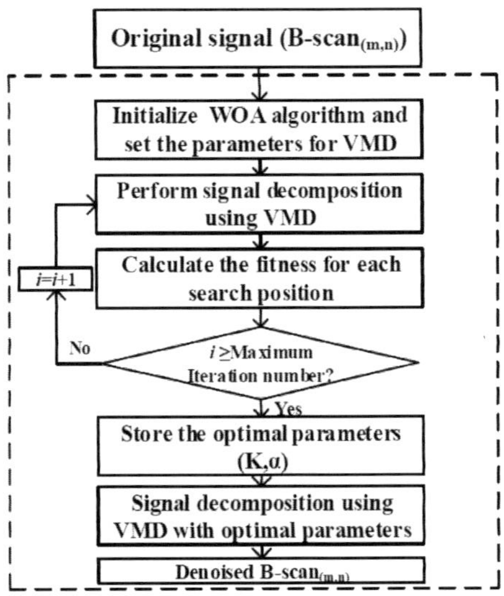

Figure 1. Processing workflow of the proposed scheme.

3. Results

The proposed scheme is applied to GPR data denoising and compared with traditional methods. Numerical simulations are conducted using gprMax [13]. The measurement area consists of air, soil, and targets, with dimensions: length L=5000 mm, width W=5 mm, height H=2000 mm. Air's dielectric constant is 1 (conductivity: 0 S/m), soil's is 6 (conductivity: 0.0004 S/m), and water's is 81 (conductivity: 0.0005 S/m). The background medium is homogeneous soil, and the target is a cylindrical water object (radius: 100 mm). Separated by 100 mm, the transmitting and receiving antennas are excited via a 400 MHz Ricker waveform. A 50 ns time window and 50 mm incremental step size are used for Bscans, generated from multiple Ascans, to simulate GPR Bscan measurements.

Denoising results for simulated Bscan are depicted in Figures 2(a)-(f). Figures 2(a) and 2(b) show the original and noise-added simulated Bscan, respectively. To acquire original simulated Bscan data, a model with detection targets is constructed to obtain its Bscan data. Subsequently, an identical model without targets is constructed. The original noise-free target Bscan data is obtained by subtracting the Bscan of the model without targets from that with targets. Figure 2(c) shows the denoising outcome with the mean filtering method, emphasizing the target signal with inadequate noise suppression. Figure 2(d) illustrates the RNMF method's denoising result, emphasizing the target signal but with poor noise suppression. Figure 2(e) displays the EMD method's denoising results, offering better noise reduction than methods in Figures 2(c)-(d), but introduces high-frequency noise in the upper left corner. Figure

2(f) presents the Adaptive VMD method's denoising outcome, optimized by WOA, achieving effective noise suppression.

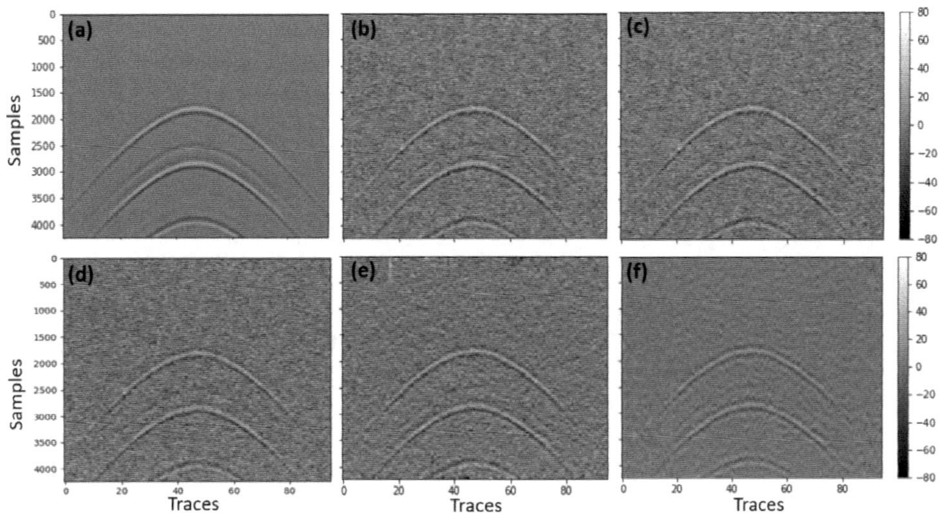

Figure 2. GPR simulated Bscan data results. (a) Noise-free GPR simulated Bscan data. (b) Noise-added Bscan data, -19.999 dB SNR. The denoised result by (c) Mean Filtering, (d) RNMF, (e) EMD, (f) the proposed WOA Based Adaptive VMD method.

To comprehensively compare denoising performances, indicators such as Signal-to-Noise Ratio (SNR), Root Mean Square Error (RMSE), and Mean Square Error (MSE) quantify the capabilities of each method. Table 1 shows the comparison results. For the noise-added Bscan data, the SNR is recorded at -19.999 dB. The proposed method increased the SNR to 0.847 dB, an improvement of 20.846 dB, outperforming the mean filtering (improvement: 0.047 dB), RNMF (improvement: 0.273 dB), and EMD methods (improvement: 14.429 dB). Moreover, the proposed method results in the most favorable outcomes for RMSE and MSE indicators.

Table 1. Comparison of SNR, MSE, and RMSE for various methods under numerical simulation data.

Method	SNR(dB)	RMSE	MSE
Noisy GPR data	-19.999	2982.724	54.614
Mean Filtering	-19.952	2951.017	54.323
RNMF	-19.726	2801.311	52.927
EMD	-5.570	107.582	10.372
Proposed WOA Based Adaptive VMD	0.847	24.541	4.953

4. Conclusions

This study proposes an adaptive VMD-based denoising scheme for GPR data. VMD parameters are adaptively optimized using WOA, followed by signal decomposition via VMD. Numerical simulations indicate that the proposed denoising scheme significantly enhances the SNR by approximately 20.846 dB, effectively suppressing noise. Simulation results validate the method's effectiveness, suggesting considerable potential for GPR data denoising. However, if noise characteristics closely resemble the signal, the scheme may result in "over-denoising," leading to loss of valuable signal information.

Moreover, this scheme involves complex nonlinear optimization, high-precision modal decomposition, and automatic parameter adjustment during execution, imposing a higher computational burden than mean filtering, RNMF, and EMD. Future work will attempt to address these issues.

Acknowledgements

This work was supported by the National Key Research and Development Program (Grant No. 2023YFB3905501).

References

[1] R. Pierri, G. Leone, F. Soldovieri, R. Persico, "Electromagnetic inversion for subsurface applications under the distorted Born approximation" Nuovo Cimento, vol. 24C, N. 2, pp 245-261, March-April 2001.
[2] R. Persico, S. D'Amico, L. Matera, E. Colica, C. De, Giorgio, A. Alescio, C. Sammut and P. Galea, GPR Investigations at St John's Co-Cathedral in Valletta. Near Surface Geophysics, vol. 17 n. 3, pp. 213-229. doi:10.1002/nsg.12046, 2019.
[3] I. Catapano, L. Crocco, R. Persico, M. Pieraccini, F. Soldovieri, "Linear and Nonlinear Microwave Tomography Approaches for Subsurface Prospecting: Validation on Real Data", IEEE Trans. on Antennas and Wireless Propagation Letters, vol. 5, pp. 49-53, 2006.
[4] R. Persico, G. Morelli, Combined Migrations and Time-Depth Conversions in GPR Prospecting: Application to Reinforced Concrete, Remote Sens. 2020, Volume 12, Issue 17, 2778, open access, DOI 10.3390/rs1172778.
[5] Yang, D., Guo, C., Wang, F., Xie, L., Zhang, K., Jin, Y, Zhao, Q. PICGAN: Conditional adversarial neural network-based permittivity inversions for ground penetrating radar data. Journal of Applied Geophysics, 217, 105164, 2023.
[6] Li, J., Yang, D., Guo, C., Ji, C., Jin, Y., Sun, H., Zhao, Q. Application of GPR system with convolutional neural network algorithm based on attention mechanism to oil pipeline leakage detection. Frontiers in Earth Science, 10, 863730, 2022.
[7] Hao, T., Jing, L., He, W. An Automated GPR Signal Denoising Scheme Based on Mode Decomposition and Principal Component Analysis. IEEE Geoscience and Remote Sensing Letters, 20, 1-5, 2022.
[8] Temlioglu, E., Erer, I. Clutter removal in ground-penetrating radar images using morphological component analysis. IEEE Geoscience and Remote Sensing Letters, 13(12), 1802-1806, 2016.
[9] Gao, Q., Zhang, P., Xia, W., Xie, D., Gao, X., & Tao, D. Enhanced tensor RPCA and its application. IEEE transactions on pattern analysis and machine intelligence, 43(6), 2133-2140, 2020.
[10] Kumlu, D., & Erer, I. Improved clutter removal in GPR by robust nonnegative matrix factorization. IEEE Geoscience and Remote Sensing Letters, 17(6), 958-962, 2019.
[11] Dragomiretskiy, K., & Zosso, D. Variational mode decomposition. IEEE transactions on signal processing, 62(3), 531-544, 2013.
[12] Mirjalili, S., & Lewis, A. The whale optimization algorithm. Advances in engineering software, 95, 51-67, 2016.
[13] C. Warren, A. Giannopoulos, I. Giannakis, gprMax: Open source software to simulate electromagnetic wave propagation for Ground Penetrating Radar, Computed Physics Communications, 209(3), pp. 163-170, 2016.

Signal Fusion of Multi-Frequency Continuous Wave Ground Penetrating Radar

Jianping Xiao[1], **Zhijing Liu**[1], **Rui Qu**[1] **and Lanbo Liu** [2]
[1]School of Geosciences and Info-physics, Central South University, Changsha, Hunan, P. R. China, 410083 (jpxiao@csu.edu.cn)
[2]Department of Earth Sciences, University of Connecticut, 354 Mansfield Road, Storrs, CT 06269-1045, USA

Abstract. Carrier-less pulse radar, with its short pulse duration, low average power, and weak anti-interference capability, poses significant limitations to the application of ground penetrating radar (GPR). Fusing continuous wave signals from multiple frequency components forms the basis for the design and data interpretation of continuous wave radar. We firstly introduce two fusion methods: the application of the Inverse Fast Fourier Transform (IFFT) to multi-frequency signals and direct superposition. And we demonstrate these methods by synthesizing three basic sinusoidal signals and analyse their time-frequency relationship. Furthermore, we perform Fast Fourier Transform (FFT) on the transmit waveform of pulse radar to obtain a set of frequency components. By superimposing three sets of frequency components with different numbers of frequency points, we obtain the fused waveform and compare it with the original Ricker wavelet. The results show that while reducing the number of frequency points may increase the error in the synthesized signal slightly, it remains capable of accurately synthesizing the Ricker wavelet. However, this reduction also decreases the time window of the synthesized signal and may result in temporal overlapping. These findings offer insights into selecting parameters for frequency components in multi-frequency continuous wave radar and lay a foundation for signal fusion in such radar systems.

1. Introduction

Currently, ground penetrating radar (GPR) operates mainly in several modes, including carrier-less pulse, frequency-stepped pulse [1], and frequency-modulated continuous wave [2, 3]. Carrier-less pulse radar, due to its simple hardware implementation and the influence of direct coupling limited to early sampling of the receiving antenna, has minimal impact from direct coupling interference, making it the mainstream operating mode for GPR [4]. Despite the peak-to-peak voltage of the ns-level short pulses reaching up to thousands of volts, the average power transmitted to the ground by carrier-less pulse radar is low due to the extremely short pulse duration. A common weakness of this mode is its weak noise resistance, requiring further improvement in detection depth and resolution [5].

Continuous wave radar, on the other hand, transmits continuous waves to the ground or target, with average power far exceeding that of carrier-less pulse radar, providing a feasible approach to improve the signal-to-noise ratio of GPR [6]. Additionally, continuous wave radar can typically adopt single-frequency stepped transmission or simultaneous multi-frequency transmission forms, both essentially being frequency-domain signals that need to be fused into interpretable signals. This paper explores the

Journal of Physics: Conference Series **2887** (2024) 012032 doi:10.1088/1742-6596/2887/1/012032

fusion of multi-frequency signals into time-domain signals using both direct superposition and inverse Fourier transform methods, validated through several case studies.

2. Theory

Consider an arbitrary signal composed of n sinusoidal components. Its analytical expression is represented by Equation (1):

$$\mathrm{x(t)} = \sum_{i=1}^{n} A_i cos(2\pi f_i + \theta_i) \tag{1}$$

where A_i , f_i , θ_i represent the amplitude, frequency, and phase of every signal component, respectively.

Taking the synthesis of three simple sinusoidal signals as an example, where the values of A_i, f_i, and θ_i are given in Table 1.

Table 1. Values of Parameters for Synthesized Signal Components.

i	f(MHz)	A	θ(rad)
1	100	0.6	0.1
2	500	0.4	-0.75
3	900	0.8	1.5

In Table 1, with $n = 3$, the transformation and synthesis results of the synthesized signal $x(t)$ are illustrated in Figure 1.

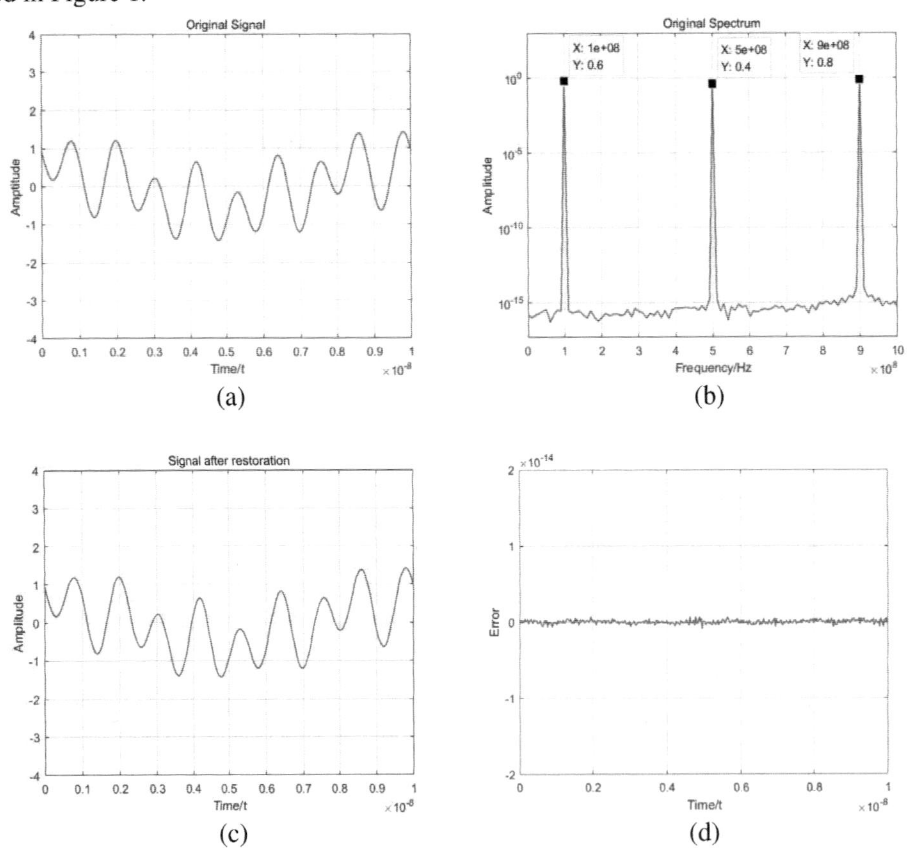

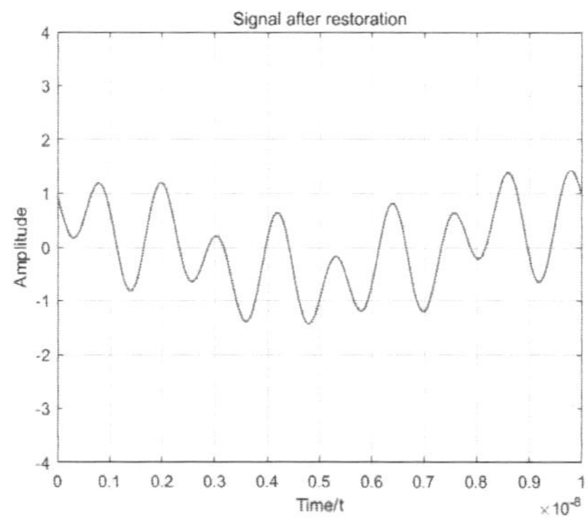

(e)

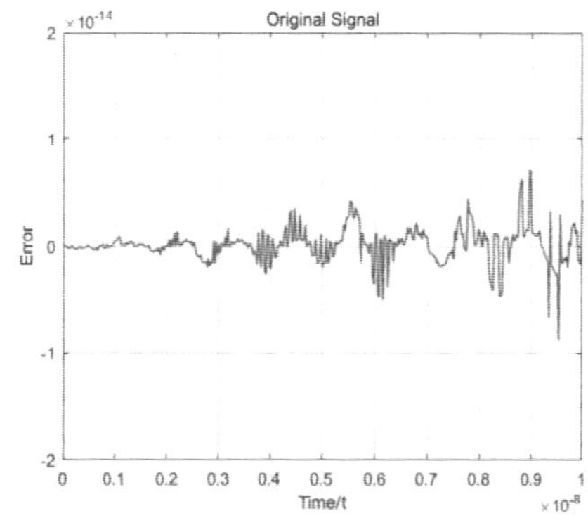

(f)

Figure 1. Numerical Error Comparison between Direct Superposition of Three Single Frequency Signals and IFFT Calculation. (a) Original Signal; (b) Spectrum of the Original Signal; (c) IFFT Synthesis; (d) Error of the IFFT Synthesized Signal; (e) IFFT Synthesis with Zero Padding except Known Frequency Points; (f) Error between Figure (e) and Figure (a).

Figure 1(a) shows the synthesized signal obtained by directly superposing the three frequency signals as illustrated in Table 1. Figure 1(b) depicts the spectrum of the signal shown in Figure 1(a). Since truncation of the signal is necessary for DFT of Figure 1(a), amplitudes at frequencies other than the three known frequency points are non-zero. Performing the inverse Fourier transform on Figure 1(b) yields Figure 1(c). The error between Figure 1(c) and the original signal in Figure 1(a) is shown in Figure 1(d). It can be observed from Figure 1(d) that the numerical error of the inverse Fourier transform can be neglected.

During the processing of multi-frequency GPR signals, the values at frequencies other than the three known frequency points in Figure 1(b) may randomly appear due to signal truncation, and their amplitudes are very small. When performing the IFFT processing, these amplitudes at these frequencies can be set to zero, resulting in the IFFT result shown in Figure 1(e). The error between Figure 1(e) and the original signal is depicted in Figure 1(f). It can be concluded from Figure 1(f) that the error can be neglected. Therefore, irrelevant frequencies can be set to zero during the inverse Fourier transform synthesis of time-domain signals using a finite number of frequency points, simplifying the processing.

3. Example Verification

3.1. Synthesis of Ricker Wavelet with Multiple Frequency Components

The most commonly used transmit waveform in time-domain pulse radar is the Ricker wavelet. In multi-frequency radar systems, the transmit waveform can be synthesized into Ricker wavelets by employing specific frequency, amplitude, and phase parameters. The resulting radar reflection signal can then be synthesized into receiving signals similar to those of pulse radar.

Taking the Ricker wavelet with a central frequency of 1GHz (as shown in Figure 2(a)) as an example, its frequency-amplitude spectrum is depicted in Figure 2(b), while the frequency-phase spectrum is illustrated in Figure 2(c). Processing the signals represented by Figure 2(b) and Figure 2(c) through IFFT yields the reconstructed signal shown in Figure 2(d), which perfectly matches the original signal in Figure 2(a).

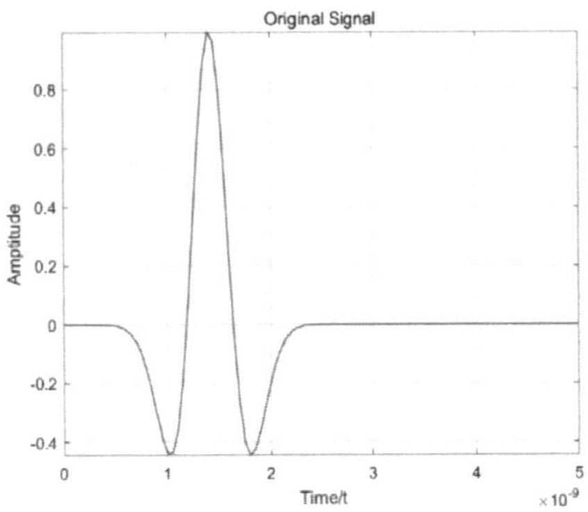
Original Signal
Amplitude
Time/t
×10⁻⁹

(a)

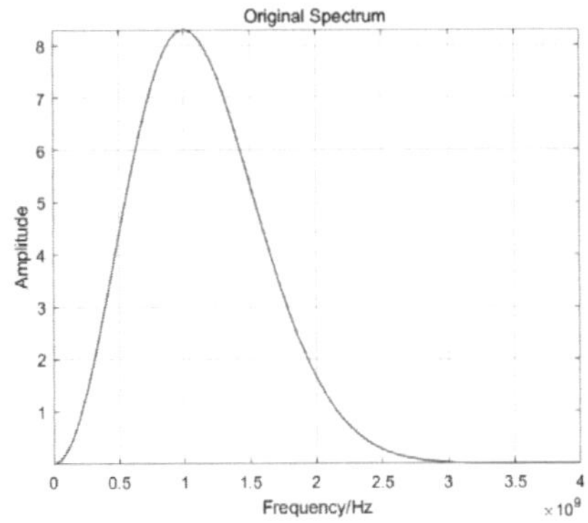
Original Spectrum
Amplitude
Frequency/Hz
×10⁹

(b)

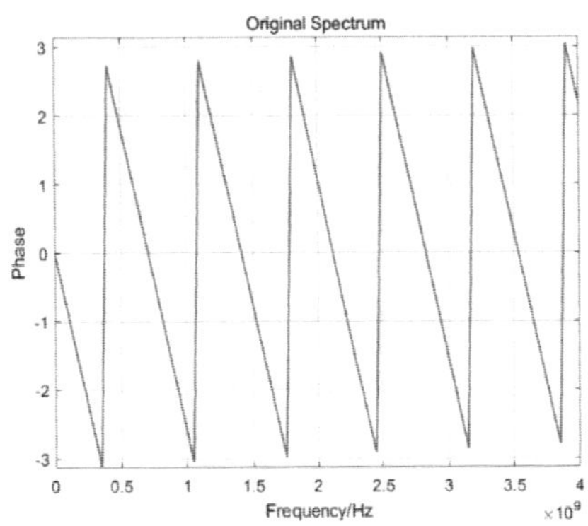
Original Spectrum
Phase
Frequency/Hz
×10⁹

(c)

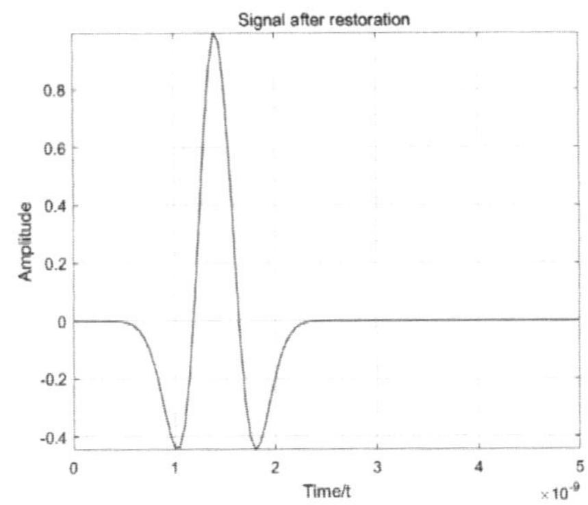
Signal after restoration
Amplitude
Time/t
×10⁻⁹

(d)

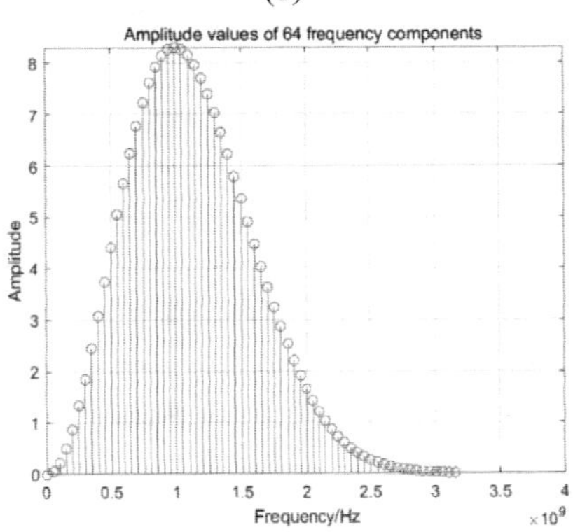
Amplitude values of 64 frequency components
Amplitude
Frequency/Hz
×10⁹

(e)

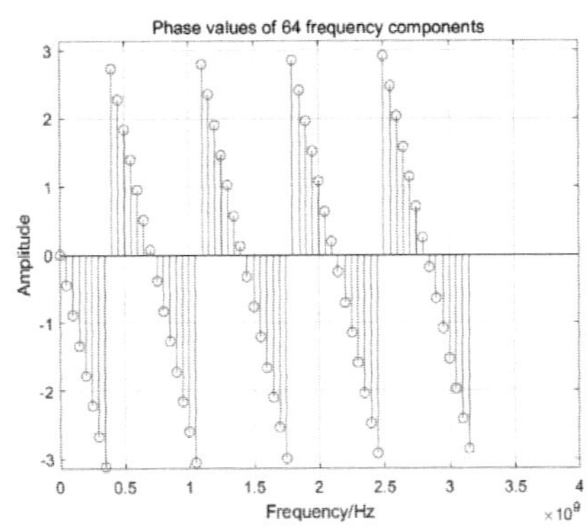
Phase values of 64 frequency components
Amplitude
Frequency/Hz
×10⁹

(f)

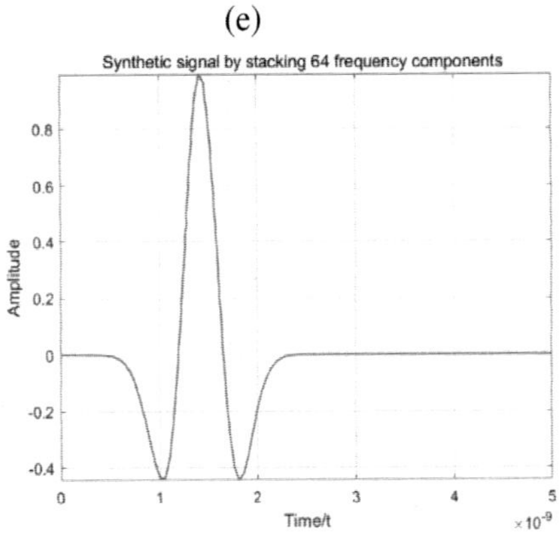
Synthetic signal by stacking 64 frequency components
Amplitude
Time/t
×10⁻⁹

(g)

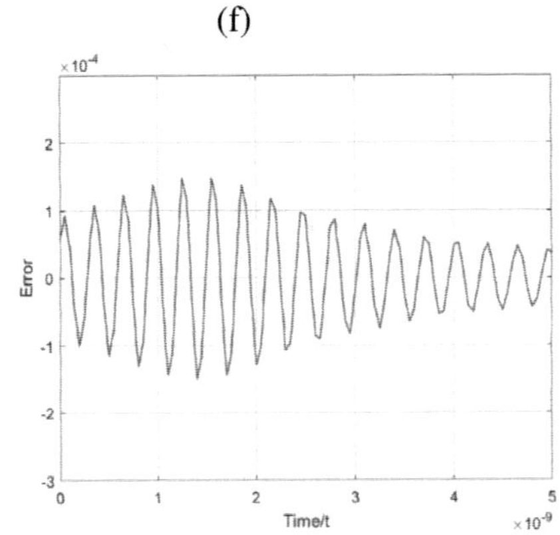
×10⁻⁴
Error
Time/t
×10⁻⁹

(h)

Figure 2. Ricker Wavelet Synthesized from Multiple Frequency Components. (a) Ricker Wavelet with a Central Frequency of 1GHz; (b) Amplitude Spectrum of the Ricker Wavelet; (c) Phase Spectrum of the Ricker Wavelet; (d) Ricker Wavelet Reconstructed; (e) Amplitudes of 64 Frequency Components; (f) Phases of 64 Frequency Components; (g) Synthesized Signal from the Superposition of 64 Frequency Components; (h) Error between the Synthesized Signal and the Original Signal.

From Figure 2(b), it can be observed that the main effective frequency components of the Ricker wavelet are concentrated within the range of 0-3.2GHz, with the amplitude of frequency components greater than 3.2GHz approaching zero. In designing multi-frequency radar systems, frequency points greater than 3.2GHz can be ignored. Simultaneously, a finite number of frequency points are selected within the frequency range of 0-3.2GHz, with 64 frequency points chosen in this example, resulting in 64 amplitude values (as shown in Figure 2(e)) and phase values (as shown in Figure 2(f)). Following the method outlined in the previous section for superimposing frequency-domain single-frequency transmission signals, the synthesized signal depicted in Figure 2(g) is obtained. The error between this synthesized signal and the original signal is shown in Figure 2(h). From Figure 2(h), it can be observed that the relative error of the synthesized signal is less than 0.1%, accurately reproducing the original signal.

3.2. Synthesis of Ricker Wavelet with Partial Frequency Components

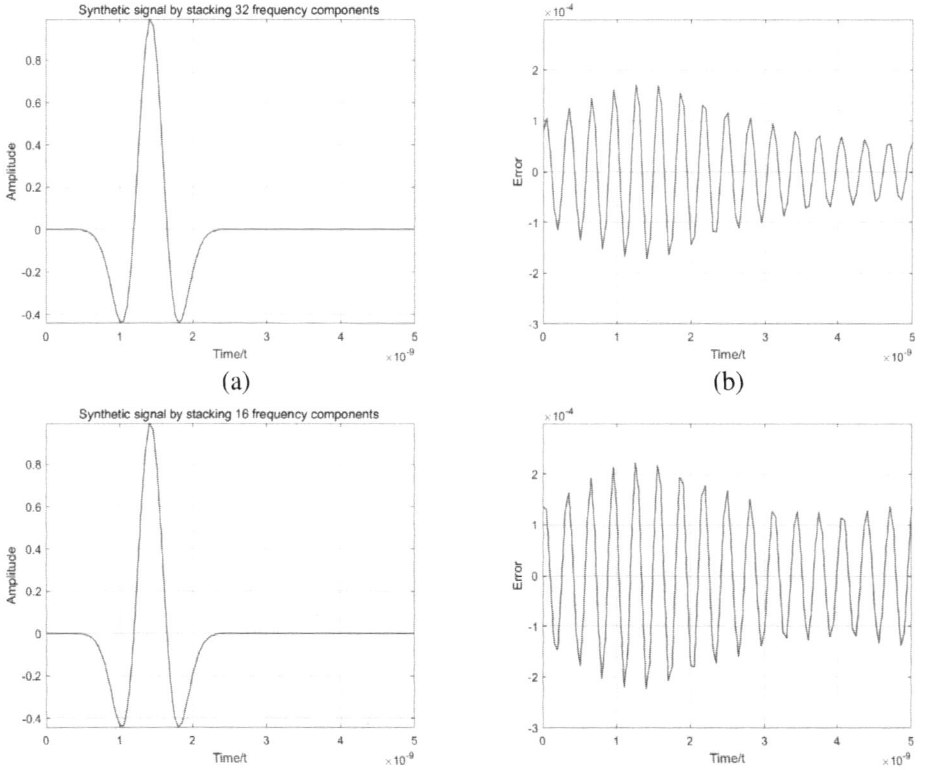

Figure 3: Ricker Wavelet Synthesized with Partial Frequency Components and its Error. (a) Synthesized Signal from the Superposition of 32 Frequency Components; (b) Error of the Synthesized Signal; (c) Synthesized Signal from the Superposition of 16 Frequency Components; (d) Error of the Synthesized Signal.

In the design of multi-frequency continuous wave radar, reducing the number of frequency points is an important approach to improving efficiency. Selecting every other frequency component from the 64 frequency components (as shown in Figure 2(e)), 32 frequency components are chosen and synthesized to obtain the synthesized signal (as shown in Figure 3(a)) along with the error compared to the Ricker wavelet (as shown in Figure 3(b)). Further, 16 frequency components are selected from the 32 frequency components, and the synthesized signal (as shown in Figure 3(c)) and its error compared to the Ricker wavelet (as shown in Figure 3(d)) are presented.

Further analysis of the errors in the signals synthesized with different frequency components reveals that the error values in Figure 3(b) and Figure 3(d) are slightly larger compared to Figure 2(h). Additionally, the error value displayed in Figure 3(d) is slightly larger than that shown in Figure 3(b). However, the absolute error values are all less than 3×10^{-4}, and the relative error values are all less than 0.1%. Therefore, within a certain range, the Ricker wavelet can be accurately recovered by the less frequency components.

On the other hand, reducing the frequency components implies a decrease in frequency resolution, leading to a reduction in the time window of the synthesized signal. In other words, if the above-mentioned 16 frequency components are further reduced by half, the time window of the synthesized signal would be half of that shown in Figure 3(c), resulting in temporal overlapping and inaccurate synthesis of the Ricker wavelet.

4. Conclusion

The results from synthesizing Ricker wavelets using multiple frequency components of continuous wave signals demonstrate that signals with a finite number of frequency points can accurately be synthesized into time-domain transmit signals for ground penetrating radar. This offers a solution for real-time signal processing or post-processing in multi-frequency continuous wave ground penetrating radar applications.

Acknowledgments

The authors would like to acknowledge the partial support from National Natural Science Foundation of China through Grant 42074171.

References

[1] L. Liu, S. Liu. Remote detection of Human vital sign with stepped-frequency continuous wave radar [J]. IEEE Journal of Selected Topic in Applied Earth Observations and Remote Sensing, 2014, 7(3):775-782.

[2] S. E. Hamran, D. P. Paige, H. E. F. Amundsen, etc. Radar imager for Mars' Subsurface Experiment—RIMFAX [J]. Space Science Reviews, 2020， 216(8):1-39.

[3] T. M. Casademont, S. Eide, E. S. Shoemaker, etc. RIMFAX ground penetrating radar reveals dielectric permittivity and rock density of shallow martian subsurface [J]. Journal of Geophysical Research-Planets, 2023, 128(5):1-13.

[4] C. Tao, H. Wang, and S. Qu. Step-sweep broadband Ground Penetrating Radar signal processing [J]. Journal of Microwaves, 2020, 36(2):67-70+78.

[5] M. Pieraccini. Noise performance comparison between continuous wave and stroboscopic pulse ground penetrating radar [J]. IEEE Geoscience and Remote Sensing Letter, 2018, 15(2):222-226.

[6] I. Nicolaescu. Improvement of stepped-frequency continuous wave ground-penetrating radar cross-range resolution [J], IEEE Transaction on Geosciences and Remote Sensing, 2013, 51(1):85-92.

Detection of Termite Nests in Dams Using Ground Penetrating Radar Based on Deep Learning

Jiao Peng[1,2,3], **Wei Xue**[1,2,3], **Wenkai Wang**[4,5], **Junlei Song**[1,2,3], **Feng Zhou**[6], **Pengchun Huang**[1,2,3], **Zhixuan Zhang**[6], **Feng Zhang**[4,5], **Xu Zhang**[4,5], **Longjia Chen**[4,5] and **Li Sun**[4,5]

[1] School of Automation, China University of Geosciences, Wuhan 430074, China
[2] Hubei Key Laboratory of Advanced Control and Intelligent Automation for Complex Systems
[3] Engineering Research Center of Intelligent Technology for Geo-Exploration, Ministry of Education
[4] Dam Safety Monitoring and Termite Control Center of Hubei Provincial Department of Water Resources
[5] Key Laboratory of Termite Control of Ministry of Water Resources
[6] School of Mechanical Engineering and Electronic Information, China University of Geosciences, Wuhan, China
(269577806@qq.com)

Abstract. Termites have always been one of the major hidden dangers to the operational safety of dams. Traditional detection methods such as manual inspection and termite baiting have limitations in terms of efficiency and accuracy. To solve the problems, paper proposes a termite nest detection method based on ground penetrating radar (GPR) and deep learning. First, the simulated GPR data are generated using gprMax software, and the real data are collected using a GPR at several dams. Subsequently, the YOLOv8 deep neural network is employed for termite nest recognition in GPR images. The experimental results show that the proposed model can achieve the mean average precision (mAP) of 0.96 and a detection speed of 53.76 frames per second (FPS).
Keywords: termite nests detection, ground penetrating radar, gprMax, YOLOv8

1. Introduction

Termite nests are characterized by their concealment, persistent nature, and destructive potential, making them a significant hazard to the safe operation of dams [1,2]. Termites rapidly construct intricate tunnel systems both the inner and outer slopes of the dam, which may cause dam failure [3].

The detection methods for termite nests in dams can be divided into invasive detection methods and non-invasive detection methods. Ground penetrating radar (GPR) is the popular non-invasive detection method for underground targets, which has been successfully used in termite nest detection [4]. Compared to other geophysical methods, GPR offers continuous scanning, real-time data collection, and higher efficiency [5]. However, the current identification of termite nests in GPR images mainly depends on artificial experience, which is time-consuming and inaccurate [6].

To address the aforementioned issue, this paper proposes a method for automatic detection of termite nests in GPR images based on YOLOv8 network. Firstly, gprMax3.0 is used to establish the numerical models to simulate GPR detection of termite nests in homogeneous and inhomogeneous media. Then, the real termite nest detection data are collected in dams using an impulse GPR system. Finally, intelligent termite nest detection is implemented using the YOLOv8 network.

2. Methodology

2.1. Forward modeling using gprMax

GprMax is the electromagnetic wave software for GPR modelling, which is based on the Finite Difference Time Domain (FDTD) method [7]. In addition to termite nests, there are stones and voids in dams. Therefore, it is essential to analyze the features of these three types of targets in GPR images.

Journal of Physics: Conference Series **2887** (2024) 012033 doi:10.1088/1742-6596/2887/1/012033

Termite nests are mainly distributed in the soil below the dam slope and above the saturation line. In addition to the main nests inhabited by the king and queen termites, there are many secondary nests and tunnels connecting the primary and secondary nests to the external environment. The main nest consists of three parts: the fungus chamber, the void and the royal family, while the secondary nest consists solely of the fungus chamber and the void. Figure 1 shows the real termite nest.

Based on the above analysis, in the geometry of termite nest model, a circular air gap outer layer is set around the fungus chamber to simulate the air gap between the fungus chamber and the surrounding soil. The FDTD simulation parameters are as follows: the size is 4×2×0.002m, the center frequency of the antenna is 400MHz, the time window is 60ns, the antenna spacing is 0.1m, the excitation waveform is Ricker wavelet, and the trace interval is 0.05m. The material parameters for each component are as follows: the relative permittivities of the fungal garden void, the air layer, and the void are all 1, and the relative permittivity of the stones is 6.

Figure 1. Termite nest. (a) The secondary nest and the primary nest, (b) The royal chamber in the primary nest.

The relative permittivity of homogeneous media is set to 9, with Figure 2 displaying the four models and the corresponding simulated GPR images before and after direct wave removal. As shown in Figure 2, each detection model contains one target. In the four models, the targets are a circular stone, a circular void, a circular termite nest, and one irregular-shaped termite nest, respectively.

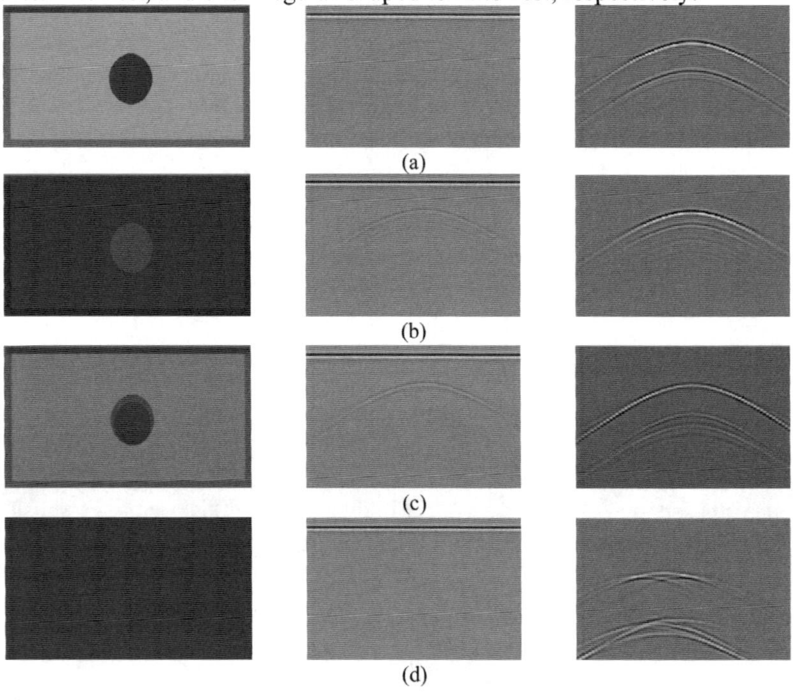

Figure 2. Simulated GPR images in homogeneous medium. (a) The circular stone, (b) The circular void, (c) The circular termite nest, (d) The irregular-shaped termite nest.

The inhomogeneous medium is a mixture of soil, air, water, limestone, and other materials with a relative permittivity range of 6-15. The simulated GPR images before and after direct wave removal are shown in Figure 3. The five models in Figure 3 contain the same targets as the models in Figure 2.

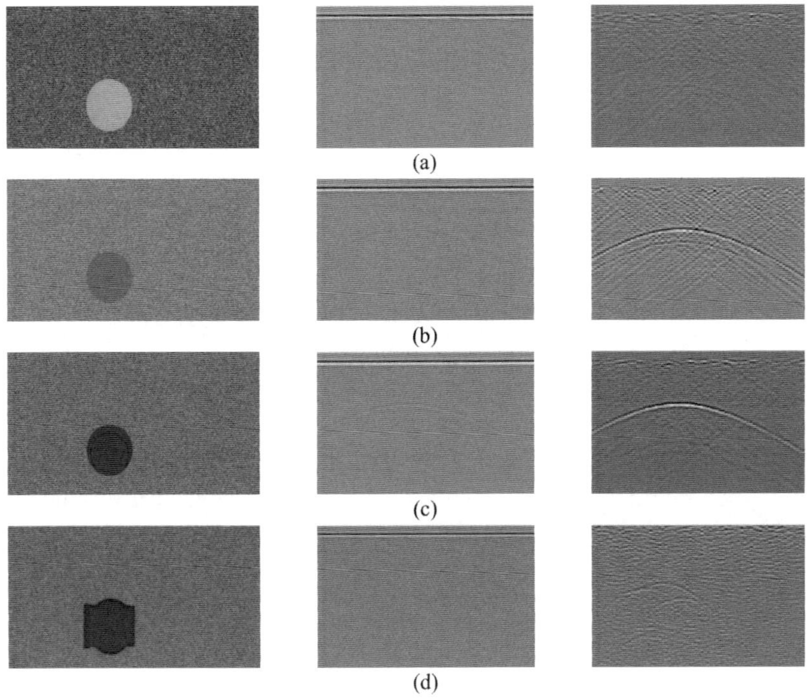

Figure 3. Simulated GPR images in inhomogeneous medium. (a) The circular stone, (b) The circular void, (c) The circular termite nest, (d) The irregular-shaped termite nest.

As shown in Figure 2, the echoes of different targets exhibit different shapes. The echoes of the stone exhibits a standard hyperbolic shape with two distinct bright hyperbolas, while the echoes of the void is composed of multiple hyperbolas that are close but have different intensities. The echoes of the circular termite nest consists of a strong hyperbola at the top and multiple relatively weak hyperbolas at the bottom. The echoes of the irregular-shaped termite nests are characterized by multiple strong hyperbolas at both the top and bottom. As shown in Figure 3, the inhomogeneous medium generates a lot of clutter, and the echoes of these targets become unclear.

2.2. YOLOv8 network model

YOLOv8 is an object detection algorithm based on previous versions of YOLO series and introduces new functionalities, offering excellent performance in detection accuracy and speed [8]. The YOLOv8 network five different versions of the network with varying widths and depths, namely YOLOv8n, YOLOv8s, YOLOv8m, YOLOv8l, and YOLOv8x. Considering the trade-off between detection speed and accuracy, this paper employs YOLOv8n as the target detection model.

3. Experimental results and analysis

3.1. Dataset

Based on the above simulation, a total of 405 GPR images are obtained, including 355 images in homogeneous medium and 145 images in inhomogeneous medium. In addition, this study utilized an impluse GPR system (ImpluseRadar CO1760) with dual frequency antennas of 170MHz and 600MHz. to collect real termite nest data in dams, and 100 real termite nest detection GPR images are obtained. To meet the training requirements of YOLOv8, various data augmentation techniques such as cropping, flipping, scaling, and color transformations are applied to the simulated and real GPR images, and finally a total of 1465 GPR images are obtained. Subsequently, the images are individually labeled, and they are divided into three sets: the training set, the validation set, and the test set at a ratio of 7:2:1.

3.2. Experimental Setup

In the experiments, the YOLOv8 model is constructed using Python 3.8.5. All the programs are performed on Windows 10 operating system equipped with the NVIDIA GeForce RTX 3060 GPU. The batch size is set to 32, and the original input image size is resized to 640 × 640.

3.3. Results

The detection results of YOLOv8n on the GPR images are shown in Figure 4.

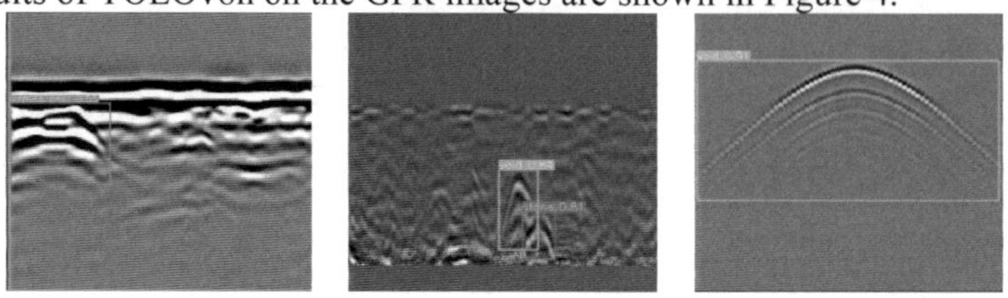

Figure 4. YOLOv8 detection results.

Table 1 lists the evaluation of detection results. As shown in Table 1, the values of average precision (AP) for the three types of targets are 0.963, 0.962 and 0.955, respectively. The value of mean average precision (mAP) is 0.96, and the detection speed reaches 53.76 frames per second (FPS). The results show that the model can realize real-time high-precision detection for termite nests in simulated and real GPR images.

Table 1. Evaluation of GPR image detection results.

	Precision	Recall	AP	mAP	FPS
Stone	0.925	0.873	0.963		
Void	0.899	0.811	0.962	0.96	53.76
Termite nests	0.916	0.921	0.955		

4. Conclusion

In this study, a method based on the YOLOv8 model for automatic detection of termite nests in dams is proposed. First, the termite nest structure is analyzed, then gprMax 3.0 is used to generate simulated GPR images including termite nests, stones, and voids. In addition, real detection data are collected from some dams using a GPR. Finally, the YOLOv8n model is employed to identify these three kinds of targets. The experimental results indicate that the proposed model achieves a mAP of 0.96 and a detection speed of 53.76 FPS. In the future, the proposed model will be validated and improved by using more real GPR images of termite nest detection.

Reference

[1] Li, Hongjie, and Chris Greening, 2020 *Termite-engineered microbial communities of termite nest structures: a new dimension to the extended phenotype* **46** 6

[2] Shi X, He X, Zhang Z, Li S, 2015 *Application of Ground Penetrating Radar (GPR) Technology in the Detection of Termites Nest* **558** 80

[3] Cosarinsky M I., 2021 *A review of micromorphological studies of ant and termite's epigean nests located in neotropical soils of Argentina* **110** 103380

[4] Xu X, Zeng Q, Li D, 2010 *GPR detection of several common subsurface voids inside dikes and dams* **111** 31-42

[5] Sham, Janet F. C., 2016 *Development of a New Algorithm for Accurate Estimation of GPR's Wave Propagation Velocity by Common-Offset Survey Method* **2** 105

[6] Deng G, Li Weichao, 2023 *Current state and prospects of detection technology for termite nests in embankments and dams* **15** 13

[7] Warren, Craig, 2016 *gprMax : Open Source Software to Simulate Electromagnetic Wave Propagation for Ground Penetrating Radar* **20** 164

[8] Talaat, Fatma M., 2023 A*n improved fire detection approach based on YOLO-v8 for smart cities* **28** 20945

Journal of Physics: Conference Series **2887** (2024) 012034 doi:10.1088/1742-6596/2887/1/012034

Spatiotemporal Localization of GPR Sequence Images With 3DResNet18

Yuchen Hu, Pengyu Zhang, Beizhen Bi, Yuwei Chen, Qin Xin and Xiaotao Huang*

College of Electronic Science and Technology, National University of Defence Technology, Changsha, Hunan 410073, China (1546221335@qq.com)

Abstract. Localizing Ground Penetrating Radar can provide a reliable support for current autonomous navigation and localization. Based on the characteristics of ground penetrating radar images, sequence-based matching and localization can effectively improve the problem of a large number of mismatched candidates in single frame image matching. Convolutional neural network is a deep model that can act directly on the original input, 2D convolutional neural network mainly extracts spatial information, we use 3DResNet18 model to extract features from both spatial and temporal dimensions by 3D convolution, so as to capture information in multiple neighbouring frames, to generate multiple channels of information from input frames, and finally combine the information from all the channels together to get a high-level feature for matching. We use the model for localizing of ground penetrating radar sequence images and it achieves better performance compared to the baseline approach.

1. Introduction

The ground penetrating radar (GPR) is mainly composed of transmitting antenna, receiving antenna and radar host. It uses the reflected and refracted waves of high-frequency waves generated on the surface of the measured object to detect the target. Due to the strong similarity of GPR images and the characteristics of vertical downward detection, the detection range of GPR in real scenes is limited, and the probability of single frame image mismatch is large. In order to improve the robustness of Localizing Ground Penetrating Radar(LGPR)[1] system in practical localizing applications, researchers explore the use of inherent sequential information.

Sequential information in LGPR is the sequence of information accumulation when the radar accesses a place again, using sequential information can improve the robustness to perceptual aliasing, because more information can be used to eliminate the ambiguity of false matching. Milford and Wyeth proposed a sequential search method SeqSLAM based on local velocity search in column normalized fractional matrix[2]. Later, researchers used models based on Network Flow[3] and KNN-DTW[4] to estimate the location matching hypothesis robustly. However, all these methods are independent of the underlying feature representation technology and do not use any learning-based mechanism.

In recent years, deep learning methods have shown obvious advantages in computer visual tasks such as image classification, target detection, and motion recognition in video.[5] Deep convolutional neural networks such as ResNet[6] can be used to extract features to get feature vectors used for matching. Garg and m Milford et al. proposed a hierarchical place recognition sequential descriptor,[7] SeqNet aggregated single-image descriptors in the sequence to obtain a sequential descriptor for matching. However, one-dimensional and two-dimensional convolutional neural networks can not extract the temporal and spatial information of sequential images at the same time.

Three-dimensional convolutional neural network(3DCNN) is a simple and effective spatiotemporal feature learning method on large-scale surveillance video data sets[8]. This paper uses 3DCNN to extract the spatiotemporal features, and combined with ResNet18[9] to realize the spatiotemporal matching and localization of GPR.

2. Algorithm and theory

2.1. 3DResNet18

3DCNN can calculate features in the dimension of space and time, so as to capture the motion information in the video stream. 3DCNN uses 3D convolution kernel in a cube by stacking multiple consecutive frames into a cube. Each feature image in the convolution layer will be connected with multiple adjacent consecutive frames in the upper layer to capture spatiotemporal information.

The main feature of ResNet is the introduction of the concept of residual block, which is used to solve the problems of gradient disappearance and gradient explosion in the deep convolution neural network. The shortcut connection can directly connect the input to the output, so that the network can learn the residual information, so as to better carry out feature extraction and processing. Figure 1 shows the basic block of ResNet18.

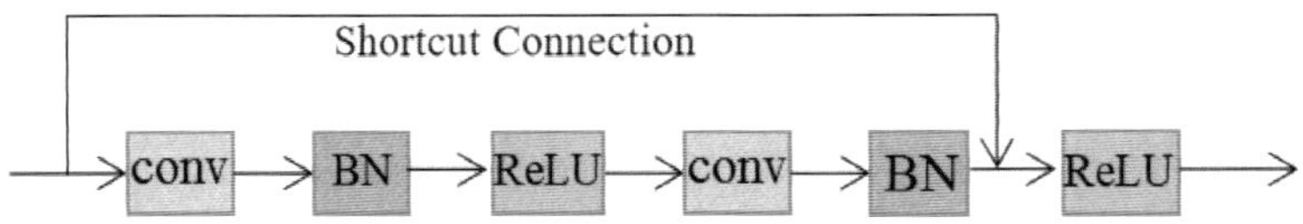

Figure 1. ResNet basic block.

3DResNet18 uses residual structures to connect 3D convolution modules. The network architecture is shown in Table 1. A $3\times3\times3$ max-pooling layer (stride 2) is located before Conv2_x for down-sampling. Spatiotemporal down-sampling is performed by conv3_1, conv4_1 and conv5_1 with a stride of 2.

Table 1. ResNet architecture.

Layer name	Conv1	Conv2_x conv2_1 ×4	Conv3_x conv3_1 ×1	Conv3_x conv3_2 ×3	Conv4_x conv4_1 ×1	Conv4_x conv4_2 ×3	Conv5_x conv5_1 ×1	Conv5_x conv5_2 ×3	FC
Outplanes	64	64	128	128	256	256	512	512	out_features=4096
Kernel_size	(7,7,7)	(3,3,3)	(3,3,3)	(3,3,3)	(3,3,3)	(3,3,3)	(3,3,3)	(3,3,3)	
Stride	(1,2,2)	(1,1,1)	(2,2,2)	(1,1,1)	(2,2,2)	(1,1,1)	(2,2,2)	(1,1,1)	
Padding	(3,3,3)	(1,1,1)	(1,1,1)	(1,1,1)	(1,1,1)	(1,1,1)	(1,1,1)	(1,1,1)	

2.2. SeqNet

The sequence-based hierarchical place recognition system first uses the global descriptor extractor NetVLAD[10] for the input query sequence to obtain the corresponding sequence of single-image descriptors. It inputs the sequence of descriptors into seqnet to generate the short learnt sequential descriptor S_{Ld}, and puts the learnt sequential descriptors from the query and reference database into the descriptor matcher based on the Euclidean distance, as in equation (1), and retrieves the sequence of the k nearest positions as matching candidates.[7]

$$p_{i,j} = \left\| S^{i}_{Ld} - S^{j}_{Ld} \right\|_2 \; \forall j \in N_{db} \tag{1}$$

Journal of Physics: Conference Series **2887** (2024) 012034 doi:10.1088/1742-6596/2887/1/012034

where N_{db} is the size of reference database and $p_{i,j}$ is the descriptor distance of query sequential descriptor at index i from the reference sequential descriptor at index j.

Then, the single-image descriptor learnt from a separate pipeline is used to perform selective sequential score aggregation, and then obtains the best matching result.

3. Experimental results

In this section, we evaluate the performance of the 3DResNet18 through experimental verification and compare 3DResNet18 with SeqNet.

3.1. Data set

We use the GPR dataset produced by Massachusetts Institute of Technology[11]. Run46 of Route4 is used as the train database, run56 (sunny day) of Route5 is used as the test map, run18 (snowy day), run57 (sunny day) and run90 (rainy day) are used as the validation query in turn, and the other two are used as the test query. The GPR data of the test map, validation query and test query are collected clockwise using the middle lane, so there is no impact of lane offset and sequence data reversal. The sequence length is set to 100. When training SeqNet, the number of positive and negative samples is set to 3. When training 3DResNet18, the number of positive samples and negative samples is respectively set to 3 and 8.

3.2. Implementation

All experiments are conducted on a computer equipped with 5GHz Intel CPU, 16GB ram and NVIDIA Geforce RTX 4060 Laptop GPU. The optimizer selects SGD, the initial learning rate is set to 0.0001, the learning step is set to 50 and the training epoch is set to 100. The batch size is set to 4 in SeqNet and 1 in 3DResNet18 limited by the graphics card. Network training uses the max-margin triplet loss[10], as in equation (2).

$$l = \max\left(\left\|X_{\alpha} - X_{p}\right\|_{2} - \left\|X_{\alpha} - X_{n}\right\|_{2} + \alpha, 0\right) \tag{2}$$

where for each query considered as an anchor X_{α}, its positives X_{p} and negatives X_{n} are generated from the reference database using a pre-defined localization radius, α is the desired margin between the positives and negatives in the descriptor space.

3.3. Results

We use Recall@K as the evaluation metric as also used by previous works.[12] It is defined as the ratio of queries that are correctly matched within its top-K retrieved images to the total number of queries.

Table 2. Performance comparison: recall@K.

	Validation query	Test query	Recall @ 1/5/10/20/40/100
3DResNet18	run18	run57, run90	0.71/0.73/0.74/0.76/0.79/0.85
SeqNet			0.53/0.54/0.55/0.56/0.59/0.62
3DResNet18	run57	run18, run90	0.47/0.50/0.52/0.57/0.64/0.73
SeqNet			0.19/0.20/0.21/0.23/0.27/0.33
3DResNet18	run90	run18, run57	0.75/0.77/0.79/0.82/0.85/0.89
SeqNet			0.52/0.53/0.54/0.56/0.60/0.64

In the process of 3DResNet18 network training, the performance gradually deteriorates from the 12th epoch, and the phenomenon of overfitting appears, which may be the reason for the simple GPR data. Table 2 shows the performance of the two networks on the GPR dataset. The reason why the recall performance of SeqNet is generally low may be that the sequence length is too long. When using a long sequence for average pooling, it will potentially reduce the discriminant ability of descriptors in the retrieval process. When the validation query is run57, the performance of the two networks is relatively poor, because the data under the same weather are similar, and the generalization ability of the learned model is weak. In general, 3DResNet18 has significantly improved the performance of each dataset compared with SeqNet, with an average increase of 20 percentage points, so it has better generalization ability.

4. Conclusion

This paper uses 3DResNet18 to realize the matching and localization of GPR sequence images. With the help of using the spatiotemporal feature learning ability of 3DCNN, 3DResNet18 can extract the features of sequence images more accurately and comprehensively, so as to improve the performance of LGPR. The experimental results on GPR data sets show that the performance of 3DResNet18 is significantly improved compared with SeqNet. 3DResNet18 still has a high recall rate in the case of long sequence, slow training, large amount of calculation and overfitting, it shows that 3DResNet18 has certain potential and advantages in GPR sequence images matching and localization.

In the future, the data of the same road section with different weather, lanes and acquisition sequences can be collected together, and the training dataset, verification dataset and test dataset can be divided to increase the diversity of data and improve the generalization ability of the model, so as to achieve higher accuracy and more robust model.

References

[1] M. Cornick, J. Koechling, B. Stanley, and B. Zhang, Localizing Ground Penetrating RADAR: A Step Toward Robust Autonomous Ground Vehicle Localization, *Journal of Field Robotics*, vol. 33, no. 1, pp. 82-102, 2016.

[2] M. J. Milford and G. F. Wyeth. SeqSLAM: Visual route-based navigation for sunny summer days and stormy winter nights. In *Robotics and Automation (ICRA), 2012 IEEE International Conference on*, pages 1643–1649. IEEE, 2012.

[3] T. Naseer, L. Spinello, W. Burgard, and C. Stachniss. Robust visual robot localization across seasons using network flows. In *Twenty-Eighth AAAI Conference on Artificial Intelligence*, 2014.

[4] A. Forechi, A. F. De Souza, C. Badue, and T. Oliveira-Santos. Sequential appearance-based global localization using an ensemble of KNN-DTW classifiers. In *Neural Networks (IJCNN), 2016 International Joint Conference on*, pages 2782–2789. IEEE, 2016.

[5] A. Krizhevsky, I. Sutskever, and G. E. Hinton, Imagenet classification with deep convolutional neural networks. In *Advances in neural information processing systems*, pages 1097–1105, 2012.

[6] K. He, X. Zhang, S. Ren, and J. Sun. Deep residual learning for image recognition. In *Proceedings of the IEEE Conference on Computer Vision and Pattern Recognition (CVPR)*, pages 770–778, 2016.

[7] S. Garg and M. J. Milford. SeqNet: Learning descriptors for sequence-based hierarchical place recognition. *IEEE Robotics and Automation Letters*, 2021.

[8] S. Ji, W. Xu, M. Yang, and K. Yu. 3D convolutional neural networks for human action recognition. *IEEE Transactions on Pattern Analysis and Machine Intelligence*, 35(1):221–231, 2013.

[9] K. Hara, H. Kataoka, and Y. Satoh. Learning spatio-temporal features with 3D residual networks for action recognition. In *Proceedings of the ICCV Workshop on Action, Gesture, and Emotion Recognition*, 2017.

[10] R. Arandjelovic, P. Gronat, A. Torii, T. Pajdla, and J. Sivic. NetVLAD: CNN architecture for weakly supervised place recognition. In *Proceedings of the IEEE Conference on Computer Vision and Pattern Recognition (CVPR)*, June 2016.

[11] Ort, Teddy; Gilitschenski, Igor; Rus, Daniela. GROUNDED: A localizing ground penetrating radar evaluation dataset for learning to localize in inclement weather.[J].*International Journal of Robotics Research*,2023,Vol.42(10): 1.

[12] M. Zaffar, S. Garg, M. Milford, J. Kooij, D. Flynn, K. McDonald-Maier, and S. Ehsan. Vprbench: An open-source visual place recognition evaluation framework with quantifiable viewpoint and appearance change. *International Journal of Computer Vision*, pages 1–39, 2021.

Denoising method of microwave earth exploration test system based on Deep-KSVD

Chen Feiyu[1], Han Shunli[1], Zhang Wenzheng[1], Gao Tianfeng[1], Hao Zeyu[1] and Li Wanrong[2]
1 Science and Technology on Electronic Test & Measurement Laboratory, Qingdao 266555, P.R. China (1023437124@qq.com)
2 Ceyear Technologies Co., Ltd, Qingdao 266555, P.R. China

Abstract. In the process of data acquisition of complex underground media in the microwave ground exploration test system, electromagnetic echoes are susceptible to various noises, resulting in the ambiguity of the in-phase axis of the diffraction hyperbolic curve, which seriously affects the subsequent interpretation and processing. Traditional denoising algorithms are difficult to meet the requirements of denoising complex data. To solve this problem, a microwave profile denoising algorithm based on Deep-KSVD is proposed in this paper. By combining the K-SVD sparse denoising algorithm with the idea of deep learning, the problem of easy loss of original signal and low computational efficiency in the original denoising algorithm is overcome. Firstly, the data is divided into small data blocks, and then an equivalent learnable scheme is used to replace the OMP algorithm in the tracking stage. Finally, each data block is denoised by tracking. The weighted average of the tracked data blocks is obtained, and the accuracy of the algorithm is verified by the actual data.

1. Introduction

As one of the highest resolution geophysical exploration methods, the microwave earth exploration test system has the characteristics of high efficiency, convenience, small size and non-destructive detection [1]. GPR is an exploration method based on exploring the propagation law of electromagnetic waves (1MHz~3GHz) in the medium, especially for the underground 0-5m area, the detection effect is very significant [2]. Using the electromagnetic waves reflected by the antenna, another antenna receives the reflected waves from the interface of the underground medium. When the electromagnetic wave propagates in the medium, its path, electromagnetic field intensity and waveform will change according to the dielectric properties and geometry through which it passes. Therefore, according to the received waveform and travel time, amplitude and waveform data, the distribution of underground media can be inferred [3].

One of the main problems in the system is clutter interference, which can come from surface direct waves, noise between different underground media, and coupling between transmitting and receiving antennas [4]. In particular, the surface direct wave, as the main component, has the characteristics of fast speed, strong intensity and early appearance, which has a serious impact on the detection signal, especially the shallow target signal, which is often buried by clutter interference [5]. At the same time, GPR uses all-pass recording to retain as much information as possible. However, due to external environment interference and hardware limitations, the inhomogeneity of underground media leads to complex electromagnetic wave propagation, and various noises and distorted waves are introduced into the echo data [6]. Due to the diversity of the operating environment of the microwave earth exploration test system, the echo signal is often small in amplitude and low in signal-to-noise ratio, so denoising the signal and improving the image quality are of great significance to improve the system performance, reduce the complexity of operation and popularize [7].

K Singular Value Decomposition (K-SVD) dictionary sparse denoising method is an improved algorithm based on the basic compressed sensing theory [8], has been widely used in image, seismic and microwave ground exploration test system and other fields of data processing. By training a large number of sample signals, this method obtains an over-complete redundant dictionary, which can use sparse basis to represent the signal adaptively, so as to achieve more accurate signal reconstruction and achieve the purpose of denoising [9]. Although KSVD dictionary sparse denoising can effectively suppress or eliminate most of the random noise in the signal, it may still leave some residual burrs at various peaks and troughs of the signal [10]. Based on the above problems, this paper proposes the application of deep-KSVD algorithm in the field of denoising in the microwave earth exploration test system. This algorithm expands K-SVD into a Deep neural network, retains the essence of sparse denoising of the original K-SVD, and uses a neural network with better fitting ability to select appropriate regularization parameters in the tracking stage. Finally, the denoising performance of this method is analyzed through the actual data.

2. Algorithm principle

K-SVD is an overcomplete dictionary design algorithm based on sparse representation: given a set of training data, it finds the best representation dictionary for each sample under the sparsity requirement [11]. By introducing the idea of Deep neural networks, Deep-KSVD automatically learns the representation and features of signals in an end-to-end manner, thus achieving more efficient dictionary learning and signal reconstruction. The specific process is shown in Figure 1.

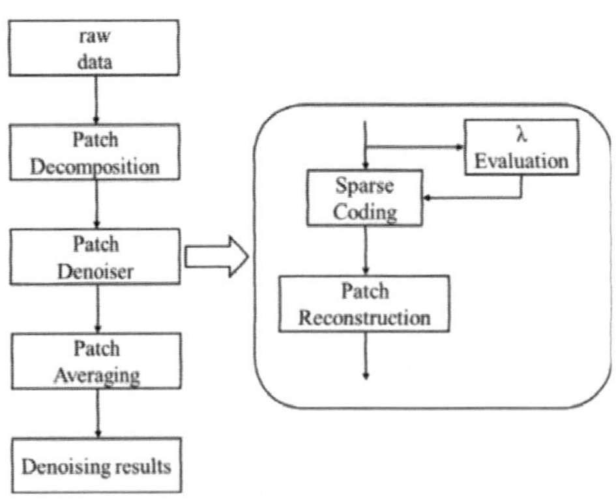

Figure 1. De-noising flow chart based on Deep-KSVD algorithm

- Data block decomposition: the microwave profile data containing noise is decomposed into y data blocks;
- Sparse coding: According to the known dictionary, sparse coding containing noise blocks is obtained iterated by gradient descent method to make it learnable;
- Regularization parameter setting: A multi-layer perceptron (MLP) network is used to estimate the regularization coefficient r for each data block y, where the MLP consists of three hidden layers. With the exception of the last layer, each layer consists of a fully connected linear map and an activation function (ReLU);
- Data block reconstruction: The denoised data blocks are reconstructed using dictionaries and sparse coding, the input data is segmented into completely overlapping databases, and then each noisy data block is processed through the above data blocks;
- Data block average: weighted average of the denoising results of these data blocks is used to obtain the final result.

3. Actual data test

In order to verify the effectiveness of deep-KSVD algorithm for denoising, actual data is used to verify the proposed method. Figure 2 shows the original microwave profile data that only eliminates direct waves at a certain location. With 300 sampling points and 500 channels of data, different algorithms are used for noise image denoising, with the main purpose of eliminating the interference of background noise on effective waveform. The experiments were processed by K-SVD method and deep-KSVD method respectively.

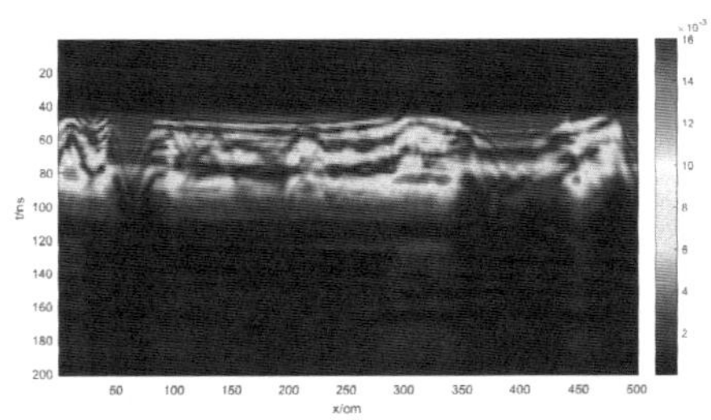

Figure 2. Only direct wave raw data is eliminated

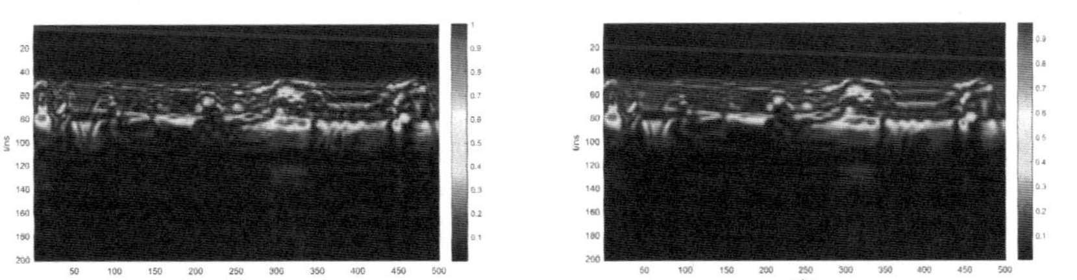

(a) (b)

Figure 3. (a) Denoising processing based on K-SVD algorithm; (b) Denoising processing based on Deep-KSVD algorithm

As can be seen from Figure 3 (a), K-SVD method can remove the background noise of microwave profile to a certain extent, but the denoising result is not thorough enough and the target signal is not clear enough. As can be seen from Figure 3 (b), the results obtained by the denoising algorithm in this paper are clearer, and the signal-to-noise ratio of the data has been significantly improved.

4. Conclusion

This paper mainly studies the application of Deep-KSVD algorithm in the field of microwave ground exploration signal denoising, firstly introduces the basic principle of K-SVD algorithm, and then designs the process of microwave ground exploration signal denoising based on Deep-KSVD algorithm. The actual data results show that Deep-KSVD can adaptively attenuate background noise and protect the continuity of effective information. Compared with K-SVD denoising method, Deep-KSVD algorithm has good adaptability and robustness in the processing of microwave profile background noise, and achieves the purpose of improving the signal-to-noise ratio of data and the accuracy of data interpretation.

References

[1] Xu X, Li J, Qiao X, et al. Fusion of multiple time-domain GPR datasets of different center frequencies[J]. Near Surface Geophysics, 2019, 17(2): 141-150.

[2] Zhou F, Miorali M, Slob E, et al. Reservoir monitoring using borehole radars to improve oil recovery: Suggestions from 3D electromagnetic and fluid modeling[J]. Geophysics, 2018, 83(2): WB19-WB32.

[3] Junoh M S A M, Sulaiman S A H, Natnan S R, et al. Estimation Diameter of Buried Pipe Using Principle of Ground Penetrating Radar and Electromagnetic Locator[J]. International Journal of Geoinformatics, 2022, 18(4): 61-72.

[4] Oskooi B, Parnow S, Smirnov M, et al. Attenuation of random noise in GPR data by image processing[J]. Arabian Journal of Geosciences, 2018, 11: 1-10.

[5] Luo J, Lei W, Hou F, et al. GPR B-scan image denoising via multi-scale convolutional autoencoder with data augmentation[J]. Electronics, 2021, 10(11): 1269.

[6] Bi W, Zhao Y, An C, et al. Clutter elimination and random-noise denoising of GPR signals using an SVD method based on the Hankel matrix in the local frequency domain[J]. Sensors, 2018, 18(10): 3422.

[7] Wang X, Feng D. Multiparameter full-waveform inversion of 3-D on-ground GPR with a modified total variation regularization scheme[J]. IEEE Geoscience and Remote Sensing Letters, 2020, 18(3): 466-470.

[8] Guo Q, Zhang C, Zhang Y, et al. An efficient SVD-based method for image denoising[J]. IEEE transactions on Circuits and Systems for Video Technology, 2015, 26(5): 868-880.

[9] Scetbon M, Elad M, Milanfar P. Deep k-svd denoising[J]. IEEE Transactions on Image Processing, 2021, 30: 5944-5955.

[10] Elad M, Aharon M. Image denoising via sparse and redundant representations over learned dictionaries[J]. IEEE Transactions on Image processing, 2006, 15(12): 3736-3745.

Journal of Physics: Conference Series **2887** (2024) 012036 doi:10.1088/1742-6596/2887/1/012036

Comparative analysis of the application of different GPR devices in the concealed karst detection of railway slopes

Zhou Chao, Jin Junjun and Zhao Siwei

Geological Prospecting & Geotechnical Engineering Design and Research Institute, CREEGC, Chengdu 610031, China
782409886@qq.com

Abstract. Due to its strong concealment and irregular development, concealed karst brings great hidden dangers to the construction and operation of railways in karst areas. It is a focus and difficulty to seek efficient and accurate detection methods. Because of its advantages of high efficiency, non-destructiveness and low cost, the ground penetrating radar（GPR）has been widely used in the detection of underground hidden karsts of railways, highways and cities in recent years. Based on the Guiyang-Nanning high-speed railway under construction, this paper applies two different kinds of GPR devices to detect the hidden karst of a rock slope. The detection results of two kinds of GPR instruments are compared and analyzed in detail, and the detection results are verified by excavation. The results show that the two different kinds of GPR instruments can obtain better detection results in the exploration of hidden karst geological bodies, which can provide some references for the selection of instruments for similar concealed karst exploration work.

1. Introduction

Karst is a geomorphological phenomenon caused by the chemical dissolution and mechanical erosion of soluble rocks by surface water and groundwater, which is widely distributed in the southwest China. Hidden karst can cause great harm to railway construction and later operation. The cave or funnel periodically floods the subgrade base, causing subsidence, churning or collapse. Even worse, sudden underground water surges can damage railway buildings. Therefore, hidden karst exploration is an important component of railway engineering geological exploration in the southwest China.

Generally speaking, the main methods of geophysical exploration for hidden karst detection include direct current sounding[1], seismic imaging[2], transient surface wave method[3], and ground penetrating radar(GPR) method[4]. The ground penetrating radar method is sensitive to water-rich or dry caves in rock masses, and has high work efficiency and low requirements for the site environment[5]. Therefore, this method has become one of the main means to detect hidden karst in rocky subgrade sections.

In this paper, the third grade rock slope of a subgrade of Guiyang-Nanning high-speed railway with strong karst development is taken as background. And two kinds of GPR devices with different antenna principles are applied to the detection work of buried karst in the same survey line and the same survey point, and the detection results are compared and analyzed according to the actual excavation of the slope. Through comparison, the two kinds of GPR devices can detect underground concealed karst well, and the comparison results also provide a more reliable method and instrument selection for the detection of subgrade concealed karst.

2. Method principle

2.1. Method principles and characteristics

GPR is a method of penetrating underground objects using high frequency electromagnetic wave technology. The basic principle of the device is that the transmitter generates a pulse electromagnetic wave signal with a center frequency of 12.5Mhz to 1200Mhz and a pulse width of 0.1ns through the transmitting antenna. When this signal meets the detection target in the rock layer, it produces a reflection signal. The direct signal and the reflected signal are input to the receiver through the receiving antenna, amplified and displayed by the oscilloscope. According to whether the oscilloscope has reflected signal, it can be determined whether there is a target being measured; Based on the arrival delay time of the reflected signal and the average reflected wave speed of the target object, the detection distance of the target can be roughly calculated.[6]

The geological radar system works in the time domain, the ordinate recorded is time, and the time can be converted into the depth information of the target directly.

$$h = \left(\frac{c*t}{2*\sqrt{\varepsilon_r}}\right) \tag{1}$$

In the formula, ε is the permittivity, μ is the permeability, σ is the conductivity, ω is the frequency, and c is the speed of light.

2.2. Instrument equipment

There are a total of two types of GPR devices for comparative testing this time: GSSI SIR-4000 and SSI pulseEKKO. These two devices adopt different theoretical basis, but both of them are improvements to varying degrees based on the original traditional GPR theory. The GSSI SIR-4000 uses an improved integrated receiving and transmitting butterfly coil to enhance detection performance, while the SSI pulseEKKO uses a multi-turn overlapping loop with transmit-receive separation. Both devices use a 100MHz transmitting antenna.

3. Engineering application

3.1. Project overview

The super-high rock slope of a certain roadbed of the Guiyang-Nanning high-speed railway under construction is located in Guangxi. The bedrock is limestone, mainly weakly weathered. In order to ensure the safety of the railway roadbed and the stability of the slope, the four-level slope design is adopted, with each level measuring 8 meters. During the excavation process of the slope, it was found that the bedrock of the slope had developed dissolution, and a total of 35 dissolution grooves or crack development areas were found, which is a suitable place to verify the detection of hidden karst by GPR. The overall view of the slope excavation is shown in Figure 1.

Figure 1. Overall view of slope excavation.

3.2. Line layout and data acquisition

The layout of the GPR detection work for this time is to arrange a geological radar scanning survey line along the third level slope platform and the fourth level slope platform, respectively numbered WT3 and WT4. This data collection work adopts the point measurement method. The collection parameters of each device are shown in Table 1

Table 1. Acquisition parameters of two kinds of GPR devices..

Model	Measurement point distance	Transmission main frequency	Sampling rate	Recording duration	Stack times per second
GSSI SIR-4000	0.2m	100MHz	2048	800ns	120
SSI pulseEKKO	0.2m	100MHz	4096	800ns	8192

4. Interpretation of detection results and comparative analysis of excavation

The built-in processing software for GSSI SIR-4000 is Radan, and the built-in processing software for SSI pulseEKKO is EKKO Project. In order to ensure the comparison with same accuracy, the data collected by the two kinds of GPR devices are processed in the same way. The process is as follows: channel head setting - channel editing - denoising - filtering - energy compensation. In order to reflect the "hyperbola" feature of the karst cave in GPR result map, deconvolution calculation was not carried out in this data processing.

4.1. Line WT3

The results of line WT3 are shown in Figures 2 and 3. As can be seen from Figure 2, there are strong hyperbolic shaped reflected waves at the mileage of 49m, 65m, and 78m, corresponding to the 22#, 21#, and 20# karst caves in Figure 1, where karst is strongly developed. There is also a hyperbolic reflection wave near the position of 12m in the small mileage section, corresponding to the 25# groove in Figure 1. The reaction in the 24# and 23# grooves is not obvious.

As shown in Figure 3, it can be seen that, except for the consistent positions of the corresponding karst caves and grooves in Figure 2, there is a hyperbolic reflection wave near the position of 21m in the SSI pulseEKKO result map, which corresponds to the 24# karst groove. The hyperbolic reflection wave of the 24# karst groove appears later than the hyperbolic reflection wave of the 25# karst groove, indicating that its burial depth is slightly larger than that of the 25# karst groove, which is consistent with the excavation morphology of the 24# and 25# karst grooves in Figure 1.

4.2. Line WT4

The detection results of line WT4 are shown in Figures 4 and 5.It can be seen from Figure 4 that there is a continuous strong reflection zone at 10-30m, which is the concentrated reaction of 33-1# cave, 33-2# cave, 34# cave and 29# groove, and karst is strongly developed in this region. The hyperbolic shape reflection wave near the position of 45m corresponds to the 28# groove that has been excavated and verified in Figure 1, and the hyperbolic shape reflection wave near the center position of 60m corresponds to the response of the 27# crack development zone. However, in the detection of GSSI SIR-4000, there was no obvious response to the 26# fracture development zone.

As shown in Figure 5, the detection results of SSI pulseEKKO and GSSI SIR-4000 are basically consistent. However, in the 65-74m region, the detection results of GSSI SIR-4000 showed a strong electromagnetic signal oscillating back and forth, indicating the influence of metal objects near the measurement line. The reason is that in order to ensure safety, the construction side arranged metal nets near the entrance of the 31# cave, so the detection effect of GSSI SIR-4000 is not good. In the detection results of SSI pulseEKKO's same line, the electromagnetic wave energy is weak in the area of 65-74m zone, but the reflection wave event is still visible, indicating that the electromagnetic wave energy emitted by SSI pulseEKKO does not oscillate back and forth between metal objects under the same influence of metal mesh.

4.3. Verification of drilling exploration

According to the drilling and measurement results completed by the construction unit at the opening of the 33-1 # and 33-2 # karst caves, this is a vertically developed karst cave that is interconnected, with

Journal of Physics: Conference Series **2887** (2024) 012036 doi:10.1088/1742-6596/2887/1/012036

a depth of about 21m. Comparing Figures 4 and 5, it can be seen that SSI pulseEKKO and GSSI SIR-4000 have both detected the bottom plate position of the karst cave.

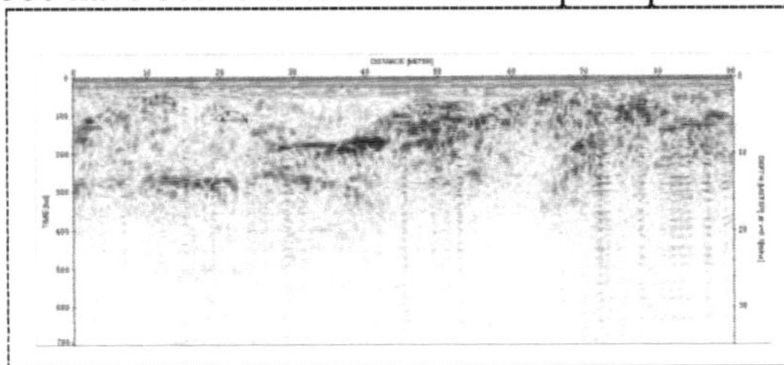

Figure 2. GSSI SIR-4000 detection results of line WT3.

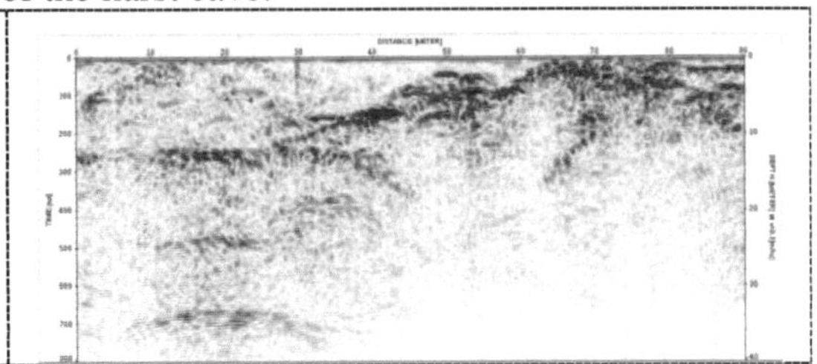

Figure 3. SSI pulseEKKO detection results of line WT3.

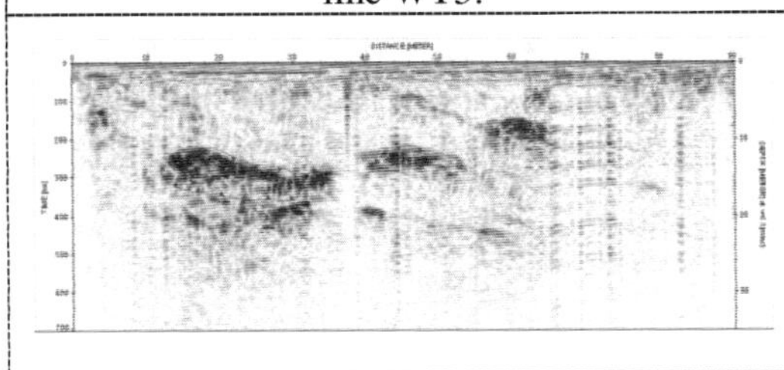

Figure 4. GSSI SIR-4000 detection results of line WT4.

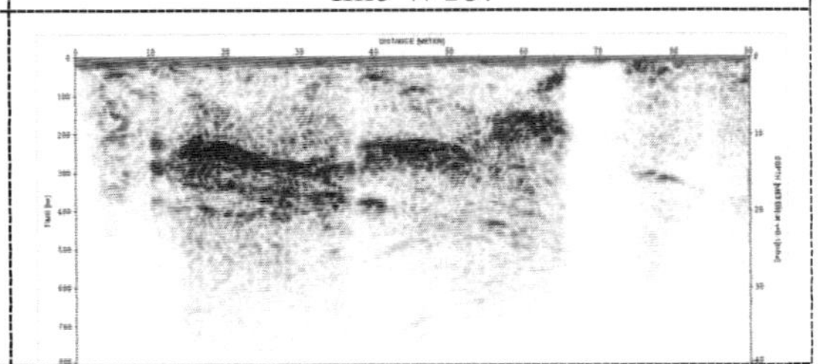

Figure 5. SSI pulseEKKO detection results of line WT4.

5. Conclusion

According to the detection results and excavation verification of two kinds of GPR equipment in buried karst, the conclusions are as follows:

(1) The two kinds of GPR devices have good detection effect on buried karst geological bodies, but the difference lies in the detection accuracy and depth.

(2) The detection depth of SSI pulseEKKO is deeper due to the high stack number and low center frequency, but the event is therefore wider, which affects the accuracy of exploration.

(3) The integrated receiving and transmitting butterfly coil in GSSI SIR-4000 is greatly affected by the interference of metal objects near the measuring line. And the electromagnetic wave energy often oscillates back and forth between the antenna and the metal objects, forming strong multiple signals on the detection results, thus drowning the effective signals.

References

[1] He G Y and Wang Y G 2019. Combined Application of High-density Electrical Resistivity Tomography and Combined Profiles Method for Hidden Karst Investigation in Transmission Line Engineering *J. Chinese Journal of Engineering Geophysics* **16(01)** 53-60

[2] Li J., Liu C, Zeng, Z., & Chen, L (2015). GPR Signal Denoising and Target Extraction with the CEEMD Method. *IEEE Geoscience and Remote Sensing Letter,* 12 (8), 1615-1619.

[3] Wang X L 2020. Application of 2D Transient Surface Wave in the Detection of Hidden Karst at the Bottom of the Tunnel *J. High Speed Railway Technology* **11(03)** 46-50

[4] Li T Y 2012. Study on Application of Ultra-low Frequency Ground Penetrating Radar in Karst Exploration *J. Journal of Railway Engineering Society* **29(03)** 6-11

[5] Song J, Wu Y Q, Hu Y B and Kang H Q 2005. Feasibility of GRP to Penetrate Hidden Cave and case Analysis *J. Journal of Henan Polytechnic University(Natural Science)* **24(05)** 355-360

[6] Yang F and Peng S P 2010 *Research on the Principles and Methods of Geological Radar Detection* (Beijing: Science Press)

Experiment on Measuring Lake Ice Thickness Using a SFCW Ground Penetrating Radar System

Hongxing Zheng, Chunlin Huang, Zhihua He, Tao Liu and Xiaoji Song
College of Electronic Science and Technology, National University of Defense Technology, Deya road 109, Changsha, China (zhenghx_1998@nudt.edu.cn)

Abstract. To ensure transportation safety and for other purposes, it becomes necessary to measure the thickness of the ice layer on natural water bodies during winter. As an efficient and non-destructive subsurface detection tool, Ground Penetrating Radar (GPR) can achieve rapid and accurate measurements of ice thickness. For this experiment, we utilize a stepped frequency continuous wave (SFCW) GPR system based on a Vector Network Analyzer (VNA) system to measure the thickness of ice layers. Additionally, based on the scenario of detecting ice layer structures with ground-penetrating radar, the formula for calculating the thickness of ice layers was revised to achieve more accurate measurements of ice thickness. Finally, we compared the calculated results with the ice thickness detection results from ice layer drilling experiments to verify the effectiveness of the system in detecting ice layer thickness.

Keywords: ice thickness, ground-penetrating radar (GPR) and VNA.

1. Introduction

Because electromagnetic waves can penetrate dielectrics and the discontinuity of electrical parameters at the interfaces of different media causes reflection phenomena, ground-penetrating radar (GPR) serves as an efficient and non-destructive geophysical surveying tool. It is used to obtain information about the internal structure of the medium[1,2]. GPR has a wide range of applications, including municipal construction[3], archaeological discoveries[4], geological surveys[5], and military engineering, etc[6].

Ground-penetrating radar has been widely applied in detecting stratified geological or engineering structures. In some cold regions, where transportation and construction on ice are necessary, ensuring the safety of traffic and construction is essential. This requires personnel to first survey the thickness of the ice layer. Fedorov M. P. et al. have used ground-penetrating radar to measure ice jams[7]. Similarly, Grab, Melchior, and others have utilized helicopters equipped with ground-penetrating radar to detect the thickness of glaciers in the Swiss Alps.[8]. Furthermore, Emmanuel Léger and colleagues have analyzed the resistivity of permafrost using ground-penetrating radar, thereby elucidating the characteristics of the Arctic permafrost region[9].

This article introduces an experiment on ice detection in freshwater lakes. We utilize a small-scale VNA system to construct an SFCW-GPR for detecting the thickness of ice layers in freshwater lakes in

Journal of Physics: Conference Series **2887** (2024) 012037 doi:10.1088/1742-6596/2887/1/012037

harsh cold environments. Additionally, based on the characteristics of experiments that use ground-penetrating radar to detect ice thickness, this article incorporates the propagation time of direct waves into the ice thickness calculation formula to achieve more accurate ice thickness calculations. Lastly, we compare the results with those from drilling exploration to verify the stability of the ground-penetrating radar system in harsh cold environments and the accuracy of the test results.

2. Theory and Methods

2.1. Theory

The SFCW-GPR system transmits a series of frequency point signals, receives echoes, and then obtains one-dimensional range profile results after performing an inverse Fourier transform[10].

The step frequency signal model is shown in Formula 1:

$$S(t)=\sum_{k=0}^{N-1}\mathrm{u}\left(\frac{t-kT_r-T_r/2}{T_r}\right)e^{j2\pi(f_0+k\Delta f)t} \tag{1}$$

In the formula, N represents the number of frequency points; Δf represents the step frequency interval; $\mathrm{u}(t)$ represents the envelope of a single frequency point signal, k represents frequency point number, T_r represents the duration of a single frequency point; f_0 represents the starting frequency point.

Based on the penetrability of electromagnetic waves, the electromagnetic waves emitted by the transmitting antenna could penetrate through the upper surface of the ice layer and propagate into the ice. When encountering an interface with discontinuous electrical constants, part of the electromagnetic waves will scatter back and be received by the receiving antenna. After performing an inverse Fourier transform on the echo signal, the Radar 1D Range Image can be obtained. In the formula, τ_i represents the echo time delay of the target, and i represents the identification number (total of M targets); Γ represents the target's scattering coefficient.

For the detection scenario, frequency domain data can be obtained by mixing the echo signal. It can be described as follows:

$$S(f_k)=\sum_{i=1}^{M}\Gamma e^{-j2\pi f_k\tau_i}, f_k=f_0+k\Delta f, k=0,1,2,\cdots N-1 \tag{2}$$

By performing an inverse Fourier transform on the mixed signal and then taking the modulus, a one-dimensional range profile can be obtained. In the formula, B represents the bandwidth of the signal; f_c repesents the carrier frequency.

$$\hat{s}(t)=IDFT\left(S(f_k)\right)=\sum_{i=1}^{M}\Gamma_i B\,\mathrm{sinc}\left(B(t-\tau_i)\right)e^{-j2\pi f_c(t-\tau_i)}$$
$$f_c=(f_0+f_{N-1})/2 \tag{3}$$

In actual detection, the propagation path of the radar waves must also be considered. To estimate a more accurate thickness of the ice layer, the actual propagation path of the radar waves, influenced by factors such as the radar volume, must be considered during the post-processing of data[8,11]. In actual measurements, since the antenna is relatively close to the ice surface, it can be assumed that the antenna is tightly adhered to the upper surface of the ice. The propagation path of the direct waves can be approximated as a straight line, and the propagation path of the echo from the ice subsurface can be approximated as the transmitted signal reflecting off the subsurface to the receiving antenna. After geometric correction, the expression for the actual ice thickness is as shown in Formula 4[12,13].

Among them, d represents the distance between the centre of the transmitting and receiving antennas. ε_{ice} represents the relative dielectric constant of freshwater lake ice, which according to the SIR-3000

 doi:10.1088/1742-6596/2887/1/012037

ground penetrating radar manual, is taken as 4; c represents the speed of light in vacuum; τ_{sur} represents the time delay of the surface echo; τ_{sub} represents the time delay of the subsurface echo; h represents the ice thickness. This formula represents an estimation of ice thickness based on the ray model, considering the transmission time of direct waves.

$$h = \sqrt{\frac{c^2((\tau_{sub} - \tau_{sur} + \frac{d}{c})/2)^2}{\varepsilon_{ice}} - (\frac{d}{2})^2} \quad (4)$$

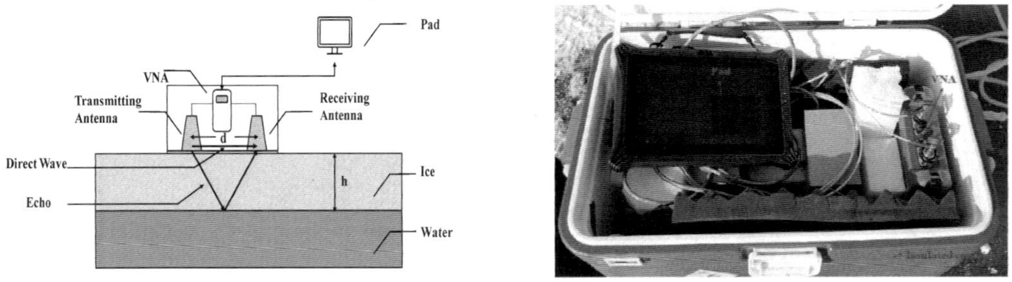

(a) Radar Detection Scenario Schematic. (b) the GPR System based on VNA

Figure 1. Radar Working Principle and System Schematic Diagram

2.2. Methods

This experiment utilized a pad to control a small-scale vector network analyser (VNA), achieving the construction of a stepped-frequency ground-penetrating radar system. The radar system, based on the VNA, is relatively compact, lightweight, and easy to control, allowing for the customization of the number of transmitting stepped-frequency points and the bandwidth [14]. The bandwidth used for this detection experiment ranges from 650 MHz to 3 GHz. The VNA system model is KEYSIGHT N9918B. The transmitting and receiving antennas used in the experiment are ultra-wideband horn antennas, which have good directivity and are conducive to conducting ice thickness detection experiments.

3. Comparison of GPR Detection Results with Drilling-Sampling Results

The ice thickness detection experiment was conducted at Cha gan Lake, located in Song yuan, Jilin Province, China, at approximately N45°. The average temperature in January at this location is below -10°C. The ice thickness at this site is suitable, and the surrounding environment is open, making it an ideal setting for GPR stratigraphic interface detection.

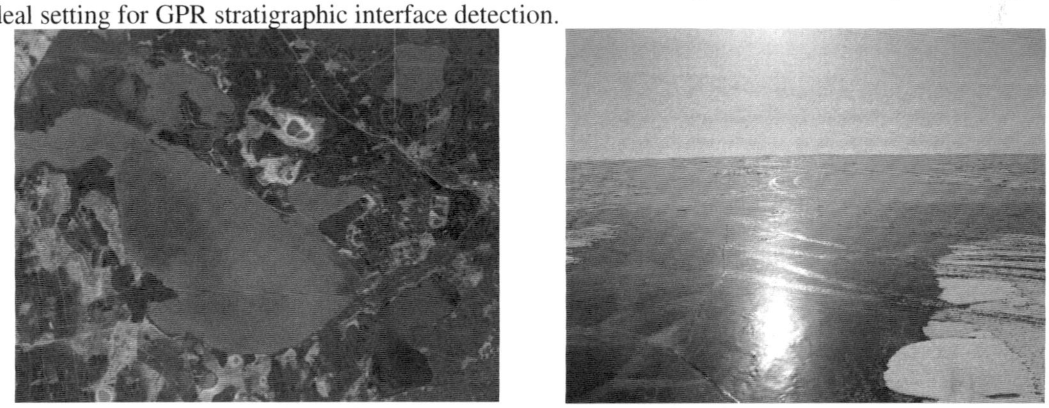

Figure 2. Cha gan Lake and Surrounding Test Area.

In this ice thickness detection experiment, we employed the system as shown in the figure. As illustrated, we installed the equipment in an insulated box and used manual dragging to measure the thickness of the ice surface. Additionally, to validate the precision of the GPR measurements, ice core samples were extracted using mechanical drilling to directly measure the ice thickness. Considering the

Journal of Physics: Conference Series **2887** (2024) 012037 doi:10.1088/1742-6596/2887/1/012037

relatively short length of the transect in comparison to the overall lake surface, and the fact that the surface of Cha gan Lake in this area can be approximately regarded as a static water body, it is assumed that these samples can serve as a representative approximation of the ice layer thickness within this specific region.

(a): Ground penetrating radar method.

(b): Drilling exploration methods.

Figure 3. GPR detection methods and drilling exploration methods.

Considering the actual situation of ice detection, we conducted multiple measurement lines at the experimental site. The uncalibrated experimental Bscan diagram is shown in Figure 4. Based on Figure 4, the ice structure appears to be a two-layered flat structure. Considering the vast area of the ice layer, an infinite two-dimensional plane model can be used as a substitute.

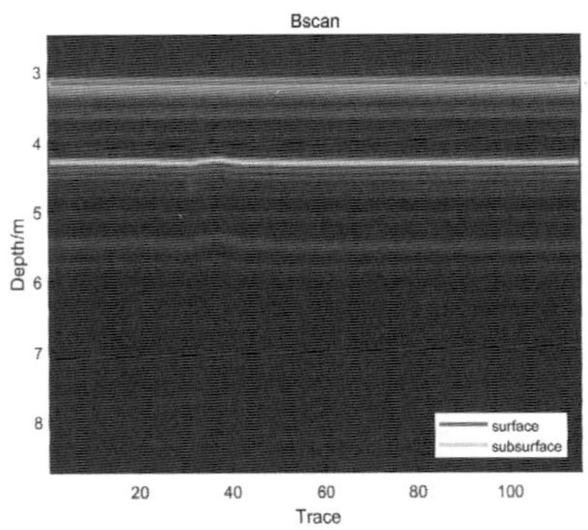

(a) Line 1.

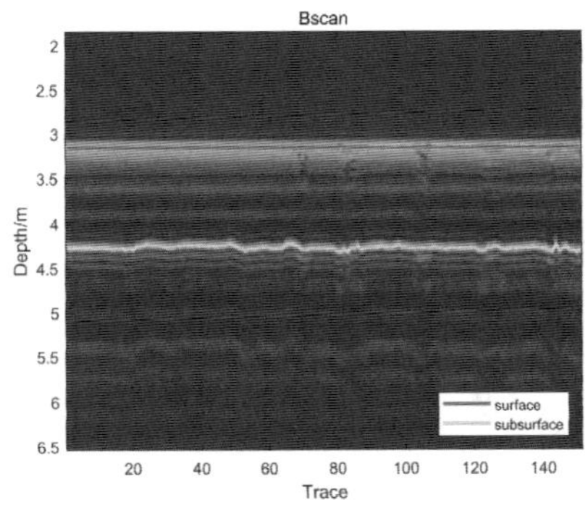

(b) Line 2

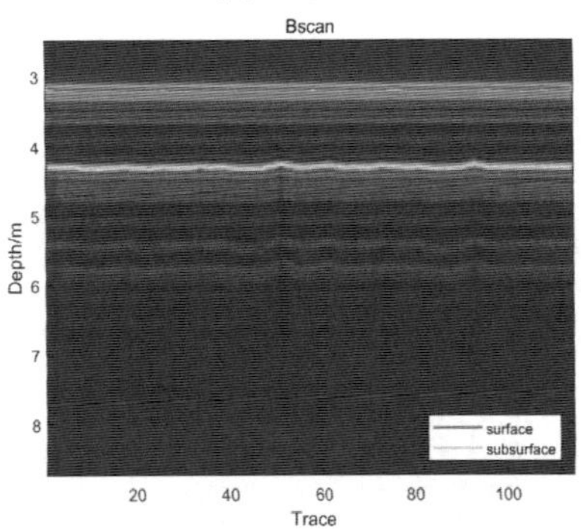

(c) Line 3

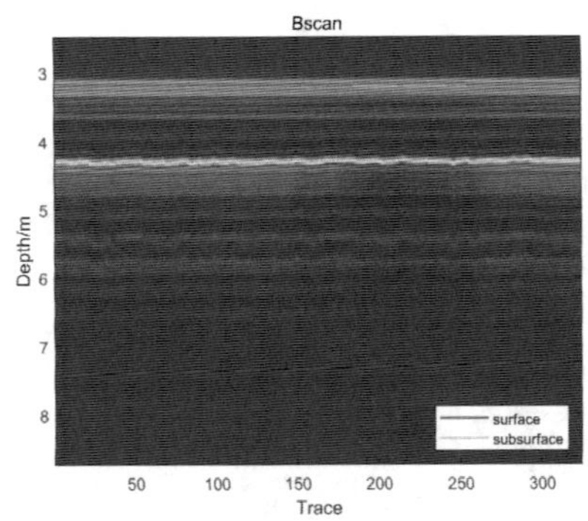

(d) Line 4

Figure 4. Test results of Bscan images of ice layers under different measurement line conditions.

4. Result

After verification through drilling experiments, the ice thickness was found to be approximately 62.8 cm. Taking Survey Line 2 as an example, we can calculate the ice thickness using GPR data by applying

Formula 4. The average thickness of the ice layer on this measuring line was 65.58 cm, with a standard deviation of 1.042 cm, and the error rate was 4.4%. If the direct wave propagation time is not considered, the average ice thickness measured by this measuring line was 55.44 cm, and the error rate increased to 11.7%. It can be clearly seen from the results that the revised formula significantly reduces the error rate. The relationship between the ice thickness calculated by the model and the ice thickness measured by drilling experiments is illustrated in Figure 5.

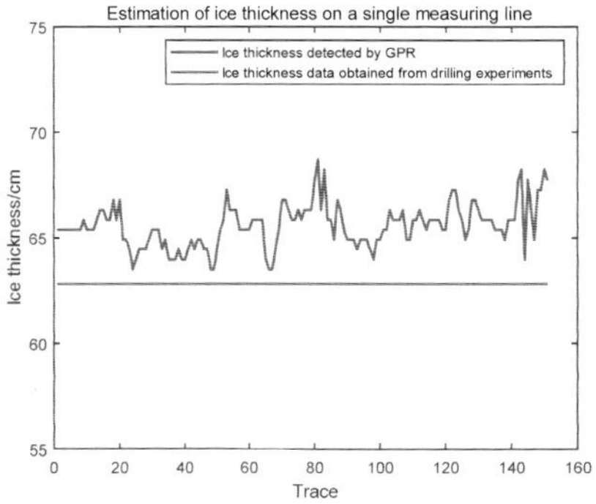

(a): Ice thickness measured by GPR

(b): Ice thickness measured by drilling method

Figure 5. Test results of Bscan images of ice layers under different measurement line conditions.

5. Conclusion

By comparing the ice thickness detection results from ground-penetrating radar and drilling exploration, it is found that the consistency between the two methods is high. However, due to the size of the antenna, we overestimate the centre distance of the antenna as the direct wave propagation path, which may result in the calculated ice thickness being slightly thicker than the actual situation. Additionally, issues such as motion jitter during measurement and the inhomogeneity of the ice sheet's dielectric constant can cause the measurement results to fluctuate relative to the actual ice sheet thickness.

In actual production and daily life, the results obtained by ground-penetrating radar and drilling methods can be considered approximately consistent. Due to its efficient and non-destructive characteristics, ground-penetrating radar can quickly detect the thickness of large areas of freshwater ice, providing technical support for construction and transportation on winter lake ice in severe cold areas.

6. References

[1] Karnena R K and Maiti S An efficient full wave model for investigating layered media using bi-static ground penetrating radar 2023 *J. Ieee Transactions On Antennas and Propagation* **71(5)** 4618-4623

[2] Smitha N G, Ullas Bharadwaj D R, Abilash S, Sridhara S N and Singh V Effect of ground undulation and mounted vehicle velocity variation on stepped frequency continuous wave gpr data 2018 *J. Exploration Geophysics (Melbourne)* **49(1)** 81-88

[3] Xu Y, Hua D, Wang J and Li J Research on application of ground penetrating radar in road inspection 2021 *J. Jop Conference Series. Earth and Environmental Science* **781(2)** 22019

[4] Barone I, Rossi M, Deiana R and Mazzariol A Evidence of attenuation and interference phenomena in GPR signals for archaeological application 2021 *C. 2021 11th International Workshop on Advanced Ground Penetrating Radar (IWAGPR)*

[5] Ullah F, Su L, Alam M, Cheng L, Rehman M U, Asghar A and Hussain G Landslide stability investigation and subsurface deformation mapping by optimizing low-frequency gpr: a mega rainfall susceptible landslide case study (gilgit baltistan, pakistan) 2022 *J. Bulletin of Engineering Geology and the Environment* **81(9)**

[6] Widodo W, Azimmah A and Santoso D 2018 Exploring the japan cave in taman hutan raya djuanda, bandung using gpr 2018 *J. Journal of Environmental and Engineering Geophysics* **23(3)** 377-381

[7] Fedorov M P, Fedorova L L, Savvin D V, Prudetskaya T P and Omelyanenko A V Investigation of ice jams using the gpr method 2018 *C. 2018 17th International Conference on Ground Penetrating Radar (GPR)*

[8] Grab M *et al* Ice thickness distribution of all swiss glaciers based on extended ground-penetrating radar data and glaciological modeling2021 *J. Journal of Glaciology* **67(266)** 1074-1092

[9] Leger E, Dafflon B, Soom F, Peterson J, Ulrich C and Hubbard S Quantification of arctic soil and permafrost properties using ground-penetrating radar and electrical resistivity tomography datasets 2017 *J. Ieee Journal of Selected Topics in Applied Earth Observations and Remote Sensing* **10(10)** 4348-4359

[10] Bukin A V, Sugak V G, Reznichenko N G, Drobna O V, Silaev Y S and Deineka I I Using sfcw gpr to search for buried objects 2020 *C. 2020 IEEE Ukrainian Microwave Week (UkrMW)*

[11] Fu H, Liu Z, Guo X and Cui H Double-frequency ground penetrating radar for measurement of ice thickness and water depth in rivers and canals: development, verification and application 2018 *J. Cold Regions Science and Technology* **154** 85-94

[12] Kowalczyk S, Lejzerowicz A and Kowalczyk B 2018 Groundwater table level changes based on ground penetrating radar images: a case study 2018 *C. 2018 17th International Conference on Ground Penetrating Radar (GPR)*

[13] Ali E, Pramudita A A and Arseno D Concrete thickness measurement model for gpr 2019 *C. 2019 IEEE Conference on Antenna Measurements & Applications (CAMA)*

[14] Jannah S, Pramudita A A and Suratman F Y Experiment of fmcw radar for small displacement detection using vna 2021 *C. 2021 International Conference on Radar, Antenna, Microwave, Electronics, and Telecommunications (ICRAMET)*

Journal of Physics: Conference Series 2887 (2024) 012038 doi:10.1088/1742-6596/2887/1/012038

Ground-penetrating radar dual-parameter full waveform inversion based on finite domain method

B Li[1], Z Fang[1], D Feng[1] and X Wang[1]

1 The School of Geosciences and Info-Physics, Central South University, Changsha, 410083 China (libingchao@csu.edu.cn)

Abstract. Ground penetrating radar (GPR) is a crucial method for shallow geophysical exploration method. GPR full waveform inversion (FWI) utilizes all available information including dynamics and kinematics, and theoretically achieving the highest imaging accuracy. However, a significant challenge for FWI is its low computational efficiency, which fails to meet the real-time requirements in actual exploration. To address this issue, we propose a finite domain method (FDM) based on the attenuation characteristic of high frequency GPR electromagnetic waves. This approach limits the calculation to a finite domain, thereby mitigating the excessive computational burden. We determine the size of the finite domain from absorption attenuation, wavefront diffusion, and scattering loss. Numerical experiments demonstrate that compared with traditional FWI, FWI based on FDM reduces the calculation time by more than 5 times.

1. Introduction

GPR is a non-destructive detection technology by its simple operation, high resolution and strong anti-interference ability. In recent years, GPR has been widely used in geological engineering, shallow exploration, and various other fields [1], [2]. Currently, the primary quantitative imaging methods for GPR include velocity analysis, migration, tomography, and FWI. While the first three methods utilize only part of the information of the data, the FWI uses all the information including dynamics and kinematics, thereby theoretically has the highest resolution.

The FWI was first proposed in the field of seismic [3], [4]. Drawing inspiration from seismic FWI, many scholars have explored the GPR FWI and applied to crosshole data [5], [6], [7], [8]. In comparison to crosshole GPR, the FWI of on-ground data is restricted by the measurement method, the inversion is more ill-posed and heavily reliant on the initial model. To enhance the inversion accuracy and efficiency for on-ground data, FWI can be optimized starting from regularization, objective function, and data encoding, etc [9], [10], [11], [12]. Regardless of whether it involves crosshole GPR FWI or on-ground GPR FWI, a significant challenge remains: the long calculation time and high memory usage, which limit its application in exploration.

The airborne electromagnetic method (AEM) typically employs a towed helicopter with the transmitter and receiver [13], and is similar to GPR in terms of the measurement method. Most effective signals originate from the induced current within a size-limited region beneath the coils, which is defined as the footprint [14]. Calculating areas beyond this footprint is inefficient, as the EM field intensity there is negligible compared to the effective signal, leading to a significant waste of computational resources [15].

Inspired by the footprint in the AEM, we have integrated the attenuation characteristics of high-frequency GPR electromagnetic waves to develop FDM. The traditional method is to excite the source at one point and simulate the propagation of the radar wave throughout the entire model. In contrast, FDM focuses on simulating radar waves within a defined 'finite domain' around the source, which will significantly reduce the calculation time. In this paper, we outline the methodology for determining the size of the finite domain and present the FWI based on finite domain. Additionally, we conduct numerical experiments to compare traditional FWI and FWI based on FDM.

2. Method

2.1. The finite domain

The wave equation of the electric field strength $\boldsymbol{E}$ can be written as:

$$\nabla^2 \boldsymbol{E} + k^2 \boldsymbol{E} = 0 \tag{1}$$

The solution is:

$$\boldsymbol{E} = E_0 e^{-jkr} = E_0 e^{-\alpha r} e^{-j\beta r} \tag{2}$$

where j is the imaginary unit, r is the distance, β is the phase coefficient, and α is the attenuation coefficient.

Therefore, the relationship between amplitude and the propagation distance r (absorption attenuation) is:

$$A = A_0 e^{-\alpha r} \tag{3}$$

where A is the amplitude, A_0 is the original amplitude.

Radar wave spread in an approximately spherical form, so the amplitude of the field intensity is inversely proportional to the distance r (wavefront diffusion):

$$A = C \cdot A_0 / r \tag{4}$$

where C is a coefficient.

The reflection and refraction also lead to amplitude changes during propagation (scattering loss). We assume that the wave incident vertically on the interface, and the refraction coefficient is:

$$T = 2Z_2 / Z_1 + Z_2 \tag{5}$$

where Z is the wave impedance.

In the 2D case, it is necessary to calculate the FDM boundaries in both directions on either side of the antenna. However, most models lack analytical solutions. To simplify the problem and avoid unnecessary costs, we employ a relatively simplified method.

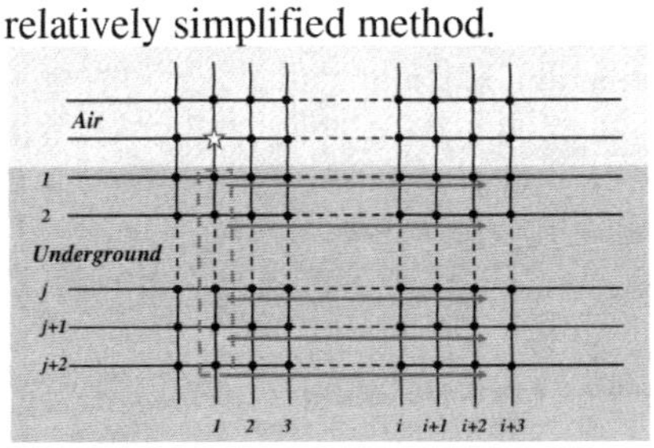

Figure 1. Calculation of the boundary of the finite domain.

As shown in figure 1, the yellow asterisk represents the source. The grid points in the same column as the source (the red dotted box) serve as the starting points (virtual source) for calculations in the horizontal direction.

Firstly, we calculate the absorption attenuation from all virtual sources to the grid (j, k) and take the maximum value:

$$D_k = \max_j \exp(\sum_{i=1}^{k} \alpha_{j,i} \Delta x) \tag{6}$$

where Δx is the space interval, and $\alpha_{j,i}$ is the attenuation coefficient.

Wavefront diffusion only involves distance, and can be calculated by:

$$W_k = \sqrt{d_{sr}} / \sqrt{d_k} \tag{7}$$

where d_{sr} is the offset, d_k is the lateral distance from the virtual source to the kth grid.

Scattering loss in underground media is quite complex, and most of the actual detected media exhibit the same permeability. For simplicity, we only consider the loss caused by radar waves as they pass through the surface:

$$T_{air} = 2\sqrt{\varepsilon_{air}} / \left(\sqrt{\varepsilon_{air}} + \sqrt{\varepsilon_1}\right) = 2/1 + \sqrt{\varepsilon_1} \tag{8}$$

where ε_{air} is the relative dielectric constant of air and equals to 1 approximately, ε_1 is the relative dielectric constant of the surface medium.

We multiply D_k, B_k, and T_{air} to obtain the attenuation ratio, and the boundaries B_k need to satisfy:

$$B_k = D_k \cdot W_k \cdot T_{air} \leqslant \eta \tag{9}$$

where η is a constant.

2.2. GPR FWI based on FDM

The GPR FWI utilizes observation data to reconstruct subsurface parameters, ensuring that the simulation data and the observed data fit optimally, and the objective function is defined as:

$$\varphi = \frac{1}{2}\sum_{s=1}^{N_s}\sum_{p=1}^{N_p}\int_0^T \left[d_{cal}\left(\boldsymbol{m}, \boldsymbol{r}_p, t|s\right) - d_{obs}\left(\boldsymbol{r}_p, t|s\right)\right]^2 dt \tag{10}$$

where N_s is the number of sources, N_p is the number of receivers, $\boldsymbol{r}_p$ is the space coordinate vector of the receiver, d_{obs} is the observed data, d_{cal} is the simulation data, t is the time, T is the duration of the observed data and $\boldsymbol{m}$ is the model parameter vector.

The gradient of the objective function can be obtained by the adjoint state method [16]:

$$\boldsymbol{g}^{\varepsilon} = \sum_{i=1}^{N_s}\int_0^T \boldsymbol{E}_i^* \frac{\partial \boldsymbol{E}_i}{\partial t} dt \qquad \boldsymbol{g}^{\sigma} = \sum_{i=1}^{N_s}\int_0^T \boldsymbol{E}_i^* \boldsymbol{E}_i dt \tag{11}$$

where $\boldsymbol{g}$ is the gradient, and * denotes the adjoint.

The FWI based on the FDM calculates the observed data of the finite field $\boldsymbol{m}_i$ for each source:

$$\varphi_{FDM} = \frac{1}{2}\sum_{s=1}^{N_s}\sum_{g=1}^{N_g}\int_0^T \left[d_{cal}\left(\boldsymbol{m}_i, \boldsymbol{r}_g, t_i|s\right) - d_{obs}\left(\boldsymbol{r}_g, t_i|s\right)\right]^2 dt \tag{12}$$

The gradient of the φ_{FDM} still can be obtained by the adjoint state method, we just need to assemble the gradients of every $\boldsymbol{m}_i$ into the gradient of $\boldsymbol{m}$ by position.

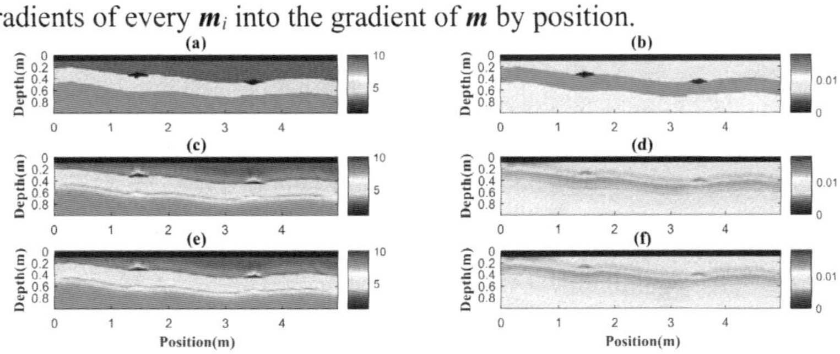

Figure 2. The FWI based on FDM. (a) and (b) are the true models; (c) and (d) are the inverted results with traditional FWI; (e) and (f) are the inverted results with FWI based on FDM

3. Experiment

We constructed a model as illustrated in figure. 2(a) and (b). The grid size is $nx \times nz = 500 \times 100$ with a space interval $ds = 0.01$ m. The time window is 16 ns, with the time interval $dt = 0.02$ ns. We evenly distributed 96 observation points in the air with the offset of 0.08m, and choose the 900MHz ricker wavelet as the source. In this experiment, we set $\eta = 0.2$.

The results of traditional FWI are shown in figure 2(c) and (d), and the results of FWI based on FDM are shown in figure 2(e) and (f). It is evident that the FWI based on FDM (FD-FWI) is almost consistent with the traditional FWI, and both of them successfully obtain the information of anomalous bodies and strata. Table 1 summarizes the reconstruction error and calculation time for both methods. The reconstruction error is assessed using structural similarity (*SSIM*); values closer to 1 indicate greater similarity between the two images and better inversion performance. Notably, compared with traditional FWI, FD-FWI exhibits a slight increase in reconstruction error (a decrease in *SSIM* value), yet it reduces computation time by more than 5 times. This experiment shows the advantage of the FWI based on FDM.

Table 1. The Reconstruction error and calculation time with different methods

Method	*SSIM* (ε_r)	*SSIM* (σ)	Time (min)
FWI	0.3984	0.6553	70.0964
FD-FWI	0.3955	0.6547	12.7295

4. Conclusion

Inspired by the footprint in AEM and integrating the attenuation characteristics of the GPR wave, we propose FDM for FWI of GPR data. This method effectively reduces the calculation time by constraining the forward modeling area. We determine the size of the finite domain by considering three factors: absorption attenuation, wavefront diffusion, and scattering loss. Numerical experiments demonstrate that, compared to traditional FWI, FWI based on FDM reduces the time by more than 5 times while maintaining accuracy. As the length of the model increases, the advantage of FDM will be more pronounced. Furthermore, FDM shows promise for 3D inversion, which will be the focus of our future work.

References

[1] Zhang L, Xu Y, Chen R, Wong H, Lai J, Meng X, Chen R, Zeng Z, Feng X and Liu C 2022 *Icarus*, **386**
[2] Feng D, Liu Y, Wang X, Zhang B, Ding S, Yu T, Li B and Feng Z 2023 *IEEE Trans. Geosci. Remote Sens.* **61** 1-15
[3] Tarantola A. *Geophysics* 1984 **49** 1259-1266.
[4] Pratt R, Shin C and Hick G 1998 *Geophys. J. Int.* **133** 341-362.
[5] Ernst J, Maurer H, Green A and Holliger K 2007 *IEEE Trans. Geosci. Remote Sens.* **45** 2807-2828
[6] Meles G, Greenhalgh S, Van der Kruk J, Green A and Maurer H 2012 *J. Appl. Geophys.* **78** 31-43
[7] Klotzsche A, Vereecken H and van der Kruk J 2019 *Geophysics* **84** H13-H28
[8] Klotzsche A, Vereecken H and van der Kruk J 2019 *J. Appl. Geophys.* **170** 103823
[9] Lavoué F, Brossier R, Métivier L, Garambois S and Virieux J 2014 *Geophys. J. Int.* **197** 248-268
[10] Ren Q 2018 *J. Appl. Geophys.* **151** 186-193
[11] Feng D, Cao C and Wang X 2019 *IEEE Trans. Geosci. Remote Sens.* **57** 9450-9465
[12] Feng D, Li B, Wang X, Ding S, Tai X, Cai L and Su X 2023 *IEEE Trans. Geosci. Remote Sens.* **61**
[13] Yin C, Ren X, Liu Y, Qi Y, Qiu C and Cai J 2015 *Geophysics* **80** W17-W31
[14] Liu R, Guo R, Liu J and Liu Z *IEEE Trans. Geosci. Remote Sens.* 2019 **16** 1809-1813
[15] Yin C, Huang X, Liu Y and Cai J 2014. *Geophysics* **79** E243-E254.
[16] Plessix R 2006 *Geophys. J. Int.* **167** 495-503

Preliminary results of the interior features inferred from radargrams, Ulanhada Volcano, Inner Mongolia, China

S Dai[1], **Y Su**[1], **C Y Ding**[2], **X P Xue**[1], **C Li**[1] and **X Y Zhu**[1]

[1] Key Laboratory of Lunar and Deep Space Exploration, National Astronomical Observatories, Chinese Academy of Sciences, Beijing 100101, China
[2] Institute of Moon-base Exploration and Observation, Institute of Advance Study, Shenzhen University, Shenzhen 518060, China

Abstract. The Ulanhada Volcano Group is located in Inner Mongolia, China. It is an important Quaternary volcanic geological relic with different types of volcanic cones, craters, lava platforms, and other structures. Ground penetrating radar (GPR) can play an important role in volcanic exploration. We used pulse EKKO radar combined with 100MHz and 250MHz antennas to detect the No.1 Redhill volcanic. The results indicate that there are significant differences in reflection characteristics in radargrams of different frequencies, which can be used to analyze the material composition, eruption period, and stratigraphic coverage relationship of volcanic eruptions. At the same time, an accurate analysis of the region will also be used as an evaluation basis for the field scientific verification experiments of China's Asteroid Core Scan Radar (ACSR) and Lunar Penetrating Radar (LPR).

1. Introduction

Ground Penetrating Radar (GPR) is a very useful non-destructive geophysical exploration technique in volcano exploration. It uses high-frequency electromagnetic wave pulses to penetrate the ground and reflect to show the underground structure and material distribution in the form of images [1]. It can be used to reveal the internal structure and characteristics of volcanoes, monitor volcanic geological changes, assess disaster risk, and more [2,3]. With the continuous development and improvement of technology, the application of GPR in volcano detection will be more extensive and deeper. At the same time, volcanoes with obvious changes in relative composition and clear bedding structure are also good objects for terrestrial planet comparison.

In this paper, the geological settings of survey sites are presented in Chapter 2, and the survey design and data acquisition are described in Chapter 3. In Chapter 4, we showed the radargrams and the discussion, and in Chapter 5 concluded.

2. Geological settings of survey sites

The Ulanhada Volcanic Group is volcanic in China, located in the Chahar Right Wing Rear Banner Area of Ulanqab City, Inner Mongolia Autonomous Region. The volcanic complex is home to several volcanoes, some of which still retain their intact morphology, and the crater is visible. The formation of volcanic groups spans a wide range of ages, from the Pliocene to the Pleistocene [4]. The geological structure mainly includes lava landforms, such as lava flows, lava waterfalls, lava tunnels, etc. At the same time, the pyroclastic debris produced during volcanic eruptions has also formed unique pyroclastic landforms, such as pyroclastic flows, volcanic ash falls, etc. The location of the research area can be seen in figure 1.

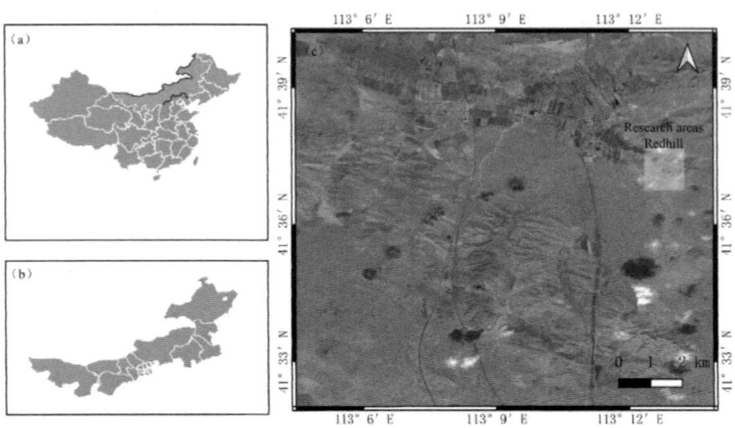

Figure 1. (a) and (b) is the volcano location in Inner Mongolia, China, (c)is the satellite image of the Ulanhada volcano areas.

3. Survey design and data acquisition

The detection is selected to be carried out at the end of December. Due to the dry and cold climate in winter, the attenuation and dispersion in the detection can be effectively reduced. The survey used a Pulse EKKO Pro™ bistatic GPR system, with the Trimble R8 GNSS™ as a real-time kinematic global positioning system (RTK GPS) which was used to acquire location and elevation information. A DJI Mavic 3 Pro™ mapping drone was used to collect high-resolution aerial multi-view image data of the target area.

GPR measurements were performed in reflection mode (single offset), operating at 100 MHz and 250 MHz and equipped with an odometer. Traces were acquired every 0.5m at 100 MHz and every 0.05 m at 250 MHz, using a time resolution of 0.5 ns with a time window of 500 ns.

There were two volcanoes in the exploration area. Fig 2(a) shows a 3D model of the detection area which was produced using high-resolution imagery collected by the mapping drone. The survey line SL1 between the two volcanoes, the survey line SL2 on the east side of the volcano named Small Redhill, and the survey lines SL3, SL4, and SL5 on the west side of the volcano named Big Redhill, which can be seen in Fig 2(b).

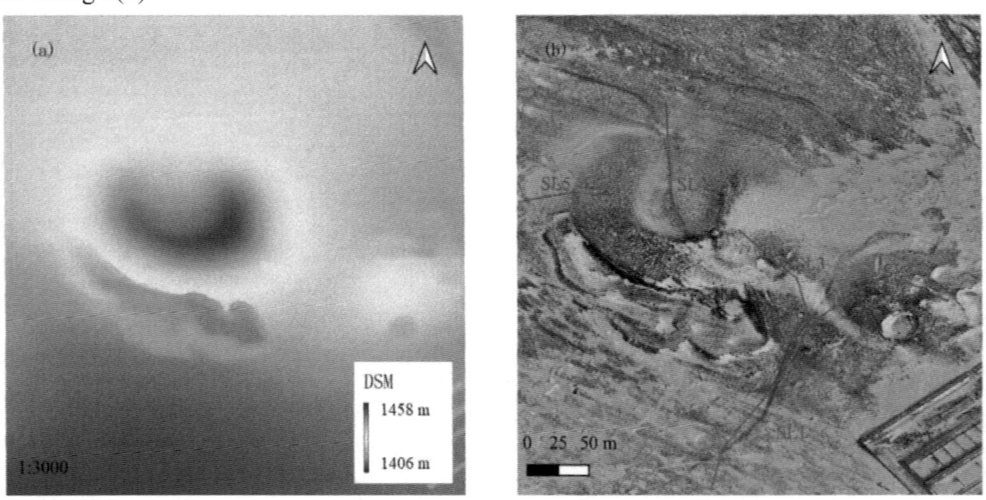

Figure 2. 3D model of the survey area and the survey liner.

Journal of Physics: Conference Series **2887** (2024) 012039 doi:10.1088/1742-6596/2887/1/012039

4. Result and discuss

4.1. Small Redhill

The diameter of the cone of a small Redhill is about 120 meters, the height is about 15 meters. It was a relatively complete volcanic cone, the centre of the volcanic cone was a relatively flat platform, and there were a small number of excavation pits around the cone. After terrain correction, the radar images are shown below using 100MHz and 250MHz antennas along the SL2 survey line across the mountain. There are fewer stray echoes in the centre of the mountain, indicating that the electromagnetic waves in this region propagate in a more homogeneous medium, which can be judged to be the cooling lava of the volcanic eruption channel. Significant stratification can be seen at the elevations of 1420 meters and 1413 meters on the left side of the volcano, indicating that the volcano has multiple eruptions, and the phenomenon of sputters overlapping with lava flows.

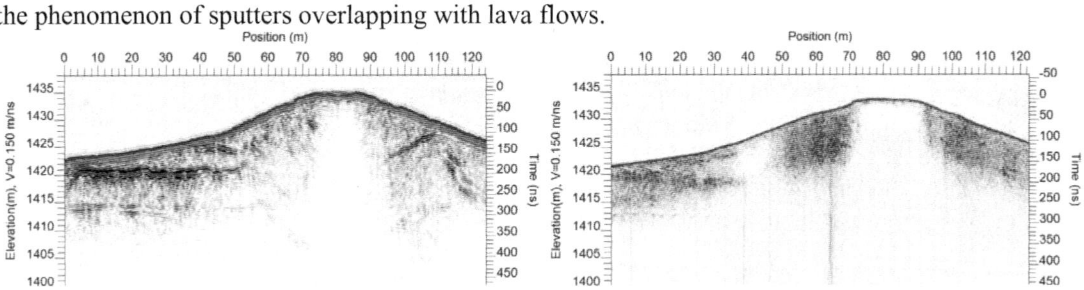

(a) 100MHz@SL2 (b) 250MHz@SL2

Figure 3. Radargrams of Small Redhill

4.2. Big Redhill

The diameter of the cone of Big Redhill is about 250 meters, the height is about 35 meters. It was a horseshoe-shaped volcanic cone, the opening facing north, the center of the volcanic cone is a relatively flat platform, and there is an obvious excavation in the south of the cone, exposing the black-red volcanic cinder. The terrain correction radar images of 100MHz and 250MHz centre frequency were obtained in different azimuths along the survey lines SL3, SL4, and SL5 as follows in Figure 4. In the eastern part of the mountain, near Small Redhill, obvious stratification can be seen at the altitudes of 1420 meters and 1413 meters in Figures 4(a), and 4(b). However, no significant stratification was found on both the western and northern survey lines in Figures 4(c), and 4(d). Except for a small amount of clutter at the top, the northern survey line is weakened by coastal uplifting. The weakened clutter in the image, compared with the topography and landform, shows that this area is the location of the lava flow after a volcanic eruption.

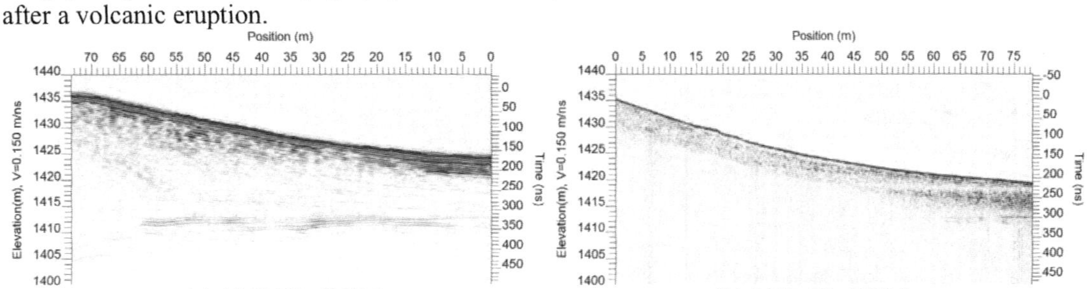

(a) 100MHz@SL3 (b) 250MHz@SL3

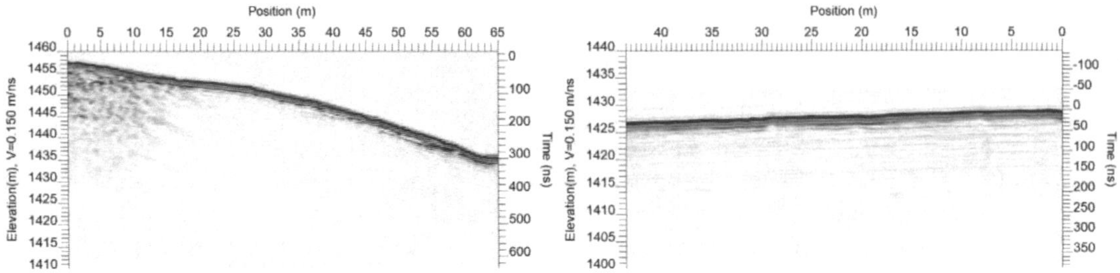

(c) 100MHz@SL4 (d) 100MHz@SL5

Figure 4. Radargrams of Big Redhill

4.3. Surrounding field

The 250MHz radar-sounding results were from the middle of the two mountains to the south, as shown in Figure 5(a). A 100MHz surrounding flown survey line SL0 can be seen in Figure 5(b). The stratification at an altitude of 1413 meters is visible. Comparing the volcanic mountain survey lines, it can be considered that the topography of this layer is a continuation of the strata at the same location in the images mentioned above. It is known that the volcano's loose sputter covers an area of about 100 meters on the surface regolith.

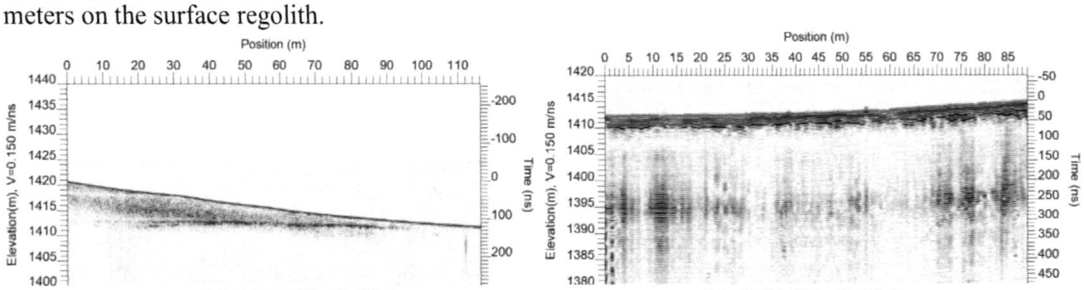

(a) 250MHz@SL1 (b) 100MHz@SL0

Figure 5. Radargrams of the surrounding field

5. Conclusion

We used pulse EKKO radar with 100 MHz center frequency and 250 MHz center frequency antennas to detect one of the No. 1 volcanic cones in the Ulanhada volcanic group. The results show significant differences in the reflection characteristics in the radar images of different frequencies, which can be used to analyze the material composition, eruption period, and stratum cover relationship of volcanic eruptions. The region has also conducted scientific verification tests of the Lunar Penetrating Radar (LPR) and Asteroid Core Scan Radar (ACSR) for deep space exploration. A precise analysis of the area will also be used to evaluate the performance of China's follow-up planetary detection radars.

6. Acknowledgments

This research was funded by the National Natural Science Foundation of China under Grant 62227901 and 12073048.

7. References

[1] H.M. Jol, 2009. *Ground Penetrating Radar. Theory and Applications*, Elsevier Science.

[2] D. Gómez-Ortiz, et.al. 2006. Characterization of volcanic materials using ground penetrating radar: A case study at Teide volcano (Canary Islands, Spain), *Journal of Applied Geophysics*, 59(1),63-78.

[3] C. Gomez, et.al. 2023. Formation Processes of Gully-side Debris-Cones Determined from Ground-Penetrating Radar (Mt. Unzen, Japan), *Journal of Applied Geophysics*, 209: 104919.

[4] Z D Bai, et.al. 2008. Quaternary Volcano cluster of Wulanhad, Right-back-banner, Chahaer, Inner Mongolia(in Chinese). *Acta Petrologica Sinica*, 24(11):2585-2594.

Design of A Linear Array Holographic Subsurface Radar System

Wei Liu, Zhihua He, Tao Liu and Xiaoji Song
College of Electronic Science and Technology, National University of Defense Technology, No.109 Deya Road, Changsha, China (1347339281@qq.com)

Abstract. This work introduces a new holographic subsurface radar (HSR) system designed with a cross-arranged linear array antenna. In non-destructive applications, holographic subsurface radar utilizes continuous wave, high-density spatial sampling, and a wide synthetic aperture to achieve high azimuth resolution. However, this results in a low data acquisition speed, which limits the detection efficiency. In this work, we designed a cross-arranged linear array antenna, where the transmitting- and receiving-antennas are layered on two parallel baselines. In this way, the space between the neighboring sampling points can still meet the spatial sampling criterion. The use of the electric-switching antenna array reduces 2-dimensional mechanical scanning to a 1-dimensional scan, significantly improving data acquisition speed while maintaining sufficient spatial sampling density. The prototype of the system was also detailed in the paper. The experiments show that the newly designed HSR system can improve detection efficiency by more than 10 times while maintaining imaging resolution.

Keywords: Subsurface Penetrating Radar; Holographic Imaging; Cross-arranged Linear Array; Non-destructive Detection.

1. Introduction

Holographic subsurface radar (HSR) achieves high azimuth resolution due to the use of high-frequency continuous wave and wide synthetic apertures. This makes it a promising technique in non-destructive testing [1-3] and evaluation, aiming at shallow-depth, high-resolution subsurface exploration, such as the detection of concealed weapons on the body and biomedical imaging [4,5]. Meanwhile, the HSR system avoids complex ultra-wideband microwave modules, leading to a more compact system design.

To solve the synthetic aperture scanning problem, researchers have proposed various system patterns. The Bauman Moscow State Technical University has developed a HSR system named RASCAN [6-8]. This system consists of a radar probe operating at C-band and a control unit. By manually moving the probe across the detecting surface along regular scanning lines, the RASCAN can generate a 2-D image of subsurface targets with a planar resolution of 1.5 cm. With the RASCAN, the team successfully detected pipes and cables marked on the floor. Hangzhou Dianzi University reported a 2-D reciprocating mechanical scanning system that completes a two-dimensional plane scan by moving the platform. The system operates within a frequency range of 6.4GHz to 6.8GHz and offers imaging resolution comparable to that of the RASCAN system. The National University of Defense Technology has developed a prototype holographic subsurface radar that utilizes a polar coordinate scanning mechanism and K-band microwave technology, achieving a resolution of 2.5m and can complete a scan in approximately 10 seconds [9].

The traditional holographic subsurface radar systems consist only a pair of transceiver antennas. To meet the requirements for high-density spatial sampling, it takes a long time to complete one scan, which limits the imaging efficiency. In this work, we designed a linear array antenna for HSR, where the transmitting- and receiving-antennas are layered on two paralleled base lines and arranged in an interleaved manner. This arrangement maintains spatial sampling standards between adjacent sampling points. RF switches were introduced to control the transmitting- and receiving-antennas, achieving time-shift switching. The use of electronically switched antenna arrays to reduce 2D mechanical scanning to 1D scanning, which significantly improves data acquisition speed. The paper also provides a detailed introduction to the prototype of this system and validates the feasibility and effectiveness of this technical scheme through penetration imaging experiments.

2. System Design

2.1. Principle of Antenna Layout

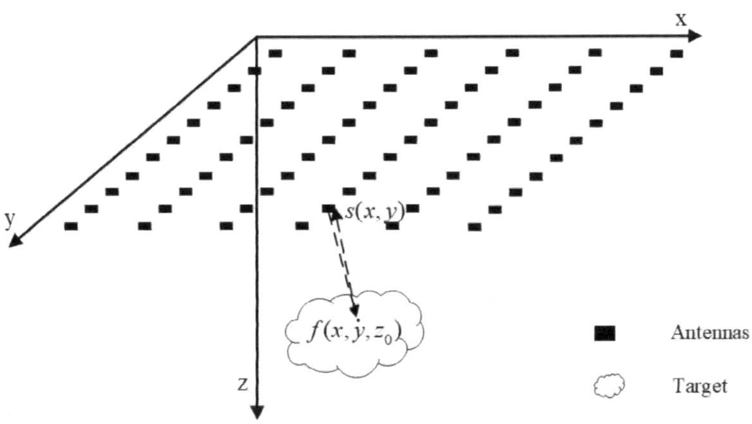

Figure 1. The regime of imaging

Assume that the echo of HSR is $s(x, y)$, where x and y are the spatial coordinates, and the target locates at the depth of z_0, then the holographic imaging algorithm can be expressed by (1):

$$f(x, y, z_0) = IFT_{2D}\left\{\left[FT_{2D}(s(x, y))\right]\exp(jk_z z_0)\right\} \tag{1}$$

where $f(x, y, z_0)$ is the imaging result at the given depth z_0, and k_z is the wavenumber of z-direction. The FT and IFT stand for Fourier transform and Inverse Fourier transform, respectively. Due to the requirement for a 2D Fourier transform, the spatial scanning must be uniformly distributed on a regular grid. [10] indicates that the sampling density must satisfy the Nyquist spatial sampling criterion to avoid aliasing effect, which means that the spatial sampling interval *dx, dy* should meet the following formula, in which k means wave-number, v means the velocity of electromagnetic wave in the medium, f is the working frequency, and λ indicates the wavelength.

$$dx \le \frac{\pi}{2k} = \frac{v}{4f} = \frac{\lambda}{4} \tag{2}$$

2.2. System Design

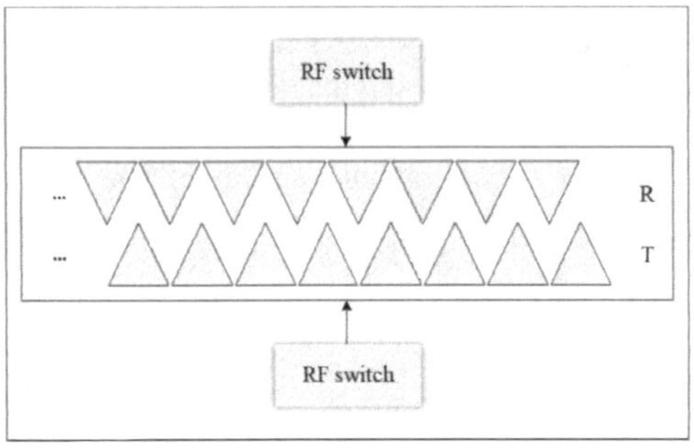

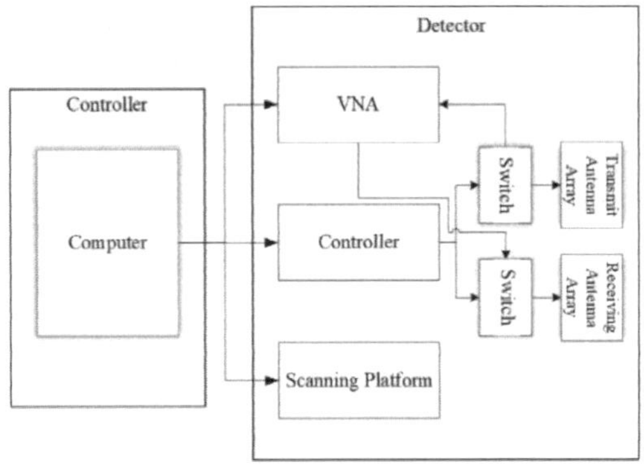

Figure 2. Design of a cross-arranged linear array. **Figure 3.** Components of the radar system.

The traditional linear antenna array is constrained by the physical dimensions of the antennas and cannot meet the principle of space sampling. And the MIMO system is also not applicable because of the near-field propagation of electromagnetic (EM) wave. By staggering the transmitting- and receiving- antennas, a cross-arranged linear array antenna was designed. As shown in Figure 2, the transmitting- and receiving- antennas are arranged on two parallel baselines. Each transmitting antenna forms a set of transceiver channels with its neighboring receiving antennas. We utilize a time-division multiplexing (TDM) mode to switch transmitting- and receiving- antennas by RF switches. The cross-arranged linear array layout maintains adequate spatial sampling density while reducing the size of the line array, and simplifying the system.

The entire system is shown in Figure 3, in which a computer was used to synchronize the vector network analyzer (VNA), the scanning platform, and the designed antennas array to accomplish the 2-D planar scanning. Hereby the VNA was used as the transmitter and receiver, which can be replaced by an RF transmitting and receiving devices or systems. The antenna array is time-division multiplexed with switching controlled by RF switches, which are controlled by a controller to perform 1-D electrical scanning. The scanning platform holds on targets and media to conduct 1-D scanning along the normal direction of the linear array.

3. Experiment

The evaluation of the rebars in the wall is in high demand, which needs a resolution of around 1~2cm and a penetrating depth of about 10cm. Based on the background, we choose the C-band microwave to design the experiment.

The experiment utilizes a total of 30 antennas, with each antenna operating at a frequency of 5.3 GHz, to create 29 transceiver channels reaching the spacing of the equivalent phase centers of 0.0125 m, which meets the principle of antenna layout. The detailed parameters are listed in the following table.

Table 1. System Configuration

Parameters	Settings
Transmit signal	Single-Frequency
Number of TR channels	29
Sampling interval	X: 0.01m, Y: 0.0125m
Frequency	5.3GHz
Scanning range	0.40m×0.35m

Before the measurement, we calibrated the inconsistencies of amplitude and phase among different channels by using a source in a known location [11,12]. A rebar was concealed beneath the surface of an ABS plastic plate with a thickness of 2.5 cm. To verify the effectiveness of the linear array holographic system, the standard measurement with a single pair of antennas is used. The imaging results of target 1 were shown in Figure 6 (b) and (c). It can be seen that the linear array holographic system gains imaging results that consistent with the standard imaging method using a pair of transmitter

Journal of Physics: Conference Series **2887** (2024) 012040 doi:10.1088/1742-6596/2887/1/012040

and receiver. Furthermore, Figure 6(a) presents the center profile of Figure 6 (b) and (c) indicated by the white dashed line, demonstrating the consistency of imaging results between the two systems.

To assess the imaging capability for complex-shaped targets, the 2nd measurement introduced a target like a * symbol, as shown in Figure 5 (b), and the imaging results are presented in Figure 7 (a) and (b).

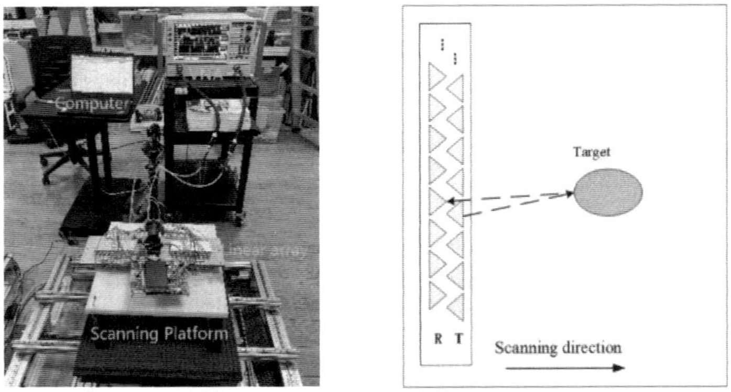

(a) Experimental scene (b) Scanning diagram of the system

Figure 4. Settings of the experiment

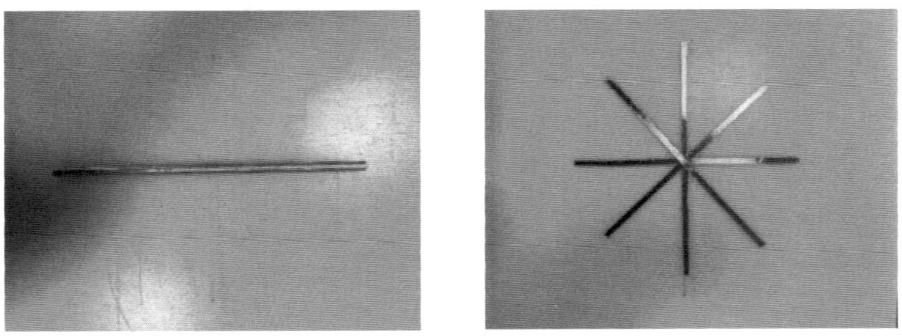

(a) Target 1 (b) Target 2

Figure 5. The deployment of targets

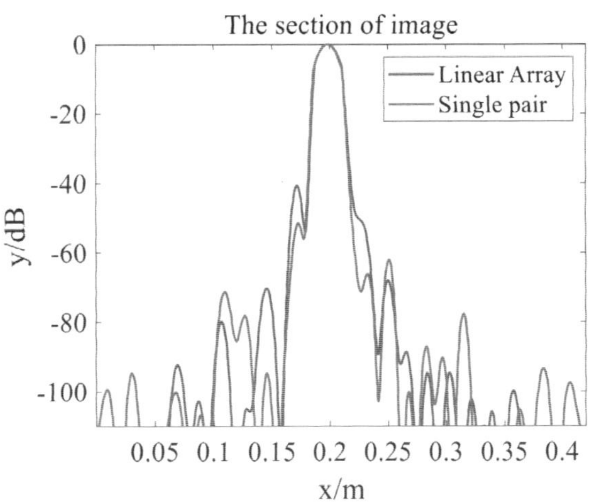

(a)1-D imaging

Journal of Physics: Conference Series **2887** (2024) 012040 doi:10.1088/1742-6596/2887/1/012040

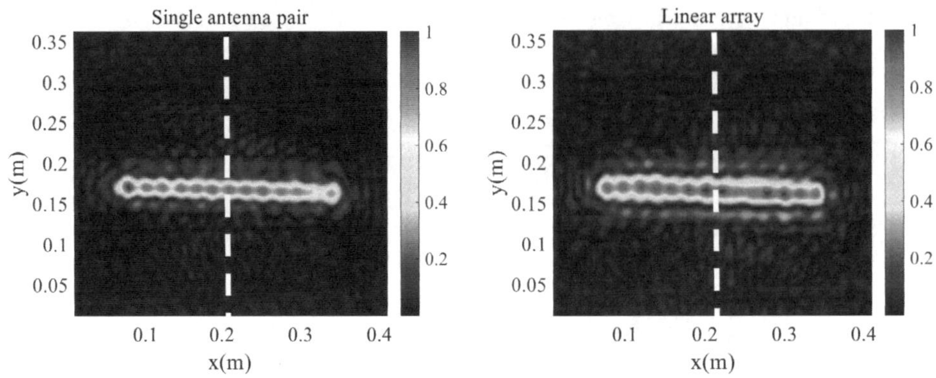

(b) 2-D imaging of single antenna pair (c) 2-D imaging of a linear array

Figure 6. The results of target 1

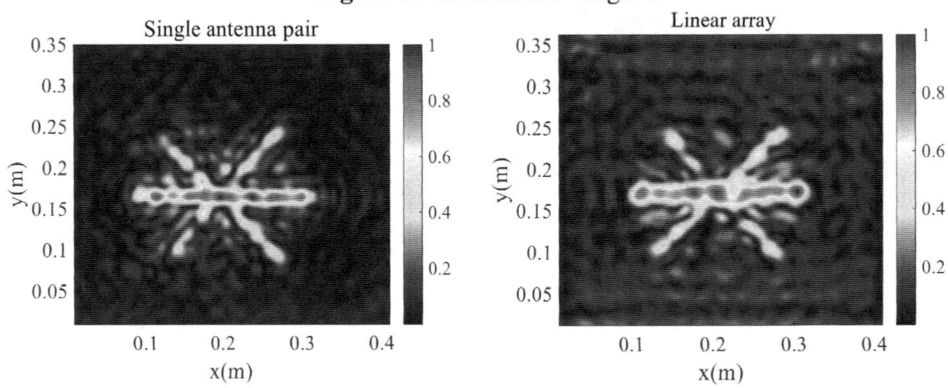

(a) 2-D imaging of signal antenna pair (b) 2-D imaging of a linear array

Figure 7. The results of target 2

The imaging results from both systems were generally consistent and the experiments show that the newly designed HSR system can improve detection efficiency by more than 10 times while maintaining imaging resolution. Note that the upper and down part of the image is incomplete resulted by the polarization effect.

4. Conclusion

In this paper, we introduce the design of a linear array holographic subsurface radar system. Through a comparison of imaging results with the antennas measurement of a single pair, it is suggested that the newly designed holographic subsurface radar system can significantly enhance detection efficiency and increase spatial sampling density while maintaining imaging resolution. The design shows promise in high-resolution subsurface exploration, non-destructive testing. Further enhancements are being pursued, particularly in optimizing the antenna design to tackle incomplete imaging results stemming from antenna polarization effects.

5. References

[1] Ivashov S I, Razevig V V, et al 2015 Microwave holography for NDT of dielectric structures *IEEE International Conference on Microwaves, Communications, Antennas and Electronic Systems (COMCAS).* pp 1-5

[2] Altuncu Y, Akduman I, et al 2007 Detecting and locating dielectric objects buried under a rough interface *Ieee Geoscience and Remote Sensing Letters* **4(2)** 251-255

[3] Ivashov S, Zhuravlev A, Chizh M, et al 2016 High resolution mw holographic system for NDT of dielectric materials and details *16th International Conference on Ground Penetrating Radar (GPR)*.pp 1-5
[4] Nikolova N K 2011 Microwave imaging for breast cancer *IEEE Microwave Magazine* **12(7)** 78-94
[5] Wang L, Simpkin R, et al 2013 Holographic microwave imaging for medical applications *Journal of Biomedical Science and Engineering* **06(08)** 823-833
[6] Chizh M, Zhuravlev A, et al 2020 Defects investigation in thermal insulation coatings with microwave imaging based on a 22 GHz holographic radar *NDT & E International: Independent Nondestructive Testing and Evaluation* **109** 102191
[7] Ivashov S I, Capineri L, et al 2021 Design and applications of multi-frequency holographic subsurface radar: Review and case histories *Remote Sensing* **13(17)** 3487
[8] Ivashov S I, Razevig V V, et al 2011 Holographic subsurface radar of RASCAN type: Development and applications *IEEE Journal of Selected Topics in Applied Earth Observations and Remote Sensing* **4(4)** 763-778
[9] Song X, Su Y, et al 2016 *16th International Conference on Ground Penetrating Radar (GPR)*. pp 1-4
[10] Sheen D M, Mcmakin D L, et al 2001 Three-dimensional millimeter-wave imaging for concealed weapon detection *IEEE Transactions On Microwave Theory and Techniques* **49(9)** 1581-1592
[11] See C M S 1995 Method for array calibration in high-resolution sensor array processing *IEEE Transactions On Microwave Theory and Techniques* **142(3)** 1581-1592
[12] Ng B C and Ser W 1992 Array shape calibration using sources in known locations *[Proceedings] Singapore ICCS/ISITA92. IEEE.* pp 836-840

An identification method of bonding state in ETIS based on GPR

C Wang[1], J Wang[2,3], M Liu[1], X Li[1], J Gong[2], X Yang[2,3], T Lan[2,3]*

[1] Beijing Building Research Institute Co., Ltd.of CSCEC, Beijing, China
[2] Chongqing Innovation Center, Beijing Institute of Technology, Chongqing, China
[3] School of Information and Electronics, Beijing Institute of Technology, Beijing, China
[4] This work was supported by the CSCEC-PT-015 CSCEC City Renovation and Intelligent Operation Engineering Research Center (Building Health Diagnosis and Treatment).
* tlan@bit.edu.cn

Abstract. The external thermal insulation system (ETIS) often has safety quality problems caused without debonding, and the classical debonding detection methods can not achieve high precision and high efficiency safety detection at the same time. Ground penetrating radar (GPR) can be one of the most efficient tools to identify the bonding state. In this paper, a methodology for bonding state detection is presented. Firstly, a modeling and simulation method for ETIS is designed, which is closer to the real form of the adhesive shape. Secondly, a 3D frequency wave-number domain filter is used to remove the low-rank direct-coupled waves and the low frequency components, then a matched filter is used to estimate the time delay arrival and identify the adhesive shape in ETIS. Finally, the effectiveness of the algorithm for the recognition of bonding state is verified by numerical results and experiments.
Keywords: External thermal insulation system (ETIS), Ground Penetrating radar(GPR), Frequency wave-number, Matched filter

1. Introduction

The utilization of external thermal insulation system (ETIS) in construction projects has become prevalent due to its capacity for enhancing building insulation performance and energy conservation.[1] In order to prevent shedding, the identification of the adhesive area of the ETIS is essential. Thus, Ground penetrating radar detection (GPR is one of the most important methods, which identifies the internal structure and characteristics of media by using high frequency electromagnetic waves.0 There are many disadvantages in other classic detection methods, such as tapping method and ultrasonic method, which have low safety coefficient and low detection efficiency, and infrared thermal image detection method is difficult to obtain high precision detection. The utilization of GPR for the identification of binding state has the advantages of high safety coefficient, easy operation, and high-efficiency detection.

This paper presents a methodology of detecting the bonding state in ETIS. Firstly, a method for modelling irregular objects is investigated. This approach to modelling and simulating adhesive area enhances the fidelity of the simulation model to the physical reality. Then, two filters are designed for suppressing the clutter and increasing the accuracy of recognition[3], including frequency-wavenumber filter and matched filter. [4] [4] Finally, the identification results of the adhesive area of the insulation board is validated through simulation and experimentation.

2. Simulation

The structure of ETIS is typically composed of four components: a plastering layer, an insulation layer, a bonding layer, and the wall. The bonding layer contains a part of concrete mortar as the adhesive between the insulation layer and the wall, while the remaining space is filled with air.

In order to simulate real situations, we designed a model with the irregular shape object that approximates the cylinder. The modeling steps are as follows:

Step1: Assume that the center of the target is fixed, and generate a random sequence $\boldsymbol{a}$ of length n which represent the polar axis in the polar coordinate system;

Step2: Extend $\boldsymbol{a}$ to $\boldsymbol{b}$, where $\boldsymbol{b} = [\boldsymbol{a}, \boldsymbol{a}]$, interpolate $\boldsymbol{b}$ nonlinearly to obtain $\boldsymbol{c}$, where the length of $\boldsymbol{c}$ is M $(M = 2nk - (k - 1))$, k is the multiple of the interpolation;

Step3: Cut a section of length $nk + 1$ in $\boldsymbol{c}$ as $\boldsymbol{d}$, where $\boldsymbol{d}$ is the polar axis after interpolation;

Step4: Generate angular coordinates $\boldsymbol{e}$ in the range of $[-\pi, \pi]$, convert $\boldsymbol{d}$ and $\boldsymbol{e}$ to the rectangular coordinate system to obtain the closed surface contour;

Step5: Fill the contour, adjust its pixels, add it to the 3D model, and import the model into gprmax for simulation.

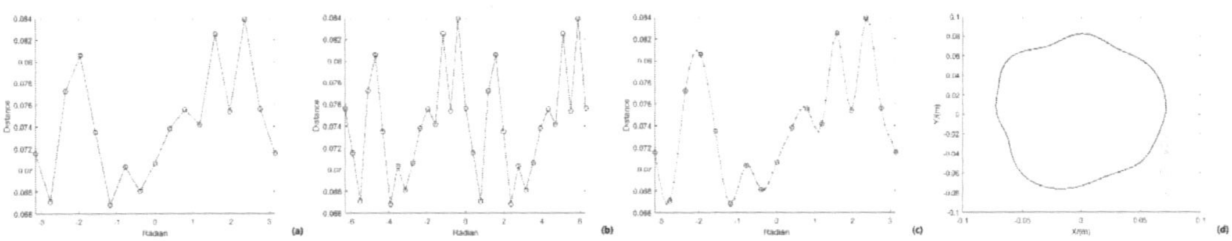

Fig 1 simulation steps (a) step1 (b) step2 (c) step3 (d) step4

3. Data processing

Frequency-wavenumber filter and matched filter are used to improve data quality.

3.1. *Frequency-wavenumber filter*

The C-scan data is transformed by 3D fast Fourier transform from time-distance domain $s(x, y, t)$ to frequency- wavenumber domain $S(k_x, k_y, \omega)$:

$$S(k_x, k_y, \omega) = \iiint s(x, y, t) \mathrm{e}^{-(ik_x x + ik_y y + i\omega t)} \mathrm{d}x\mathrm{d}y\mathrm{d}t \tag{1}$$

where k_x and k_y correspond to the spatial frequencies in x and y dimensions and ω is the temporal frequency. For the direct coupled wave with strong correlation, its energy is predominantly concentrated in the $k_x = 0$ and $k_y = 0$ regions in the frequency- wavenumber domain. For the target signal, it shows different forms in the time-distance domain, and its energy is relatively diffused in the frequency- wavenumber domain. Therefore, a three-dimensional filter is designed as:

$$S'(k_x, k_y, \omega) = S(k_x, k_y, \omega) H(k_x, k_y, \omega) \tag{2}$$

$H(k_x, k_y, \omega)$ is the three-dimension filter, it can achieve different degrees of processing effect by selecting the bandpass width. $S'(k_x, k_y, \omega)$ is the frequency-wavenumber domain data after filtering, and its 3D inverse fast Fourier transform represents the data that removes the direct coupled wave. In this study, the $k_x = 0$ and $k_y = 0$ regions in $S(k_x, k_y, \omega)$ are mainly zeroized.

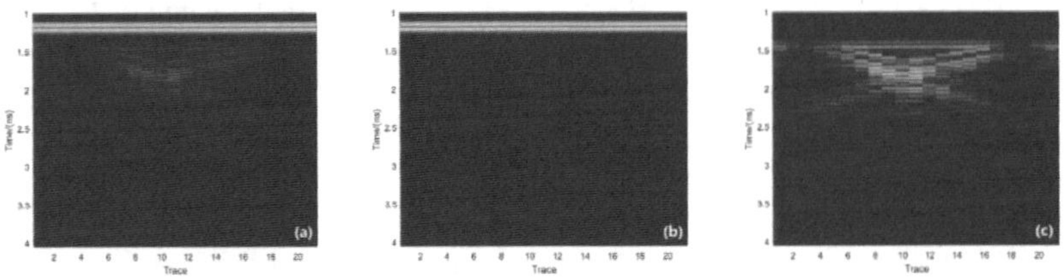

Fig 2 B-scan taken from 3D data (a) raw data (b) direct coupled wave (c) filtered data

The raw data contains a strong direct coupled wave and target echoes. Frequency- wavenumber filter can remove the direct coupled waves with low rank attributes and retain the target echoes.

3.2. Matched filter

The matched filter, considered the optimal detector, can be used to determine the presence of a target signal $s(t)$ embedded in Gaussian variance white noise. The identification problem of adhesive area can be regarded as a stratified media structure, with its signal model $r(t)$ depicted as follows0:

$$r(t) = \sum_{i=1}^{N} A_i s(t - \tau_i) + n(t) \tag{3}$$

Where A_i and t_i represent the amplitude and delay of the reflected pulse, respectively. $s(t)$ is spurious responses and $n(t)$ is a random noise process. The normalized system signal response is convolved with the reflected signal response from the interface to determine the corresponding peak position of the interface. In this study, the system function of the filter $x(t)$.

$$r'(t) = r(t) * h(t) = r(t) * x(T - h) \tag{4}$$

Where T is the duration of signal $x(t)$ introduced to make the filter causal. After the signal is matched, the bonding layer state can be identified by selecting the peak position t_i.

4. Validation

In order to ensure the effectiveness and performance of the algorithm, numerical simulation and experimental model are used to validate the algorithm performance.

4.1. Numerical validation

The simulation model designed in section 2 was adopted. The model shape and result are as follows. Ten sets of models were then used for verification, with an average morphological accuracy of 92.95%.

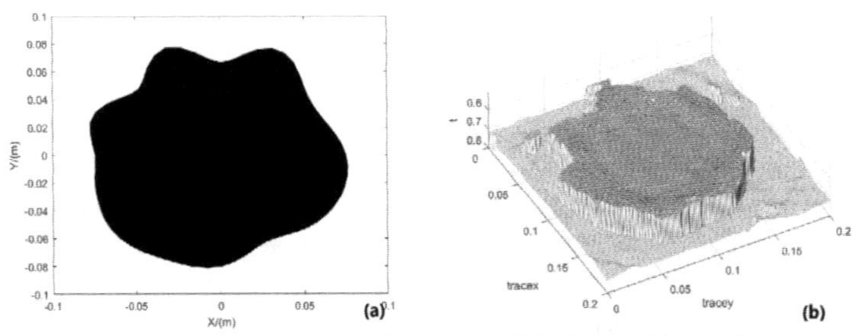

Fig 3 Simulation model (a) model shape (b) result

Fig 3 (a) shows the adhesive shape of the ETIS simulation model. From the cross-section view, it has a nearly circular structure with random shape distribution. (b) is the recognition result of the algorithm for (a) which accurately reflects its shape characteristics.

4.2. Experiment validation

In order to further validate the authenticity of the model and the effectiveness of the algorithm, an external insulation system was investigated by GPR. The adhesive structure located in the bonding layer. Beyond this bonding layer is a thermal insulation layer with a thickness of about 30 mm, followed by a plastering layer with a thickness of 5 mm.

Fig 4 Experimental environment and scanning mode

The GPR equipment is Proceq GP8800, and the raw data is scanned from left to right by the antenna in turn upward, as shown in figure 4. The raw data has three dimensions: x, y, t, in which 33 traces are spaced 50mm on the x axis and 161 traces are spaced 5mm on the y axis. 0 Figure 5 shows the adhesive morphology in the two regions tested in the experiment and their identification results:

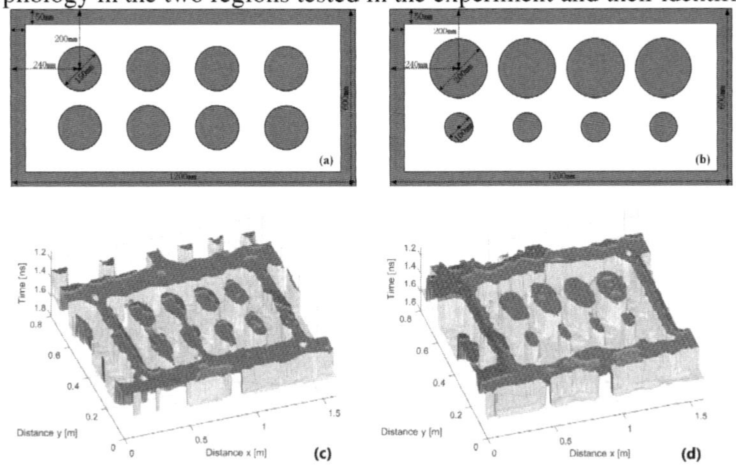

Fig 5 Adhesive morphology (a) region 1 (b) region 2 (c) result 1 (d) result 2

From the recognition result in the image, the algorithm avoids the influence of direct coupled wave, interface reflection wave and other clutter, and the adhesive structure can be clearly recognized.

5. Conclusion

The primary focus of this study is to evaluate the bonding state in ETIS. The numerical simulation based and experiments are verified.

Acknowledgment

This work was supported by the CSCEC-PT-015 CSCEC City Renovation and Intelligent Operation Engineering Research Center (Building Health Diagnosis and Treatment).

References

[1] Yin X Q, Li H Q and Li M *2011 Int Conf on Electric Technology and Civil Engineering (ICET CE), Lushan, 2011*, pp. 1599-1602, https://doi.org/10.1109/ICETCE.2011.5776202
[2] Jol, H. Radar, Ground Penetrating. "Ground penetrating radar: theory and applications." *Jol ISB N* (2009): 978-0.
[3] Jean-Luc Starck, E. J. Candes and D. L. Donoho *IEEE Trans on Image Processing*, vol. 11, no. 6, pp. 670-684, June 2002, https://doi.org/10.1109/TIP.2002.1014998
[4] Kurtz J L, Fisher III J W, Skau G, et al. *Radar Sensor Technology II. SPIE*, 1997, 3066: 11-21. https://doi.org/10.1117/12.276101
[5] Lahouar S, Al-Qadi I L. *NDT and E International: Independent Nondestructive Testing and Eva luation*, 2008(2):41. https://doi.org/10.1016/j.ndteint.2007.09.001

[6] Sun M, Le Bastard C, Pinel N, et al. *Signal Processing*, 2017, 132: 272-283. https://doi.org/10.1016/j.sigpro.2016.05.029
[7] Liu T, Zhu Y, Su Y *Sensors* 2018, 18, 1366. https://doi.org/10.3390/s18051366

Application of ground-penetrating radar based on U-Net in the statistics on the bonded area of external wall insulation

J Wang[1], Y Li[2], Z Jiang[1], M Liu[1], J Gong[2,3], X Yang[2,3], T Lan[2,3*]

[1] Beijing Building Research Institute Co., Ltd.of CSCEC, Beijing, China
[2] College of Information Science and Technology, Beijing University of Chemical Technology, Beijing, China
[3] Chongqing Innovation Center, Beijing Institute of Technology, Chongqing, China
[4] School of Information and Electronics, Beijing Institute of Technology, Beijing, China
[5] This work was supported by the CSCEC-PT-015 CSCEC City Renovation and Intelligent Operation Engineering Research Center (Building Health Diagnosis and Treatment).

Abstract. The insulation of external walls of buildings often suffers from various diseases. Therefore, it is necessary to accurately detect the bonding area of the insulation layer. Ground-penetrating radar is a non-destructive detection method that can effectively determine the bonding status. In this paper, FDTD forward simulation of the exterior wall insulation layer by ground-penetrating radar is carried out, and the GPR images of the 4GHZ and 6GHZ antennas are simulated, respectively, and field experiments are conducted to collect the stepped frequency vector net data of 2.5-8GHZ and 6-8GHZ. The data obtained are processed by using the improved SP-ICA clutter suppression algorithm, followed by the improved loss function of the U-Net and training. The F1_score of the obtained model can reach 0.8399, and the error of bonding area statistics is 0.23%, which realizes more accurate semantic segmentation and area statistics.
Keywords: Ground-penetrating radar (GPR), Building envelope insulation, Clutter suppression, Semantic segmentation

1. Introduction

Diseases such as hollowing and other diseases of the external wall insulation layer of the building will lead to the lack of the adhesive area of the insulation board, which will have a negative impact on its thermal insulation effect and energy consumption. The standard of external wall thermal insulation engineering technology stipulates that the minimum bonding area of XPS thermal insulation board is 50%. Therefore, timely and accurate testing of the bonding status of the building exterior wall insulation layer is of great significance for guaranteeing building safety and prolonging service life.

As an important non-destructive detection method, ground-penetrating radar (GPR) plays an important role in the fields of pipeline localization, road detection, and tunnel lining [1]. In recent years, ground-penetrating radar technology has also been applied to the defect detection of external wall insulation system, which verifies the feasibility of ground-penetrating radar to detect the external wall insulation system [2]. By scanning the external wall of the building, ground-penetrating radar can obtain information such as the thickness and internal structure of the insulation layer. Due to its non-destructive and high resolution characteristics, it makes the calculation of adhesive area possible.

In this paper, numerical simulation and vector-net measured data were collected for the adhesive of ground-penetrating radar exterior wall insulation, the resulting data were processed using the improved SP-ICA clutter suppression algorithm, and the collected ground-penetrating radar C-scan data were used to train the U-Net model after the improvement of the loss function, which realized a more accurate pixel segmentation and area statistics.

2. Method

The U-Net is a classical convolutional neural network structure which consists of an encoder and a decoder. The encoder is responsible for gradually decreasing the size of the feature map and increasing the number of channels, while the decoder is responsible for gradually restoring the original image size and decreasing the number of channels.The U-Net is characterized by jump connections, which allow the utilization of feature information from different layers, thus improving the accuracy of pixel segmentation and detail preservation. Through forward propagation, the U-Net model classifies each pixel and determines it as either adhesive or background. It has been widely used in medical image segmentation [3], nature image segmentation and other fields.

In this paper, we use the U-Net model with an improved loss function for pixel-level semantic segmentation of ground-penetrating radar C-scan image slices. In order to obtain clear boundaries of the segmented image with high accuracy, a hybrid loss function is used to reduce the gap between the segmented image and the labeled image, which is defined as:

$$l = \lambda_1 l_{BCE} + \lambda_2 l_{Iou} + \lambda_3 l_{Focal}$$

Where l_{BCE} is the BCEWithLogitsLoss binary cross-entropy loss function, l_{Iou} is the Iou loss function, and l_{Focal} is the FocalLoss loss function. $\lambda_1, \lambda_2, \lambda_3$ are the hyperparameters of each loss function. The model is set to：

$$\lambda_1 = 0.4; \quad \lambda_2 = 0.3; \quad \lambda_3 = 0.3$$

BCEWithLogitsLoss targets the pixel-by-pixel binary classification problem; Iou Loss aims to maximize the intersection area of the prediction frame and the true target frame while minimizing the concatenation area; and the FocalLoss loss function focuses on samples that are difficult to classify, as well as the problem of weighting positive and negative samples.

3. Data acquisition and processing

The data used in this paper includes both simulated and measured data. The simulation data is carried out with the help of GprMax for the orthogonal simulation of multiphase random medium for insulation board bonding agent with 3GHZ and 6GHZ antennas. Some of the results of the simulation experiments are shown in Figure 1, from top to bottom, air layer, plaster layer, insulation board, adhesive, and wall. Meanwhile, the vector net scanning field measured ground-penetrating radar data of irregular bonding agent of insulation boards with 2.5-8 GHZ and 6-8 GHZ step frequencies were collected. After processing, C-scan slice plots of the simulated and measured data were obtained. Some of the results of the real measurement experiments are shown in Figure 2.

Figure 1 Simulation data and corresponding ground-penetrating radar imaging (6GHZ)

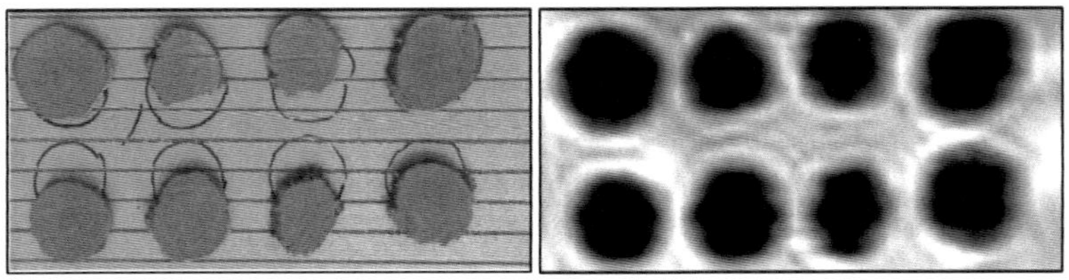

Figure 2 C-scan slices of field experimental and measured data (6-8 GHZ)

Simulated data has better signal-to-noise ratio and target image entropy than measured data. The measured data is seriously disturbed, it should be preprocessed. Ground-penetrating radar data preprocessing can suppress the interference of clutter and noise, and can effectively improve the quality of data. In this paper, the preprocessing process of the measured data includes background removal, band-pass filtering, inverse Fourier transform, and gain adjustment, in addition to the clutter suppression process before the simulation and measured data are used.
The data preprocessing steps are:

- Averaging method of background removal.
- Hamming window bandpass filtering.
- Inverse Fourier Transform.
- Gain Adjustment.
- Independent Component Analysis (SP-ICA) clutter suppression. The proposed method improves on ICA by using SVD-based PCA as a front preprocessing in the FASTICA iteration [4]. After homogenization, it will be decomposed by SVD, the reconstructed matrix is selected, and the reconstructed matrix is then whitened, and finally the fixed-point iteration. The flow of the improved algorithm is shown in Figure 3. The comparison of the proposed method with other clutter suppression algorithms is shown in Table 1.

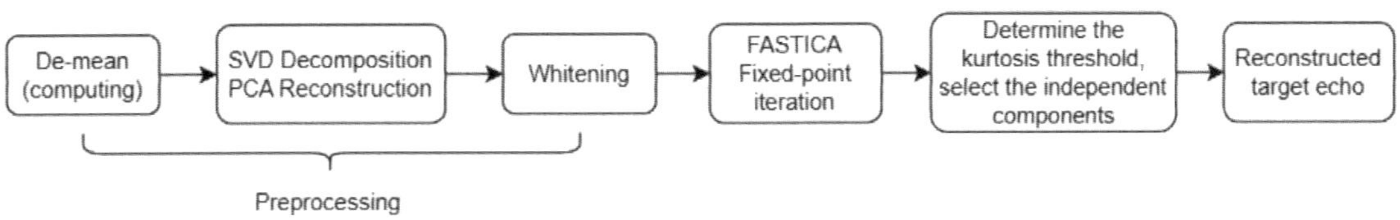

Figure 3 Improved SP-ICA algorithm for clutter suppression

The ground-penetrating radar image cannot be used directly as an input dataset, and the radar data corresponding to the adhesive has to be labeled and transformed into a binary image. Considering the convenience of practical application, we limit the background pixels to 255 (white) and the adhesive pixels to 0 (black). Here we perform the labeling and binarization of the dataset with the help of Labelme tool. The training and validation sets are randomly divided by 8:2.

Table 1 Target image entropy for the proposed method and other clutter suppression algorithms

Arithmetic	Mean square elimination	Localized energy 75% auto PCA	Global energy 99% auto PCA	SVD	SVD-based PCA	ICA	SVD-based ICA	SP-ICA
Image entropy	40.473	43.199	44.191	44.735	40.402	40.541	40.361	40.345

4. Experimental results

Our approach is to combine the data preprocessing proposed above with the U-Net network to train a model capable of segmenting the C-scan data slices of the insulation board adhesive ground-penetrating radar at different frequencies. By analyzing the various quantitative metrics and qualitative results obtained from the U-Net model, we can draw the following conclusions:

Table 2 Evaluation indicators

Evaluation indicators	IOU ↑	Precision ↑	Recall ↑	Accuracy ↑	F1_score ↑
Numerical	0.792	0.7999	0.9877	0.9916	0.8839

- The individual metrics used in this paper are IOU, Precision, Recall, Accuracy, F1_score. through training, the results are shown in Table 2. Among them, the segmented F1 index reaches 88.39%, which basically meets the engineering requirements.
- We use the best model in training to segment the ground-penetrating radar C-scan image, and the segmented binary map for the calculation of the area share of the bonding agent, the tested image with the base wall of the bonding area share of 33.02%, the actual bonding area of the thermal insulation board is calculated as 33.25%, with an error of 0.23%. The error is controlled within the range of $\pm 0.5\%$, and the result is more accurate.
- The proposed algorithm includes both identification and localization of adhesive targets. It is possible to frame the range of each adhesive more accurately and to determine the location of the adhesive. The recognition results are shown in Figure 5.

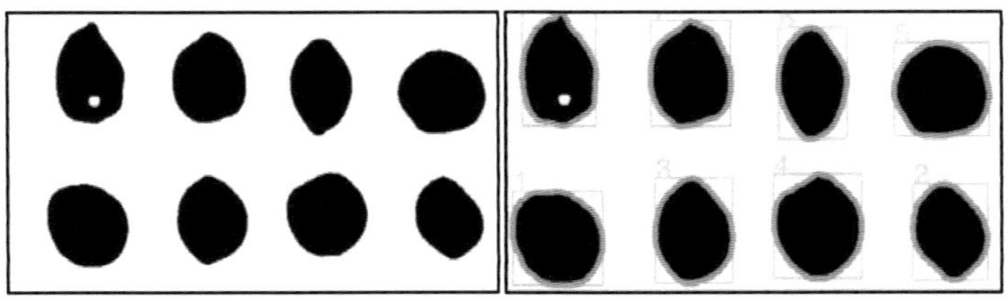

Figure 5 U-Net segmentation results (left) and adhesive recognition results (right)

5. Conclusion

First, we use the SP-ICA algorithm to display the target signal clearly from the cluttered background and improve the target image entropy. Second, we use U-Net with improved loss function to predict the resulting ground-penetrating radar image and obtain a semantic segmentation map. Finally, the resulting segmented image is utilized for adhesive area counting, as well as adhesive detection and location identification. The method used provides better performance and performs well on both simulated and measured data.

Acknowledgments

References

[1] Yang T, Tangqing J, Zhang G, et al. *2023 International Annual Conference on Complex Systems and Intelligent Science (CSIS-IAC), Shenzhen, China, 2023*, pp. 779-783. https://doi.org/10.1109/CSIS-IAC60628.2023.10364184

[2] Zhang D B, et al. *International Conference on Smart Monitoring, Assessment and Rehabilitation of Civil Structures*, 2021, pp. 177-193. https://doi.org/10.1007/978-981-99-3362-4_15

[3] Islam S S and Hossain A B M A. *2021 IEEE 2nd International Conference on Applied Electromagnetics, Signal Processing, & Communication (AESPC), Bhubaneswar, India, 2021*, pp. 1-5. https://doi.org/10.1109/AESPC52704.2021.9708467

[4] Kumlu D and Erer I. *2017 4th International Conference on Electrical and Electronic Engineering (ICEEE), Ankara, Turkey, 2017*, pp. 323-328. https://doi.org/10.1109/ICEEE2.2017.7935843

Journal of Physics: Conference Series **2887** (2024) 012043 doi:10.1088/1742-6596/2887/1/012043

A compact air-coupled SFCW ground penetrating radar and its application in the earthen sites

Shengbo Ye[1,2,5], Xu Li[1,2,3], Xinhui Li[1,2,3], Xiaojun Liu[1,2], Guangyou Fang[1,2,3], Hejun Jiang[4,5] and Rongqi Lyu[4]

[1]Aerospace Information Research Institute, Chinese Academy of Sciences, Beijing, 100190, China
[2]Key Laboratory of Electromagnetic Radiation and Sensing Technology, Chinese Academy of Sciences, Beijing, 100190, China
[3]School of Electronic, Electrical and Communication Engineering, University of Chinese Academy of Sciences, Beijing, 100049, China
[4]Science and Technology on Near-surface Detection Laboratory, Wuxi, 214035, China
[5]Corresponding author's e-mail: yesb@aircas.ac.cn, jhj68@126.com

Abstract. As a non-destructive testing technology, ground-penetrating radar (GPR) has been widely used in archaeological investigations of the earthen sites. Traditionally, a GPR antenna is ground-couple. It is attached to the measured object and thus may cause damage to the earthen sites. This paper develops a lightweight and compact air-coupled stepped frequency continuous waveform (SFCW) GPR system with frequency range from 1GHz up to 6GHz. We suppress the nonlinear variables of the radar system by calibration. And we perform tests on different weathering samples from the earthen sites. Our system is verified that can discriminate layered structures in samples and the results of tests are consistent with the simulations of the gprMax model. In general, this radar system can assess the weathering degree and provide technical support for the preservation of the earthen sites.

1. Introduction

As an important part of cultural heritage resources, the earthen sites contain abundant historical information. However, the earthen sites have low strength and poor ability to resist weathering. The trend of global warming and extreme weather phenomena threaten the long-term stable conservation and utilization of the earthen sites. Therefore, the analysis of the weathering degree and reinforcement of the earthen sites is a crucial mission of archaeological work.

GPR is a versatile non-invasive method that can quickly obtain high resolution imaging of underground targets at a low cost. It has wide applications in the field of archaeology. In 1975, Roger S. Vickers and others used GPR systems to detect ancient buried Indian structures at Chaco Canyon, New Mexico[1]. The applications of GPR in the field of archaeology has become diversified. According to S. P. Lukyanov and O. V. Stukach, GPR can be used for underground environmental monitoring in archaeology[2]. In addition, GPR has high sensitivity to the variations in water content. I. Morris and others quantitatively analyse the characteristics of Roman archaeological sites respectively under wet and dry scans, which improves the interpretation of the GPR archaeological survey under only one set of conditions[3].

In this paper, we develop an air-coupled SFCW GPR system which conforms to the trend of miniaturization and multi-functionality of the GPR systems. The system uses a horn antenna to achieve transceiver sharing to reduce the cost. To verify the reliability and practicability of the system, we perform tests to measure the thickness of weathering layer and the permittivity of the weathering samples. By comparing and analysing the performance characteristics of the methods of reinforcement, we obtain the adaptability and durability of different treatments of weathering.

This paper is organized as follows: in Section 2, we introduce our radar system and the proposed system model solution. In Section 3, we analyse the thickness of weathering layer and measure the permittivity of the weathering samples. Finally, the conclusion is drawn.

2. Radar system

The compact air-coupled SFCW GPR system proposed in this paper consists of an ultra-wideband double-spine horn antenna, Vector Network Analyzer (VNA) module and data acquisition module. The basic structure of the system is shown in Figure 1.The horn antenna realizes the sharing of transceiver and receiver to reduce cost, which has the advantages of wide impedance bandwidth and high directivity. Using a remote-control slide rail module, the radar system operates in air-coupled mode with the frequency range from 1 GHz to 6 GHz, and can be used in side-looking mode or down-looking mode.

As shown in Figure 2, a system function of the far-field radiation model is proposed by Lambot[4], which is used for layered media detection applications. Assuming that the electric field measured locally by the antenna in the far-field region tends to a plane wave, radar in a single-station configuration can be modeled by a frequency-dependent transfer function.

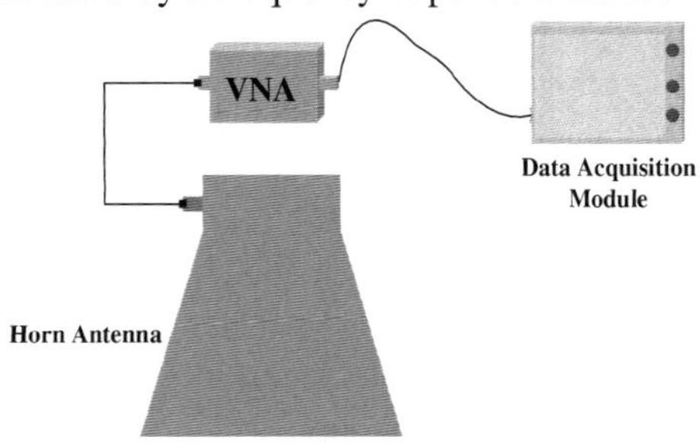

Figure 1. The basic structure of the radar system.

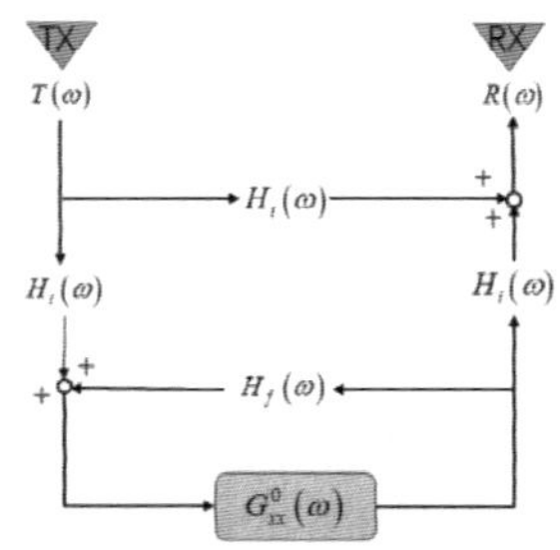

Figure 2. The system function of the far-field radiation model.

Simplifying Figure 2, the received signal of the radar can be denoted as in equation (1).

$$S_{11}(\omega)=\frac{R(\omega)}{T(\omega)}=H_i(\omega)+\frac{H(\omega)G_{xx}^0(\omega)}{1-H_f(\omega)G_{xx}^0(\omega)} \tag{1}$$

$$H(\omega)=H_r(\omega)H_t(\omega) \tag{2}$$

where $T(\omega)$ and $R(\omega)$ are the transmitted and received signals; $H_t(\omega)$, $H_r(\omega)$ and $H_i(\omega)$ represent transmitting, receiving and return loss transfer functions of the antenna; and $G_{xx}^0(\omega)$ is the reflected electric field of the electric dipole in the layered medium.

The effect of the antenna in the model is only related to frequency, and the equivalent zero-bias Green's function G_{xx}^0 can be converted into the original waveform of the antenna. From the signal expression, the extraction of the Green's function can effectively remove the direct wave and compensate the zero point of time. Equation (3) shows the extraction formula of Green's function. The waveforms of the echo at the distance of 52 cm and its Green's function in time domain are as shown in Figure 3 and Figure 4.

$$G_{xx}^0(\omega)=\frac{S_{11}(\omega)-H_i(\omega)}{H(\omega)+H_f(\omega)\left(S_{11}(\omega)-H_i(\omega)\right)} \tag{3}$$

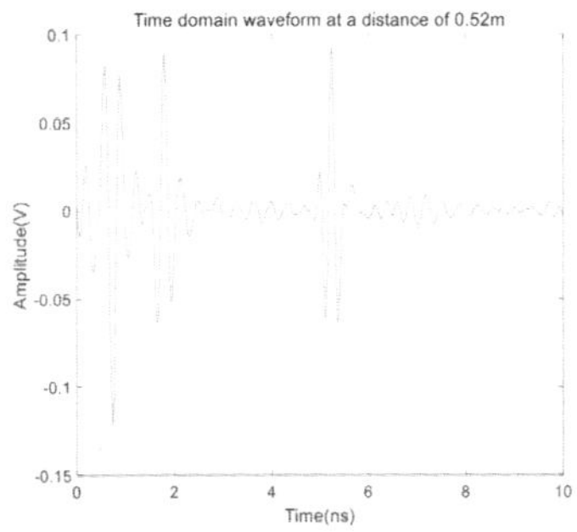

Figure 3. Measured waveform of the metal plate at 0.52 m.

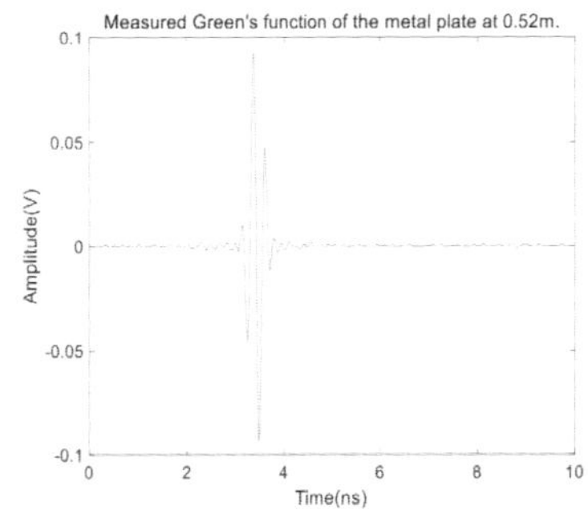

Figure 4. Measured Green's function of the metal plate at 0.52 m.

3. Experiment

The weathering samples of the earthen sites provided by the Dunhuang Academy in China are as shown in Figure 5, including cube samples with the edge length of 30 cm and samples with trapezoidal vertical profile. We establish the equally proportional simulation models of these two kinds of samples respectively in gprMax, and compare with the measured results. Figure 6 shows the measurement scene of a cube samples, and it's under the radar system. The radar transmits electromagnetic waves vertically downward for horizontal detection of the samples in vertical profiles and fixed-point detection of the central points of the samples.

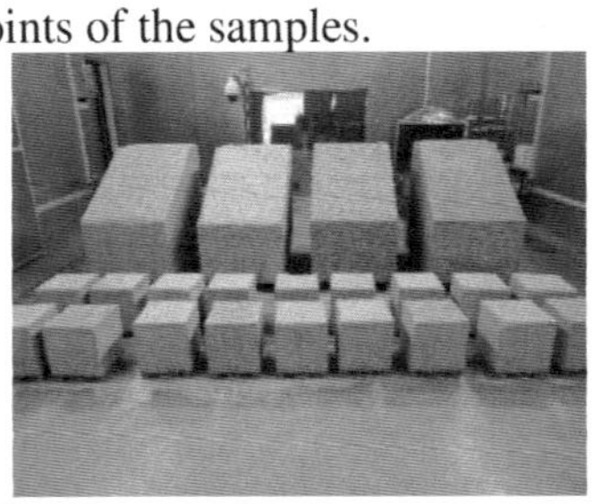

Figure 5. The weathering samples of the earthen sites.

Figure 6. The measurement scene of a cube samples.

The simulation model of the cube samples with five layers is as shown in Figure 7, and the gap between each layer is 1 mm. The weathering layer has the thickness of 2 cm, and others are 6.9 cm. From top to bottom, the permittivity of each layer is 3.2, 3.0, 3.2, 3.1 and 3.2. We can clearly distinguish the five layers of the model from the B-scan results as shown in Figure 7. And there is an edge effect as the measured sample is above the ground. For the cube sample, after signal compensation four layers can be roughly detected from the B-scan results as shown in Figure 8. As the electromagnetic waves can penetrate the sample, the strong reflection signal received from the bottom metal plate can be seen.

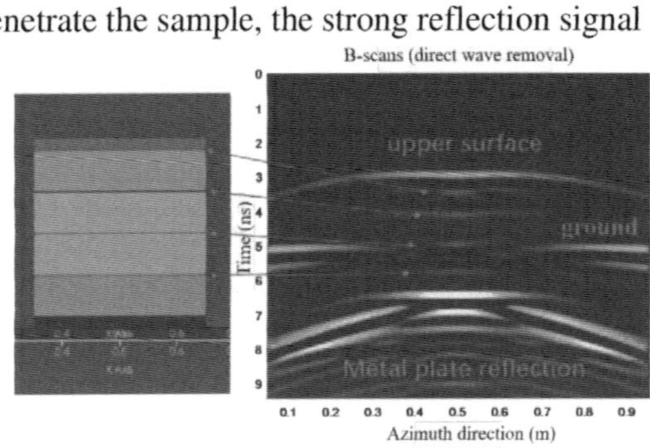

Figure 7. The simulation model and B-scan result of the cube sample.

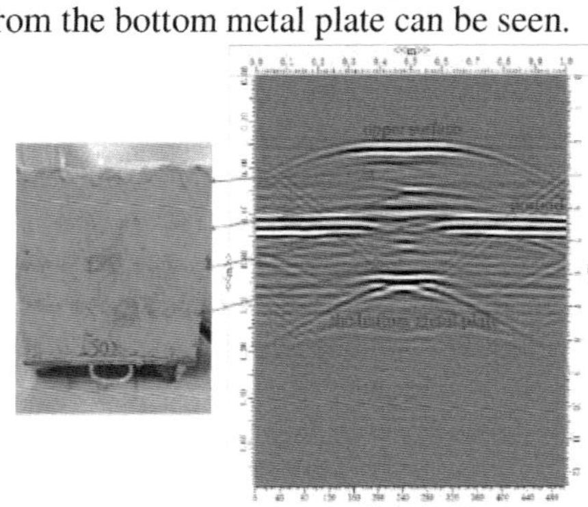

Figure 8. The B-scan result of the cube sample.

As shown in Figure 9, the simulation model in gprMax of the sample with trapezoidal vertical profile has two 2 mm air cracks below the surface at 10.6 cm and 21.2 cm respectively. The transceiver antenna in multiplexing mode moves along the x-axis from right to left with the step of 4 mm to collect 111 A-scans. As shown in Figure 10, the trend of the reflected echoes of the top surface and air cracks for the model is consistent with the sample.

Figure 9. The sample with trapezoidal vertical profile and the simulation model in gprMax.

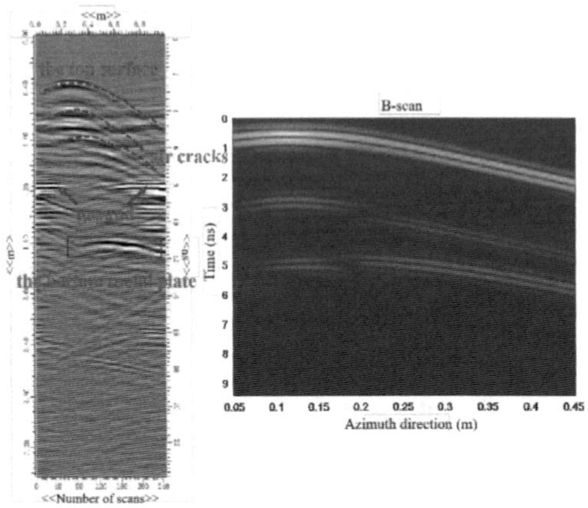

Figure 10. The B-scan results of the sample and the corresponding gprMax model.

4. Conclusion

A compact air-coupled SFCW GPR system is presented in this paper, with frequency range from 1 GHz up to 6 GHz. The system is applied to the reinforcement tests of the weathering samples in Dunhuang in China. The results show that our system can noninvasively penetrate the weathering samples with a thickness of 30 cm. And it can distinguish the weathering layer and stacking layers of the detect objects. In conclusion, this system realizes non-contact detection for the permittivity and the weathering degree of the weathering layers. It will help the protection work of the earthen sites in the future.

Acknowledgement

This research work is supported by the Strategic Priority Research Program of the Chinese Academy of Sciences under Grant XDA28050302, and is funded by Science and Technology on Near-surface Detection Laboratory under Grant 6142414220710.

Reference

[1] Vickers R S and Dolphin L T 1975 A communication on an archaeological radar experiment at Chaco Canyon, New Mexico *Masca Newsletter.*

[2] Lukyanov S P and Stukach O V 2009 Archeological researches of the ancient fortress by GPR *2009 International Siberian Conference on Control and Communications* pp. 324–31.

[3] Morris I, Glišió B and Gonciar A 2018 Archaeological Ground Penetrating Radar Surveys Under Variable Soil Moisture: Visual and Numerical Results *2018 17th International Conference on Ground Penetrating Radar (GPR)* pp. 1–4.

[4] Li J., Liu L., Zeng, Z. , & Liu, F. . (2014). Advanced signal processing for vital sign extraction with applications in UWB radar detection of trapped victims in complex environments. IEEE Journal of Selected Topics in Applied Earth Observations & Remote Sensing, 7(3), 783-791.

[5] Li J., Zeng Z., Sun J., Liu F., (2012) Through-wall detection of human being's movement by UWB radar. IEEE Geoscience and Remote Sensing Letter, 9(6), 1079-1083.

A Ground Penetrating Radar Array Antenna Designed for Small Targets Detection

Xinghua Shi [1,2], Shaolong Li [1], YueMeng Yin[1] and Zhao zhang[1]

[1] Xi'an JiaoTong University, Xi'an 710049, China (ruhu43@qq.com)
[2] China Research Institute of Radiowave Propagation, Qingdao 266107, China

Abstract. The detection of small targets, especially those made of non-metallic materials, presents significant challenges due to their small size and the difficulty of being detected by traditional metal detectors. Additionally, environmental factors and soil conditions further complicate their detection. Traditional Ground Penetrating Radar (GPR) operate with a single-line detection method, resulting in a low number of scan lines and inefficient detection. This approach makes it difficult to generate slice images and can lead to high missed detection. A miniaturized GPR array antenna suitable for handheld and unmanned platforms has been developed. Utilizing a planar dipole antenna with multiple resistive and capacitive loading, a horizontal oblique polarization array comprising five transmitters and six receivers has been constructed, achieving dual-reception from a single transmission across ten channels, significantly improving upon the limitations of single antenna systems in terms of data volume and sampling interval. Experimental tests demonstrate that the designed system can successfully detect the echo signals of metallic and non-metallic small targets like buried up to 20 cm deep in typical soil conditions, and generate clear slice images, proving its capability for high-resolution imaging detection of shallow buried small targets.

1. Introduction

Ground Penetrating Radar (GPR) plays a crucial role in engineering applications, particularly in the detection of shallow-depth small targets. Due to the typically small size of these targets, high resolution is required from GPR. GPR finds extensive applications in various domains, especially in detecting buried target and other hazardous materials[1].

The application of GPR in safety inspections is also paramount. In major event venues or critical facilities, GPR is used to detect concealed objects within walls or beneath floors to prevent potential security threats. During renovation and construction inspections, GPR is employed to locate internal wiring and pipelines before drilling into walls, thereby avoiding accidental damage during construction. Additionally, GPR is widely utilized for quality inspections, such as measuring thickness during the construction of highways, airport runways, and tunnels, as well as for routine maintenance checks. By detecting the quantity and distribution of reinforcement bars within building materials, GPR provides comprehensive quality assessment data[2-4]. GPR can also be used to inspect concrete slabs for relative thickness, deterioration, and structural issues. By utilizing GPR technology, the concrete surface can be scanned, and the variations in thickness of the slab can be determined by analysing the reflected signals. Deterioration and structural problems such as voids, cracks, and rebar corrosion can also be detected using GPR[5].

However, GPR technology also exhibits certain limitations in shallow-depth detection. Firstly, the detection depth of GPR is restricted by soil type and moisture content. In damp or highly viscous soils, electromagnetic wave attenuation occurs at a faster rate, resulting in reduced detection depth. Secondly, natural obstacles such as tree roots and rocks can interfere with signals, increasing detection complexity and false alarm rates. Furthermore, the high cost of GPR equipment and data processing software limits its application in resource-constrained environments. Professional training of operators is required to ensure accurate data interpretation and effective utilization.

This paper introduces a compact array antenna design aimed at improving the detection of shallow-buried targets and enhancing the safety of manual detection methods. The design incorporates loaded dipole antennas as transceivers within an array configuration of five transmitters and six receivers, tailored for detecting small-scale targets within a detection range of 20cm. This approach markedly increases detection precision and efficiency. Experimental validations of the antenna design confirm its capability to meet the detection requirements for small targets. Furthermore, the array data facilitates the slicing imaging of small targets, significantly improving the accuracy of detection.

2. Methodology

2.1. the Antenna Selection

In radar technology, the lateral resolution is often related to the size of the Fresnel zone. The size of the Fresnel zone depends on the distances along the propagation path, the wavelength, and the Fresnel zone number under consideration. The lateral resolution for a single target can approximately be half the radius of the first Fresnel zone. Therefore, the lateral resolution ΔR for a single target can be approximated with the following formula:

$$\Delta R = \sqrt{\frac{\lambda * depth}{2} + \frac{\lambda * \lambda}{16}} / 2 \tag{1}$$

Where ΔR is the lateral resolution. λ is the wavelength of the electromagnetic wave. $depth$ is the total propagation distance.

Under conditions of a 20 cm depth and a dielectric constant of 8 the lateral resolution for detecting a single target needs to be at least 4 cm to meet the requirements for small targets detection. Setting the antenna's center frequency at 1500 MHz addresses multiple operational requirements, including penetration depth, compactness, and high-quality imaging, yielding a satisfactory lateral resolution of 3.1 cm for targets up to 10 cm deep, or 4.3 cm for targets up to 20 cm deep.

In response to the requirements of a compact multi-channel array project, a planar dipole antenna with multiple resistive and capacitive loadings is implemented, with the antenna unit confined to dimensions of 50 mm × 100 mm as shown in Figure 1. A balun positioned at the antenna's center facilitates the conversion from a balanced to an unbalanced signal, crucial for minimizing direct coupling between the transmit and receive antennas and reducing multiple signal oscillations. The inclusion of layered absorptive material within the shielding case's interior further aids in mitigating interference[6].

For the shielding enclosure, an aluminium shell design is typically utilized. In this project, considering the need for portability and lightweight in portable devices, a non-magnetic aluminium shell is chosen to be crafted into a cuboid with dimensions of 100 mm in length, 50 mm in width, and 60 mm in height, serving as the radar antenna's shield. Moreover, based on practical requirements, a circular hole approximately 10 mm in diameter is made at the center of the shell's top surface for cable routing.

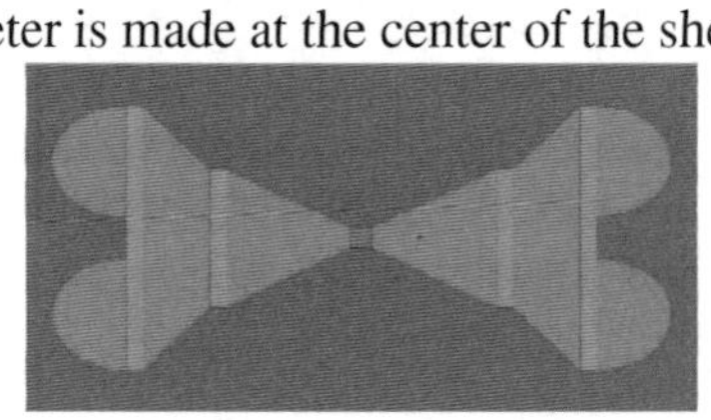

Figure 1. The diagram of the radar antenna model

2.2. The layout of Antenna array

To accurately capture the surface and structural features of small targets, extensive scanning with multiple measurement lines over small targets is necessary. The denser the measurement lines, the tighter the detection grid on the surface of the small targets, resulting in clearer imaging results, more pronounced features, and a lower false alarm rate. Extensive research and experimentation on the imaging characteristics of targets and the radiation efficiency of radar antennas have shown that smaller intervals between channels lead to higher imaging and resolution, improved recognition rates of small targets, and reduced false alarms. The maximum distance between the transmitting and receiving antennas should not exceed 5cm; exceeding this distance may increase the risk of missed detections and false alarms. To ensure effective detection and imaging of targets while considering the size of the antenna array and a certain level of radiation efficiency, the distance between the center of the transmitting and receiving antennas is set to approximately 5 cm[7].

Considering the requirements for miniaturization and measurement line spacing, this paper innovates on the traditional antenna arrangement by implementing a 90 degree rotation strategy to reduce the line spacing. Specifically, two different antenna layout schemes are proposed.

The first scheme arranges all transmitting and receiving antennas in a linear configuration, significantly reduces the size of the device and shortens the distance between transmission and reception, thereby enhancing the strength of the echo reflections. However, this approach superimposes direct-coupled waves with ground waves, which for very shallowly buried small targets, may result in their signals overlapping with those of the ground waves, making them difficult to distinguish

The second scheme employs a staggered array arrangement of transmitting and receiving antennas, aimed at capturing reflected and scattered signals from various angles. This arrangement not only optimizes the distance between antennas but also enhances the system's capability to capture signal diversity, which is crucial for improving the resolution and accuracy of subsurface detection as shown in Figure 2.

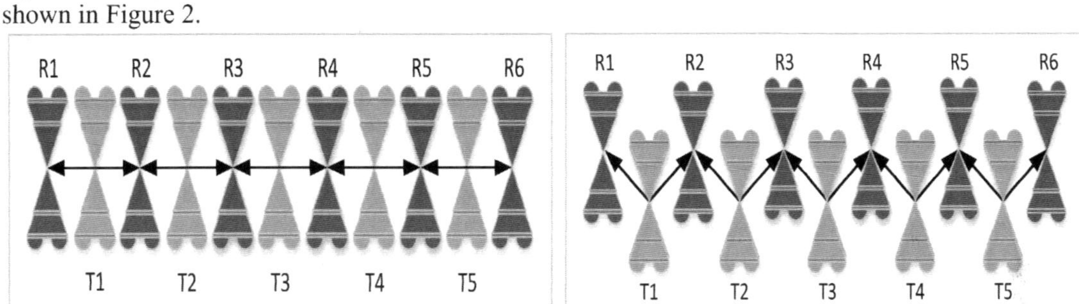

(a) Linear arrangement (b) Alternating arrangement

Figure 2. The arrangements of Antenna array

2.3. Simulation

- Single-Channel Simulation

For a single antenna, we simulated the scenario of a typical non-metallic cylindrical object with a burial depth of 60 mm and a height above ground of 70 mm, using a ground model measuring 600 × 600 × 300 mm as shown in Figure 3.

(a) Side of the antenna (b) Beneath the antenna

Figure 3. Single-channel simulation model

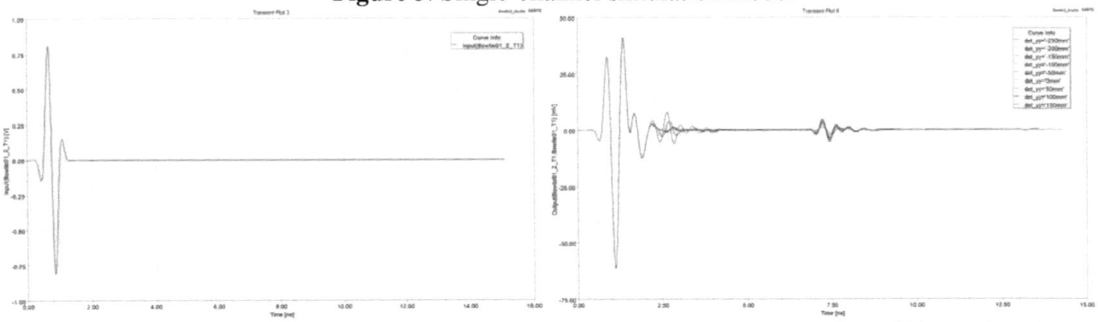

(a) Input Signal wave (b) Received Signal wave - different locations

Figure 4. A-Scan Waveform in different position

From the received Signal wave, it is observed that the strongest echo is received when the antenna is directly above the cylinder, and the echo becomes very weak at the edges of the ground model. In the measurement lines at different positions, the direct coupling wave and the ground reflection wave are completely consistent, forming a relatively clear fixed wave in the B-scan. This signal can be used to corroborate the antenna's lift-off height and the burial depth of the cylinder.

- Array Simulation

To verify which of these two array configurations is more effective, we conducted simulations for both scenarios separately. The simulation with each antenna individually enclosed in a shielding case. The antennas are positioned at a height of 6 cm above the ground. The objective is to simulate a plastic cylinder buried at a depth of 10 cm with a diameter of 7 cm. The center frequency of the antenna is set to 1500 MHz. We collected data from the transmission T4 channel directly above the cylinder, and from the three receiving channels R2, R3, and R4. Figure 5 and Figure 6 present the two types of antenna array arrangement methods and the radar waveforms received by different receiving antennas.

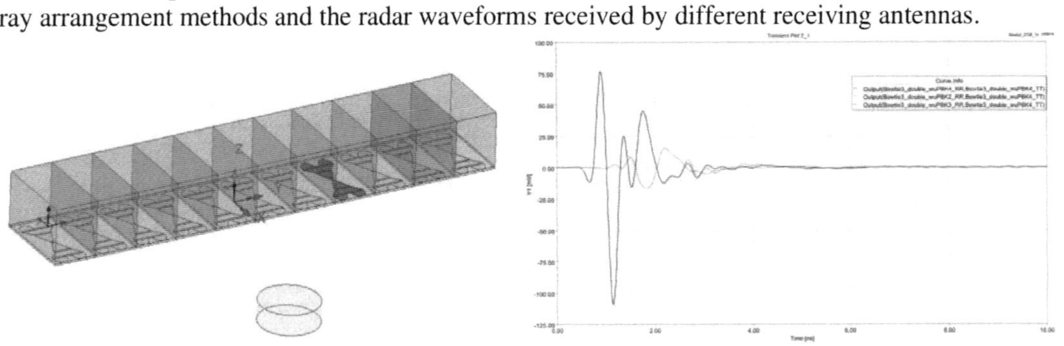

(a) line array arrangement (b) Received waves

Figure 5. The line array arrangement and the T4 transmission R2, R3, R4 receive signal waves

Journal of Physics: Conference Series **2887** (2024) 012044 doi:10.1088/1742-6596/2887/1/012044

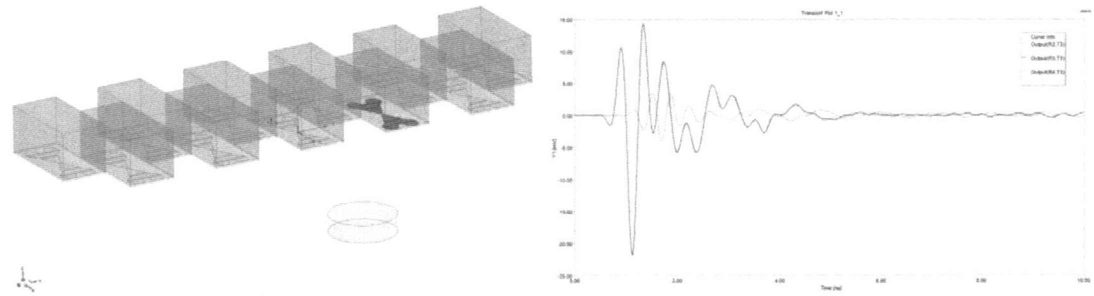

(a) Staggered array arrangement (b) Received waves

Figure 6. The Staggered array arrangement and the T4 transmission R2, R3, R4 receive signal waves

From the simulated radar waves, it is evident that the energy of the direct wave in the linear arrangement is significantly greater than that in the cross arrangement. However, for the detection of cylinder targets, the echo response from the cylinder target in the cross arrangement is not smaller than that in the linear arrangement. Therefore, we normalized these two waveforms and compared them under the same conditions.

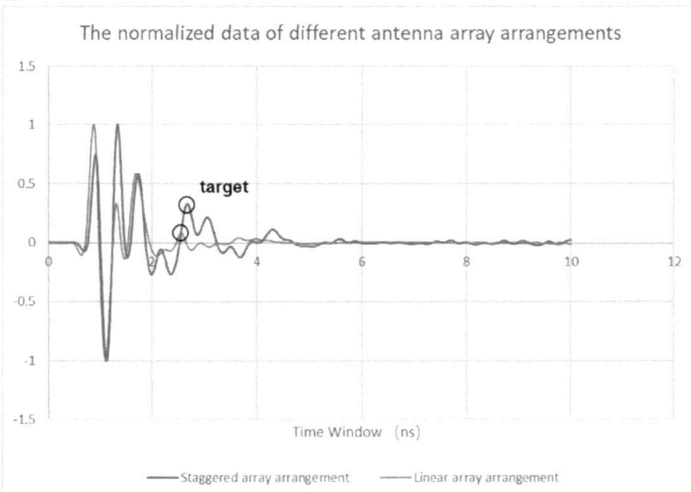

Figure 7. The normalized data of different antenna array arrangements

From the Figure 7, it can be observed that after normalizing the two datasets, the target response is greater with the staggered array arrangement, making the targets more pronounced. Therefore, although the staggered arrangement can reduce the amplitude of direct coupling, it facilitates better performance for shallow targets. Hence, we have chosen the staggered array configuration for the antenna arrangement. In subsequent experiments, the staggered arrangement method was also adopted.

3. Experimental device and Results

To verify the effectiveness of the antenna, tests were conducted on array antennas at the CRIRP target detection test field as shown in Figure 8. The dimensions of the field were 6m in length and 4m in width. The soil at the site was sandy, containing interference objects (tree roots) at various depths. In this experiment, the antenna operated at a frequency of 1500 MHz with 5 transmit and 6 receive channels, spaced 10 channels apart. The inter-channel spacing was 5 cm. To facilitate measurements in accordance with the established methodology, the surface of the test site was covered with foam noodles for ease of walking and measuring the height above ground. Each sampling interval was 2 cm, with a time window of 8ns and 256 sampling points. The width of the measurement line was designed to be 50 cm. Two targets were buried along the measurement line at depths of 10 cm and 8 cm, respectively, for detection purposes.

Figure 8. The experimentation field and detection site condition

The dielectric constant of the sandy soil was 6, with a wave velocity of 12.2 cm/ns. Radar images were obtained through multiple channels after ground correcting, as shown in Figure 9.

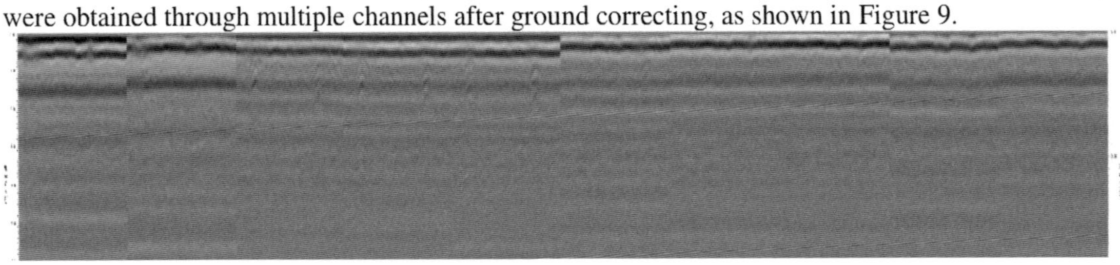

Figure 9. The GPR profile data collected from 10 channels

The data from the 10 channels can be used to generate slice images. After background elimination of the raw data, slice images of cylinder targets at different depths can be obtained as shown in Figure 10.

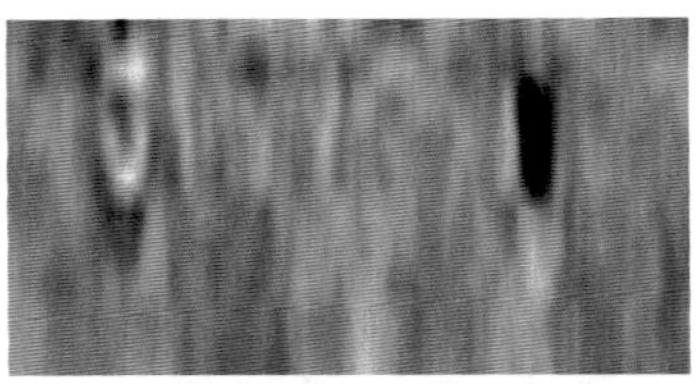
(a) depth 6 cm

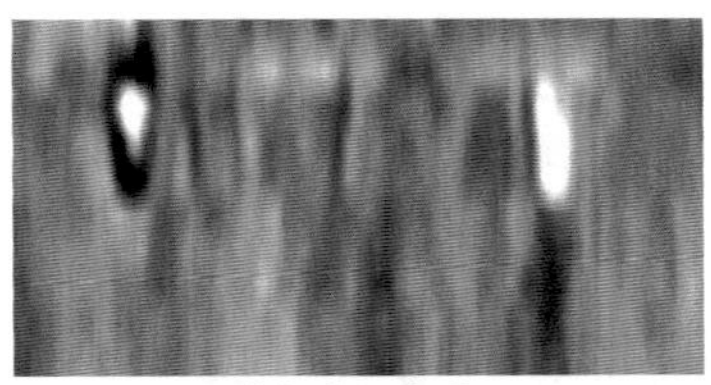
(b) depth 8 cm

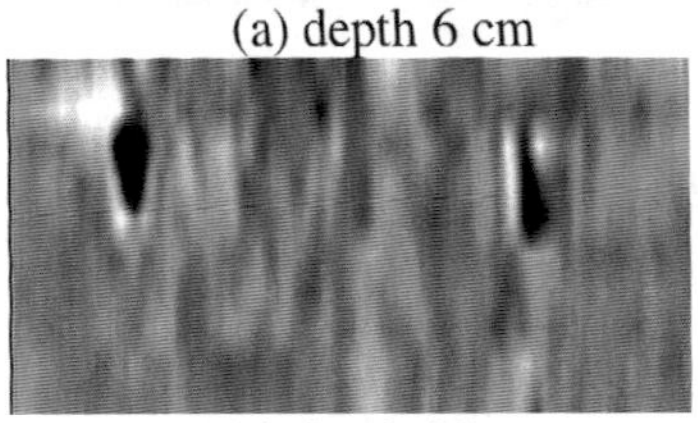
(c) depth 10 cm

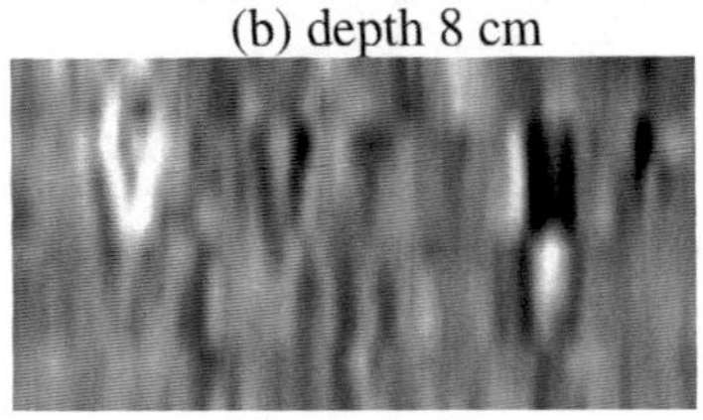
(d) depth 12 cm

Figure 10. The slice images of cylinder targets at different depths

The proposed antenna array offering higher resolution and more pronounced responses to anomalies, making it more resistant to interference compared to other methods. By analysing multi-channel radar data, three-dimensional slicing imaging technology can clearly display radar images. In the slice images, the response characteristics of plastic cylinder target is clearly distinguishable, allowing for the exclusion of tree roots or other objects.

4. Conclusion

Using high frequency GPR array for the detection of small targets is an effective method. the proposed system comprising a planar dipole antenna with resistive and capacitive loading in a horizontal oblique polarization array enhances detection capabilities. The array, featuring five transmitters and six receivers, enables dual-reception from a single transmission across ten channels, markedly improving data volume and sampling intervals over single antenna systems. Experimental results confirm the system's efficacy in detecting echo signals from cylinder targets buried up to 10 cm deep under typical soil conditions and in producing clear slice images for high-resolution imaging detection of shallowly buried small targets. These findings underscore the potential of the developed GPR array antenna in enhancing the detection and imaging of hazardous or metallic targets, contributing significantly to engineering application and safety detection.

Reference

[1] Baek S-H, Kim S-S, Kwon J-S and Um E S 2017 Ground penetrating radar for fracture mapping in underground hazardous waste disposal sites: A case study from an underground research tunnel, South Korea *Journal of Applied Geophysics* **141** 24-33

[2] Amran T S T, Ahmad M R, Haron A H, Masenwat N A, Mustapha I, Osman M F, Abd Razak N, Adnan M A K, Rahman A H and Adnan A A 2024 Underground utility inspection using ground penetrating radar. In: *IOP Conference Series: Materials Science and Engineering*: IOP Publishing) p 012021

[3] Guo S, Xu Z, Li X and Zhu P 2022 Detection and characterization of cracks in highway pavement with the amplitude variation of GPR Diffracted waves: Insights from forward modeling and field data *Remote Sensing* **14** 976

[4] Boldrin P, Fornasari G and Rizzo E 2024 Review of Ground Penetrating Radar Applications for Bridge Infrastructures *NDT* **2** 53-75

[5] Almalki M A and Almutairi K F 2024 Inspection of reinforced concrete structures using ground penetrating radar: Experimental approach *Journal of King Saud University - Science* **36** 103140

[6] Scheers B, Plasman Y, Piette M, Acheroy M P and Vander Vorst A 2000 Laboratory UWB GPR system for land mine detection. In: *Eighth International Conference on Ground Penetrating Radar*: SPIE) pp 747-52

[7] Shi X, Cheng D, Song Z and Wang C 2018 A real-time method for landmine detection using vehicle array GPR. In: *2018 17th International Conference on Ground Penetrating Radar (GPR)*: IEEE) pp 1-4

Image Enhancement Algorithm for Ground Penetrating Radar Based on Nonlinear Technology

Guo Xian Wu[1] **, Ying Liu**[2]

[1] School of Mechanical and Electrical Engineering, Tianjin Modern Vocational Technology College, Tianjin, China

[2] School of Electronic Engineering, Tianjin University of Technology and Education, Tianjin, China (liuying@tute.edu.cn)

Abstract. Ground penetrating radar images typically have characteristics such as weak target return signals, high background noise, and strong direct wave return signals. In order to obtain images with acceptable quality, a non-linear technology based image enhancement method was proposed. The method used LIP model and CDF-HSD function as the core to test multiple ground penetrating radar images and compared them with three common image enhancement algorithms. Finally, six image quality evaluation methods were used to evaluate the output images, verifying that this method can improve the contrast of target return signals while suppressing background noise in ground penetrating radar image processing.

Keywords: ground penetrating radar, image enhancement, LIP model, CDF-HSD function

1. Introduction

Ground penetrating radar (GPR) is a commonly used instrument for non-invasive underground detection. In urban areas, ground penetrating radar can not only detect and characterize hazards, but also monitor underground resources, including underground infrastructure, historical sites, and agricultural exploration. Ground penetrating radar provides features of shallow underground layers, making it particularly suitable for detecting non-metallic objects. The basic principle is that when electromagnetic wave pulse signals propagate underground, if there are no other objects underground and the dielectric constant is basically the same, then there is no other return signal except for the direct component returned from the surface. However, due to the existence of various different media objects underground, which are the targets we want to detect, there is discontinuity in the underground dielectric constant. The pulse signal is reflected or refracted, and the corresponding return signal is obtained. Based on the return signal, the corresponding characteristics of the underground object can be calculated [1]. However, due to the fact that actual ground penetrating radar signals are contaminated by various types of noise, or when the target is a small object, the return signal is often weak and contains more noise. Therefore, in order to more accurately detect underground targets, an image enhancement method that can enhance the return signal and attenuate the background area signal is needed.

The basic image enhancement algorithm is based on histogram equalization algorithm [2]. These algorithms are simple, fast, and widely used, and can achieve results that are generally acceptable in most cases. However, histogram equalization is a global processing of the image, which is not suitable for preserving local details and has certain losses in visual effects and image quality. Afterwards, people proposed some adaptive histogram equalization techniques, as well as Other image enhancement methods, such as gamma correction [3], SVD [4], DWT [5], DCT and other different techniques, which brought certain improvements. This article proposes a ground penetrating radar B-scan image enhancement algorithm, which is based on nonlinear image transformation technology to solve the problems of weak target signals and large background noise in ground penetrating radar B-scan images.

2. Nonlinear image transformation technology

The main purpose is to obtain acceptable image quality effects from various ground penetrating radar B-scan images, which are mostly grayscale images characterized by overall darkness, high noise, and low contrast between the target and background signal, as shown in Figure 1.

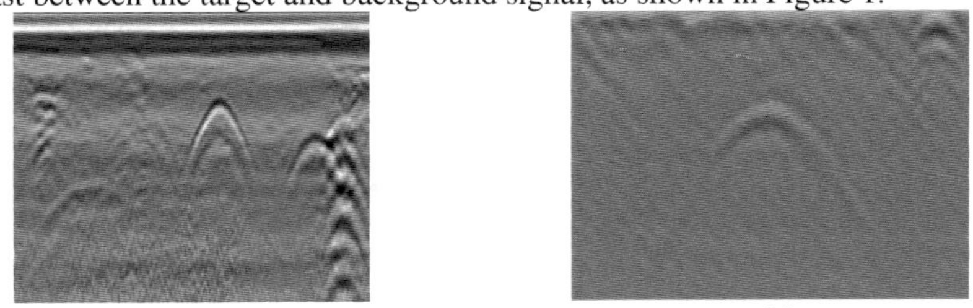

(a) (b)

Figure 1. Original image of ground penetrating radar: (a) Original Image 1; (b) Original Image 2

For GPR images, the proposed image enhancement algorithm is generally required to improve overall brightness and contrast while reducing noise. When these elements are processed well, enhancing the image usually presents better results. This technique uses the following method for calculation, first performing logarithmic transformation:

$$I_1 = \log(I + 1) \tag{1}$$

In the formula, I is the normalized GPR grayscale image input, with a numerical range of 0~1. I_1 is the result image of logarithmic transformation.

Then, an exponential function is used to operate on the input GPR image I to enhance image contrast, with the following equation:

$$I_2 = 1 - \exp(-I) \tag{2}$$

After obtaining the logarithmic operation result image I_1 and the exponential operation result image I_2, we can try using an appropriate LIP model to combine the result images I_1 and I_2. Considering the low brightness and contrast of the image, the LIP mode [6] used can be calculated using the following equation:

$$I_3 = \left(1 - \frac{(1 - I_1^2) * (1 - I_2^2)}{1 - (I_1^2 * I_2^2)}\right)^{1/2} \tag{3}$$

After the first three steps of processing, the quality of the ground penetrating radar image has improved, but after testing, the brightness of the image is still not high, and the details of underground target objects are not clearly displayed. Therefore, the Modified Cumulative Distribution Function (CDF-HSD) [7]with hyperbolic secant distribution is used to improve the brightness of GPR images. Meanwhile, in order to achieve better performance of the function, modifications were made. The newly modified function has the following formula:

$$I_4 = erf(\lambda * \arctan(\exp(I_3)) - 0.5 * I_3) \tag{4}$$

In the equation, I_4 is the graph of the corrected function result. Adding an error function (*erf*) to enhance the curve transformation of this function can better improve the brightness of dark areas in the image. Meanwhile, there is a parameter λ in the equation that controls the enhancement effect, which is the key to the enhancement effect of this function. Finally, normalize the resulting image.

3. Parameter selection and algorithm comparison

3.1. Parameter selection

By relying on the nonlinear image transformation technique proposed in Section 2, the ground penetrating radar B-scan image was processed, and the results were obtained under different variable λ values as follows, The original image is shown in Figure 1 (a).

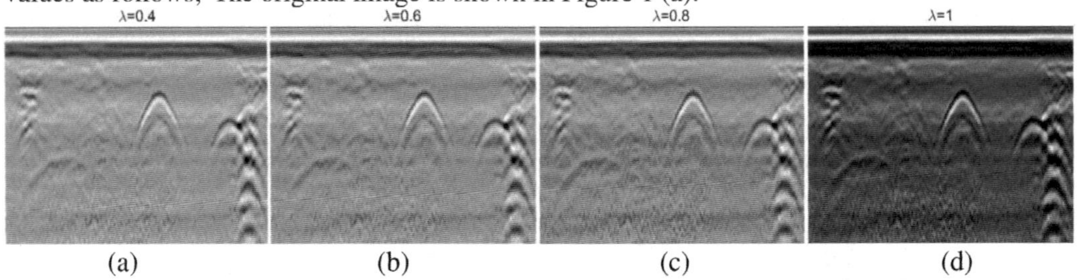

Figure 2. Processing results under different λ: (a) 0.4; (b) 0.6; (c) 0.8; (d) 1.0

If the processed image allows the observer to better perceive the expected information in the image, it can be said that the output image is enhanced compared to the original image. Generally speaking, the improvement of enhanced images is difficult to measure, and there is no universal method to measure the objective and intuitive effectiveness of image enhancement methods. Of course, in practice, there are many quality evaluation methods for image enhancement. In order to measure the quality of image enhancement and select the optimal processing parameters, we used several common image quality evaluation methods [8], namely SRCC, KRCC[9], PLCC, RMSE, PSNR, and SSIM. The results are shown in Table 1.

Table 1. Evaluation results

Evaluation methods	λ =0.4	λ =0.6	λ =0.8	λ =1.0
SRCC	1	1	1	1
KRCC	1	1	1	1
PLCC	1	1	1	0.9998
RMSE	1.6241 e-04	1.1719 e-04	6.1683 e-05	0.0025
PSNR	17.5160	19.3739	24.3540	15.1202
SSIM	0.9647	0.9763	0.9919	0.8663

Among them, SRCC, KRCC, PLCC, and SSIM, the closer their values are to 1, the smaller the RMSE value, and the larger the PSNR value, indicating better image quality.

Both from the visual representation in Figure 2 and the evaluation results in Table 1, it can be seen that the algorithm performed better at that time λ=0.8.

3.2. Algorithm comparison

Below, we will compare the nonlinear transformation technique proposed in Section 2 with several commonly used algorithms, including contrast limited adaptive histogram equalization (CLAHE), unsharpening mask (UM), and histogram equalization (HE) algorithms.

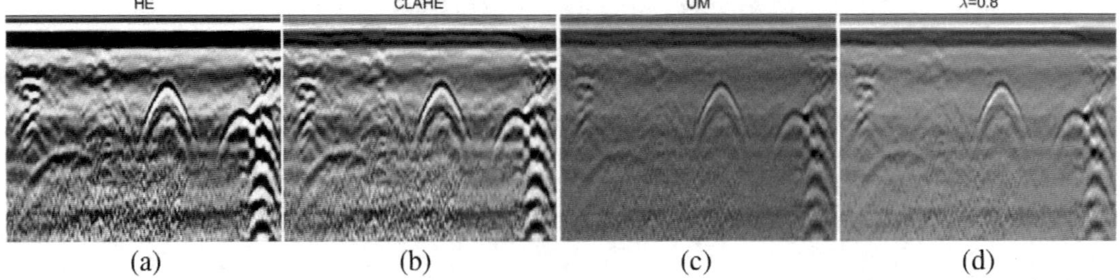

Figure 3. Different algorithm processing results: (a) HE; (b) CLAHE; (c) UM; (d) Proposed Method

Table 2. Evaluation results of different algorithms

Evaluation methods	HE	CLAHE	UM	Proposed Method
SRCC	0.9999	0.9668	0.9817	1
KRCC	0.9976	0.8467	0.9134	1
PLCC	0.8974	0.9026	0.9831	1
RMSE	0.0538	0.0525	0.0223	6.1683 e-05
PSNR	14.3601	18.3166	21.5818	24.3540
SSIM	0.5061	0.6704	0.9538	0.9919

4. Conclusion

From the evaluation results in Table 2, it can be seen that the algorithm proposed in this paper performs the best on the data, followed by the UM algorithm. There is a significant gap between the HE and CLAHE algorithms and the previous two algorithms. From the intuitive representation in Figure 3, the same conclusion can also be drawn.

This article proposes a ground penetrating radar image enhancement method, which is mainly based on nonlinear transformation technology and has been validated through various common image quality evaluation methods, including SRCC, KRCC, PLCC, RMSE, PSNR, and SSIM methods. The experimental results show that the algorithm significantly improves the contrast and overall brightness of ground penetrating radar images, and reduces noise, verifying the effectiveness of the algorithm.

Acknowledgments

This work was supported by the National Natural Science Foundation of China (Grant No.62241109), and Scientific Research Fund of Tianjin Education Commission (Grant No. 2021KJ005).

References

[1] M. Kim, S.D. Kim, J. Hahm, etc. GPR Image Enhancement Based on Frequency Shifting and Histogram Dissimilarity, IEEE Geoscience and Remote Sensing Letters,2018,15,(5): pp. 684 - 688.

[2] R.C. Gonzalez and R.E. Woods, Digital Image Processing, 3rd ed., Prentice Hall, Englewood Cliffs, NJ, 2007.

[3] P. Sujith Reddy, Himanshu Singh, Anil Kumar, L. K. Balyan and Heung-No Lee. Retinal Fundus Image Enhancement Using Piecewise Gamma Corrected Dominant Orientation Based Histogram Equalization, 2018 International Conference on Communication and Signal Processing (ICCSP), April 3-5, 2018.

[4] Aditi Sharma and Ajay Khunteta. Satellite image contrast and resolution enhancement using discrete wavelet transform and singular value decomposition, 2016 International Conference on Emerging Trends in Electrical Electronics & Sustainable Energy Systems (ICETEESES), March 11-12, 2016.

[5] P Megha, M Swarna, V Sowmya and K P Soman. Low contrast satellite image restoration based on adaptive histogram equalization and discrete wavelet transform, 2016 International Conference on Communication and Signal Processing (ICCSP), April 6-8, 2016.

[6] Florea, C. and Florea, L.. Parametric logarithmic type image processing for contrast based auto-focus in extreme lighting conditions, Int. J. Appl. Math. Comput. Sci., 2013, 23, (3), pp. 637–648.

[7] Zohair Al-Ameen. Nighttime image enhancement using a new illumination boost algorithm, IET Image Processing,2019,13,(8), pp. 1314-1320.

[8] Zicheng Zhang, Wei Sun, Xiongkuo Min, Wenhan Zhu, Tao Wang, Wei Lu and Guangtao Zhai. A No-Reference Evaluation Metric for Low-Light Image Enhancement, 2021 IEEE International Conference on Multimedia and Expo (ICME), July 05-09, 2021.

Journal of Physics: Conference Series **2887** (2024) 012045 doi:10.1088/1742-6596/2887/1/012045

[9] H. Yeganeh and Z. Wang, Objective assessment of tone mapping algorithms, in IEEE ICIP, 2010, pp. 2477-2480.

Journal of Physics: Conference Series **2887** (2024) 012046 doi:10.1088/1742-6596/2887/1/012046

Application of Ground Penetrating Radar in Rapid and Non destructive Detection of Urban Tunnels

Beiqiang Zhao, Zhihua He, Tao Liu, Xiaoji Song and Hongxing Zheng
College of Electronic Science and Technology, National University of Defense Technology, Deya road 109, Changsha, China (z_025227@163.com)

Abstract. Due to the rapid advancement of modern transportation, tunnels have emerged as an important component of transportation infrastructure. A large number of pipelines and tunnels makes the urban underground environment extremely complex. Detecting underground tunnels and pipelines can provide guidance for urban construction. Ground penetrating radar (GPR) possesses attributes such as swift detection speed, no loss, convenient and flexible operation, thereby playing a crucial role in numerous domains. The aim of this research is to verify the feasibility of detecting tunnels with depths on the order of 10 meters, by evaluating tunnel dimensions and depth parameters through a combined approach of simulation and experimental measurements. In this investigation, gprMax was utilized to model and simulate the underground tunnel to establish the correlation between echo data and target position. Additionally, this research employed an 80MHz GPR system to detecting a undergound tunnel and acquiring data which facilitated successful determination of the depth of the underground tunnel based on empirical evidence.

1. Introduction

Tunnels and underground engineering, with their prominent advantage of not occupying surface land resources, are playing an increasingly important role in addressing the three major challenges of population growth, resource scarcity, and environmental degradation facing society today. They are an important component of modern transportation infrastructure. Due to the rapid development in China, as of the year 2021, the number of highway tunnels in China has exceeded 23000, making it necessary to understand the usage of underground space to effectively utilize infrastructure space and avoid conflicts.

Ground-penetrating radar (GPR) is a geophysical method that uses electromagnetic waves to detect the distribution of materials in the subsurface[1]. The amplitude, phase, and waveform information of the GPR echo signals can reflect the position, shape, and electromagnetic properties of underground targets. With wide application prospects in urban development [2][3], infrastructure construction[4][5], geological exploration[6], mine detection[7][8], and many other fields[9][10], GPR has become an important research method in the field of geophysics.

This research first conducts numerical analysis, analysing the mathematical models of echo simulation and imaging. Utilizing electromagnetic simulation, the feasibility of GPR detecting tunnels with depths on the order of 10 meters is verified. Finally, an 80MHz GPR system is employed to detect and analyse the echoes of the Nyanja Lake Tunnel, obtaining information on tunnel depth.

Journal of Physics: Conference Series **2887** (2024) 012046 doi:10.1088/1742-6596/2887/1/012046

2. Imaging processing

Assuming that the signal received by GPR is $s(x,z,t)$, according to the wave theory[11], it should satisfy the following scalar wave equation.

$$\left(\frac{\partial^2}{\partial x^2}+\frac{\partial^2}{\partial z^2}-\frac{1}{(v/2)^2}\frac{\partial^2}{\partial t^2}\right)s(x,z,t)=0 \tag{1}$$

where x represents the azimuth coordinate, z represents the distance coordinate, t represents the distance time, and v represents the wave velocity in the medium. Performing a three-dimensional Fourier transform on equation (1) yields.

$$\left(k_x^2+k_z^2-\frac{\omega^2}{(v/2)^2}\right)S(k_x,k_z,\omega)=0 \tag{2}$$

where ω represents the frequency, k_x represents the wave number corresponding to the coordinate x, and k_z represents the wave number corresponding to the coordinate z. Equation (2) can derive the following relationship between frequency and wavenumber domain variables.

$$k_x^2+k_z^2=\frac{\omega^2}{(v/2)^2} \tag{3}$$

The wavefield reconstruction process of the FK algorithm commences with performing a two-dimensional Fourier transform on the received echo in the x and z directions, thereby obtaining the frequency spectrum of the echo signal in the $k_x-\omega$ domain.

$$S(k_x,z=0,\omega)=\iint s(x,z=0,t)\exp\left[-j\left(k_x x-\omega t\right)\right]dxdt \tag{4}$$

Based on the wave field extrapolation theory, the echo signal $s(x,z=0,t)$ is characterized by a plane wave field located within a homogeneous space and perpendicular to the propagation direction. Therefore, by applying phase shifts to the echo signal in the frequency domain, one can obtain the wave field at any point within the plane.

$$S\left(k_x,z,\omega\right)=S\left(k_x,z=0,\omega\right)\exp\left(jk_z z\right) \tag{5}$$

Performing a two-dimensional Fourier inverse transform on equation (5) converts the echo signal from the $k_x-\omega$ domain back to the time domain.

$$s\left(x,z,t\right)=\iint S\left(k_x,z=0,\omega\right)\exp\left(jk_z z+jk_x x-j\omega t\right)dk_x d\omega \tag{6}$$

The wave field $s(x,z,t=0)$ at the moment when the explosion source is just activated is the final imaging result, so taking t as 0 yields the final imaging result.

$$s\left(x,z,t=0\right)=\iint S\left(k_x,z=0,\omega\right)\exp\left(jk_z z+jk_x x\right)dk_x d\omega \tag{7}$$

By interpolating equation (7) to equation (8) (where the relationship between k and ω is given by equation (3))[12].

$$s\left(x,z,t=0\right)=\iint S\left(k_x,z=0,k_z\right)\exp\left(jk_z z+jk_x x\right)\left(-\frac{v}{2}\frac{k_z}{\sqrt{k_x^{\ 2}+k_z^{\ 2}}}\right)dk_x dk_z \tag{8}$$

The equation (8) illustrate the process of the FK algorithm. The echo signal undergoes a two-dimensional Fourier transform into the $k_x-\omega$ domain. Then, by interpolating, the signal is transformed from the $k_x-\omega$ domain to the k_x-k_z domain. After multiplying by certain coefficients and then performing the Fourier inverse transform, the resulting signal represents the imaging output.

3. Numerical simulation

The Nyanja Lake Tunnel is located in the Kaifu District, east of the Xiang River in Changsha City, Hunan Province, China. It starts at the intersection of Yingpan Road and Dongfeng Road to the east and passes under the Nyanja Lake in Martyrs Park, ending at the intersection of Yingpan Road and Chezhan North Road to the west. The tunnel consists of double tunnels with four lanes in each direction, with a single tunnel width of 10.35 meters and height of 4.5 meters. The detection point is located at the intersection of Martyrs Park and Nyanja Lake, with a depth of approximately 11 meters at the intersection.

Figure 1. (a) The image of Nyanja Lake Tunnel; (b) Nyanja Lake Tunnel in googel earth. The solid line is the detection track and the dotted line is the tunnel under the ground.

Before actual measurements, a forward simulation is conducted to obtain the relationship between the depth of the tunnel and the simulated echoes. Build the following model using gprMax simulation software: the background is the underground medium with a relative dielectric constant of 6, the target is a circular tunnel with a diameter of 10.35 meters. The tunnel contains a lane with a width of 7.25 meters and a hight of 4.5 meters. The distance between the top of the tunnel and the ground surface is 10 meters.

Table 1. Simulation parameter setting

Type of media	Relative permittivity	Electrical conductivity	Magnetic permeability	Magnetic loss
Air	1	0	1	0
Soil	6	0	1	0
Concrete	10	0	1	0

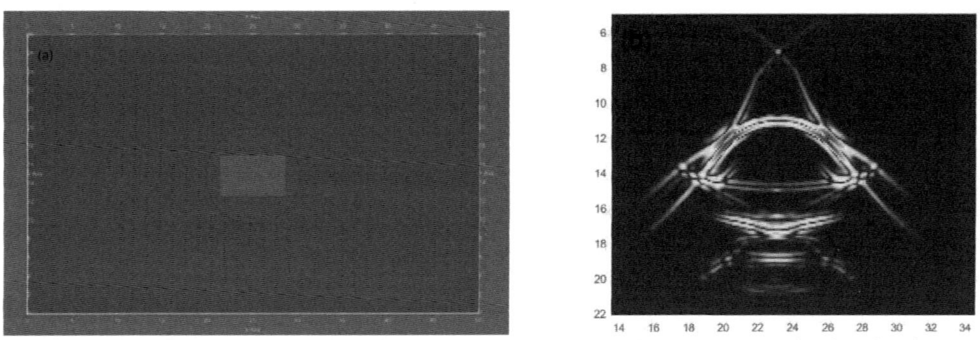

Figure 2. (a) Simulation model. The circular area is concrete, the square area in the circular area is air, and the rest of the area is soil; (b) Simulation result after migration. The displayed section is limited to the centre only.

From the simulation results and imaging analysis, it can be observed that GPR is capable of detecting the tunnel with depths on the order of 10 meters. The result after migration reveals a distinct curve, corresponding to the upper part of the concrete in the model. The diameter of the curve is approximately 10 meters, which closely matches the actual dimensions. Below the curve, two additional curves represent the upper and lower surfaces of the tunnel, with a height of about 4 meters, also consistent with real dimensions. This indicates that in the simulation, GPR can not only detect tunnels with depths on the order of 10 meters but also provide specific dimensional information about the tunnel.

4. Data measurement

In this research, the center frequency of the radar system is 80MHz, the bandwidth is 120MHz, the frequency range is 20-140MHz, the dynamic range of the radar system is 196dB, and the sampling frequency is 640MHz.

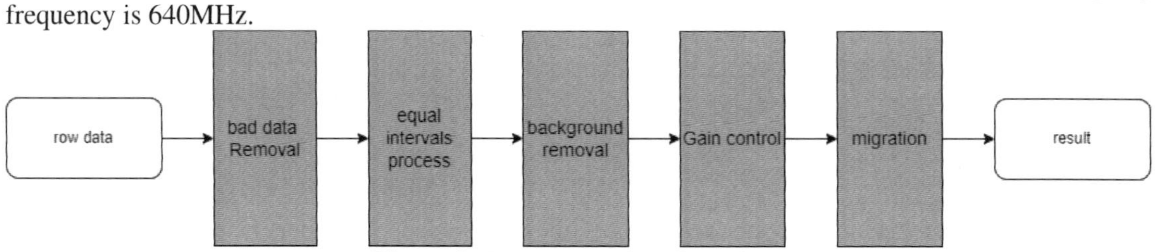

Figure 3. Data processing flow chart

The presence of direct waves, noise, clutter, and other interferences make the target echoes in the echo image was completely obscured by the background. To address this issue, a series of data preprocessing steps are required. Subsequently, the pre-processed data is subjected to imaging processing for further analysis and target identification.

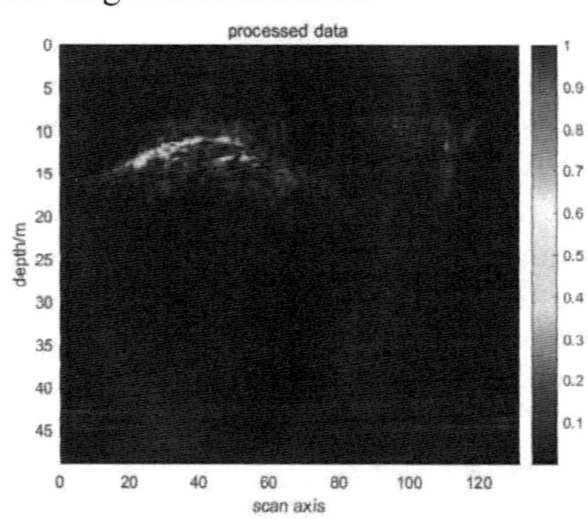

Figure 4. Results of processing data

As the Nyanja Lake tunnel is a twin-bore tunnel, two hyperbolic curves can be observed in the processed data image. The depth of the Nyanja Lake tunnel at the measured location is estimated to be around 11 meters based on the depth at the vertex of the hyperbolic curve, which aligns closely with the actual dimensions. From the image, it can be observed that the heights of the two echoes are inconsistent, indicating that there may be differences in the heights of the two tunnels above the ground surface in the measurement area. However, compared to the simulation results, the measured data obtained are unable to focus on the interior of the tunnel due to the attenuation of electromagnetic waves.

5. Conclusion

This research conducted numerical simulations and collected measured data on a section of the Nyanja Lake tunnel. The performance of GPR in detecting tunnels at a depth of 10 meters has been validated through numerical simulation. Under ideal simulation conditions, GPR not only discerns the tunnel's depth but also provides dimensional information.

The actual data collection was carried out using an 80MHz GPR system. After preprocessing the echo data, hyperbolic curves were obtained. By combining the results of numerical simulations with the echo data analysis, the depth of the tunnel was determined and found to be consistent with the actual dimensions. However, in the current field measurements, due to the attenuation of electromagnetic waves, accurate estimation of the tunnel's dimensions is impeded, and the detection of internal voids within the tunnel is unfeasible. Therefore, the utilization of GPR with enhanced penetrating capability is warranted.

Reference

[1] Harry m. Jol - ground penetrating radar - theory and applications
[2] Liang X, Hu D, Li Y, Zhang Y and Yang X 2022 Application of gpr underground pipeline detection technology in urban complex geological environments *Geofluids* 2022 1-9
[3] Harseno R W, Lee S, Kee S and Kim S 2022 Evaluation of air-cavities behind concrete tunnel linings using gpr measurements *Remote Sensing (Basel, Switzerland)* 14(21) 5348
[4] Rasol M, Pais J C, Pérez-Gracia V, Solla M, Fernandes F M, Fontul S, Ayala-Cabrera D, Schmidt F and Assadollahi H 2022 Gpr monitoring for road transport infrastructure: a systematic review and machine learning insights *Construction and Building Materials* 324 126686
[5] Cheng L, Lu J and Zhou C 2023 Non-destructive compaction quality evaluation of runway construction based on gpr data *Nondestructive Testing and Evaluation* 1-28
[6] Zhang L, Zhang S, Deng Z and Ling T 2024 Geological detection of hard rocks by gpr and signal time-frequency characteristics analysis in urban underground trenchless construction *Measurement Science & Technology* 35(4) 45406
[7] Song X, Liu T, Xiang D and Su Y 2019 Gpr antipersonnel mine detection based on tensor robust principal analysis *Remote Sensing* 11(8) 984
[8] Song X, Xiang D, Zhou K and Su Y 2019 Fast prescreening for gpr antipersonnel mine detection via go decomposition *Ieee Geoscience and Remote Sensing Letters* 16(1) 15-19
[9] Forte E, Santin I, Ponti S, Colucci R R, Gutgesell P and Guglielmin M 2021 New insights in glaciers characterization by differential diagnosis integrating gpr and remote sensing techniques: a case study for the eastern gran zebrù glacier (central alps) *Remote Sensing of Environment* 267 112715
[10] Bornik A and Neubauer W 2022 3d visualization techniques for analysis and archaeological interpretation of gpr data *Remote Sensing* 14(7) 1709
[11] Xu X, Miller E L and Rappaport C M 2003 Minimum entropy regularization in frequency-wavenumber migration to localize subsurface objects *Ieee Transactions On Geoscience and Remote Sensing* 41(8) 1804-1812
[12] Zhou L and Su Y 2011 Gpr imaging with rm algorithm in layered mediums *Ieee Geoscience and Remote Sensing Letters* 8(5) 934-938

Journal of Physics: Conference Series **2887** (2024) 012047 doi:10.1088/1742-6596/2887/1/012047

Numerical parametric study for ballast assessment using SVM applied to GPR data

Ernest Mbubia T[1,5], David Guilbert[2], Jérome Tissier[1,4], Antoine Martin[5], Théo Dezert[6], Yannick Fargier[3], Amine Ihamouten[1]

[1] Lames, Mast department, Gustave Eiffel University, Bouguenais, France (ernest.mbubia@univ-eiffel.fr)
[2] GeoEnd, Gers department, Gustave Eiffel University, Bouguenais, France
[3] Rro, Gers department, Gustave Eiffel University, Bron, France
[4] Eseo, engineering school, Angers, France
[5] Technical Department, ETF, Vinci Construction, Rueil-Malmaison, France
[6] FI-NDT, Start-up of Gustave Eiffel University, Bouguenais, France

Abstract. This study explores novel radar-based methodologies for ballast fouling evaluation and structural anomalies in railway structures. Ground Penetrating Radar (GPR), coupled with numerical modeling, is utilized to provide insights into ballast condition and structure. Various data processing methods, including the Matrix Pencil Method (MPM) and Full Waveform Inversion (FWI), are investigated for their effectiveness in detecting fouling. Support Vector Machine (SVM)-based machine learning approaches are employed to enhance classification accuracy. The results demonstrate that integrating raw GPR signals with MPM yields the most accurate classification results, facilitating efficient assessment of ballast condition and contributing to improved railway maintenance strategies.

1. Introduction

Ground Penetrating Radar (GPR) technology significantly enhances railway infrastructure maintenance, especially in adressing ballast fouling. This fouling, marked by the accumulation of particles and water infiltration, leads to operational disruptions and escalated maintenance expenses [1, 2]. Current methodologies, such as visual inspection, are recognized for their limitations in precision and time-consuming nature. While Selig and Waters' approach [3] effectively addresses granular aspects, its lack of continuity and high cost renders it inadequate and insufficient for quantifying fouling.

In response to this gap, novel radar-based methodologies for detecting fouling and structural anomalies are being investigated, with GPR coupled with numerical modeling providing valuable insights into ballast condition and structure [4, 5, 6]. This study introduces a numerical approach utilizing gprMax simulations [7], evaluating methods like the Matrix Pencil Method (MPM) and Full Waveform Inversion (FWI), alongside a Support Vector Machine (SVM)-based machine learning approach. A hybrid approach combining MPM, SVM, and FWI is aimed at enhancing classification accuracy between fouled and clean zones. These methods are tested on numerical data with the objective of comprehensively comparing their accuracy and efficiency in detecting ballast fouling.

Journal of Physics: Conference Series **2887** (2024) 012047 doi:10.1088/1742-6596/2887/1/012047

2. Data processing methods

The investigation of the SVM method and its application to raw and processed GPR data is proposed. The delayed response of the GPR signal enables feature identification based on its geometry and physical properties. This signal can be dimensionally reduced using processing methods such MPM or inversion method (FWI) while preserving its essence. The SVM is effectively applied to both raw and processed data, as well as to the combination of raw and processed signals, to study the influence of the two previously mentioned methods on classification.

2.1. Support Vector Machine (SVM)

Conventional radar techniques can estimate subsurface characteristics but struggle with detecting subtle dielectric variations. Machine Learning, particularly SVM, offers a solution by classifying data based on labels established during GPR data training.

SVM maximizes the decision boundary to separate variables and classify them, using support vectors and selecting an optimal hyperplane [8].

Key parameters in SVM include the kernel function and hyperparameters C and γ, which control regularization and error penalty during classification.

The model outputs automatic classification results for the evaluated zone, providing information on the ballast's condition.

2.2. Matrix Pencil Method (MPM)

The MPM is a linear algebra technique used for analyzing signals composed of exponentially damped sinusoids. It separates observations into signal and noise components and is commonly applied in adaptive filtering and parameter estimation. MPM involves computing a Hankel matrix to determine the number and values of poles, utilizing eigenvalue and singular value decompositions. For a detailed understanding, refer to [9]. MPM enables the reconstruction of the original signal from obtained poles and residues, enhancing precision and robustness of inverse calculations while significantly reducing input data dimensionality.

2.3. Full Waveform Inversion (FWI)

The electromagnetic model presented here describes radar wave propagation in a multi-layered medium, considering the frequency-dependent nature of EM properties. The objective is to compute causal parameters from the GPR signal, involving solving an inverse problem [10]. FWI allows reconstruction of the original signal from Green's functions with limited parameters, enhancing precision and robustness of the inverse calculation while reducing input data dimensionality. Therefore, we will explore this method's potential for replacing raw GPR signals with a reduced dimensionality of Green's function parameters while incorporating all relevant signal components.

3. Parametric numerical study

Following extensive experimental validation to determine the ranges of dielectric properties of the media under study, a framework is developed for the numerical investigation presented in this paper. GprMax is employed to produce multiple signals that are representative of real-world scenarios. The modeling characteristics and the study database are elaborated upon in the subsequent section. Additionally, the aim is to model the propagation of electromagnetic waves through multi-layered structures with well-defined physical and geometric properties, with the objective of distinguishing between clean and fouled ballast in railway structures.

Journal of Physics: Conference Series **2887** (2024) 012047 doi:10.1088/1742-6596/2887/1/012047

3.1. Modeling

To effectively address the problem, we model EM wave propagation through different configurations considering the number of layers and the ballast's state. Simulation domain parameters are set as follows:

- Domain size: 0.051 m, 0.425 m, 0.001 m with a spatial discretization of 0.001 m.
- Theoretical Hertzian dipole source with a Ricker waveform (1.5 GHz center frequency).
- Antenna position at (0.026 m, 0.405 m) (5 mm above the surface).
- Time window: 12 ns.
- Total thickness of the ballast layer: 0.4 m.

We create a database containing different configurations, defining four ballast conditions (Table 1): Clean, Moderate 1, Moderate 2, and Fouled. Each condition exhibits distinct dielectric permittivity and conductivity characteristics, with variable thickness ranging between 0.04 m and 0.4 m depending on the specific configurations. In the results, Moderate 2 and Fouled states are considered as Fouled, while Clean and Moderate 1 states are assimilated to Clean. For each state, we consider single-layer, bi-layer, tri-layer, and four-layer configurations, resulting in over 3700 A-scans (3200 corresponding to various levels of ballast fouling and 500 to a clean ballast).

Table 1. Ballast Layer State.

Ballast Layer State	Permittivity ε	Conductivity σ ($\mathrm{S\,m^{-1}}$)
Clean	[2.5 - 3.5]	10^{-5}
Moderate 1	[5.5 - 6.5]	10^{-4}
Moderate 2	[8.5 - 9.5]	10^{-3}
Fouled	[11 - 12]	10^{-2}

3.2. Processing step

A comparative study of SVM performance between global and local approaches, utilizing raw signals and features extracted from preprocessing methods (MPM and FWI), is conducted. We employ a two-class SVM method to identify ballast fouling anomalies, separating data into "Clean" and "Fouled" classes. The inputs injected into the SVM model include: case 1: Global approach (raw signal only); case 2: Local approach (MPM only); case 3: Mixed approach (Raw signal + MPM); case 4: Raw signal + FWI.

A 70/30 split is employed for training and test sets, with 70 % used for training (2590 A-scans) and 30 % for testing (1110 A-scans).

3.3. Results

Through a parametric numerical study, we explored various configurations and data processing methods to improve accuracy and efficiency in detecting fouling. The SVM-based machine learning approach, coupled with features extracted from MPM and FWI, provided valuable insights into the ballast's condition, further enhancing the model's performance in distinguishing between clean and fouled samples. The confusion matrix (Table 2) highlights the effectiveness of each approach, with the "Predict Raw signal + MPM" scenario achieving the highest classification accuracy across both clean and fouled samples. This approach significantly improves prediction and classification while efficiently reducing computational costs. It may be necessary to incorporate some limitations into our study. These could include factors such as noise in the measurement not being accounted for, as well as the binary nature of our approach (fouled/clean). This underscores the importance of integrating advanced data processing techniques with machine learning algorithms to achieve reliable and accurate classification results. This choice is reinforced by the continuously improving capabilities of learning-based methods compared to FWI. Nonetheless, the unoptimized SVM kernels and 'limited' database suggest that higher-quality results could be achieved with optimization in the future.

Table 2. Confusion matrix for the various approaches using SVM.

Real	Clean	106 (73 %)	40 (27 %)	75 (51 %)	71 (49 %)	135 (92 %)	11 (8 %)	129 (88 %)	17 (12 %)
	Fouled	6 (1 %)	958 (99 %)	1 (0 %)	963 (100 %)	1 (0 %)	963 (100 %)	0 (0 %)	964 (100 %)
		Clean	Fouled	Clean	Fouled	Clean	Fouled	Clean	Fouled
		Predict Case 1: Raw signal		**Predict** Case 2: MPM		**Predict** Case 3: Raw signal + MPM		**Predict** Case 4: Raw signal + FWI	

4. Conclusion

In conclusion, our study contributes to the development of an innovative radar-based methodology for detecting ballast fouling and structural anomalies in railway infrastructure. The proposed approach offers a viable solution for efficient and accurate assessment of ballast condition, ultimately leading to improved maintenance strategies and enhanced railway operations. Future research should focus on further optimizing SVM kernels, expanding the database, and refining hyperparameters to achieve even higher-quality results in real-world applications.

References

[1] Sussmann T R, Ruel M and Chrismer S M 2012 *Transportation Research Record* **2289** 87–94
[2] Bruzek R, Stark T D, Wilk S T, Thompson H B and Sussmann Jr T R 2016 Fouled ballast definitions and parameters *ASME/IEEE Joint Rail Conference*
[3] Selig E T and Waters J M 1994 *TRACK GEOTECHNOLOGY and SUBSTRUCTURE MANAGEMENT* (Thomas Telford Publishing)
[4] Fontul S, Fortunato E, De Chiara F, Burrinha R and Baldeiras M 2016 *Procedia Engineering* **143** 1193–1200 ISSN 1877-7058 advances in Transportation Geotechnics III
[5] Wang S, Liu G, Jing G, Feng Q, Liu H and Guo Y 2022 *Sensors* **22** ISSN 1424-8220
[6] Kingsuwannaphong T, Bräu C and Heberling D 2022 *Fouled railway ballast modeling using rigid body simulation* (Society of Exploration Geophysicists) pp 63–66
[7] Warren C, Giannopoulos A and Giannakis I 2016 *Computer Physics Communications* **209** 163–170 ISSN 0010-4655
[8] Xie X, Li P, Qin H, Liu L and Nobes D C 2013 *Journal of Geophysics and Engineering* **10** 034002
[9] Drissi K E K and Poljak D 2015 *WIT Transactions on Modelling and Simulation* **59** 13–24
[10] Guan B, Ihamouten A, Dérobert X, Guilbert D, Lambot S and Villain G 2017 *IEEE Journal of Selected Topics in Applied Earth Observations and Remote Sensing* **10** 4328–4336

Journal of Physics: Conference Series **2887** (2024) 012048 doi:10.1088/1742-6596/2887/1/012048

Calculation of the reflected wave from an inclined conducting cylinder for the array-type singlehole borehole radar

S Ebihara, R Tsukuda and S Yamanaka

Department of Electrical and Electronic Engineering, Osaka Electro-Communication University, Neyagawa, 572-8530, JAPAN (ebihara@osakac.ac.jp)

Abstract. This paper uses MoM/Green's function method to model the reflected wave from an inclined conducting cylinder efficiently for the array-type singlehole borehole radar. Using the exact analytical solution of the point electric source field around the conducting cylinder requires efficient modeling of the source in a borehole. We proposed replacing the source with the vertical electric dipole and magnetic one in a homogeneous medium. Numerical calculations with the proposed method gave the time domain signals of the reflections from the inclined conducting cylinder. In this simulation, the used antennas were the dipole and loop antennas. We ran the field experiment in the soil to measure the inclined conducting cylinder, similar to the one used in the numerical simulation. Observing this experimental data, we confirmed that the difference between the dipole and loop antenna signals was identical to the numerical simulation.

1. Introduction

In Japan, buildings are often rebuilt in response to frequent earthquakes. At this time, the foundation piles of the old building must be removed. Measurement of these foundation piles is required when construction documentation sometimes has gone [1]. Measurement tools such as borehole magnetic and borehole sonic can explore the foundation piles near it [2]. Borehole radar is also a promising technique for the measurement. The radar in an array antenna in a borehole would estimate the azimuth angle of the direction of the foundation pile [3]. An infinitely long conducting cylinder may model the foundation pile in interpreting the borehole radar signals.

The cylindrical layers, such as the borehole wall and medium inside the borehole, influence the borehole radar signals. Interpreting these radar signals requires modeling the cylindrical layers in electromagnetic analysis. The analytical exact solution for electromagnetic analysis can synthesize the array signals measured by an antenna in a borehole of the reflected waves from the infinitely long cylinder parallel to the borehole [4]. In this case, the borehole and cylinder cylindrical coordinates are in a parallel-shifted relationship. This situation allows the use of the addition theorem for the cylindrical coordinates, which helps express the wave functions [5]. However, when the axis of the borehole is twisted to that of the cylinder, we cannot use the addition theorem. This case corresponds to the scenario where the antennas in an inclined borehole measure vertical foundation piles [6].

In this paper, we introduce Green's function to represent the reflected fields in multiple cylindrical layers, which are reflected from other multiple cylindrical layers. The source is inside the same multiple cylindrical layers as the observation point. We assume that the two axes of the cylindrical layers are twisted, and the distance between the two is large. We compare the numerical data with the measured borehole radar in the last part of this paper.

2. Construction of Green's functions

Fig. 1 shows the coordinate systems representing the reflected fields from an infinitely long conducting cylinder for the borehole radar. There are cylindrical layers with the cylindrical coordinate system (ρ, ϕ, z). The rectangular coordinate system (x, y, z) shares the same origin. The axial unit vector is defined like $\mathbf{z}$ when it directs the z-axis. These layers represent the borehole, inside which the source and observation points exist. The other cylindrical layers are in the coordinate system (ρ'', ϕ'', z''), and an infinitely long conducting cylinder exists around the z''-axis. The shortest line segment of ones that connects these two axes of the z and z''-axis is on $z=z_l$, and the length of this line segment is L. Note that the z-axis is twisted to the z''-axis. This means that the angle between the vectors $\mathbf{z}$ and $\mathbf{z}''$ is θ_l, which is not zero. A point electric dipole is $\boldsymbol{J}=J_\phi\boldsymbol{\phi}+J_z\mathbf{z}$ at ($\rho=\rho'$, $\phi=\phi'$, $z=z'=0$) on the x-y plane in the innermost layer around the z-axis in Fig. 2(a). This wave source radiates the wave in the direction of the reflection point at ($\rho=\rho_0$, $\phi=\phi_0$, $z=z_0$), from which the reflected wave makes the electric fields E_ϕ and E_z at (ρ, ϕ, $z=z_r$) in the innermost layer of the cylindrical layers around the z-axis, where the ϕ and z components of the electric field are defined as E_ϕ and E_z, respectively.

If all the permittivity ε_i and permeability μ_i (i=1, 2, ..., N) in the cylindrical layers around the z-axis were the same values ε_N and μ_N, respectively, *i.e.*, the medium around the wave source and observation points was homogeneous, we would obtain the exact analytical solution of E_ϕ and E_z from Chap. 3 in [5]. If the z-axis and z''-axis were in parallel, *i.e.*, $\theta_l=0$, even when all the permittivity ε_i were different, usage of the addition theorems in Appendix D in [5] would give the exact solution for the electric fields. Assuming L is sufficiently large, in this paper, we would give the solution for the cylindrically inhomogeneous ($\varepsilon_i\neq\varepsilon_N$, $\mu_i\neq\mu_N$) around the z-axis with $\theta_l\neq0$ as follows.

2.1. Replacing the current in a cylindrical layer with a virtual current in a homogeneous medium

We propose the replacement of the point electric dipole $\boldsymbol{J}$ in Fig. 2(a) with another point electric dipole $\boldsymbol{J}_e=J_{ez}\mathbf{z}$ in Fig. 2(b) and magnetic dipole $\boldsymbol{J}_m=J_{mz}\mathbf{z}$ in Fig. 2(c). These dipoles $\boldsymbol{J}_e$ and $\boldsymbol{J}_m$ are in homogeneous media with the permittivity ε_N and μ_N, and the summation of the far fields radiated by $\boldsymbol{J}_e$ and $\boldsymbol{J}_m$ in the direction of the reflection point is the same as that by $\boldsymbol{J}$. In this case, the currents J_{ez} and J_{mz} for large L should be

$$J_{ez}=\frac{\varepsilon_N}{\varepsilon_1 k_{N\rho 0}^2 H_0^{(1)}(k_{N\rho 0}\rho_0)}\sum_{n=-\infty}^{\infty}F_{1,n} \quad \text{and} \quad J_{mz}=\frac{\mu_N}{\varepsilon_1 k_{N\rho 0}^2 H_0^{(1)}(k_{N\rho 0}\rho_0)}\sum_{n=-\infty}^{\infty}F_{2,n}, \tag{1}$$

where

$$\begin{bmatrix}F_{1,m}\\ F_{2,m}\end{bmatrix}=e^{in(\phi_0-\phi')}H_n^{(1)}(k_{N\rho 0}\rho_0)\tilde{\mathbf{T}}_{1N}^{(n)}\left\{J_z J_n(k_{1\rho 0}\rho')\begin{bmatrix}k_{1\rho 0}^2\\ 0\end{bmatrix}+J_\phi\begin{bmatrix}-\dfrac{nk_{z0}}{\rho'}J_n(k_{1\rho 0}\rho')\\ i\omega\varepsilon_1\dfrac{\partial}{\partial\rho'}J_n(k_{1\rho 0}\rho')\end{bmatrix}\right\}, \tag{2}$$

$k_{z0}=k_N\cos\theta_0$, $k_{i\rho 0}=\sqrt{k_i^2-k_{z0}^2}$, $k_i=\omega\sqrt{\varepsilon_i\mu_i}$ (i=1,2,...,N), the Bessel function is $J_n(\cdot)$, and the Hankel function of the first kind is $H_n^{(1)}(\cdot)$. The generalized transmission matrix $\tilde{\mathbf{T}}_{1N}^{(n)}$ is given in (3.2.11a) in [5].

2.2. Calculation of the reflected wave from the infinitely long conducting cylinder

After the replacement of the wave source in Fig. 2, we may calculate the reflected fields E'_ϕ and E'_z at (ρ, ϕ, $z=z_r$) from the conducting cylinder around the z''-axis, assuming that the medium is homogeneous, *i.e.*, $\varepsilon_i = \varepsilon_N$ and $\mu_i = \mu_N$. Here, we can use the exact solution of (3.3.21) in [5] to obtain the electromagnetic fields of the reflected wave. Following the reciprocity theorem, we can transfer the fields (E'_ϕ, E'_z) in the homogeneous media to those ones (E_ϕ, E_z) in the cylindrical layers in Fig. 1.

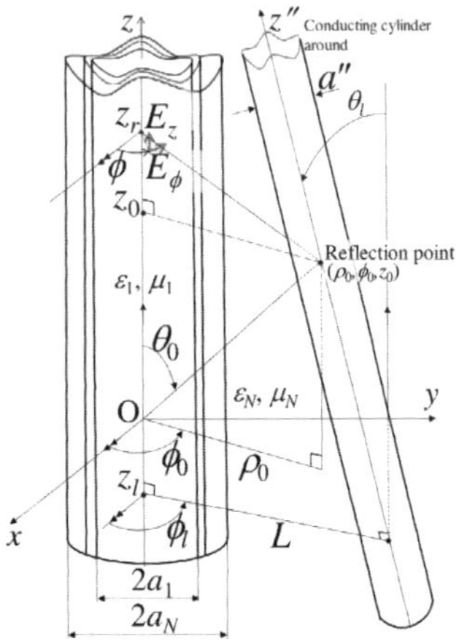

Figure 1. The fields (E_ϕ, E_z) excited by the point currents in the same cylinders and reflected by the inclined infinitely long conducting cylinder.

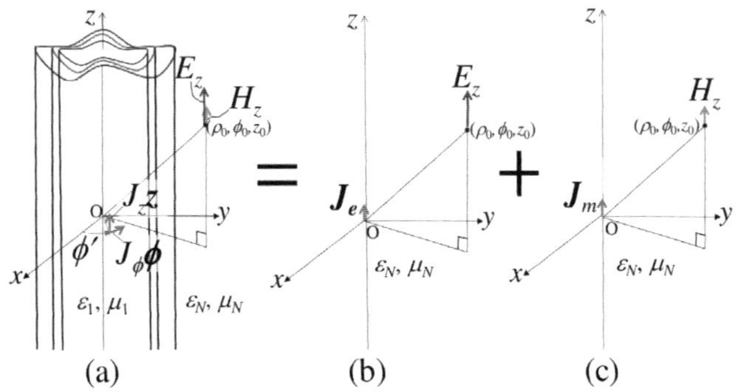

Figure 2. Replacement of the point electric dipole $\boldsymbol{J}= J_\phi\boldsymbol{\phi}+ J_z\boldsymbol{z}$ in cylindrical layers with the point electric dipole $\boldsymbol{J}_e$ and magnetic dipole $\boldsymbol{J}_m$ in homogeneous media.

3. Radar signals of the reflected wave from an inclined infinitely long conducting cylinder

We apply the method of moment (MoM) with Green's function to synthesize the borehole radar signals of the reflected wave from the conducting cylinder. We introduced the point electric dipole's fields proposed in Section 2 when calculating the Green's function. Fig. 3 shows the antenna model in the cylindrical layers. The four cylindrical layers imitate the fiber-reinforced plastic housing and fresh water, which is medium inside a borehole. The loop antenna is a transmitter, while the two dipole and one loop antennas are receivers. There is the inclined infinitely long conducting cylinder, like in Fig. 1, from which the reflected wave arrives at the receivers. The MoM outputs the transfer functions representing a transmission between the transmitter and each receiver in the frequency domain. We applied the bandpass windowing function and the inverse Fourier transform to the transfer functions and obtained the time domain signals of the receivers. Fig. 4(a) shows the exited wave in the time domain.

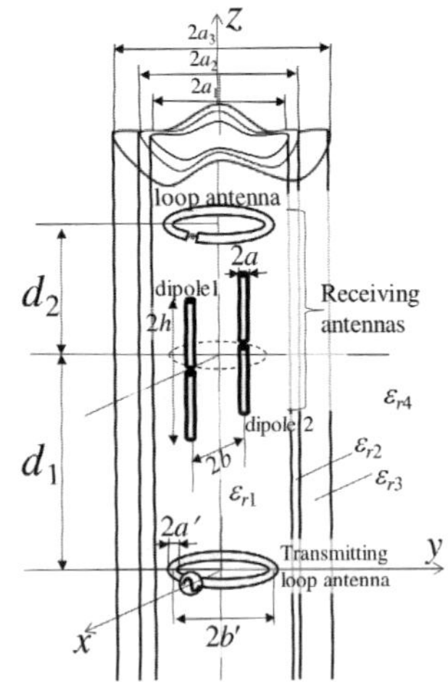

Figure 3. The model for the antennas in a borehole. d_1=57 cm, d_2=21 cm, $2a'$=2 mm, $2b'$=3.6 cm, $2h$=20 cm, $2a$= 2 mm, $2b$=4 cm, $2a_1$=4.7 cm, $2a_2$=5.7 cm, $2a_3$=12 cm, $\varepsilon_1=3\varepsilon_0$, $\varepsilon_2=5\varepsilon_0$, $\varepsilon_3 = 81\varepsilon_0 - j4.1\times10^{-3}/\omega$, $\varepsilon_4 = \varepsilon'_4 = 31.0\varepsilon_0 - j9.06\times10^{-2}/\omega$ at $\omega = 2\pi\times$ 100 MHz.

Fig. 4 (b) shows the simulated time domain signals. The reflected waves arrive at about 32 ns at all three antennas. Although the transmitting loop antenna radiates a horizontally polarized wave whose electric field oscillates in the horizontal direction, the dipole antenna sensing the vertically polarized wave receives the signals. This is because the inclined conducting cylinder changes the wave polarization. A phase delay exists between the first and second dipole antenna signals. This is because the second dipole antenna's position is closer to the conducting cylinder than the first one. The antenna efficiency of the dipole is larger than that of the loop antenna, which causes the larger amplitude of the dipole antenna.

We conducted a field experiment with antennas, boreholes, and conducting cylinders similar to those assumed in the numerical simulation. Fig. 4(c) shows the measured signals of the reflected waves. The direction and inclination of the conducting cylinder are also similar to the numerical simulation case. The test site was where we experimented in [3]. They used the dipole only, but we used both the dipole and loop in this paper. The reflected wave arrives between 32 and 35 ns at all the antennas. We can see the phase delay between the two dipole array signals and the amplitude of the dipole antenna is larger than the loop antenna. This is as we saw the numerical simulation result in Fig. 4(b).

4. Conclusions

When calculating the reflected wave from the conducting cylinder for the array-type borehole radar, we proposed replacing the point electric dipole in the cylindrical layers with an alternative electric and magnetic dipole in homogeneous media. These virtual wave sources would help calculate the array signals with the MoM. We compared the MoM and experimental signals and found similarities, confirming the accurate calculations of the MoM.

5. References

[1] Santos M C de S, Mendonça C A, Souza T de J, Hemsi P S, Gandolfo O C B 2022 Borehole magnetometry to determine the depth of reinforced foundations in telecommunication towers *GEOPHYSICS* **87** 4 pp EN33-43

[2] Coe J T, Kermani B, and Nyquist J 2019 Evaluation of Unknown Bridge Foundations Using Borehole-based Nondestructive Testing Methods: A Case Study in Urban Settings *Journal of Environmental and Engineering Geophysics* **24** 2 pp 299-315

[3] Ebihara S, Toyozawa K, Hasegawa T, Tsujikawa Y and Koyama K 2024 Influence of a Horizontally Polarized Wave Incidence on DOA Estimation by a Dipole Array Antenna in a Borehole *IEEE Transactions on Geoscience and Remote Sensing* **62** pp 1-16

[4] Ebihara S, Wada K, Karasawa S, and Kawata K 2017 Probe rotation effects on direction of arrival estimation in array-type directional borehole radar *Near Surface Geophysics* **15** pp 286-297

[5] Chew W C 1995 *Waves and Field in Inhomogeneous Media*. New York, NY, USA: IEEE Press

[6] Yamashita Y and Toshioka T 2002 Application of borehole radar to buried object survey for civil engineering purpose *Proc. 64th EAGE Conf. Exhib.* p. P041

Acknowledgments

We appreciate R Aoki and T Orihashi's assistance with the 2023 experiments. The Japan Society for the Promotion of Science (JSPS) KAKENHI Grant Number 21K04969 supported this work.

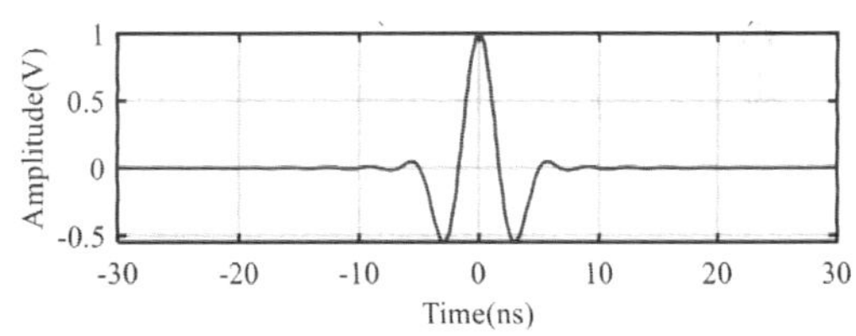

(a) Excited wave

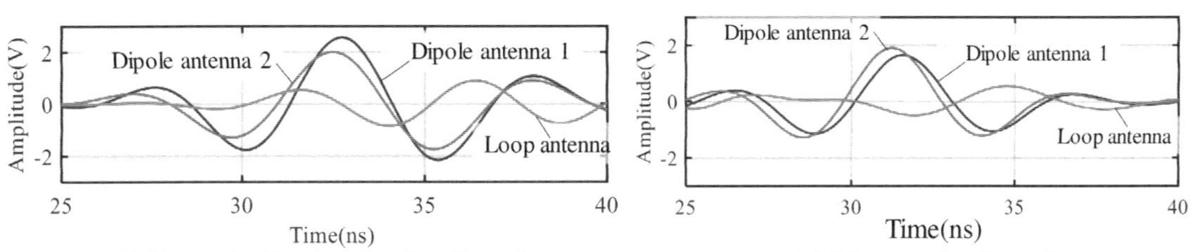

(b) Numerically simulated reflected waves

(c) Measured reflected waves

Figure 4. The excited wave and reflected waves from the conducting cylinder. The parameters in Fig. 1: L=72 cm, $\theta_l = -44°$, ϕ_l=236°, $2a''$=6 cm.

Journal of Physics: Conference Series **2887** (2024) 012049 doi:10.1088/1742-6596/2887/1/012049

Adding the outer pipe to a thin borehole radar probe in a thick water-filled borehole in rock to reduce direct coupling

S Ebihara, R Masui, K Koyama, Y Tsujikawa and Y Nishida

Department of Electrical and Electronic Engineering, Osaka Electro-Communication University, Neyagawa, 572-8530, JAPAN (ebihara@osakac.ac.jp)

Abstract. We show the experimental evidence that the additional outer pipe attached to a borehole radar probe reduces the direct wave's power in single-hole measurement and emphasizes the reflected wave from subsurface objects in hard rock. This pipe replaces the water layer with the air one when the difference between the diameter of the radar probe and that of the borehole is large, for example, by 4 cm. In the field experiment, the radar probe is scanned in the water-filled borehole in the rock in the two cases, with and without the additional outer pipe. The comparison between these two cases confirmed that the direct wave's power difference increases as the frequency gets higher and reaches 20 dB at 300 MHz. The radar probe with the additional outer pipe detected the reflected wave from the objects in the time domain. This contrasted with the fact that the radar probe without the additional outer pipe observed only the strong direct waves rather than the reflected waves.

1. Introduction

The geological disposal of radioactive waste is an important subject for humans. This disposal requires obtaining site descriptive models that capture geological characteristics [1]. One piece of this information is the detailed three-dimensional distribution of fractures in the rock over tens of meters. For this purpose, the borehole radar, which radiates electromagnetic waves of 30 – 300 MHz in a borehole, may be applied to this application.

The thin radar probe is preferred since this probe may be applied to boreholes having various diameters. The diameters of the commercial radar probes are usually several centimeters, while those of the boreholes for resource exploration may be more than 10 cm. These boreholes are usually drilled in rock and filled with water. In this case, the thickness of the water layers may work as dielectric materials having large permittivity. Some authors pointed out that the water layer influences the direct wave signals in the cross-hole in water-saturated sediments [2] and single-hole radar measurement in wet soil [3]. This effect may be increased in the water-filled borehole in hard, dry rock having small permittivity. This is because the permittivity contrast between the water and dry rock is larger than the one between the water and soil. An additional outer pipe (AOP) attached to a radar probe may reduce the influence of the water layer. The effect of the AOP has already been examined in the case of the single-hole measurement in wet soil [3]. We expect this additional outer cylinder to reduce the direct wave's power more effectively in rock than in wet soil.

In this paper, we experimentally investigate the effect of the additional outer pipe on the borehole radar probe on reducing a direct wave in the water-filled borehole drilled in the hard rock. We demonstrate that this additional outer pipe emphasizes the reflected waves from the object with the field experiment data.

2. Radar probe with the additional outer pipe

Fig. 1 shows the AOP attached to the fiber-reinforced plastics (FRP) vessel, which houses the thin borehole radar probes. The outer surfaces of the AOP are made of acrylic plastic, while the inside is filled with air. This means the AOP's permittivity is much smaller than water's. The inner diameter of the AOP is almost the same as the outer one of the radar probe. Furthermore, the outer diameter of the AOP is close to the diameter of the borehole in which the radar probe will be inserted. This AOP is removable from the radar probe.

3. Field experiments in a water-filled borehole in rock

We conducted the experiments at the field test site inside a gallery at the Nakatatsu mine in Fukui Prefecture, Japan. The host rock at this site is Skarn. The fracture zone is observed on the wall of the gallery. An inclined borehole with a diameter of 10 cm drilled from the gallery is filled with fresh water. This borehole intersects the fracture zone and sphalerite vein. In past measurements, the thick borehole radar probe received reflected waves from these objects [4]. The main difference between the experiments here and in [3] is the host medium of the site. The permittivity's contrast between the water and hard rock here is larger than that between the water and the wet soil. In this case, we expect the AOP's larger effect in the reduction of the direct wave's power than in the case of the wet soil. Fig. 2 shows the positional relationship between the radar probe and the objects. One of the loop antennas is for the transmitter, and the other is for the receiver. The loop antenna radiates and receives the horizontally polarized wave. A vector network analyzer measures the transfer function between the transmitter and the receiver in the frequency domain. The operating frequencies are 0.3 – 300 MHz.

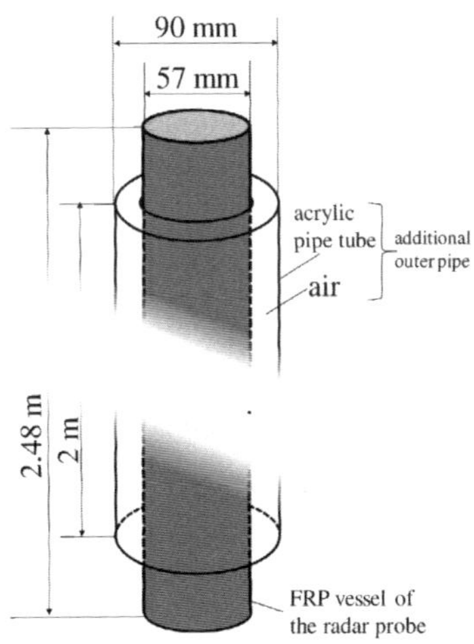

Figure 1. The additional outer pipe attached to the FRP vessel of the radar probe.

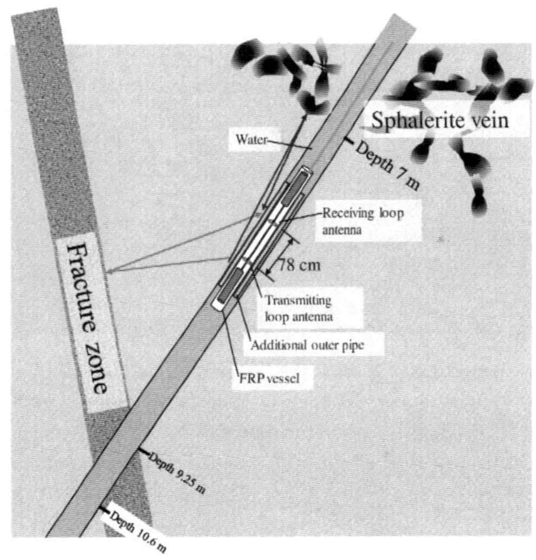

Figure 2. The position relationship between the radar probe and the objects.

Figure 3 shows the power spectrums of the transfer functions between the transmitter and the receiver in the frequency domain. At this depth, the power of the direct wave from the transmitter to the receiver approximately accounts for the total power spectrums shown in the figure since that of the reflected wave from the objects is small. It should be noted that the power difference between with and without the pipe increases as the frequency gets higher. This implies that the existence of the thin water layer increases the power of the direct wave. Figure 4 shows the time domain signals obtained from frequency domain data via inverse Fourier transform. The direct wave arrives at about 10 ns. This figure confirms that the direct wave's power without the additional pipe is much larger than that with the pipe.

Figure 5 shows the time-domain signals received at multiple depths. Note that comparing the amplitudes with and without the additional pipe is impossible since the amplitudes are normalized in these figures. Fig. 5 (a) includes the reflected waves from the objects. For example, the reflected waves from the sphalerite vein arrive from 12 ns at 830 cm to 20 ns at 885 cm in Fig. 5 (a), whose arrival times get late as the depth goes deep. Other reflected waves arrive from the fracture zone, whose arrival times change from 11 ns at 1000 cm to 20 ns at 900 cm. These waves' arrival times get late as the depth goes shallow. In Fig. 5 (b), we see only the strong direct wave, whose arrival times do not change from a depth of 796 to 980 cm. This means the strong direct wave shades the reflected waves from the sphalerite vein and fracture zone. This confirms that the AOP reduces the direct wave's power and highlights the reflected waves. It should be noted that the AOP reduces the powers of the direct waves rather than those of the reflected waves. This differs from the effect of the AOP for the measurement in the case of the wet soil [3], in which the AOP reduced both the powers of the direct wave and the reflected wave in the wet soil.

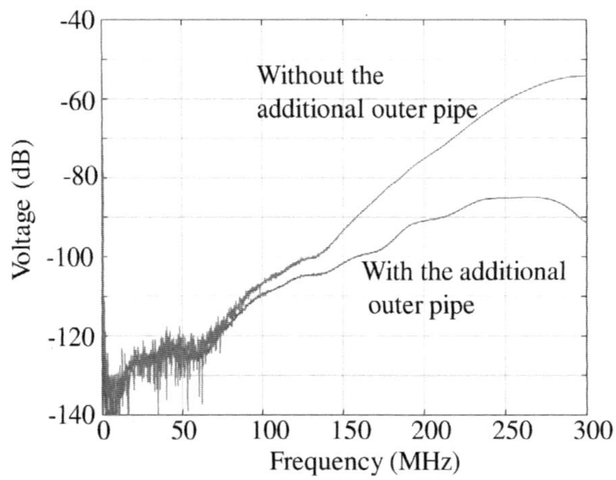

Figure 3. Comparison of power spectrums. The depth of the radar probe is 800 cm with the additional outer pipe and 804 cm without the pipe.

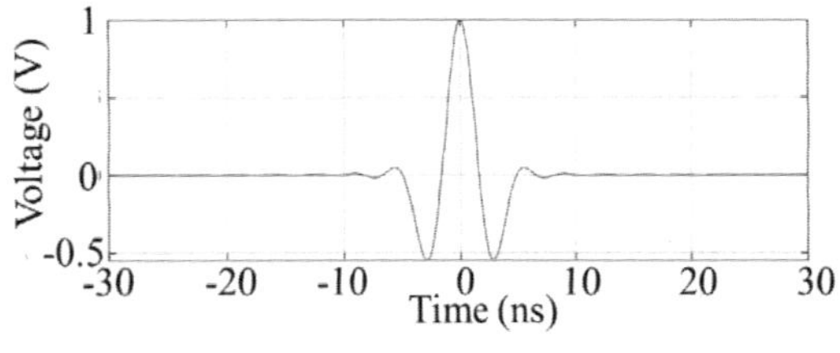

(a) Excited wave at the transmitter's feeding point.

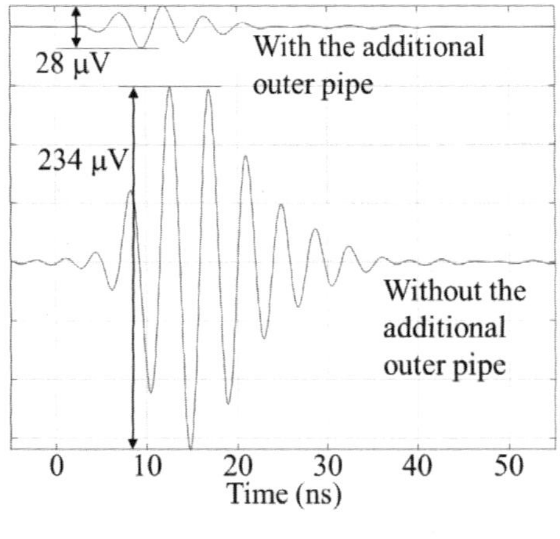

(b) Time domain signals measured at the same depths in Fig. 3.

Figure 4. Received time domain signals with the additional outer pipe.

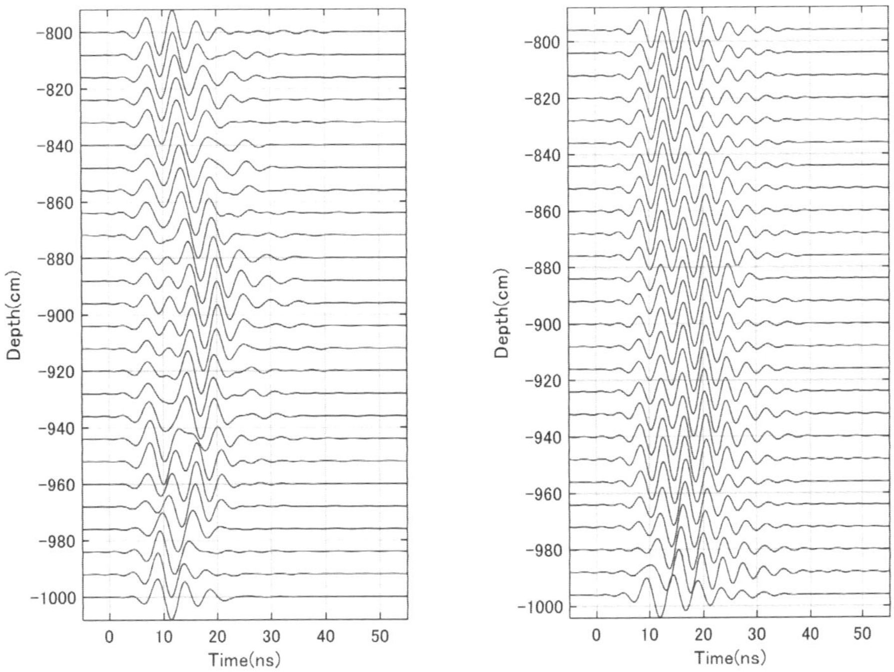

(a) with the additional outer pipe. **(b)** without the additional outer pipe.

Figure 5. Received time domain signals with the additional outer pipe.

4. Conclusions

We experimentally demonstrated that the additional outer pipe reduced the direct wave and highlighted the reflected waves from the subsurface objects. This pipe's effect in the hard rock in this paper differed from that in the wet soil, which was investigated in the previous work. The field experiment confirmed the direct wave's power reduction by 20 dB at 300 MHz when the additional outer pipe was attached to the radar probe.

5. References

[1] Nuclear Waste Management Organization of Japan (NUMO) 2021 *The NUMO Pre-siting SDM-based Safety Case* NUMO-TR-21-01

[2] Tronicke J and Holliger K 2004 Effects of gas- and water-filled boreholes on the amplitudes of crosshole georadar data as inferred from experimental evidence *Geophysics* **69** 5 pp 1255-1260

[3] Ebihara S, Nakamoto Y and Kawata K 2015 Cylindrical Air Layer to Reduce the Influence of the Direct Wave in Singlehole Borehole Radar *Proceedings of the 8th International Workshop on Advanced Ground-Penetrating Radar* (Florence) pp 1-6

[4] Ebihara S, Yamamoto S, Inada K, Ishida K, and Kawasaki A 2016 Experiments in Skarn with a Polarization-sensitive Directional Borehole Radar *Proceedings of the 16th International Conference of Ground Penetrating Radar (GPR 2016)*, Hong Kong, pp 13-16

Acknowledgments

The authors thank Kazuhiko Yamanaka of Mitsui Mining & Smelting Co., Ltd. for his permission to perform the experiment in the Nakatatsu mine. We appreciate Seiichi Ikeda at Nakatatsu Mine for his assistance with the experiments in September 2022. This work was supported by the Japan Society for the Promotion of Science (JSPS) KAKENHI Grant Number 21K04969.

Journal of Physics: Conference Series **2887** (2024) 012050 doi:10.1088/1742-6596/2887/1/012050

Russian RES and GPR research in Antarctica

S V Popov[1,2]

[1]Polar Marine Geosurvey Expedition (PMGE), 24, Pobedy str., Lomonosov, St. Petersburg, 198412, Russia (spopov67@yandex.ru)
[2]Institute of Earth Sciences, Saint Petersburg State University, 31–33 10-ya Liniya V.O., 199178 St. Petersburg, Russia

Abstract. Russian (former Soviet Union) investigations of Antarctica by radio-echo sounding (RES) were commenced in 1964. In two years later, ice-penetrating radar has been fixed on board the airplane IL-14 opening the door to future airborne studying of the Antarctic ice sheet and bedrock topography. By present the ground-based profiles and mostly airborne RES flights have covered more than 5.5 mln. sq. km of icy continent. Since 1985 Russian airborne research concentrated to study 60E – 110E sector of East Antarctica carrying out 5-km apart surveys. In the period of 1990s – 2010s ground-based RES research in scientific traverses were concentrated to study Lake Vostok and the Antarctic inland. Systematic Russian ground-penetrating radar (GPR) investigations was commenced in 2010s. Mostly it was a reason to study and revealing of crevasses in the area of logistic traverse route and other safety supply of the logistic operations. As well GPR engineering research used to find a right place to construct a new snow-runway at Mirny Station, also to investigate icy-runway at Novolazarevskaya Station and new snow airfield Zenith, near Progress Station, East Antarctica.

1. The initial stage of systematic geophysical research: 1950s - 1970s

An important old stage of Antarctic research finished with the beginning of the First International Geophysical Year (IGY), which took place in 1957-58. It lasted more than a hundred years, and had a pronounced geographical focus. Despite the fact that a large amount of research was carried out during that time, knowledge about the continent as a whole was fragmentary and descriptive. The most significant achievement of that stage was the determination of the coastline of the icy continent, the mapping of large glaciers and mountain ranges, as well as the creation of a system of geographical names for lands, coasts and seas. New stage which commenced with IGY brought active study of Antarctica, its ice sheet and subglacial relief [1,2]. The importance of the geophysical, glaciological and geological data collected in that time cannot be overestimated for related fields of science and lights are the basis of most constructions. In particular, the maps of ice thickness and subglacial relief are the base for creation of structural-tectonic schemes and geomorphological maps, also for glaciological and climatic reconstructions.

In February 1956, Russia (Soviet Union in that time), opened its first Antarctic station, Mirny, on the Davis Sea coast. This remarkable event opened the doors to study of the icy continent in the 20th century, including the ice sheet and subglacial topography. At that time, only very complex seismic methods were used for this [1,2]. The turning point came in February 1964, when new radio-echo sounding (RES) technique was carried out in the area of Mirny station for the first time in the history of Russian Antarctic research. The GUYS-1M4 naval radar was used for these first experiments [1,2]. Two years later it was installed on the IL-14 aircraft, and the first aviation tests were carried out. This opened wide prospects for using a new effective mobile geophysical method to study the glaciers and also subglacial relief. In a year later, in 1967, the first special Russian ice-penetrating radar was developed in the Arctic and Antarctic Research Institute (AARI) [1,2]. In February 1968, the first Russian airborne RES survey from IL-14 aircraft was carried out on Enderby Land (East Antarctica) with space between profiles of about 50 km. The GUYS-1M4 radar was used. The total length of the flights was about 11,000 km [1,2]. The location of all RES and seismic research is shown in fig. 1.

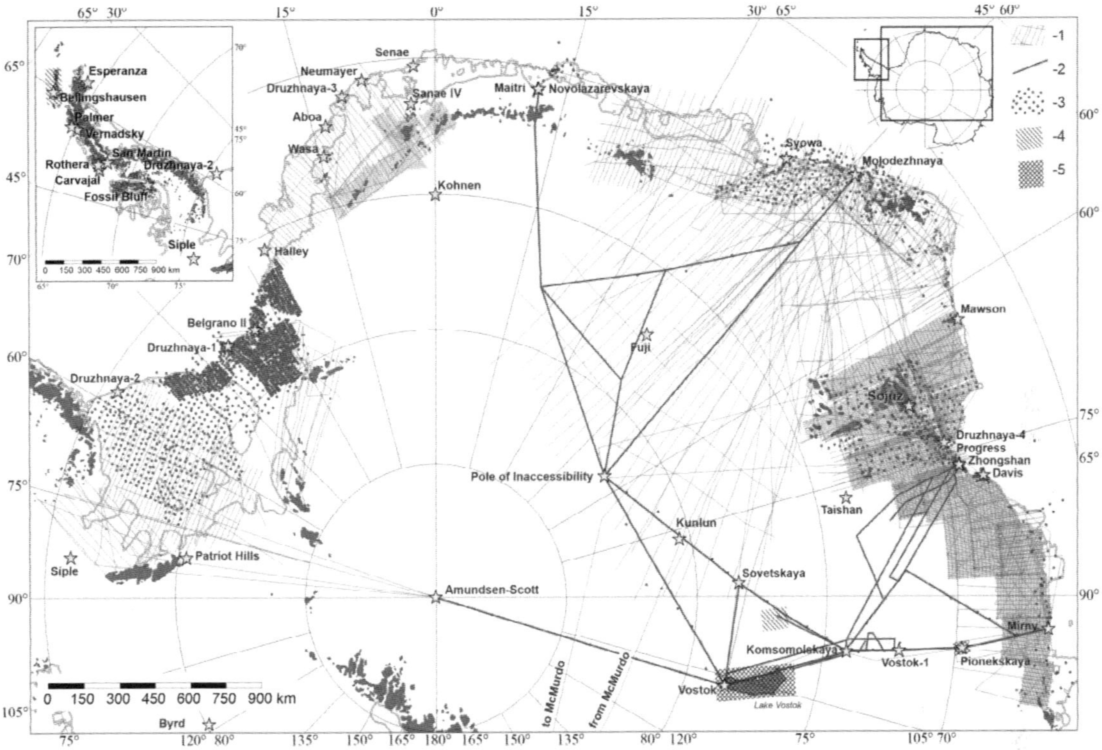

Figure 1. Location of Russian RES and seismic research. 1- airborne RES; 2- ground-based RES; 3- reflection seismic soundings; 4- area of ground-based RES survey; 5- area of ground-based RES and reflection seismic survey.

2. Stage of systematic airborne survey: 1970s – 1990s

The period from the beginning of 1970s to the end of 1980s is a “Golden Age of Russian Geophysics”. Over these two decades, complex airborne geophysical surveys, including RES, covered more than 5 million sq. km of Antarctica [1,2]. Operation “Amery”, which was carried out in 1971-74, can rightfully be considered the widest Russian project of that time which carried out by the Polar Marine Geosurvey Expedition (Polar Geophysical Expedition in that time) in the sector of 60E – 110E (fig. 1). The main tasks of operation "Amery" included a comprehensive study of the area of the Lambert-Amery glacier system and Lambert rift valley. Airborne RES flights were carried out from IL-14 aircraft along a regular network with the space between profiles of about 20 km. Special designed ice-penetrating radar RLS-60-67 with a main frequency of 60 MHz was used. Reflection seismic soundings were also carried out on a network of 30×30 km to measure the sea depths under Amery Ice Shelf (fig. 1) [1,2]. The collected data was taken as the base for subsequent structural and tectonic constructions for this region [3–5]. After a significant interruption, in 1985, geophysical research in this area was resumed by carrying out systematic airborne RES and magnetic surveys with the space between profiles of 5 km (fig. 1) [1,6].

The next important result of this stage was wide airborne geophysical research in the Antarctic inland. These investigations were carried out over three field seasons of 1987-90 using the flying laboratory based on IL-18D GAL airplane with space between profiles of 50 km [1,2]. One of the main results of these research was the revealing the main features of the subglacial Gamburtsev mountains, the largest mountain country, covered by ice which discovered in 1958 during the scientific traverse of the 3rd Soviet Expedition [7]. The next time the study of this difficult to access region was carried out under umbrella of the international project AGAP only two decades later [8,9].

The main result of this stage was the accumulation of plenty radio-echo and seismic data as Russian as by international expeditions. As following, this was a good opportunity to start ambitious international project Bedmap to compile all the data on the ice thickness and bedrock which collected in Antarctica.

3. Recent RES research: 1990s – 2020s

During the current stage, Russian research is focused on studying the sector of 60E – 110E (fig. 1). At the same time, there was a change in the scientific priorities: the main topic transforms to more deep study of the glacier structure and the subglacial processes including hydrological ones. Airborne investigations were carried out using a flying laboratory based on AN-2 aircraft. See details at [6].

The most grandiose scientific event of the end of the last century was the Lake Vostok revealing [10,11]. Due to the geographical location and logistics capabilities since 1995 Russian researchers began a systematic study of this area using ground-based remote geophysical methods. They included reflection seismic (since 1995) and RES (since 1998) [12]. These investigations were yielded to determine of the lake coast and the ice thickness in the drilling site vicinity. That was extremely important for safe penetration into Lake Vostok. Thise geophysical research was resulted in the area of the water surface of Lake Vostok is 15,790 km^2, eleven islands have been identified inside the lake, and 56 isolated subglacial reservoirs were also revealed. The total volume of Lake Vostok is estimated about 6100 km^3 [12].

In the same time during the field seasons of 2004-13 ground-based RES mostly with glaciology and geodesy were carried out along the "Mirny – Vostok" and "Progress – Vostok" traverse routes (fig. 1). They gave the first new information about glacier structure and bedrock topography of a large gap between Argus Dome and Lake Vostok. In addition, these researches discovered a subglacial lake located in 821-st km from Mirny to Vostok and one more was revealed exactly under Pionerskaya Station [1,2].

4. GPR research

Russian GPR research in Antarctica has been commenced in 2012 for safety supply of transport operations of the Russian Antarctic Expedition mostly [1,2]. Now they include searching right places for airfields, testing airfields, engineering surveys for the construction of new buildings of the Vostok station, choosing a site for unloading ships and others. The most important GPR engineering research was three-year investigations at Mirny station to construct new airfield for ski-chassis airplane. On February 10, 2016 the DC-3T (BT-67) "Turbobasler" airplane by ALCI Company was accepted. Thus, aviation communication with Mirny station, interrupted for two decades, was restored. After that GPR research was carried out at icy runway at Novolazarevskaya Station and new snow runway Zenith at Progress Station to check the internal structure there.

One of the most significant recent fundamental GPR research was the study of a vast depression 183×220 m, with a depth of 20-30 m in the Dålk glacier (Larseman Hills, Princess Elizabeth Land, East Antarctica). It formed after outburst of an interglacial lake [13,14]. Thus, researchers have a real opportunity to see and simulate the processes that occur under a very thick ice and multidisciplinary scientific research, including GPR, made it possible to create a mathematical model which describe this kind of processes [14].

Russian RES and GPR research continue as a systematic investigation to study glacier structure, bedrock, and subglacial processes especially subglacial hydrological processes.

Acknowledgments
The author thanks his colleagues from the Polar Marine Geosurvey Expedition, Arctic and Antarctic Research Institute, Russian Antarctic Expedition and Saint Petersburg State University with whom he worked in the field and processed RES and GPR data.

References

[1] Popov S 2020 Fifty-five years of Russian radio-echo sounding investigations in Antarctica *Ann Glaciol* **61(81)** 14–24
[2] Popov S V 2021 Six decades of radar and seismic research in Antarctica *Led i Sneg* **61(4)** 587–619
[3] Craddock C 1982 Geologic map of Antarctica *Am Assoc Pet Geol Bull* **66(7)** 962–963
[4] Elliot D H 1975 Tectonics of Antarctica: a review *Am J Sci* **275** 45–106
[5] Grikurov G E and Mikhalskii E V 2002 Tectonic structure and evolution of East Antarctica in the light of knowledge about supercontinents *Russian Journal of Earth Sciences* **4** 247–257
[6] Popov S 2022 Ice Cover, Subglacial Landscape, and Estimation of Bottom Melting of Mac. Robertson, Princess Elizabeth, Wilhelm II, and Western Queen Mary Lands, East Antarctica *Remote Sens* **14(1)** 241
[7] Sorokhtin O G, Avsyuk Yu N and Kopteev V I 1959 Results of determining the thickness of the ice sheet in East Antarctica *Inform bull SAE* **11** 9–13
[8] Bo S, Siegert M J, Mudd S M, Sugden D, Fujita S, Xiangbin C, Yunyun J, Xueyuan T and Yuansheng L 2009 The Gamburtsev mountains and the origin and early evolution of the Antarctic Ice Sheet *Nature* **459** 690–693
[9] Bell R E, Ferraccioli F, Creyts T T, Braaten D, Corr H, Das I, Damaske D, Frearson N, Jordan T and Rose K 2011 Widespread persistent thickening of the East Antarctic Ice Sheet by freezing from the base *Science* **331** 1592–1595
[10] Ridley J K, Cudlip W and Laxon S W 1993 Identification of subglacial lakes using ERS-1 radar altimeter *J Glaciol* **39** 625–634
[11] Kapitsa A P, Ridley J K, Robin G D Q, Siegert M J and Zotikov I A 1996 A large deep freshwater lake beneath the ice of central East Antarctica *Nature* **381** 684–686
[12] Popov S V, Masolov V N, Lukin V V and Popkov A M 2012 Russian seismic, radio and seismological investigations of subglacial Lake Vostok *Ice and Snow* **52(4)** 31–38
[13] Popov S V, Pryakhin S S, Bliakharskii D P, Pryakhina G V and Tyurin S V 2017 Vast ice depression in Dålk Glacier, East Antarctica *Ice and Snow* **57(3)** 427–432
[14] Boronina A, Popov S, Pryakhina G, Chetverova A, Ryzhova E and Grigoreva S. 2021 Formation of a large ice depression on Dålk Glacier (Larsemann Hills, East Antarctica) caused by the rapid drainage of an englacial cavity *J Glaciol* **67(266)** 1121–1136

Journal of Physics: Conference Series **2887** (2024) 012051 doi:10.1088/1742-6596/2887/1/012051

Characterization of GPR radiation pattern in the upper and lower hemisphere (theory and experiment)

P Morozov, I Prokopovich, M Lazarev, A Popov, V Sakhterov and D Edemsky

Pushkov Institute of Terrestrial Magnetism, Ionosphere and Radio Wave Propagation, Russian Academy of Sciences (IZMIRAN), 108840, 4 Kaluzhskoye Hwy, Troitsk, Moscow, Russia (prokop21@mail.ru)

Abstract. The directivity pattern of a powerful GPR has been studied in archeological works. The methods used: return pulse measurements by passing a compact underground object, multiple passages with changing antennas azimuth for estimating the horizontal directivity, use of a prolonged overhead object in order to estimate spurious radiation into the upper hemisphere.

1. Introduction

Ground penetrating radar (GPR) [1-5] is widely used in archeology, geology, civil engineering and safety issues. The directional characteristics of GPR radiation play an important role in planning measurements in nature or urban conditions. On the one hand, a wide radiation pattern increases the field of view of the device. On the other hand, especially in urban environments, "aerial" reflections from the upper hemisphere make it difficult to detect and identify the subterranean objects of interest. This article presents the results of theoretical and experimental studies of GPR resistively loaded dipole antennas directional patterns in the subsurface environment and upper half space, obtained by the field research team of a Russian Science Foundation project.

2. Equipment and methods

In this work, we studied the directional patterns of Loza GPR developed and mass-produced in Russia. Their antennas, made as dipoles of different length, have a resistive load approximately corresponding to the Wu-King law [6], in order to minimize "ringing" at the central frequency of the probing pulse The main design features of a mid-frequency version of this GPR are: receiver frequency band: 50-300 MHz, pulse voltage applied to the antenna: 5 kV, dynamic range: 120 dB, sampling frequency: 1 GHz, antenna central frequencies: 100, 200, 300 MHz.

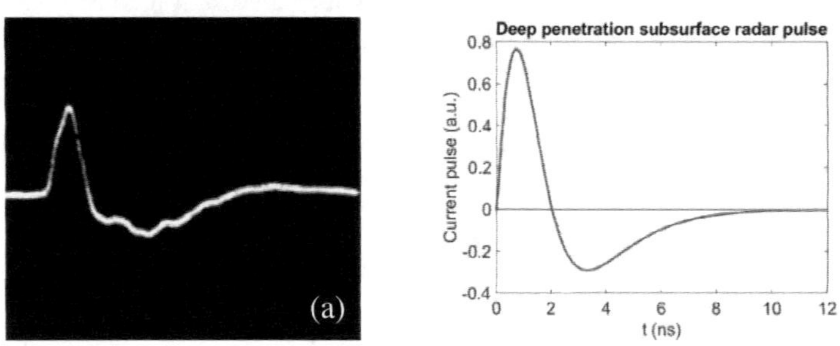

(b)

Figure 1. (a) Measured current pulse in a resistively loaded dipole antenna (arbitrary units) [2]; (b) Theoretical model of aperiodic GPR pulse [5].

3. Theoretical estimates

The GPR antenna emits an aperiodic pulse into the subsurface medium (figure 1). The directional pattern is determined by the soil parameters and antenna position. When the radar is located on the ground surface, the radiation pattern is substantially oriented towards the medium with a higher dielectric constant (figure 2). According to estimates [1, 4, 5], n^2 times more energy is emitted into the lower hemisphere (ground) than upwards (air), where n is the soil refractive index.

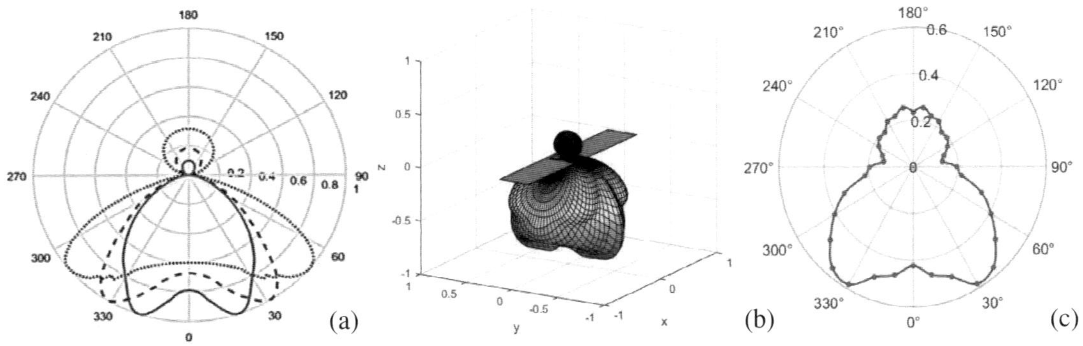

Figure 2. (a) Radiation patterns of an infinitely long dipole antenna placed on the surface of a dielectric half-space with $n=2$ (···); 4 (- - -); 9 (——); (b) Radiation pattern of a strip dipole at the air-ground interface ($n=4$, numerical simulation—gprMax software); (c) same, measured on a semi-cylindrical experimental stand (IZMIRAN).

Ground properties determine the angular width of the beam pattern. The higher the dielectric constant of the lower medium, the narrower is the radiation pattern, and more energy is emitted into the lower hemisphere (ground) (figure 2a). The beam angle in the lower hemisphere can be estimated as $\Theta = \arcsin(1/n)$. A 3D radiation pattern of a planar dipole GPR antenna in the lower half-space, calculated with gprMax open source software [7], is shown in figure 2b.

4. Experimental studies

The directivity of a GPR set is a product of the transmitter and receiver antenna patterns [8]. Three modes of measurements have been carried out in course of the expedition works:

(A)—2D vertical section of the GPR directivity pattern ("radar image" of the object) is constructed as the reflected signal amplitude when passing over a contrast underground object.
(B)—2D horizontal section of the radiation pattern in the upper hemisphere is obtained by several measurements of the GPR pulse scattered by a linear overhead object (e.g. electric power line) from one point underneath, with changing the antennas azimuth from 0° to 360°.
(C)—2D horizontal section of the radiation pattern in the lower hemisphere is measured by a series of passes above a local contrast object along the same trajectory, with different antenna skew angle on each pass.

4.1. Vertical radiation pattern in the lower hemisphere

The vertical section of the GPR directivity pattern has been obtained from the "hyperbola"—a characteristic reflected signal when crossing a compact archaeological object (two pipes of an ancient water supply in this case)—method A. The amplitude variation of the reflected signal over the passage reaches 20 dB (figure 3a).

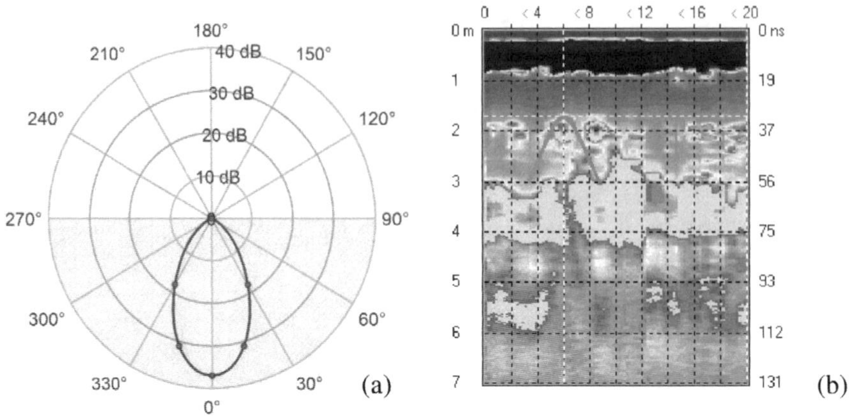

Figure 3. (a) Vertical directivity pattern in the lower hemisphere; (b) GPR B-scan across two antique water pipes.

4.2. Horizontal section of the GPR directivity pattern in the upper hemisphere

A horizontal section of the GPR directivity pattern has been obtained by measuring the reflected signal from a 10 kV aerial electric power line wire at a 15 meter from the GPR position (method B)—see figure 4.

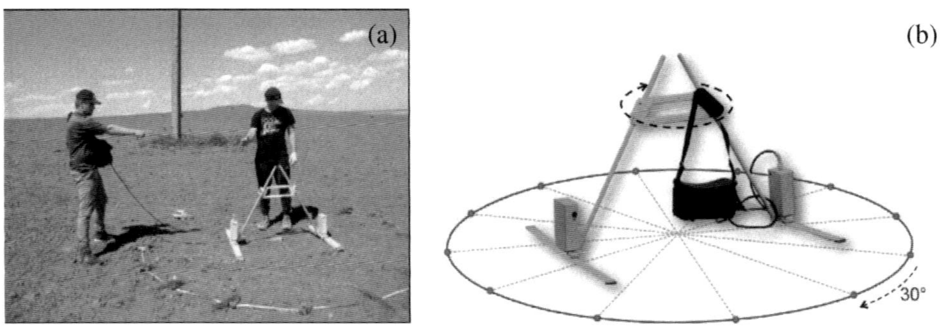

Figure 4. Varying GPR antennas orientation for directivity pattern measurements.

The amplitude difference of the signal reflected from an upper hemisphere object, between the maximum and minimum of the directivity pattern, reaches 40 dB (figure 5).

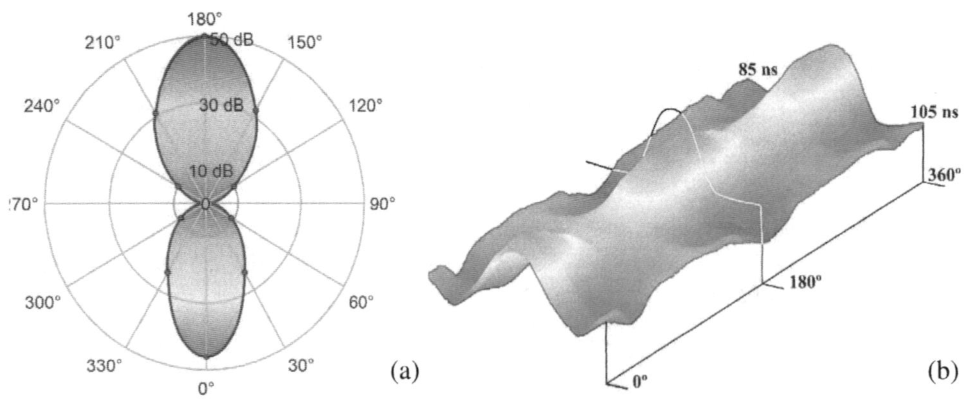

Figure 5. (a) Horizontal section, (b) 3D sweep of the GPR directivity pattern in the upper hemisphere.

4.3. Horizontal section of the GPR directivity pattern in the lower hemisphere

A horizontal section of the GPR directivity pattern has been obtained from the reflected signal trace (hyperbola clearly seen in figure 6c) from a compact archeological object (method C). The polar plot (figure 6a) has been derived from the amplitude values of the GPR return signal from the same compact subsurface object obtained by twelve passes along the 001 GPR profile (figure 6b) with changing by 30 degrees the orientation of the antennas from 0° to 360°. As shows figure 6a, the difference in the amplitude of the reflected signal reaches only 10 dB, which characterizes almost round GPR horizontal directivity pattern.

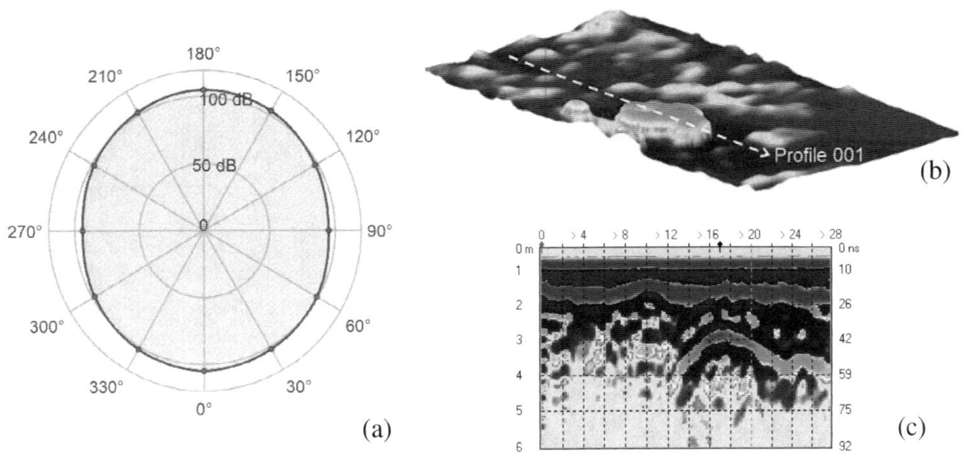

Figure 6. (a) Horizontal cross-section of the GPR directivity pattern in the lower hemisphere; (b) Horizontal slice of GPR directivity pattern at the 3 m depth; (c) GPR B-scan along the 001 profile across an ancient archeologic object.

5. Conclusion

GPR dipole antennas placed on the ground surface have a pronounced directivity pattern aimed to the subsurface medium. The shape of the directivity pattern depends on the subsurface medium properties. The methods of GPR directivity characterization in the lower and upper hemisphere are illustrated by practical examples of GPR studies of archeological objects.

6. References

[1] Daniels D J 2004 *Ground Penetrating Radar, 2nd ed.* (London: IET)
[2] Engheta N, Papas C H and Elachi C 1981 *Appl. Phys. B.* **26** 231
[3] Edemsky F, Popov A and Zapunidi S 2011 *IJET* **57** 407
[4] Arcone S 1995 *J. Appl. Geophys.* **33** 39
[5] Berkut A I, Edemsky D E, Kopeikin V V, Morozov P A, Prokopovich I V and Popov A V 2017 *Proc. Int. Workshop on Advanced Ground Penetrating Radar (Edinburgh)* pp 1–6
[6] Wu T T and King R W P 1965 *IEEE Trans. Antennas Propag.* **13** 369
[7] Giannopoulos A 2005 *Constr. Build. Mater.* **19** 755
[8] Morozov P, Morozov F, Lazarev M and Popov A 2023 *Remote Sensing* **15** 5452

Acknowledgments

Experimental research, as a geophysical support for Crimean expeditions of 2022 (State Hermitage Museum and RAS Institute of Archeology of Crimea) has been supported by the Russian Science Foundation, grant № 22-12-00083 "Deep GPR: theory, methods, experiment".

A Comprehensive Detection and Evaluation Method for the Internal Health Status of Highways

W Y Feng[1], D D Cheng[1,2], C H Wang[1], X Cheng[1], Y J Yin[1], H Y Geng[1]

[1]China Research Institute of Radio Propagation, Qingdao (fengwenya3@aliyun.com)

[2]College of Geoexploration Science and Technology, Jilin University, Changchun

Abstract: It is necessary to regularly diagnose highway health status. The current evaluation system pays less attention to internal damages, which makes the evaluation results not comprehensive and accurate enough. Therefore, Multi frequency array ground penetrating radar and falling weight deflectometer are adopted to detect the thickness of structural layers, underground disease, and structural bearing capacity. Establish evaluation criteria, quantify and grade the detection results, comprehensively evaluate health status of highways.Experimentresults demonstrate the feasibility and effectiveness of this method.

1. introduction

Some highways have not strictly met the design standards. Even for highways with strict requirements, long-term effects of water, heat, force, and other factors may have a certain negative impact on their structural performance [1-2]. Therefore, it is necessary to regularly diagnose the internal health status of highways[3-4].

At present, more attention is paid to surface disease and structural bearing capacity in the evaluation of highway. There is a lack of comprehensive diagnosis [5-6]. Therefore, ground penetrating radar and falling weight deflectometerare used to evaluate the internal structural layer, diseases, and structural bearing capacity. Statistical methods are used to establish evaluation criteria and classify levels.More scientifically judge of a highway is gotten[7-8].

2. Ground penetrating radar internal inspection and evaluation

Ground penetrating radar is used for internal detection of highways. Its principle is that the radar antenna sends high-frequency electromagnetic waves in the form of pulses to the underground. When encountering underground targets with electrical differences, the electromagnetic waves are reflected and received by the antenna [9].

2.1. Structural layer

Structural layers can be seen in the radar spectrum. Based on the propagation speed V of electromagnetic waves in the highway and the arrival time t of layer reflections, the thickness of structural layer$\boldsymbol{h}$ can be determined.

The structural layer index PTI is calculated by the qualification rate of layerthickness at all sampling locations. At the sampling locationi the thickness measurement value $\boldsymbol{h_{i1}}$ and the design thickness value $\boldsymbol{h_{i0}}$ are compared.If$\boldsymbol{h_{i1}} \geq \boldsymbol{h_{i0}}$, the position will be judged as qualified.The evaluation levelof structural layer is obtained According to the table1.

2.2. *Underground disease*

Underground disease including cavity, void, loose and water-rich can be seen in the radar spectrum. The area and the depth of underground disease can be detected. The risk index PRI for underground disease should be calculated according to the following equation:

$$PRI = 100 - \sum_{i=0}^{N} K_i(0.7P_{Ai} + 0.3P_{Bi}) \quad (1)$$

In the equation: $\boldsymbol{PRI}$——Risk index of underground disease, $0 \leqslant \boldsymbol{PRI} \leqslant 100$, no diseased $\boldsymbol{PRI} = 100$;N——total number of underground disease;$\boldsymbol{K_i}$—— Coefficient of the disease type. Such as 1.0 for cavity, 1.0 for void, 0.8 for loose, and0.7 for water-rich;$\boldsymbol{P_A}$—— Underground disease area index, $10 \leqslant \boldsymbol{P_A} \leqslant 40$, the larger the area, the larger the value;$\boldsymbol{P_B}$—— Depth index of underground disease, $10 \leqslant \boldsymbol{P_B} \leqslant 40$, the shallower the depth, the larger the value;The evaluation level of underground disease is obtained According to the table1.

3. Load bearing capacity detection and evaluation of falling weight deflectometer

A heavy hammer is lifted and dropped by the falling weight deflectometer.A pulseloadis applied to the road surface and the impact of wheel loads is simulatedduring driving [10]. A deformation of the road surface is causedby the applied load. The deformations are measured by multiple sensors and collected by a computer system to obtain the deflection values of each measuring point.

The structural strength index PSSI is calculated by the deflection value, using the following equation:

$$PSSI = \frac{100}{1+\alpha_0 e^{\alpha_1 SSR}} \quad (2)$$

$$SSR = \frac{l_0}{l} \quad (3)$$

In the equation: $\boldsymbol{PSSI}$——Structural strength index;SSR——Strength coefficient of pavement structure;$\boldsymbol{l_0}$—— Standard value of pavement deflection (0.01mm);**l**—— The measured deflection value of the road surface (0.01mm);$\boldsymbol{\alpha_0}$—— Model parameters, using 15.71;$\boldsymbol{\alpha_1}$—— Model parameters, using -5.19.The evaluation level of bearing capacity is obtained According to the table1.

4. Comprehensive detection and evaluation of internal health status of highways

The comprehensive inspection and evaluation process for the internal health status of highways is shown in the following figure:

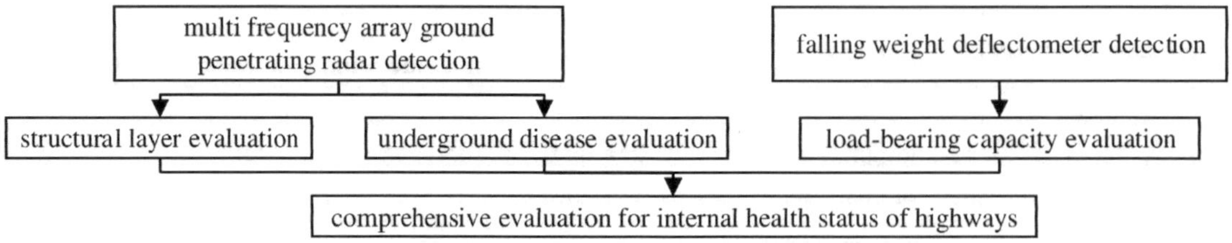

Figure 1. Comprehensive inspection and evaluation process for internal health status of highways.

The comprehensive health index PHI of highways is proposed as follows:

$$PHI = \omega_{PTI} \times PTI + \omega_{PRI} \times PRI + \omega_{PSSI} \times PSSI (6)$$

In the equation: PHI——Comprehensive indicator of internal health status of highways; PTI——Structural layer index;$\boldsymbol{PRI}$——Risk index of underground disease;$\boldsymbol{PSSI}$—— Structural strength index; $\boldsymbol{\omega_{PTI}}$——Weight coefficient of structural layer; $\boldsymbol{\omega_{PRI}}$——Weight coefficient of underground disease;$\boldsymbol{\omega_{PSSI}}$——Weight coefficient of structural strength; l

According to the following evaluation criteria, obtain the evaluation level of each indicator.

Table 1. Evaluationcriteria.

Evaluation level	1	2	3	4	5
PTI, PRI,PSSI, PHI	≥90	≥80,<90	≥ 70,<80	≥60,<70	<60

5. Detection and evaluation of experimental road

Select the K0+000~K1+000 section of the Hebei Avenue as the test section, with one evaluation unit per 200m. 1500M and 400M array ground penetrating radars were used to detect the structural layers and underground disease, with sampling intervals of 5cm. Using a falling weight deflectometer for bearing capacity testing, with a sampling interval of 50m. The above comprehensive evaluation method for the internal health status of highways was applied for evaluation, and the results are shown in Table 2.

According to the traditionalevaluation method, the pavement disease condition index *PCI* was calculated, which is also listed in Table 2.The structural layer index is generally high. The structural strength are also relatively high. In the unit K0+600~K0+800, two cavity were found. One was verified by core sampling, and the results are shown in the following figure. The index of underground disease was down to 70.0. Therefore, the comprehensive index in the K0+600~K0+800 unit was affected, with low values, while the other units remained high.Due to only minor disease such as cracks and ruts appearing on the test road surface, the traditional index isgenerally high.

Table 2. Comparison of internal health status and road surface damage status of highways.

Pile number	*PTI*	*PRI*	*PSSI*	*PHI*	*PCI*
K0+000~K0+200	95.89	100.0	87.34	94.97	96.61
K0+200~K0+400	94.48	100.0	87.10	94.47	92.42
K0+400~K0+600	93.34	90.5	86.97	90.29	91.88
K0+600~K0+800	81.38	70.0	82.05	77.03	88.30
K0+800~K1+000	97.34	100.0	92.98	97.10	94.26

(a)Detection results of cavity disease (b)Verification results

Figure 2. Detection and verification results of cavity disease in K0+600-K0+800 evaluation units.

The experiment found that the internal structural layers, underground diseases, and bearing capacity of highway can be detected by ground penetrating radar and falling weight deflectometer. Compared to the traditional index, the degradation of the internal health status of highways can be accurately reflected with the comprehensive evaluation method.2

6. conclusion

The traditional index *PCI* cannot accurately evaluate highways. Because the road cover isn't damaged by the underground diseases in the early stage of development, and the road surface is good. But if these diseases are not treated timely and effectively, surface diseaseseven driving accidents will be occured.

Therefore, multi frequency array ground penetrating radar and falling weight deflectometer are combined to detect Structural layer, underground disease, and structural bearing capacity. Acomprehensive evaluation of the highwaysinternal health status is used. It's a supplement to the traditional method and accurate to identify the causes of diseases.

References

[1] Ma D and Yang B W2022*Theory and Application of Highway Launch Site Modeling*(Beijing: Science Press)p 19

[2] Sang F, Luo R and Chen Y2023*J Construction and Building Materials.***408**370–5

[3] Liu S2023 *Evaluation of the performance of asphalt pavement on highways and research on preventive maintenance strategies*(Shijiazhuang:Shijiazhuang Railway University)p 15

[4] Jun L, Chen XL and Chun L X2023 *JCase Studies in Construction Materials.* **19** 279–83

[5] Highway Science Research Institute of the Ministry of Transport2019*JTG5210-2018 Evaluation Standard for Highway Technical Conditions*(Beijing: People's Transportation Press)p 43

[6] Beijing Survey Design and Research Institute Co. Ltd 2018*JGJ/T437-2018 Technical Standard for Comprehensive Detection and Risk Assessment of Urban Underground Disease*(Beijing: China Construction Industry Press)p 130

[7] Konstantinos G, Christina P and Andreas L 2023*J Sustainability.* **15** 3622–7

[8] Csaba T, Peter P2022*J Coatings.* **12**1944–50

[9] Calhoon T, Zegeye E and Velasquez R2022*J Construction and Building Materials.* **351** 831–7

[10] Edgar R 2023*Characterization and modeling of unbound and strategically stabilized materials for structural analysis of multilayered payment systems* (Texas: The University of Texas) p59

An intelligent power management system design of ground penetrating radar

Hongyan Geng[1 3] ,Yijun Gengv[1 2] ,Huiying Che[1] ,Qiuyuan Chen[1] ,Shanshan Huang[1] ,Minggang Ding[1] and Yun Liu[1]
[1] No.22 Research Institute of CETC, Qingdao 266107
[2]Corresponding author,email:373358603@qq.com
[3]Email: 8326580@qq.com

Abstract: A method for the design of an intelligent power management system for Ground Penetrating Radar (GPR) is presented in this article. For the mobile devices which are dependent on their battery power supply, the power management system is vital. The circuit design adopts a voltage stabilizer chip with low-loss and high-efficiency, greatly improving the equipment's continuous operating time. The power management system also provides the functions as follows: battery charging and discharging automatically switching; providing stable, intelligent and controllable power supply to subsequent circuits; the power differential mode, common mode filtering and π filtering significantly decrease the power noise and enhance the stability of the devices; the power supply circuit is designed with anti-reverse insertion and overheat protection functions, which can prevent the equipment from working within the fault and abnormal temperature range and reduce the risk of equipment failure.
Keywords: intelligent; GPR; power management system; low-loss

1. Introduction

In recent years, Ground Penetrating Radar technology has been widely used in fields such as landmines, geology, bridges, tunnels, pipeline detection, and soil assessment due to its high resolution, convenient assembly, and non-destructive detection [1-3]. The improvement of radar technology and the demand for intelligentialize and mobility have brought new challenges to power supply design. The traditional power module is inefficiency, large in size and cannot interact with the system in real time [4]. It cannot complete various functional control and information transmission. In order to meet the needs of new GPR, it is urgent to design a high -efficiency, low-loss and intelligent power management system [5].

2. Design of power management system

2.1. Requirements and Function Analysis

Ground Penetrating Radar (GPR) is a kind of activated electro-magnetic device, mainly detecting the targets by means of high-frequency electro-magnetic waves and receiving the return signals returned by the underground medium, analyzing and processing based on the echo signals, further exploring the measured targets and characteristics [6]. The GPR consists of a radar main unit and an aerial. The radar host is connected to the antenna through a comprehensive cable to provide working power supply and trigger pulses for the antenna. The power management system, the DSP board, the STEP board, the PULSE board, the capture and control components, the mainframe battery, and other components make up the radar mainframe. GPR's intelligent power management system is called the power management system. It can be thought of as the equipment's heart because it supplies the circuit with steady, continuous energy and can react quickly to prevent or minimize equipment loss. The DSP board mainly completes the AD acquisition, processing and uploading of echo; the primary output of the STEP board is a ps-level high-precision step signal; the PULSE board is a pulse formation circuit that provides trigger pulses and working power to the antenna; the acquisition and control component mainly realizes human-machine interaction and post-processing algorithm; the host battery is a portable rechargeable battery, when there is no adapter power supply, the host battery provides power to the GPR [6-8].

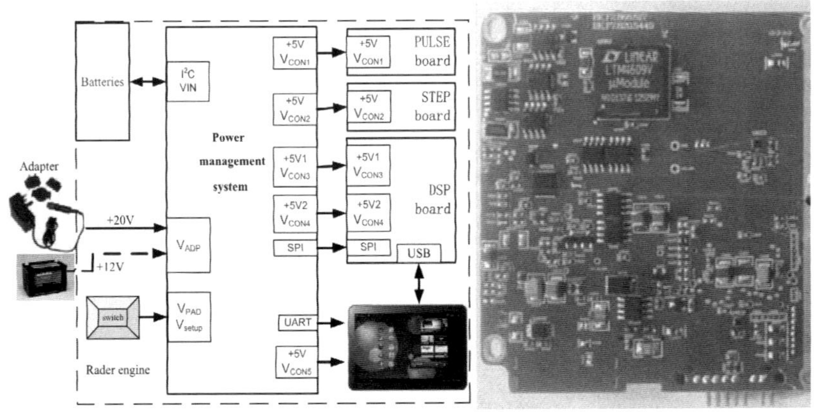

Figure 1. Power management system functional block diagram and circuit board

The DSP board, STEP board, PULSE board, and the capture and control component require an independently regulated +5V high-current power supply from the power management system. The power management system in general shall also be enabled to perform mainframe battery charging and to switch

Journal of Physics: Conference Series **2887** (2024) 012053 doi:10.1088/1742-6596/2887/1/012053

between battery and adapter power automatically. That means that when an adapter is available, the adapter powers the system and automatically charges the mainframe battery; when there is no adapter to power, the host battery will automatically provide power for the system. To realize intelligent control, the power management system uses the I^2C interface to read the battery level of the main unit, checks the GPR's inside temperature in real time, and uploads the relevant information to the acquisition control component and DSP board to realize the battery under voltage alarm, temperature super threshold alarm and other functions. In order to protect the GPR equipment, the adapter and battery end are designed with anti-reverse circuits; for the convenience of users, work indicator lights are also designed [9].

Six independent controlled high-current power sources are output by the power management system, as seen in Figure 1. These include two +5V power supplies for the DSP board, +5V power supplies for the PULSE and STEP boards, +5V power supplies for the acquisition and control component, and +13.1V power supply for the host battery. The power management system has three communication interfaces: it can communicate with the DSP board through the SPI port, the acquisition and control component through the UART port, and the host battery power can be read through I^2C interface.

2.2. System Circuit Design

Figure 2 displays the design circuit for the power management system:

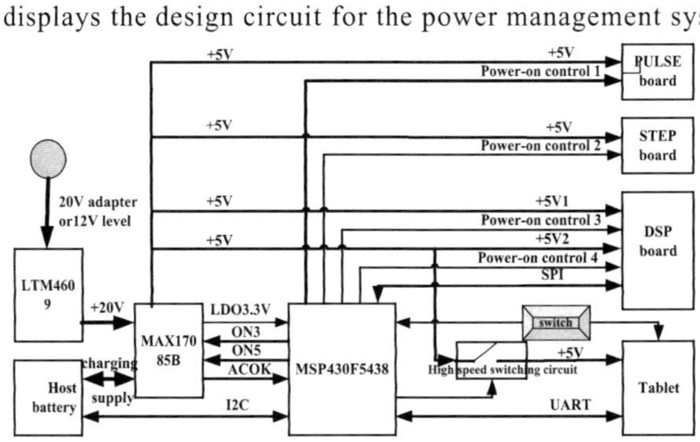

Figure 2. Circuit design diagram of the power management system

The MCU serves as the system's central control unit, managing the output and data exchange across different power supplies. Usually operating at low power, it can be awakened by an interruption.

When the adapter is not plugged in, the MAX17085B chip automatically selects the host battery to power the system. The MAX17085B chip uses the adapter power supply to power the system and automatically regulates the charge of the host battery when the adapter is placed into charging mode. The switch button causes the MCU to wake up when the host goes on. The MCU then flips on the high-speed switch to power the acquisition and control component, which then gets going. The MCU detects the host battery power through the I^2C port, detects whether there is an adapter charging, and uploads the data information through the UART port. The start-up instructions sent by the acquisition and control component, control the power-on 1, power-on 2, power-on 3 and power-on 4 in turn to power to PULSE board, STEP board and DSP board.

The stop operation instructions sent by the acquisition and control component, turn off the power-on 1, power-on 2, power-on 3 and power-on 4 in turn to power off the the PULSE board, STEP board and DSP board.

3. Conclusion

An MCU serves as the primary control chip in the GPR's power management system, enabling communication with the DSP board, acquisition and control component, and output controllable power supply and other intelligent controls. It takes MAX17085B as the power management module to realize the automatic switching of battery and adapter power supply; to realize the charging of the host battery; to provide LDO 3.3V, LDO 5V and +5V/8A, +3.3V/8A to supply power to the MCU, PULSE board, STEP board, DSP board and acquisition and control component. The first-level voltage regulator expands the input voltage range to +12V to +20V, which can be a +20V/3A adapter, or a +12V storage battery or a car battery input. Anti-reverse connection circuits, real-time monitoring of the system temperature, battery under-voltage alarm and working indicator lights are added in the circuit design, and this design improves the system's capacity for protection and real-time monitoring.

This power management system has been batch applied to GPR equipment, truly realizing the intelligent power management of the GPR. It is a great progress in power management technology.

Reference

[1] Pambudi A D, Fauss M, Ahmad F, et al. Minimax robust landmine detection using forward-looking ground-penetrating radar[J]. IEEE Transactions on Geoscience and Remote Sensing, 2020, 99: 1-10

[2] D. Pasculli, A. Natali, W. Salvatore, F. Morelli and D. Morandi. Investigation of reinforced concrete bridges by using a dual-polarized high-frequency GPR[C]. 2018 17th International Conference on Ground Penetrating Radar, 2018:1-5.

[3] Ebrahim S M, Medhat N I, Mansour K K, et al. Examination of soil effect upon GPR detectability of landmine with different orientations[J]. NRIAG Journal of Astronomy and Geophysics, 2018, 7(1): 90-98.

[4] Yang T C, Dai S X, Deng H Y, et al. Application research of GPR in quality detection of urban underground jacked pipes construction[J]. IOP Conference Series Earth and Environmental Science, 2019, 295: 042069.

[5] Tess X H L, Wallace W L L, Ray k W C, et al. GPR imaging criteria[J]. Journal of Applied Geophysics, 2019, 165: 37-48.

[6] Shiau J K, Ma D M, Yang P Y, et al. Design of a solar power management system for an experimental UAV[J]. IEEE Transactions on Aerospace and Electronic Systems, 2009, 45(4): 1350-1360.

[7] Kotra S, Mishra M K. A supervisory power management system for a hybrid microgrid with HESS[J]. IEEE Transactions on Industrial Electronics, 2017, 64(5): 3640-3649.

[8] Benini L, Bogliolo A, De Micheli G. A survey of design techniques for system-level dynamic power management[J]. IEEE Transactions on Very Large Scale Integration (VLSI) Systems, 2000, 8(3): 299-316.

[9]Li J., Zeng Z., Sun J., Liu F., (2012) Through-wall detection of human being's movement by UWB radar. IEEE Geoscience and Remote Sensing Letter, 9(6), 1079-1083.

Journal of Physics: Conference Series **2887** (2024) 012054 | doi:10.1088/1742-6596/2887/1/012054

Accurate real-time ANN classification of cavities and utilities using 3D GPR data in the MFCC domain

J N Ogunbo[1,2], S H Baek[1] and S W Kim[1]

[1]Isung Corporation, GPR Division, South Korea
[2]nosajide@yahoo.com

Abstract. Real-time identification of serviceable utilities and cavities is critical for the safety of urban settlements because delay has been reportedly costly and destructive. For utilities, maintaining their integrity is the goal, while for cavity imaging, prevention or mitigation becomes the focus. Several existing research has recommended some methodology for classification of these utilities and cavities, but their approaches are either geometrically restrictive or cumbersome. We therefore present a practical methodology that provides accurate real-time classification of cavity, box, manhole, patch, and pipe. 3D GPR data is reshaped into 1D supertrace and a 2D mel-frequency cepstrum coefficients (MFFC) feature are computed, serving as an input pattern, paired up with a utility class. The ANN is subsequently trained and tested during training to avoid overfitting of data. 99% model testing accuracy was achieved after 100 epochs, translating to only 3 misclassifications out of 1000 3D GPR data inputs used for computing a confusion matrix.

1. Introduction

At the heart of urban settlement's environmental ecosystem are the serviceable utilities, although unwanted inutilities may occasionally leech on to the ecosystem. Serviceable utilities are manmade infrastructures that support urban's socioeconomic convenience, for example the manholes, pipes, box and so on. However, the unwanted inutilities, like the cavities that form naturally, stemming from hydrologic and geologic factors [1]. While the regular monitoring of the integrity of the utilities is essential, the timely detection of inutilities cannot be overemphasized as well, especially since they pose serious safety hazards in urban settlements. Attendant consequences of collapsed subsurface cavities are the sinkhole-induced lakes and ponds. Indeed, safe, and timely identification of sinkholes have far-reaching benefits at saving both cost and human lives, and even preventing mining accidents [1].

Safe detection of utilities and inutilities have been reportedly done by non-intrusive geophysical procedure like the high-frequency ground penetrating radar (GPR) method. At a high frequency, the GPR works on electromagnetic induction principles, mapping the dielectric contrasts between different materials in the subsurface. GPR A-scan and B-scan, respectively 1D [2] and 2D images, have been applied to identify cavities but the final interpretation is geometrically limited and non-unique. However, the 3D (C-scan) GPR survey offers a more reliable characterization of the subsurface object, lending to unique interpretation. The use of drilled boreholes for mapping subsurface utilities and inutilities are reportedly accurate but not practical in terms of cost and time [1]. Timely detection is key to adequate planning, monitoring, and mitigation of integrity of utilities, as well as the formation of inutilities. GPR interpretation has truly evolved from the primordial manual visualization of GPR data, to traditional data inversion workflow involving multiple calls to forward modelling of synthetic data and computation of sensitivity matrix until a prescribed convergence value is reached. This approach is not only inefficient but also cumbersome, considering the setting values for several hyperparameters required for optimum performance [3].

Nowadays, artificial intelligence (AI) has been deployed to produce results in real-time. AI feeds on big data, apparently needed for establishing patterns common to the same class of target objects which are different from other patterns classifying other types of target objects. Large number of misclassifications are inevitably reported from GPR experiments conducted on 1D and 2D geometry due to the inadequate characterization by geometrical extent. Khudoyarov et al. [4] performed Convolutional Neural Network (CNN) classification of underground objects from 3D GPR data sets. Their methodology is cumbersome, containing restrictive conditions of the presence of the peak of hyperbolic shape inside the sliding 3D box. The success of their classifications dictates that clear hyperbolic pattern is non-negotiable, and a fixed data volume is compulsory ingredient for a good performance. Such approach will fail if the data acquisition misses a part of the hyperbola, therefore their results are suspects.

We will present the results of ANN classification of utilities and inutilities, using the 2D features provided by mel-frequency cepstrum coefficients (MFFC). The approach uses 3D GPR data to overcome the limitation of geometrical extent, providing more flexibility to not focus on centralizing the peak of the hyperbola, that uses variable volume of data set. It is believed that a part of the data for a particular object is sufficient to provide a distinguishing feature that differentiates it from other objects. Although MFCCs, as features, have been especially used for voice signal processing and found relevant for industrial and medical analysis [5]. Salamon et al. [6] classified urban sounds (gunshot, car horns etc) in the MFCC domain. We particularly adapted the MFCC sound application in [6] to solving ANN classification of 3D GPR data set. The 3D GPR data is reshaped into 1D supertrace, which is converted into 2D MFCC features, serving as input pattern pairing up with classification flags for cavities and the other utilities.

2. Methodology

The workflow sets up for executing the project involves.

2.1. 3D GPR data

High resolution GPR data is acquired based on the principle of electromagnetic induction. 3D Radar signals sent into the subsurface from the transmitter are reflected and recorded at the receivers. The radargrams indicate dielectric contrasts between the host rock and the target geologic materials. Since dielectric properties of utilities and inutilities produce different characteristic radargrams, the GPR method is readily helpful for their classification.

2.2. Mel Frequency Cepstral Coefficients (MFCC)

MFCCs are features which have extensive application in speech recognition. Its computation involves 5 basic steps [5] namely: signal framing, power spectrum computation, mel filter bank application, finding the logarithm of filter banks and applying the discrete cosine transform. The output is a 2D cepstral of MFCCs at different frequencies.

2.3. Artificial Neural Network (ANN)

The unique pairing of input patterns with output targets (classes) is the strength of the supervised feedforward artificial neural network, ANN [7]. The 3D GPR data may contain complicated features in the time-space domain, which may limit the performance of the ANN trained machine. Therefore, in the MFCC domain, a unique feature with a reduced dimension is an utmost preference.

2.3.1. MFCC of supertrace. The traces in the 3D GPR data volume are combined into a single trace called supertrace. This 1D supertrace is converted to the 2D MFCCs cepstral, serving as the input pattern for a particular class of object (cavities or utilities).

Journal of Physics: Conference Series **2887** (2024) 012054 doi:10.1088/1742-6596/2887/1/012054

3. Classification by ANN

The sequence of classification includes these basic steps:

- Supply a signal-processed 3D GPR data set. Reshape the 3D GPR data set into 1D supertrace.
- Generate a 2D MFCC feature of the 1D supertrace,
- Feed forward the MFCC feature as an input pattern to ANN machine, pairing up with preassigned class labels. Train and validate the ANN for a specified number of epochs.

5 classes of objects are the targets namely: cavity, box, manhole, patch, and pipe. An average of 2000 3D GPR data sets were prepared for each class, 20% of which will be randomly used for testing during the ANN training.

Figure 1 shows the typical 3D GPR data acquired over cavity, box, manhole, patch, and pipe structures. Their corresponding supertraces and the 2D MFCC features are respectively presented in figures 2 and 3.

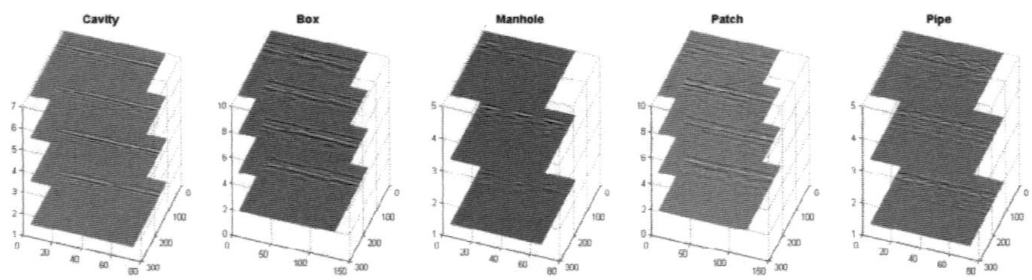

Figure 1. Typical 3D GPR data for cavity, box, manhole, patch, and pipe.

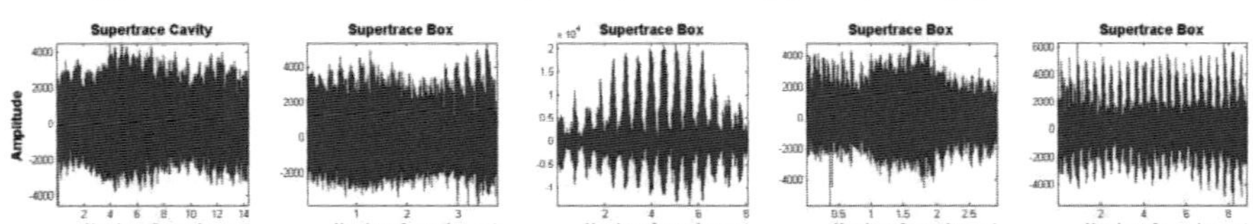

Figure 2. Supertraces for typical 3D GPR data for cavity, box, manhole, patch, and pipe.

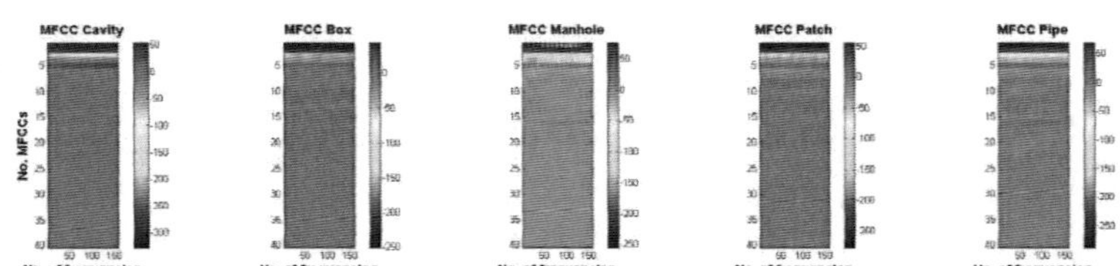

Figure 3. 2D MFCC features for typical cavity, box, manhole, patch, and pipe supertraces.

A 3-layer (including the input layer) ANN architecture is designed for the training, testing and validation with respective activation functions of ReLU, ReLU and softmax. 40 MFCCs are used for this training and the hidden layers have 256 neurons. The training of the machine lasted for one hundred epochs updating the weights and biases by the Adam's optimizer. Figures 4 and 5 presents the percentage accuracy and loss with epoch. For both training and testing, the model accuracies of the increase from 35% and 50% to 97% and 99% respectively with corresponding decrease in the model loss values.

The ANN machine was finally tested on 200 3D GPR for each class. The result is presented as a confusion matrix in Table 1. 3D GPR for actual cavity, box, manhole and pipes were accurately predicted without any misclassification. However, actual patch has 3 misclassifications, 2 as box and 1 as manhole.

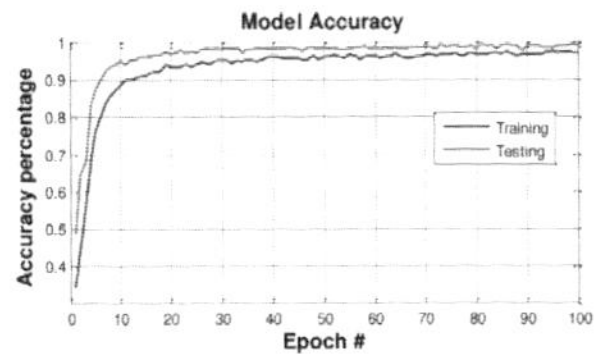

Figure 4. Model accuracy with epoch.

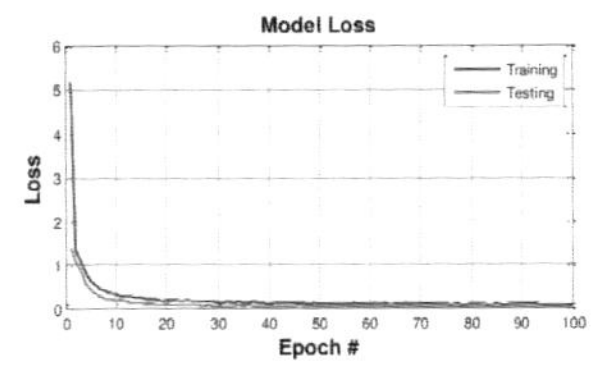

Figure 5. Model loss with epoch.

Table 1. Confusion Matrix from ANN classification

Predicted **Actual**	Cavity	Box	Manhole	Patch	Pipe
Cavity	200	0	0	0	0
Box	0	200	0	0	0
Manhole	0	0	200	0	0
Patch	0	2	1	197	0
Pipe	0	0	0	0	200

4. Conclusions

The mel-frequency cepstral coefficients (MFCC) feature has been effectively used as an input pattern for training an ANN which accurately classifies cavity, box, manhole, patch, and pipe. 2D MFCC is computed from 1D supertrace reshaped from 3D GPR data set. 99% accuracy of the model tested is reported during training of the ANN and confirmed in a computed confusion matrix, potentially offering an accurate real-time prevention and mitigation of urban settlement utilities and inutilites.

Acknowledgments

This research was supported by a grand (20024996) of Development of social complex disaster response technology funded by the Ministry of Interior and Safety (MOIS, Korea).

References

[1] Kuhns G L 1983 *Application of Ground Penetrating Radar to the Detection of Subsurface Cavities* (Florida: University of Central Florida) p 87

[2] Ogunbo N J 2022 Algorithm for accurate velocity analysis of 1D acoustic reflection and transmission responses by projection constrained multi-physics inversion *J. Applied Geophysics* jappgeo(199)104593

[3] Ogunbo N J 2018 MATLAB code for data-driven initial model of 1D Schlumberger sounding curve *Geophysics* GEO(83)2

[4] Khudoyarov S, Kim N and Lee J 2020 Three-dimensional convolution neural network-based underground object classification using three-dimensional ground penetrating radar data *Structural Health Monitoring* SHM(00)

[5] Abdul Z K H, Al-Talabani A K 2022 Mel frequency Cepstral Coefficient and its Applications: A Review *IEEE Access* Access(10)

[6] Salamon J, Jacoby C and Bello J P 2014 *A dataset and Taxonomy for Urban Sound Research* (USA: Orlando USA) p 103

[7] Xiong H., Li J., Li Z., & Zhang Z., (2024). GPR-GAN: A ground-penetrating radar data generative adversarial network. *IEEE Transactions on Geoscience and Remote Sensing* 62, 1-14.

The GPR and ERT detection for road hazard in loess region

Jiahe Yan[1], Zhaofa Zeng[1], Baizhou An[2], Boyang Sun[2,3], Zhilian Li[1], Haoyu Wang[1] and Jianwei Zhao[1]

[1] College of Geo-Exploration Science and Technology, Jilin University, Changchun 130026, China
[2] Geophysical and Geochemical Exploration Institute of Ningxia Hui Autonomous Region (Autonomous Regional Deep Earth Exploration Center), Yinchuan 750001, China
[3] School of Resources and Geosciences, China University of Mining and Technology, Xuzhou 221116, China

Abstract. The road hazards have frequently happened in the loess region in Ningxia autonomous region. The reason for this phenomenon is that the loess under the roads has always changed because of its properties, such as collapsibility, and leads to the instability of the geological structure. Detection of the loess under the roads can obtain the information about the geological structure and identify potential road hazard regions, which has great significance. Therefore, in this paper, we conducted a field experiment about road hazard in loess region using ground penetrating radar (GPR) and electrical resistivity tomography (ERT). The field experiment demonstrated that each method can be used in the detection and identification of potential road hazard but limited in their different detection capabilities, but better results can be obtained by using both methods comprehensively.

1. Introduction

In recent years, the road hazards have already been characterized by a wide range, high frequency and severe losses, and have been a serious problem in Ningxia autonomous region. This is because there are large areas of loess located in Ningxia autonomous region. However, the loess has the property of collapsibility [1]. It will cause a large area of collapse when suffering long-term rainfall and accumulated water [2]. It will also cause serious road hazards when it happens under the roads. Therefore, in order to enable early warning and prevent the road hazards, it is of great significance to detect the loess structure under the roads.

Geophysical detection is a non-invasive technique used to study the subsurface structures. Among all geophysical methods, ground penetrating radar (GPR) and electrical resistivity tomography (ERT) have already been widely used in loess detection and road hazard detection [3][4]. However, these two methods have different resolution in different region. Therefore, it is necessary to detect the road hazard in loess region by comprehensively using these two methods.

In this paper, we conducted a field experiment to detect the road hazards in loess region using GPR and ERT in Ningxia autonomous region. We measured GPR data using the antenna with the frequency of 200 MHz and 400 MHz, and measured ERT data using Wenner and Dipole-Dipole array. After that, we processed the GPR and ERT data, and comprehensively analysed the road hazards in the experiment field.

Journal of Physics: Conference Series **2887** (2024) 012055 doi:10.1088/1742-6596/2887/1/012055

2. Field experiment in Ningxia autonomous region

2.1. Experiment site

The experiment site is located in Guyuan, Ningxia autonomous region, as shown in figure 1. A GPR and an ERT survey line have been carried out in the experiment field, with a length of 30 m. The GPR data was measured on the road using the antenna with the frequency of 200 and 400 MHz, respectively. There is a total of 60 electrodes in the ERT survey line, with 0.5 m electrode spacing. The Wenner and Dipole-Dipole array were used to measure ERT data, each of which contains 570 data. In the field, there is a culvert located between 8 and 10 m of the survey line, which would produce a large anomaly in the GPR and ERT images.

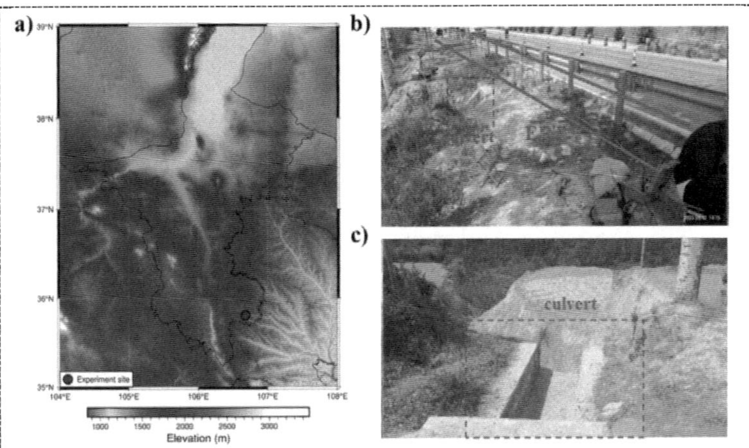

Figure 1. a) The location of the field experiment site. b) ERT and GPR survey line in the field. c) The culvert under the road.

2.2. GPR and ERT Results

The GPR images measured by the antenna with the frequency of 200 and 400 MHz were processed by removing direct current component, applied gain by power and amplitude decay, as shown in figure 2.

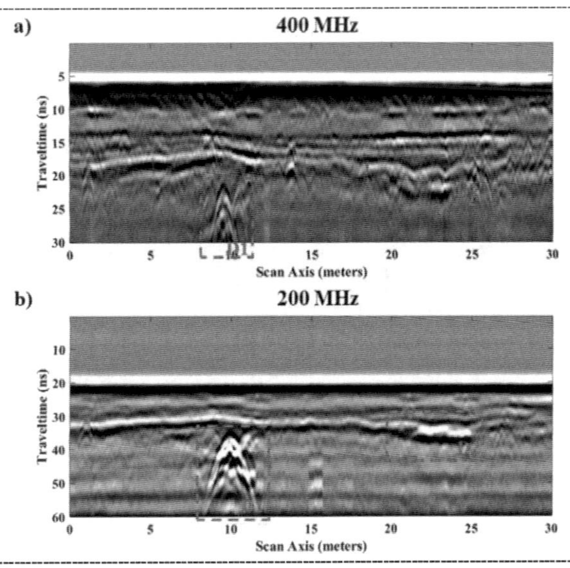

Figure 2. The processed GPR images. a) 400 MHz; b) 200 MHz.

There are two reflection anomalies can be seen in GPR images in figure 2, named D1 and D2. D1 is a local strong reflection signal located in X =10 m. D2 is also a regional strong reflection under the lowest in-phase axis of the GPR images.

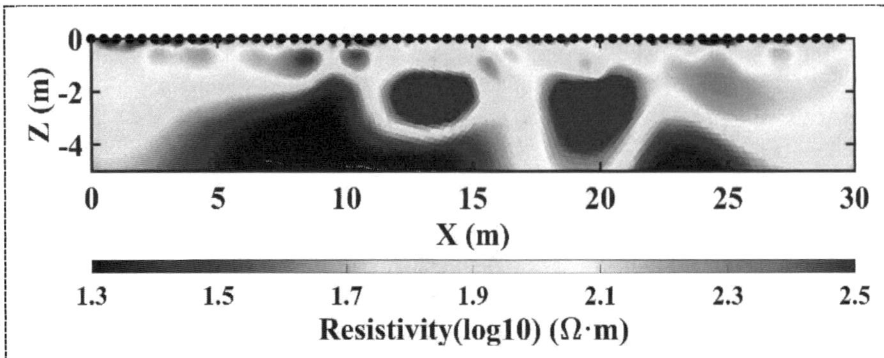

Figure 3. The resistivity image inverted from ERT data.

Figure 3 shows the resistivity image inverted from ERT data using Wenner and Dipole-Dipole array. As can be seen in figure 3, the left and right part of the shallow part of the resistivity image are dominated by high resistivity, while the middle region shows low resistivity. There is also a large area of low resistivity located in the deep left and right part of the image. In addition, there are two high resistivity regions at the depth more than 2m in the resistivity image, which can reflect the property of the loess under the road at that depth to some degrees.

3. Discussion

As important near-surface geophysical methods, GPR and ERT are usually used together in unsaturated zones and collapse region [5]-[7]. Therefore, we applied these two methods to test the imaging effect of these two methods in road hazard detection in loess region.

In figure 2, it is reasonable to infer that D1 is a reflection anomaly caused by the culvert in figure 1c. This is inferred according to the field situation. In figure 3, the culvert is a high- permittivity structure because it usually contains steel reinforcing bars and it is used for drainage with a high-water content. There is a low resistivity region on the left part of the culvert, which is inferred as seepage region. However, it cannot be seen in GPR image because its depth has already been more than the detection depth of the GPR.

D2 is a regional strong reflection in GPR image, which has the characteristics of B-scan profile of road subsidence [8]. This is inferred because its in-phase axis is shifted downward at X=20-24m. Therefore, it is reasonable to infer that D2 is a road subsidence region. There is a high-resistivity region under D2, which is inferred to be region with high air content.

To verify the reliability of the GPR interpretation, a validation experiment was carried out. We established a road permittivity model and conducted GPR simulation using gprMax. The experiment showed that the synthetized B-scan profile in figure 4 is similar to the GPR profile measured in field experiment, which proves the reliability of interpretation.

The comprehensively analysis of GPR and ERT image shows that both GPR and ERT method can be used in road hazard detection in loess region but there are still some limitations. The GPR images show that it has high resolution in shallow region but its detection depth is limited. In addition, it is easily affected by the low resistivity region, which means the structure under the low region is hardly imaged. Compared to GPR, ERT shows a greater capability in detecting low resistivity region but its resolution is definitely lower than that of GPR. Therefore, in order to fully utilize the advantages of both methods, a combination of these two methods is necessary in road hazard detection in loess region, as indicated in the field experiment.

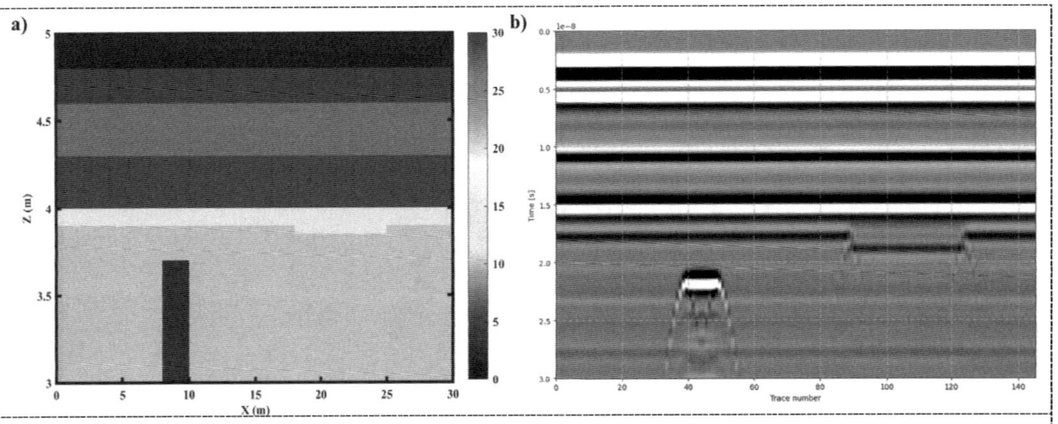

Figure 4. GPR validation experiment with frequency of 400MHz. a) permittivity model; b) synthetized B-scan profile.

4. Conclusions

In this paper, a field experiment was conducted to evaluate the validity of GPR and ERT in road hazard detection in loess region. GPR can be used in the detection of reflection interfaces underground but limited in detection depth and capability in low resistivity regions, while ERT has a deeper detection depth and higher resolution in low resistivity regions but is weak in resolution. Both methods have their limitations, but the comprehensively use of these two methods can maximize their advantages in road detection in loess regions, which can provide better references for road hazard warnings. Therefore, it is recommended that both two methods are used in road hazard detection in loess region.

5. References

[1] Wang L Q, Shao S J and She F T 2020 A new method for evaluating loess collapsibility and its application *Eng Geol* **264** 105376.

[2] Meng X F, Liao H J and Zhang J W 2021 Research on the collapsibility of loess after water immersion *Nat Hazards* **109** 303-28.

[3] Lv G, Yang J, Li N, Hu D X, Zhang Y and Zhao F 2018 Dielectric Characteristics of Unsaturated Loess and the Safety Detection of the Road Subgrade Based on GPR *J Sensors* **2018**.

[4] Wu J Q, Dai F C, Liu P, Huang Z Q and Meng L C 2023 Application of the electrical resistivity tomography in groundwater detection on loess plateau *Sci Rep-Uk* **13**.

[5] Tao M, Chen X, Cheng Q B and Binley A 2022 Evaluating the joint use of GPR and ERT on mapping shallow subsurface features of karst critical zone in southwest China *Vadose Zone J* **21**.

[6] Carrière S D, Chalikakis K, Sénéchal G, Danquigny C and Emblanch C 2013 Combining Electrical Resistivity Tomography and Ground Penetrating Radar to study geological structuring of karst Unsaturated Zone *J Appl Geophys* **94** 31-41.

[7] Gómez-Ortiz D and Martín-Crespo T 2012 Assessing the risk of subsidence of a sinkhole collapse using ground penetrating radar and electrical resistivity tomography *Eng Geol* **149** 1-12.

[8] Pan J W, Shi Z S, Meng X, Yue Y P, Lin C D, Chen J, Liu H and Cui J 2023 Reflection characteristics of typical road defects in 3D GPR images for collapse mitigation *J Appl Geophys* **217**.

Acknowledgments

The research was funded by National Key Research and Development Program of China (grant number: 2022YFC3003402) and 2022 Science and Technology Benefiting the People Project of Ningxia Hui Autonomous Region (grant number: 2022CMG03006).

Journal of Physics: Conference Series **2887** (2024) 012056 doi:10.1088/1742-6596/2887/1/012056

GPR circumferential survey line radius's effect on shield machine tunnel look-ahead detection

W Duan[1,2], X Xie[1,2,3], T Zhu[4], Y Yang[4], H Wu[5] and Y Pei[1,2]

[1] Department of Geotechnical Engineering College of Civil Engineering Tongji University, Shanghai, 200092, China (2310025@tongji.edu.cn)
[2] Key Laboratory of Geotechnical & Underground Engineering of Ministry of Education Tongji University, Shanghai, 200092, China
[3] State Key Laboratory of Disaster Reduction in Civil Engineering, Shanghai, 200092, China
[4] Department of Road and Bridge Shanghai Tunnel Engineering Quality Inspection Co., Ltd, Shanghai, 201109, China
[5] Office of the Chief Engineering Shanghai Tunnel Engineering Co., Ltd, Shanghai, 200032, China

Abstract. In southern cities of China shield machines often encounter obstacles such as large boulders in complex strata which may cause more abrasion to the shield machine's cutterhead and even make the shield machine stuck. Therefore, it is important to know distribution of obstacles in front of excavation face in advance. Ground penetrating radar (GPR) is an efficient and accurate detection method to ahead detect large boulders when GPR antennas are put in the cutter head. Synthetic simulations of GPR look-ahead detection are carried out based on finite difference time domain mothed. Antennas are put at two different radii of 4 meters and 7 meters within a 15-meter diameter cutter head. GPR response of boulders in various positions can be clearly found in B-scan images and effect of survey line radius is discussed. Reflection wave in circumferential survey line has continuous sinusoidal wave feature which is different from hyperbola curve feature in conditional straight survey line. The reflection waves of boulders in GPR B-scan images show same characteristics at two different radii of the survey line. Numerical simulation results provide evidences for GPR in accurate tunnel forward-looking detection in complex strata.

1. Introduction

The problem of boulders detection in composite strata has become more and more common in shield tunnel construction in recent years [1]. Many scholars have proposed different methods to detect isolated rocks, such as micromotion detection, opposing coils transient electromagnetic method, cross-hole detection, etc [2, 3, 4]. These methods detect boulders from the ground surface along the tunnel route before excavation. With the advancement of detection equipment technology, detection accuracy of these methods has been improved. According to field engineering experience, when diameter of boulders is larger than 80cm, these boulders will cause damage to shield machines [5]. As the designed depths of tunnels gradually increase, it is more and more difficult to meet the detection accuracy requirement by these surface detection methods.

Ground Penetrating Radar (GPR) is also used to detect isolated stones from the surface as a high-precision method [6]. However, even using the low-frequency ground-penetrating radar, it is still difficult to detect the shield tunnelling strata from the surface when the tunnel depth is very large. On the contrary, by installing the ground-penetrating radar on the shield machine, it is possible to horizontally and forward detect the shield tunnelling strata. GPR starts to work simultaneously with the cutter rotating and the measuring orbit is a circle. When obstacles in the stratum are located directly in front of the cutter centre, GPR with 3-meter radius measuring line can detect these obstacles [7]. However, obstacles are randomly distributed in space and diameters of some shield tunnels are close to 15 meters. GPR detection effect in large diameter tunnels needs to be further analysed.

In this paper, based on the principle of time-domain finite difference, GPR forward numerical simulations are carried out for a 15-meter-diameter shield tunnel. GPR antennas are put in two measurement lines with 4-meter and 7-meter radii. Boulders are randomly distributed in the shield excavation section. Influence of GPR antenna's radius on the GPR look ahead detection is compared and discussed. Simulation results aims to provide guidance for the reasonable placement of GPR antennas.

2. Forward numerical simulation method

GPR works by following the basic principles of electromagnetic waves, which is the set of Maxwell's equations. As a method of solving Maxwell's equations using differential grid, time-domain finite-difference principle has now become a common numerical method for GPR simulation [8]. Numerical simulations in this paper are carried out by using gprMax which is an open-source simulation software based on the time-domain finite-difference principle [9].

The size of a three-dimensional model is 10*16*16 meters. The 3D model consists of a 15-meter diameter excavation section with 1m thickness, soil layer and boulders. Length of differential grid is 0.05 meters. The absorbing boundary conditions is first order Complex Frequency Shifted Perfect Matched Layer. The emission source is a Ricker waveform and the polarisation direction is x-direction. The centre frequency of the waveform is 50MHz. The distance between the receiving point and the transmitting point is 2 meters considering the antennas' offset. The time window is set to 400 ns with 4156 iterations. The electromagnetic properties of materials are listed in table 1.

Table 1. Electromagnetic properties of materials.

Materials	Relative Permittivity	Conductivity/ (S/m)	Relative Permeability
Soil	16	0.04	1
Rock	6	0.004	1
Free space	1	0	1

A simulated boulder shape combines a 1.2-meter-diameter sphere and a 1-meter-long cube to create the inhomogeneous surface of actual boulders. The positions of boulders are designed to be deviated from the centre of shield excavation surface.

The measurement step is 0.2 meters. Rotation direction of GPR antennas is counter-clockwise. When the radius of the measurement line is 4 meters, one circle of measurement line includes 126 traces of A-scan and the total length is 25.2 meters. When the radius of the measurement line is 7 meters, a B-scan includes 220 traces of A-scan and the total length is 44 meters.

3. Simulation result and discussion

The coordinates of the centre of the shield excavation surface in the 3D model are (1,8,8). The simulated shield excavation direction is x-direction. Transceivers are put at 4-meter radius and 7-meter radius circles separately.

3.1. One boulder close to the tunnel central axis

The centre coordinate of one boulder is (5.5, 6, 6). The projection of this boulder in the excavation surface is 2.83 m from the tunnel centre point. The locations of the transceiver points are 4 meters and 7 meters separately away from the centre of the shield excavation surface. The positional relationships of the boulder projection point and GPR circumferential survey lines are as shown in figure 1.

Journal of Physics: Conference Series **2887** (2024) 012056 doi:10.1088/1742-6596/2887/1/012056

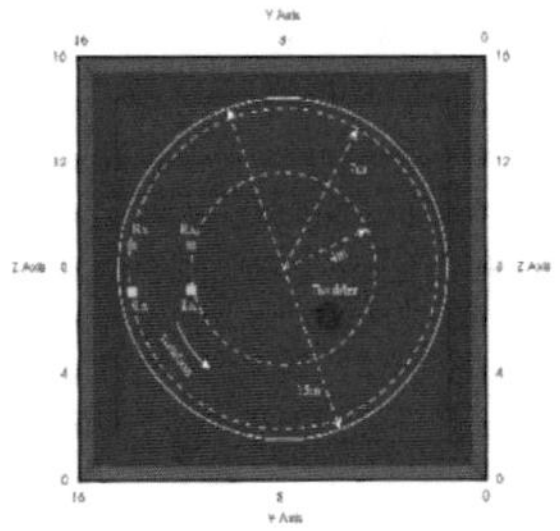

Figure 1. The positional relationships of the boulder projection point and GPR circumferential survey lines.

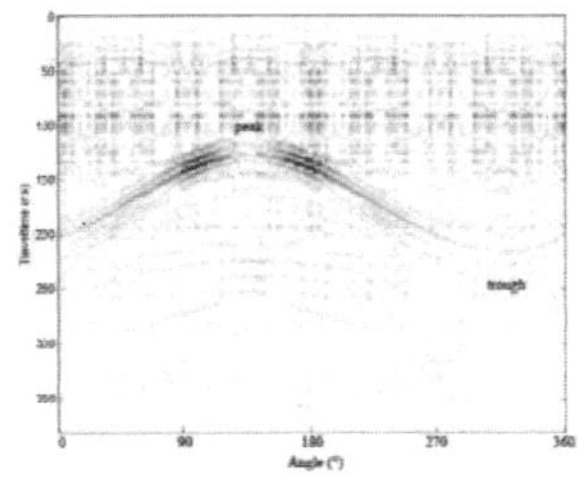

Figure 2. GPR B-scan image of 4-meter radius survey line of one boulder's projection point close to the centre.

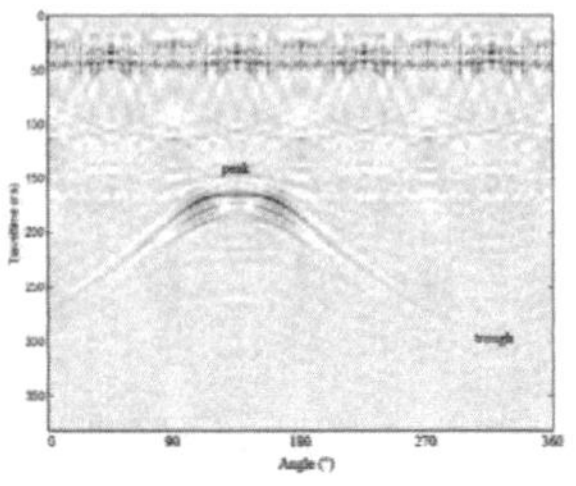

Figure 3. GPR B-scan image of 7-meter radius survey line of one boulder's projection point close to the centre.

The transceiver antenna was placed on a 4-meter radius circle and rotated one cycle to form the B-scan image as shown in figure 2. The location of this boulder remains unchanged and GPR B-scan image of 7-meter radius survey line is as shown in figure 3. From figure 2 and figure 3 the reflection wave of the boulder can be clearly found. Its form is a curved line with clear peak and slight trough.

Comparing figure 2 and figure 3, it can be found that shape of refection wave is similar. However, the vertical coordinate value of the reflection wave peak in Figure 3 is larger. The increase of the vertical coordinate value is due to the fact that the distance between the 7-meter radius transceiver point and the stone is farther away than the 4-meter radius.

3.2. One boulder close to the tunnel edge

The boulder's centre coordinate is arbitrarily changed to (7.5,3,10). Distance between the boulder and the tunnel central axis is 5.39 meters. The projection point of the boulder is between the 4-meter radius measuring line and the 7-meter radius measuring line, as shown in the figure 4.

GPR B-scan image of 4-meter radius survey line is as shown in figure 5. Though this boulder is located outside the range of the survey line, the scattering wave profile of the lone rock can still be seen clearly in figure 5. Compared with figure 2, peak point of scattering wave in figure 5 is changed with change of the boulder's position.

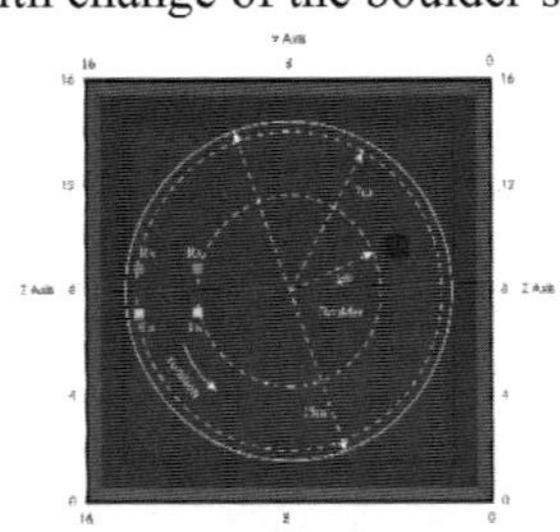

Figure 4. One boulder's projection point close to the tunnel edge.

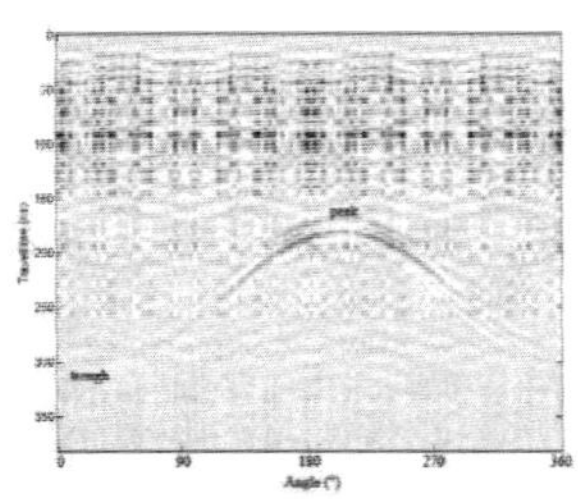

Figure 5. B-scan image of 4-meter radius detection line of one boulder's projection point close to the tunnel edge.

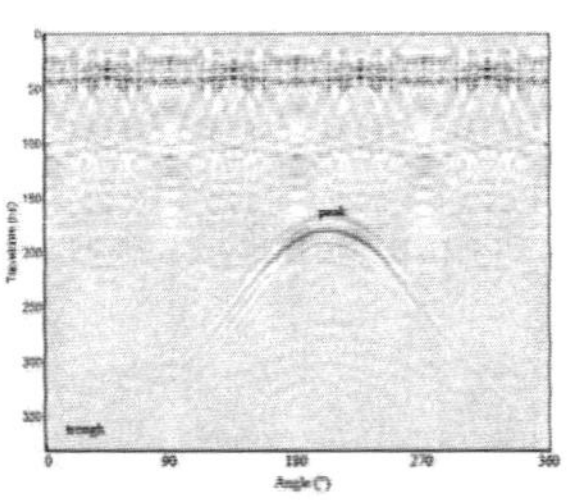

Figure 6. B-scan image of 7-meter radius detection line of one boulder's projection point close to the tunnel edge.

The location of the boulder remains unchanged. GPR B-scan image of 7-meter radius measurement line is shown as figure 6. Compared with figure 5, the scattering wave has a similar peak point position but greater slope around peak point.

4. Conclusions

Based on time-domain finite-difference principle, forward numerical simulations of boulder detection were carried out. The target can be detected by the ground penetrating radar regardless of its projection point close to the tunnel centre or tunnel edge. The scattering wave shows sinusoidal characteristics with peaks and troughs and position of wave peak corresponds to the position of the

Journal of Physics: Conference Series **2887** (2024) 012056 doi:10.1088/1742-6596/2887/1/012056

boulder. The reflection waves of boulders in GPR B-scan images show same characteristics at two different radii of the survey line. However, as a matter of fact when ground-penetrating radar antennas radiate electromagnetic waves, the electromagnetic energy is focused within a certain angle. Further research needs to be carried out considering actual ground-penetrating radar antennas to determine whether only a group of ground-penetrating radar antennas can detect the entire tunnel boring section.

5. References

[1] You S and Sun J. Deformation and Control of Super-Large-Diameter Shield in the Upper-Soft and Lower-Hard Ground Crossing the Embankment 2022 *Appl. Sci.* **12** 4324

[2] Xu P, Shi W, Ling S, et al. Mapping spherically weathered "Boulders" using 2D microtremor profiling method: a case study along subway line 7 in Shenzhen 2012 *Chinese J. Geophys.* (in Chinese) **55**(6) 2120-2128

[3] Xi Z, Long X, Zhou S, et al. Opposing coils transient electromagnetic method for shallow subsurface detection 2014 *Chinese J. Geophys.* (in Chinese) **59**(9) 3428-3425

[4] Cao Q, Xiang W, Jia H, et al. Application of cross-hole ultra-density resistivity method to detection of spherically weathered granite 2013 *J. Engineering Geology* (in Chinese) **21**(5) 730-735

[5] Li B. Key technology of super large diameter slurry pressure balanced shield construction in marine composite strata 2021 *Tunnel and rail transit* (in Chinese) **2021**(01) 19-23

[6] Zhao D, Zhang B, Zong Q, et al. Application of integrated geophysical method to detection of boulder in subway shield zone 2022 *Progress in Geophysics* (in Chinese) **37**(03) 1360-1370

[7] Duan W, Xie X, Yang y, et al. Forward modelling of ground penetrating radar in look-ahead detecting of slurry balance shield machine 2022 8th International Conference on Hydraulic and Civil Engineering: Deep Space Intelligent Development and Utilization Forum (ICHCE): 181-184

[8] Yee K. Numerical solution of initial boundary value problems involving maxwell's equations in isotropic media 1966 *IEEE Trans. Antennas Propagat.* **14**(3) 302-307

[9] Warren C, Giannopoulos A, Giannakis I. gprMax: Open source software to simulate electromagnetic wave propagation for Ground Penetrating Radar 2016 *Comput. Phys. Commun.* **209** 163-170

Acknowledgments

Authors thanks to the Shanghai Digital Shield Engineering Technology Innovation Center (grant no. 21DZ2220700), National Natural Science Foundation of China (grant no. 52038008, grant no 51978431), Science and Technology Innovation Plan of Shanghai Science and Technology Commission (grant no. 22dz1203004), and State Grid Shanghai Municipal Electric Power Company (grant no. 52090W220001) for their founding.

Journal of Physics: Conference Series **2887** (2024) 012057 doi:10.1088/1742-6596/2887/1/012057

A Study on the Characteristics of Time-lapse Full-Polarimetric GPR for Ice Fractures

Yafei An[1], Xuan Feng[1*], Qi Lu[1], Tseeulam Khuut[2], Jiarun Yang[1], Zejun Dong[1], Haoqiu Zhou[1], Minghe Zhang[1], Yinghao Zhang[1], ENHEDELIHAI[1]

[1] College of Geo-Exploration Science and Technology, Jilin University, Changchun 130026, China
(Corresponding author: Xuan Feng. E-mail:fengxuan@jlu.edu.cn)
[2] Mongolian University of Science and Technology, Ulaanbaatar 17025, Mongolia

Abstract. Real time detection of ice fractures is crucial for the safety of Antarctic scientific research and understanding the ice cracking process caused by avalanches. Using a full-polarimetric ground penetrating radar (FP-GPR) system to detect ice fractures, the scattering matrix obtained is processed using polarization decomposition. Compared to traditional single polarization radar, we can obtain more comprehensive and intuitive information about ice fractures. This study uses the Time-lapse FP-GPR method to detect ice fractures, and extracts the polarimetric attributes of fracture targets using Pauli decomposition and Freeman decomposition methods. The results indicate that polarimetric target decomposition is an effective method for identifying fractures, and the results in this paper provide important theoretical support and technical guidance for using time-lapse FP-GPR to detect surface ice fractures in practice.

1. Introduction

Ground Penetrating Radar (GPR) is a non-destructive geophysical method that utilizes electromagnetic wave propagation to detect underground environments. Due to the electrical differences between different media interfaces underground, the characteristics of electromagnetic waves emitted underground will change with the electromagnetic and physical properties of the media. By analysing the electromagnetic waves received by the receiving antenna, the purpose of detecting underground targets can be achieved. The propagation and scattering of electromagnetic waves in media are vector phenomena, and polarization can be used to study the vector characteristics of electromagnetic waves. In 1965, Huynen proposed the theory of polarized scattering, revealed the scattering mechanism of targets, which is the foundation of polarimetric decomposition theory[1]. In 1970, Williams et al. used radar to observe the polarimetric phenomenon of electromagnetic waves under the influence of targets, which received widespread attention from scholars[2]. Since then, FP-GPR has achieved rapid development.

At present, FP-GPR has been developed for mapping fine structures underground[3-5]. Fracture detection is one of the important applications of FP-GPR. In 1999, Miwa et al. designed a polarized borehole radar for detecting underground fractures[6]. In 2006, Tsoflias used multi- polarimetric GPR data to characterize the characteristics of vertical fractures[7]. With the development of polarimetric synthetic aperture radar (SAR) target decomposition methods, polarimetric decomposition methods have been applied to the processing of FP-GPR data. In 2017, Feng et al. combined H-Alpha decomposition, Pauli decomposition, and Freeman decomposition techniques with migration imaging techniques to achieve high-precision classification and recognition of underground targets[4, 8]..

Detecting fractures in ice layers is an important research topic. A common method of material unloading for Antarctic exploration in various countries around the world is to use transportation vehicles such as snow trucks to unload sea ice on the Antarctic ice surface. The presence of ice fractures poses a serious threat to this type of operation on the ice surface. These potential hazards are not easily detected by satellite photos and visual observations due to long-term snow cover. Therefore, real-time detection of ice fractures formed during the melting process of Antarctic sea ice is necessary. This study pre-processed and polarized the FP-GPR data of ice fractures before and after water injection, using two polarimetric decomposition methods: Pauli decomposition and Freeman decomposition, and analysed the results in detail.

2. Methodology

This section introduces the methods used in this article, containing the polarimetric target decomposition.

2.1. Pauli decomposition

Pauli decomposition is a polarimetric coherent decomposition method for deterministic targets, which decomposes the scattering matrix S into a weighted sum of complex forms of each Pauli matrix. Each Pauli basis matrix corresponds to a scattering mechanism, which is given by formula (1):

$$S = \begin{bmatrix} S_{HH} & S_{HV} \\ S_{VH} & S_{VV} \end{bmatrix} = \frac{a}{\sqrt{2}}\begin{bmatrix} 1 & 0 \\ 0 & 1 \end{bmatrix} + \frac{b}{\sqrt{2}}\begin{bmatrix} 1 & 0 \\ 0 & -1 \end{bmatrix} + \frac{c}{\sqrt{2}}\begin{bmatrix} 0 & 1 \\ 1 & 0 \end{bmatrix} \quad (1)$$

The values of a, b, and c are Pauli bases for polarimetric decomposition, corresponding to sphere scattering, dihedral scattering, and scattering other than sphere scattering and dihedral scattering, respectively. The total power is given by formula (2):

$$Span = |S_{HH}|^2 + 2|S_{HV}|^2 + |S_{VV}|^2 = |a|^2 + |b|^2 + |c|^2 \quad (2)$$

2.2. Freeman decomposition

Freeman decomposition is a three-component decomposition method based on physical models, where the mechanism of each scattering point is composed of surface scattering component, double rebound scattering component, and volume scattering component. Define the scattering vector of the following equation using the elements of the scattering matrix:

$$k_L = \begin{bmatrix} S_{HH} & \sqrt{2}S_{HV} & S_{VV} \end{bmatrix}^T \quad (3)$$

The covariance matrix can be calculated using k_L ,and C is composed of three basic scattering mechanism matrices:

$$C = C_s + C_d + C_v \quad (4)$$

Among them, C_s, C_d, and C_v represent the covariance matrices of surface scattering, double bounce scattering, and bulk scattering. Solve the equation system to obtain a, b, f_s, f_d, and f_v, and calculate the power of the three scattering types as follows:

$$P_s = f_s(1 + b^2); P_d = f_s(1 + a^2); P_v = f_v \quad (5)$$

f_s, f_d, and f_v represent the energy of the corresponding S_{VV} signals for these three types of scattering, a and b represent the values of S_{HH}/S_{VV} for double bounce scattering and surface scattering.

3. Field experiments

The experimental site is located on the surface of Jing Yue Lake in Changchun City, Jilin Province, China. The 1.6GHz MALA ground penetrating radar shown in Figure 1 (a) was used in the experiment. In order to obtain FP-GPR data from a single polarization GPR, we set the angle between the antenna and the measuring line to 0°, 45°, and 90°. The ice fractures on the lake surface are shown in Figure 1 (d). In the experiment, a measuring line perpendicular to the fractures was selected, with a length of 0.6m. We conducted water injection experiments in the fractures to obtain time-lapse data. Figure 1 (e) can be clearly observed that water is filled with ice fractures.

Figure 2 shows the pre-processed radar profile. From the radar profile, a series of obvious diffracted waves can be seen at positions 0.3m on the horizontal axis and 2ns on the vertical axis, corresponding to the location of ice fractures. We infer that the diffracted wave at a depth of 11ns and a horizontal axis of 0.28m is at the bottom of the fracture.

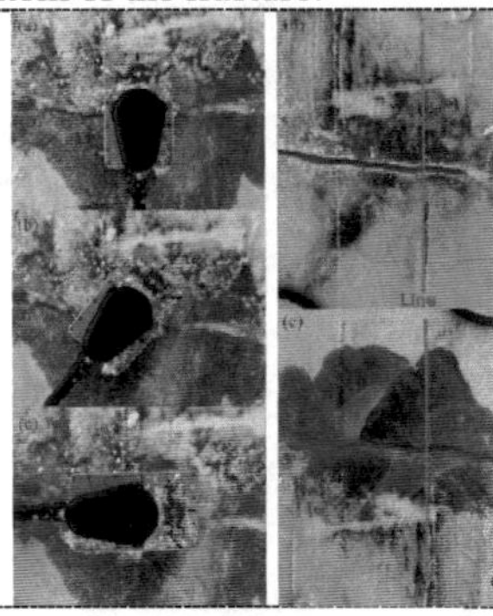

Figure 1. Measurement line diagram of the experimental site; (a)-(c) Three measurement angles for obtaining FP-GPR data, namely (a) 0 °, (b) 45 °, and (c) 90 °; (d) Line before water injection experiment; (e) Measurement line after water injection experiment.

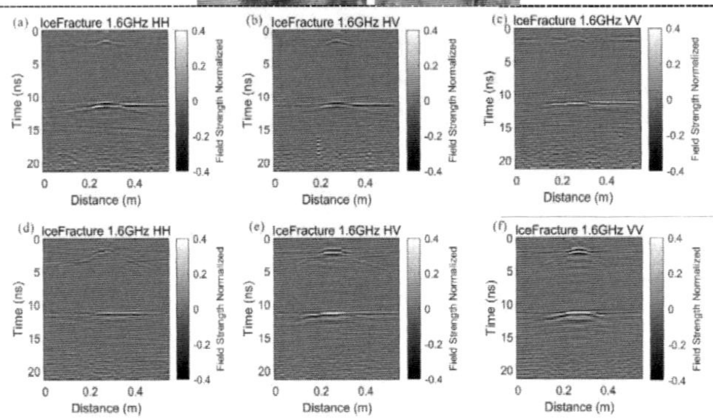

Figure 2. Profile of measured data; (a)-(c) The measured data profile before water injection in the fracture of line; (d)-(f) The measured data profile after water injection in the fracture of line.

4. Polarimetric attribute analysis

Applying Pauli decomposition to the pre-processed FP-GPR data before and after water injection on the line can calculate three parameters: *a*, *b*, and *c*. The energy intensity of the three scattering mechanisms in the profile can be characterized by $|a|^2$, $|b|^2$, and $|c|^2$. Usually, using blue to represent single scattering ($|a|^2$), red to represent secondary scattering ($|b|^2$), and green to represent volume scattering ($|c|^2$) yields the results in Figure 3. The results show that both the stratigraphic interface and interlayer reflectors exhibit strong single scattering, which can be used to locate abnormal bodies or layers below the lake surface. After water injection, due to the sharp increase in the dielectric constant of the fracture medium, the scattering energy intensity is significantly higher than before water injection, and the polarimetric component of the surface fracture can be seen after water injection.

The Freeman decomposition was applied to the pre-processed FP-GPR data of the Line before and after water injection. The decomposed Ps, P_D, and P_V components are shown in Figure 4, where Ps represents the surface scattering component, P_D represents the dihedral angle scattering component, and P_V represents the volume scattering component. The results in Figure 4 indicate that the P_V scattering component dominates, followed by the Ps scattering component, and the P_D scattering component is the weakest. After the water injection experiment, the scattered energy of surface cracks was enhanced.

5. Conclusion

This paper uses Pauli decomposition and Freeman decomposition methods to identify fractures and underground layers. By using a FP-GPR system, the polarimetric information of underground targets is received, and scattering features are extracted, which can effectively classify targets. The time-lapse water injection experiment on ice layer fractures achieved good results.

Acknowledgments

Supported by the National Key Reserach and Development Program of China（2022YFC3003402）

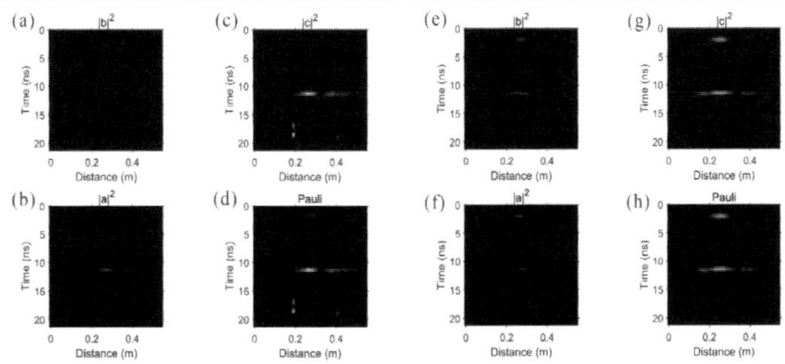

Figure 3. Pauli decomposition results; (a)-(d) Pauli decomposition results before water injection of line; (e)-(h) Pauli decomposition results after water injection of line.

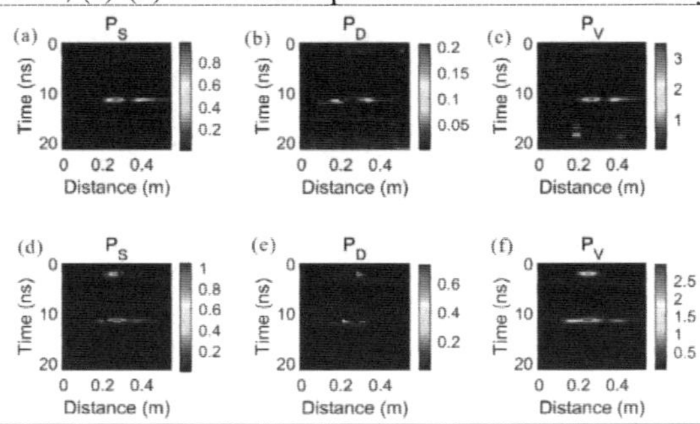

Figure 4. Freeman decomposition results; (a)-(c) Freeman decomposition results before water injection of line; (d)-(f) Freeman decomposition results after water injection of line.

References

[1] J. R. Huynen, "Measurement of the target scattering matrix," *Proceedings of the IEEE,* vol. Vol.53, no. NO.8, pp. 936-946, 1965.

[2] C. S. Williams, A. Sandia Laboratories, N. Mex., J. A. Cooper, and J. R. Huynen, "Antenna-polarization and terrain-depolarization effects on pulse-radar return from extended areas at the near vertical," *Proceedings of the IEEE,* vol. Vol.58, no. NO.9, pp. 1322-1328, 1970.

[3] M. Lualdi, and F. Lombardi, "Significance of GPR polarisation for improving target detection and characterisation(Article)," *Nondestructive Testing and Evaluation,* vol. Vol.29, no. No.4, pp. 345-356, 2014.

[4] X. Feng, Y. Yu, C. Liu, and M. Fehler, "Subsurface polarimetric migration imaging for full polarimetric ground-penetrating radar," *Geophysical Journal International,* vol. Vol.202, no. No.2, pp. 1324-1338, 2015.

[5] Z. Dong, X. Feng, H. Zhou, M. Zhang, Q. Liu, Y. An, Y. Zhang, J. Yang, W. Liang, Y. Yu, and C. Liu, "Super-Resolution Detection of Millimeter-Scale Fractures With Fluid Flow Using Time-Lapse Full-Polarimetric GPR and Anisotropy Analysis," *IEEE Transactions on Geoscience and Remote Sensing,* vol. Vol.62, pp. 1-16, 2024.

[6] M. Takashi, N. Hiroaki, and S. Motoyuki, "Subsurface fracture measurement with polarimetric borehole radar," *IEEE Transactions on Geoscience and Remote Sensing,* vol. Vol.37, no. No.2Part1, pp. 828-837, 1999.

[7] G. P. Tsoflias, and A. Hoch, "Investigating multi-polarization GPR wave transmission through thin layers: Implications for vertical fracture characterization," *Geophysical Research Letters,* vol. Vol.33, no. No.20, pp. 0, 2006.

[8] X. Feng, W. Liang, C. Liu, E. Nilot, M. Zhang, and S. Liang, "Application of Freeman decomposition to full polarimetric GPR for improving subsurface target classification," *Signal Processing,* vol. Vol.132, no. No.C, pp. 284-292, 2017.

Journal of Physics: Conference Series **2887** (2024) 012058 doi:10.1088/1742-6596/2887/1/012058

Antenna Comparison of Different Antennas in Through-Wall Radar Human Vital Signs Recognition

Tieyu Liu [1], **Jing Li** [1, 2*], **Zhaofa Zeng** [1,2] **and Hai Liu** [3]

[1] College of Geo-Exploration Science and Technology, Jilin University, 130026, China.
[2] Key Laboratory of Applied Geophysics, Jilin University, 130026, China
[3] Guangzhou University, Guangzhou, China

Abstract. Through wall radar (TWR) is an essential technology in the post-disaster rescue of buried individuals. The human vital signs analysis of breathing, heartbeat, and other periodic motions is the information used to identify and classify the human target. In the TWR system, selecting a suitable antenna and processing weak signals due to wall clutter and weak human vital signals is essential. In this abstract, we compare the performance of different antennas in TWR human vital feature recognition. We introduce a threshold pixel-connectivity signal processing method to locate the human body efficiently and accurately and extract the human respiratory frequency. The conclusion is that the Horn antenna has strong directionality, radiation energy, and a high resolution for recognizing the vital features of stationary human targets.

1. Introduction

Through wall radar (TWR) technology holds significant applicative value in the post-disaster rescue of buried individuals and military counterterrorism. TWR antennas usually have a wider bandwidth and lower frequency for the electromagnetic waves to penetrate the wall better .

Many different TWR systems have been built based on various antennas. Anirudh et al. designed an ultra-wide-band (UWB) Horn antenna with a gain of 21.5dB for excellent detection performance [1]. Horn antennas have the advantages of high gain and narrow beamwidth, but they are rarely used in TWR systems due to its cumbersome. Lu et al. designed a butterfly antenna for TWR with added reflectors for unidirectionality and wide beams [2]. Kumar et al. designed a slotted Vivaldi antenna with a return loss response better than -10dB over the operating bandwidth and an antenna gain greater than 8dB for good detection performance [3]. The Vivaldi antenna is extensively used owing to its portability, high gain, and ultra-wide bandwidth. As one of the essential components of the TWR, the performance of the antenna has a great impact on the TWR. Therefore, choosing a suitable antenna for various scenarios is crucial for through-wall detection.

Therefore, based on the synchrosqueezing wavelet transform (SWT) and micro-Doppler effect, combined with the actual TWR data, this paper evaluates the effect of multiple types of antennas (Double-Ridged Horn antenna, Spiral antenna, Vivaldi antenna, and Spiral-Vivaldi combination) in detecting respiratory frequency, which provides a practical solution for selecting antenna types in actual scenarios. Based on the threshold pixel connectivity, the TWR data is processed, which can completely adaptively and accurately extract the respiration rate, which improves the effect of through-wall detection.

2. Methods

2.1. SWT and Threshold -pixel connectivity

Daubechies et al. proposed the theory of the SWT method, which focuses the energy of the spectrogram onto the intermediate frequency curve [4]. For any (a, b) and $Ws(a, b) \neq 0$.

$$\omega_s(a,b) = -\frac{i}{\left(W_s(a,b)\right)}\frac{\partial}{\partial b}W_s(a,b) \tag{1}$$

After the vital sign signals processing by the SWT method, dispersed noise signals with low pixel connectivity (PC) can be seen in the time-frequency spectrum with high-frequency resolution. PC means that if the edges of the pixels are in contact with each other, then these pixels have connectivity. Since the image will have many wavelet coefficients with high noise dispersion, the PC method is used at this point for further image processing. Before the PC processing, the time-frequency spectrum $T_x(t, f)$ of the signal needs to be thresholded binarized for the wavelet coefficients using thresholding as shown in the algorithm (1)

Algorithm 1 Binarization of time and frequency images.

1. **for** i=1 to N **do**
2. $C_i = \max(T_x(i,f))$
3. **End**
4. $c_{there} = \min(C_i)$

After that PC is taken to remove the noise dispersed by wavelet coefficients.

In determining the effective region, a threshold must be selected to remove as many invalid wavelet noise coefficients as possible. Zeng et al. proposed an adaptive pixel connectivity thresholding method to remove the residual noise in [5], with the thresholding function

$$S_{\text{thre}} = \sqrt{2\,ln(c)}MAD \tag{2}$$

where $MAD = median(S - mean(S))$, S is the sum of all PC components, S_{thre} is the threshold value, and c is the number of PC components in the time-frequency domain. When the PC component is greater than the threshold value, the wavelet coefficient of the point is retained. Otherwise, it is set to zero so that the PC method can remove the dispersed noise component and the signal-to-noise ratio can be improved.

2.2. Signal-to-Noise Ratio estimation

In the field data of TWR, most of the noise is caused by antennas, wall scattering, and still clutter interference, so it is difficult to separate the noise from the signal completely. In the actual calculation of the signal-to-noise ratio (SNR), since the frequency of the respiration signal is mostly between 0.2 Hz and 0.4 Hz, the signal in this range is considered a valid signal, and the other signals are noise. The SNR is calculated using equation (3)

$$SNR = 10\log_{10}\left(\frac{p^2(g)}{\sum p^2(f) - p^2(g)}\right) \tag{3}$$

Where $p(g)$ is the value of the life signal spectrum and $\sum p^2(f)$is the total energy of the life signal spectrum.

Journal of Physics: Conference Series **2887** (2024) 012058 doi:10.1088/1742-6596/2887/1/012058

3. Results and Discussion

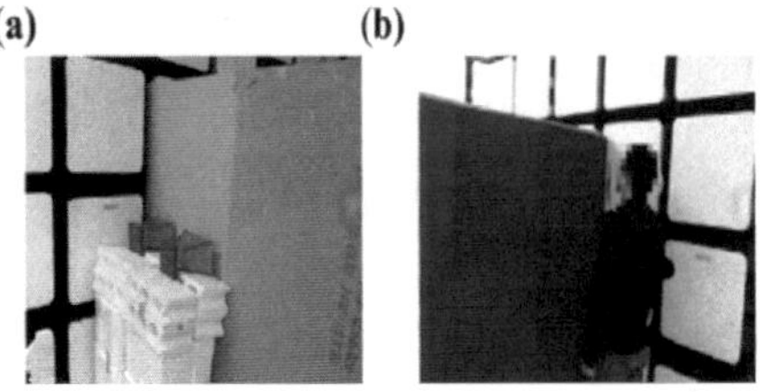

Figure 1. TWR experimental system.

Four commonly used TWR antennas were connected to a vector network analyzer (VNA) for the field TWR test. The TWR human vital signal experiment and antenna position are shown in Figure 1. We use Spiral, Vivaldi, Horn and Spiral-Vivaldi antenna respectively (Figure 2 (a-d)). There is an obvious signal of the periodic activity of the human body between 5-7 ns (Figure 2 (e-h)). The short-time Fourier transform (STFT) can obtain the micro-doppler effect of the pulmonary contraction of the human body transform (Figure 2 (i-l)), the apparent change of pulmonary contraction can be seen.

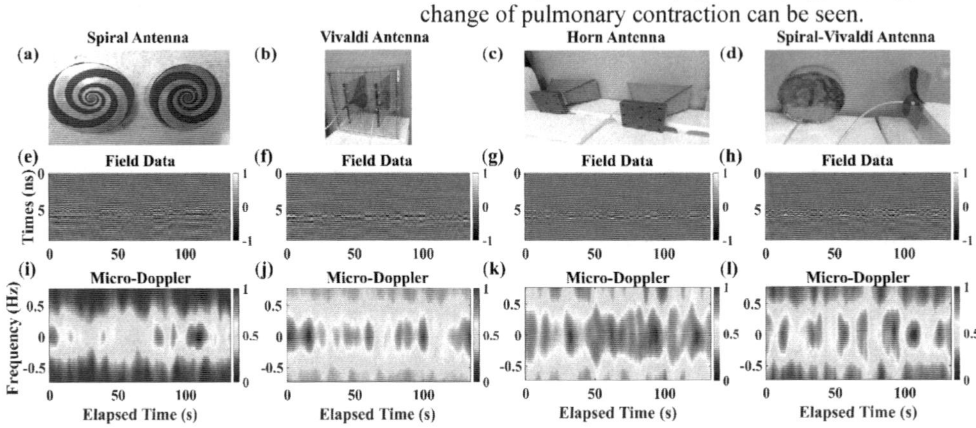

Figure 2. (a)-(d) Four antenna configurations; (e)-(h) TWR data after background removal;(i)-(l) Micro-Doppler effect.

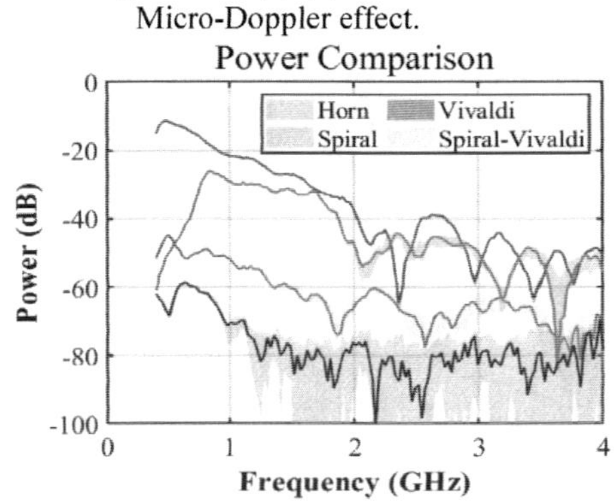

Figure 3. Comparison of gain curves when through-wall detection, the orange curve is the change of the gain curve when the human body is breathing.

Comparing the changes in pulmonary contraction detected by the four antenna combinations, the detected micro-Doppler of the periodic activity of the human body is not apparent because the spiral antenna is circularly polarized and usually has a lower gain. Compared with the Spiral antenna and the Vivaldi antenna, the Spiral-Vivaldi combination antenna has a higher resolution in detecting the contraction of the human pulmonary because the circularly polarized waveform will be converted into more vertically polarized waves after being reflected by the wall and the human body, which the Vivaldi antenna will receive with less power loss.

Table 1. Comparison of SNR before and after processing.

	Horn	Spiral	Vivaldi	Spiral-Vivaldi
SNR(Original Data)	-8.4655	-11.1812	-10.5453	-9.8309

SNR(Processed)	-4.6950	-6.8781	-7.3881	-7.6990

The Horn antenna has a higher gain due to its better directionality, and the detected microdoppler of pulmonary contraction is the most obvious among these four antennas.

The time-frequency spectrum is then obtained by SWT on the sampling timeline of the signal in the time domain (as shown in Section 2.1). As can be seen from the figure, the four antennas can accurately obtain the respiratory rate and the position of the human body. In contrast, the horn antenna can accurately describe the front chest and back wave changes when the human breathes with a higher resolution. The position of the Spiral antenna and the Spiral-Vivaldi combined antenna on the chest is more evident than the back position. The positioning of the human body with a Vivaldi antenna is less accurate than the other three antennas. According to power results, the Vivaldi antenna is not sensitive enough to human respiration when it changes, which is limited by its polarization type and small gain.

The respiratory rate can be accurately extracted in the Horn antenna result by comparing the time-frequency spectrum before and after the threshold pixel connectivity. For computing efficiency, the CPU using Intel(R) Core(TM) i5-8500 has a processing time of only 3.7s, and the respiratory rate of the human and the corresponding position behind the wall can be obtained.

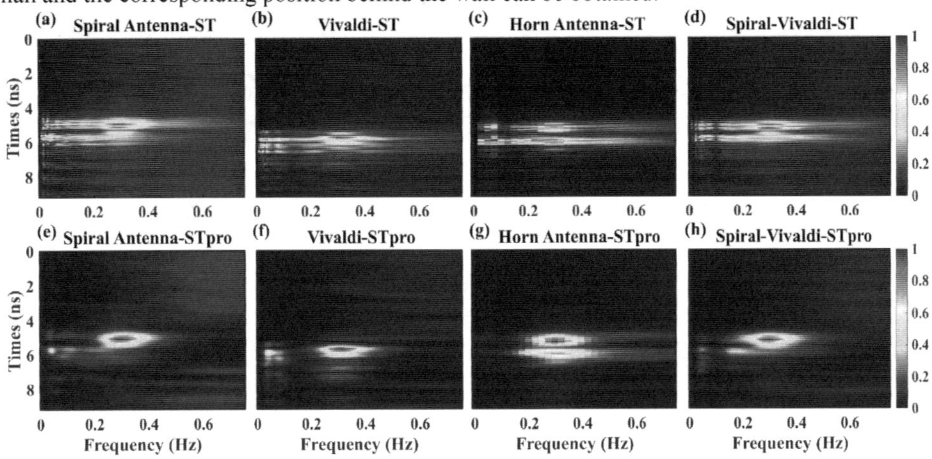

Figure 4. (a)-(d) Time-frequency spectrum of the original data; (e)-(h) Time-frequency spectrum after thresholded pixel connectivity processing.

4. Conclusion

In this abstract, we compare the performance of different antennas in TWR human vital feature recognition. We used the SWT and threshold pixel connectivity to extract the human respiratory frequency and micro-Doppler effect, respectively. The results show that the Horn antenna has a better effect on detecting the human respiratory frequency with a small detection range. The Spiral-Vivaldi combination antenna has a higher resolution in detecting the contraction of the human pulmonary with the micro-doppler effect. Although the Vivaldi and Spiral antennas have the advantage of portability, high gain, easy-to-make array, and ultra-wide bandwidth, they are not sensitive to get human respiration. The threshold pixel connectivity can quickly and accurately adapt the positioning of the human body behind the wall and extract the human respiratory rate. Our work provides a valuable tool for human signal detection using TWR.

Acknowledgments

This work was supported in part by the Natural Science Foundation of China under Grant 42222407 and Grant 42174065

Journal of Physics: Conference Series **2887** (2024) 012058 doi:10.1088/1742-6596/2887/1/012058

References

[1] A. Karanth, N. Onkar, N. S. N. Smitha, V. Singh 2017 Through-wall imaging system using horn antennas *Vipula Singh 2017 4th International Conference on Advanced Computing and Communication Systems (ICACCS)* 1-6

[2] Q. L. Lu, L. Y. Zhou, C. M. Tan 2010 A novel wide beam UWB antenna design for Through-the-Wall radar *2010 IEEE Antennas and Propagation Society International Symposium*. 1-4

[3] B. Kumar, K. A. Vardhan and P. Sharma 2016 Design of slotted Vivaldi antenna for through wall imaging system *2016 11th International Conference on Industrial and Information Systems (ICIIS)* 704-709

[4] Daubechies, I., Lu, J., Wu, H.T 2011 Synchrosqueezed wavelet transforms: An empirical mode decomposition-like tool. *Applied and Computational Harmonic Analysis* **30** 243–261.

[5] Zhiyi Zeng and others 2023 Microseismic data denoising in the sychrosqueezed domain by integrating the wavelet coefficient thresholding and pixel connectivity, *Geophysical Journal International*, **232** 1113–1128

DL-GPRd: Deep Learning for Ground Penetrating Radar Deconvolution

G.Roncoroni[1], E. Forte[1], M. Pipan[1]

[1]Department of Mathematics Informatics and geosciences, University of Trieste, Trieste, Italy.

Abstract.

Ground-penetrating radar (GPR) is a vital tool for subsurface investigation, but traditional deconvolution methods have limitations. This study explores the integration of Deep Learning (DL) techniques to enhance GPR data processing. The model utilizes Bi-Directional LSTM neurons and synthetic data generation for training. Results show improved resolution and reduced noise compared to classical methods. Computational analysis highlights the efficiency of the approach, with faster prediction times. This DL-based deconvolution offers promising advancements for GPR imaging, enhancing subsurface characterization and exploration capabilities.

1. Introduction

Ground-penetrating radar (GPR) is a commonly used geophysical method for subsurface investigation due to its non-invasive nature and high-resolution imaging capabilities. Deconvolution in GPR entails estimating or removing the impulse response of the GPR system from the recorded data, with the goal of enhancing resolution and interpretability of subsurface features (Economou & Vafidis, 2011).

Traditional deconvolution methods for GPR data have historically relied on deterministic approaches, often implemented in the time-frequency domain (Economou & Vafidis, 2011). These methods typically make assumptions about the GPR wavelet's characteristics and the subsurface media. However, recent advancements have seen the incorporation of Deep Learning (DL) techniques into the deconvolution process, providing a data-driven approach capable of adapting to complex signal variations and noise patterns (Rohatgi et al., 2022). DL-based deconvolution methods has demonstrated potential in enhancing the accuracy and efficiency of GPR data processing, particularly in scenarios with challenging subsurface conditions or noisy environments (Rohatgi et al., 2022). This approach is particularly beneficial in scenarios with challenging subsurface conditions or high levels of noise (Kang & An, 2020). This study delves into the integration of DL-based deconvolution techniques within GPR data processing. The objective is to explore the capacity of deep learning models to enhance the resolution, clarity, and interpretability of GPR images, thereby advancing the capabilities of subsurface imaging and geological mapping through GPR technology. By harnessing artificial intelligence and neural networks, we aim to expand the horizons of GPR data processing and unlock new opportunities for subsurface characterization and exploration.

2. Methods

To build the model we used Bi-Directional LSTM neurons (Hochreiter and Schmidhuber, 1997), due to their capability to manage temporal dependencies and predict varying temporal lengths, courtesy of their recurrent nature.

Synthetic data is generated through a convolutional model comprising reflection coefficient time series (Roncoroni et al., 2023):

$$T(t) = w(t) * (R(t) + n_1(t)) + n_2(t) \quad (1)$$

where R(t) is a uniformly distributed reflection coefficient time series with random values, $n_1(t)$ and $n_2(t)$ are random noise with uniform distribution, w(t) is the wavelet. $n_1(t)$ has been added to simulate real data noise due to wave propagation or scattering and $n_2(t)$ represent instrumental or environmental noise. In our case, both and , randomly vary between 2 to 10% of the maximum amplitude of the signal. Specifically, two types of wavelets are employed: a Ricker wavelet with a 200MHz dominant frequency for the first trace and a broadband Ricker wavelet with frequencies spanning from 100 to 800MHz, for the target trace, see Figure 1. The training dataset is

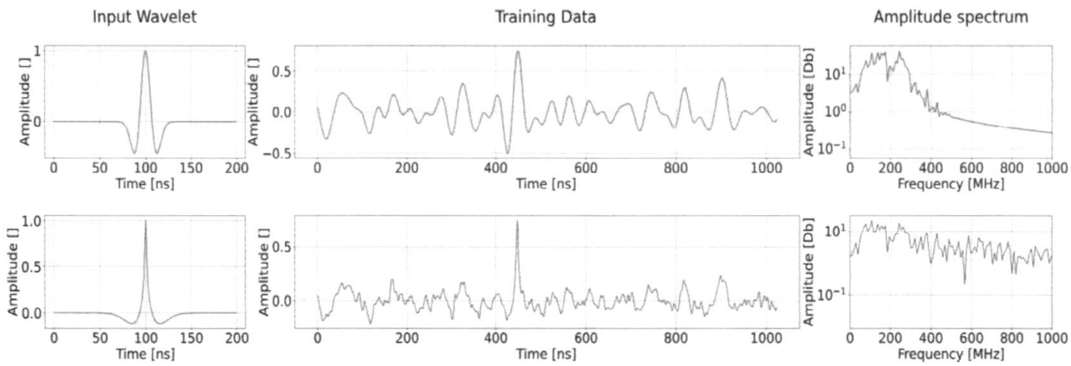

Figure 1: General description of the NN training: we train a LSTM-based model to map the change between red signal (top) and green signal (bottom). The model maps not only the amplitude domain, but also the amplitude-frequency spectrum in the loss function, see Eq. 2.

engineered to maintain maximum phase to mitigate potential phase shifts that could disrupt the placement of reflections during prediction.

Lastly, the choice of the neural network involves selecting a configuration with six hidden layers with 32, 16, 8, 4, 2, and 1 Bi-Directional neurons, respectively. This configuration was determined following experimentation with various complexities, where excessively complex geometries were found to necessitate significant increases in training and prediction time without appreciable reductions in errors. The loss function involves summing the

Mean Squared Error (MSE) between real and predicted traces, considering both time and frequency domains:

We trained the NN using 100,000 randomly generated traces, using as optimizer Adamax (Kingma and Ba 2014).

$$loss(X,Y) = \| X - Y \|_2 + \| \mathbb{R}(\hat{X}) - \mathbb{R}(\hat{Y}) \|_2 \qquad (2)$$

3. Results

3.1. Comparison with classical methods

We used the GPR profile from Ercoli et al., 2022, Figure 5 therein, where sub-horizontal and laterally continuous reflections in the range of approximately 6-16 nanoseconds. Strong reflections are prominent between 15 and 20 meters, with a deeper wavy reflection observed between 50 and 60 nanoseconds (approximately 5-12 meters).
The deconvolution workflows applied to the data in aim to improve the vertical resolution and sharpen the reflections, particularly within the dim area representing the faulted zone. The use of deconvolution techniques, such as Gabor deconvolution, helps enhance the clarity and quality of the GPR data, enabling better imaging and interpretation of subsurface structures, including seismogenic faults in the study area, in Figure 2 we compare the results of the proposed methodology Figure 2B with input data Figure 2A, and the results from the classical approaches presented in Ercoli et al., 2022, Gabor deconvolution and Spiking deconvolution, Figure 2C and D, respectively.

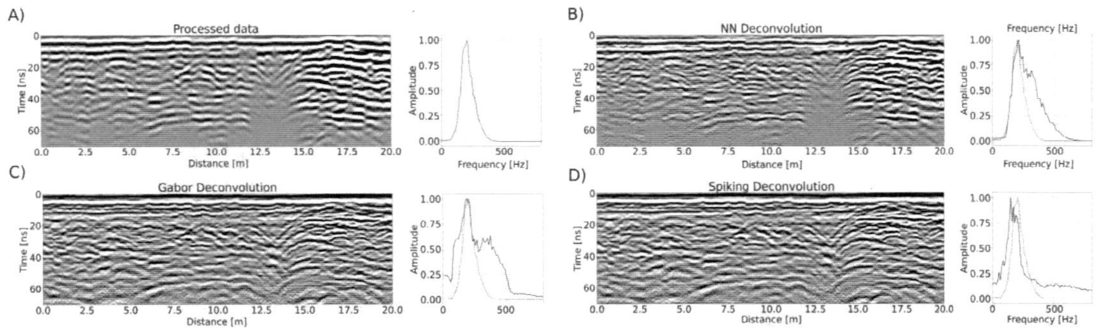

Figure 2: Comparison of the methodology with classical approach: A) input data used in Ercoli et al., 2022, B) result of the proposed approach, C) result of the Gabor deconvolution (Ercoli et al., 2022) and D) results of the Spiking deconvolution (Ercoli et al., 2022).

As we can see the results are comparable to the Gabor deconvolution, with less high frequency noise introduced in the last 20ns. If we focus on the wedge t 0-10ns we can see that the NN approach better resolve the waveforms, making it possible to perform an easier interpretation of the structure.

3.2. 2.5-D Examples

Data are taken from Pipan et al., 2002, Figure 7 therein shows a 200 MHz GPR data. The single-fold section displays a prominent bedding joint in the central part of the section.

These data show various geological feature including bedding joints, fractures in the limestone study area. In Figure 3A we present the processed data from Pipan et al., 2002, while in Figure 3B we display the results of the NN deconvolution. As we can see we have a huge improvement in the resolution of the data, especially in the first 10ns we can better resolve the dipping structures.

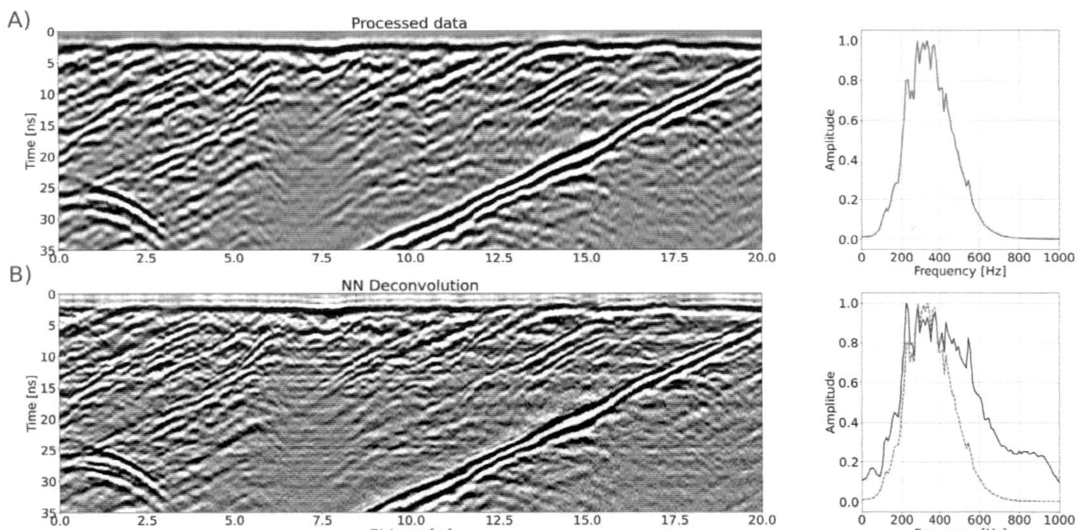

Figure 3: Comparison of the application of the presented methodology on a data from Pipan et al., 2002. The processed data is shown in A, while the deconvolution results is shown in B. The increase in resolution is remarkable in the first 10ns, where the dipping events are better resolved by the proposed methodology.

To further prove the stability of the approach we applied the NN deconvolution on the 3-D dataset, Pipan et al., 2002, Figure 8 therein. In Figure 4A we display the processed data, while in Figure 4B the results of our methodology are depicted. The application of the NN deconvolution on the whole 3-D volume prove the stability of the approach and its robustness.

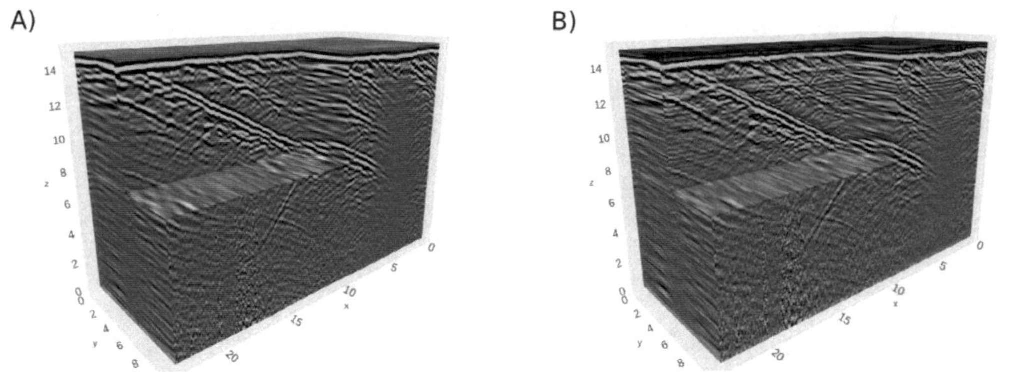

Figure 4: Comparison of the application of the methodology on the 3-D volume from Pipan et al., 2002. The processed volume is shown in A, while the deconvolved one is shown in B. We can appreciate the low amount of high frequency noise introduced by the deconvolution process.

3.3. Computational Complexity

The methodology is computationally profitable: the training procedure has been done on 100,000 traces on a cluster with a Nvidia A100, 18Gb GPU, 128Gb RAM for 1000 Epochs, lasting 14 hours.
After this training: the same NN has been applied to produce all the results in this work. Prediction time time B-scan shown in Figure 2 takes 0.78s, definitely faster than the Gabor deconvolution that took 34.7s on the same machine using the function proposed in Ercoli et al., 2022

4. Conclusions

This study demonstrates the efficacy of DL-based deconvolution in advancing GPR imaging for subsurface investigation. By integrating Bi-LSTM neurons and synthetic data generation, our approach significantly improves resolution and reduces noise compared to traditional methods. The computational analysis underscores the efficiency of DL-based deconvolution, with substantially faster prediction times. These findings signify a notable leap forward in GPR data processing, offering enhanced imaging capabilities crucial for subsurface characterization and exploration.

5. Acknowledgments

The authors would like to thank Dr. Maurizio Ercoli for sharing the data used in this work.

6. References

Economou, N. and Vafidis, A. (2011). Deterministic deconvolution for gpr data in the t-f domain. Near Surface Geophysics, 9(5), 427-433. doi: 10.3997/1873-0604.2011020

Ercoli, M., & Ferguson, R.J. (2022), Ground-penetrating radar attenuation compensation by Gabor deconvolution: Seismogenic fault imaging at Castelluccio di Norcia (Central Italy), *GEOPHYSICS* 87: WB55-WB67, doi: 10.1190/geo2021-0354.1

Hochreiter, S., & Schmidhuber, J. (1997), Long Short-Term Memory, Neural Computation 9, 1735-1780, doi: 10.1162/neco.1997.9.8.1735.

Kang, M. and An, Y. (2020). Frequency–wavenumber analysis of deep learning-based super resolution 3d gpr images. Remote Sensing, 12(18), 3056. doi: 10.3390/rs12183056

Kingma, D., & Ba, J., (2014) Adam: A method for stochastic optimization. International Conference on Learning Representations, doi: 11245/1.505367.

Pipan, M., Forte, E., Guangyou, F., & Finetti, I. (2003). High resolution GPR imaging and joint characterization in limestone. Near Surface Geophysics. 1. doi: 10.3997/1873-0604.2002006.

Rohatgi, N., Ghoshdastider, U., Baruah, P., Kulshrestha, T., & Skanderup, A. (2022). A pan-cancer metabolic atlas of the tumor microenvironment. Cell Reports, 39(6), 110800. doi: 10.1016/j.celrep.2022.110800

Roncoroni, G., Forte, E., Bortolussi, L., & Pipan, M. (2022). Efficient extraction of seismic reflection with Deep Learning, Computers & Geosciences, 166, 105190, doi: 10.1016/j.cageo.2022.105190.

Journal of Physics: Conference Series 2887 (2024) 012060 doi:10.1088/1742-6596/2887/1/012060

Numerical simulation and dispersion inversion of ground penetrating radar waveguide

Kaiwen Zhang[1], Jing Li[1,2]*, Ruizhe Sun[1], Tieyu Liu[1], Lige Bai[1]

[1] College of Geo-Exploration Science and Technology, Jilin University, 130026, China. (kwzhang23@mails.jlu.edu.cn)

[2] Key Laboratory of Applied Geophysics, Jilin University, 130026, China

Abstract. Ground penetrating radar (GPR) is widely used in near-surface detection due to its non-destructive and high-efficiency advantages. Radar waves are continuously reflected in the waveguide structure, and dispersion will be formed. In order to find out the propagation regularity and characteristics of radar waves in waveguides, we use the simulation and field GPR waveguide data to test the radar wave dispersion characteristic. Then, the Dix-type inversion method is used to estimate the permittivity from the radar wave dispersion curve. The synthetic and field data tests show that the GPR waveguide dispersion inversion can accurately estimate the permittivity result.

1. Introduction

Dispersion is a typical feature of seismic surface wave data caused by the change in spatial waveform due to the different phase velocities of the waves in each component, and dispersion contains abundant stratigraphic information. Compared with seismic waves, the dispersion of radar waves only occurs in specific structures. However, radar data processing has excellent limitations, so studying the dispersion of radar waves is significant.

In ground penetrating radar (GPR), the transmitting and receiving antennas are usually placed on the ground, so the air layer always covers the waveguide. Total reflection occurs at the interface when the substrate layer has a low permittivity. For example, when wet soil covers gravel, the low permittivity of the base layer leads to total reflection. Arcane et al. researched the waveguide characteristics of GPR through wet soil-dry sand, gave the thickness of the waveguide layer, and deduced the permittivity of two layers [1]. Van der Kruk proposed the derivation of dispersion curves from GPR data and the subsequent inversion of subsurface media properties [2]. When the substrate layer has a low permittivity, it can only provide strong reflection. However, there will still be some energy transmission at the substrate interface besides strong reflection, called leakage waveguide. Arcane researched the electromagnetic properties of ice over water and determined the permittivity of the ice layer by assuming that the dominant air-ice layer is the waveguide cut-off [3].

Based on the GPR waveguide theory, we study the propagation characteristics of radar waves in waveguides using the finite difference time domain (FDTD) method. Dispersion inversion experiments of simulated and field GPR waveguide data are carried out to obtain two-dimensional underground structures and physical parameters. These provide a theoretical basis for applying GPR waveguides in unique structures such as ice layers and frozen earth.

2. GPR waveguide theory

The GPR waveguide is a unique physical phenomenon based on the significant difference in physical parameters of the underground medium. If the permittivity of the overlying surface layer and the substrate layer differ significantly, the radar wave will be reflected. Suppose the covered surface layer is very thin, and the thickness is less than the wavelength of a GPR electromagnetic wave. In that case, the reflected energy may be trapped in the thin surface layer, which can be considered a waveguide.

When the thin surface layer is covered on the substrate with low permittivity, the electromagnetic wave is completely reflected outside the critical angle of the upper and lower interfaces. At this point, the thin layer can be considered a low-velocity waveguide. When a thin surface layer is covered on a substrate with a much higher permittivity than the surface layer, total reflection occurs only at the upper interface, some energy will be transmitted downwards through the interface. At this point, the thin layer can be considered a leaky waveguide [2, 4].

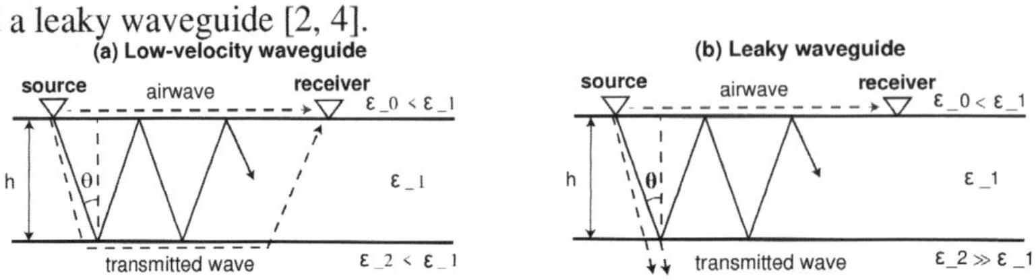

Figure 1. Diagram of two waveguide structures (a) low-velocity waveguide, (b) leaky waveguide

3. Numerical simulation

In this paper, a horizontally layered model and a dipping interface model are set up, as shown in figure 2. The upper layer is air, the middle layer is a waveguide layer with a relative permittivity of 21, the permittivity of base layer is set to 7, the thickness of the layered model is 0.32 m, the thickness of the dipping interface model is 0.24 m at the left and 0.48 m at the right.

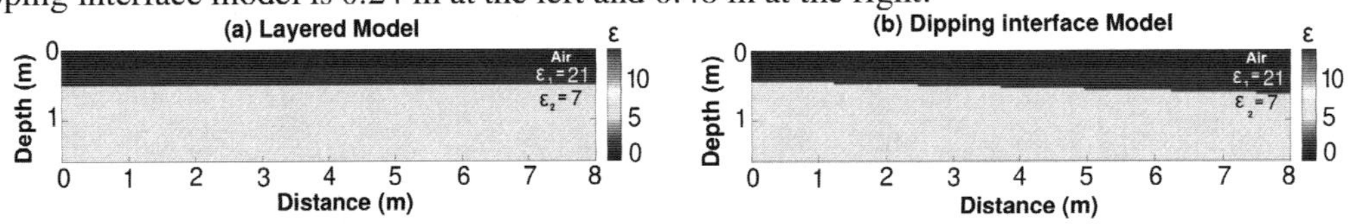

Figure 2. (a) layered model, (b) dipping interface model

The simulation uses a wide-angle method with a forward model. A 100MHz antenna is positioned at the air-waveguide interface, with receiving antennas evenly spaced at 0.08m intervals along the interface. The resulting CMP data is depicted in figure 3(a).

The dipping interface model was used to study the influence of the transverse waveguide thickness variation on the results and to obtain the two-dimensional structure of the waveguide layer. The forward results of three positions corresponding to different depths at the left end, the middle, and the right end of the model are shown in figure 3 (b), (c), and (d).

Convert common middle point (CMP) data E (X, T) to the frequency domain using the following equation [1]:

$$\hat{E}(x,f) = \int E(x,t)exp(-2\pi jft)dt \tag{1}$$

The corresponding phase correction is performed within the specified sum phase velocity range so that constructive interference occurs when a certain phase velocity is moved. The phase velocity spectrum is calculated by:

$$D(v,f) = \left| \sum_x \frac{\hat{E}(x,f)}{|\hat{E}(x,f)|} exp\left(j\frac{2\pi f}{v}x\right) \right| \tag{2}$$

The corresponding phase velocity spectrum is shown in figure 3 (e) ~ (h) determination of a single dispersion curve of phase velocity with variation by selecting maximum:

$$v_{data}(f_i) = max[D(v,f_i)] \tag{3}$$

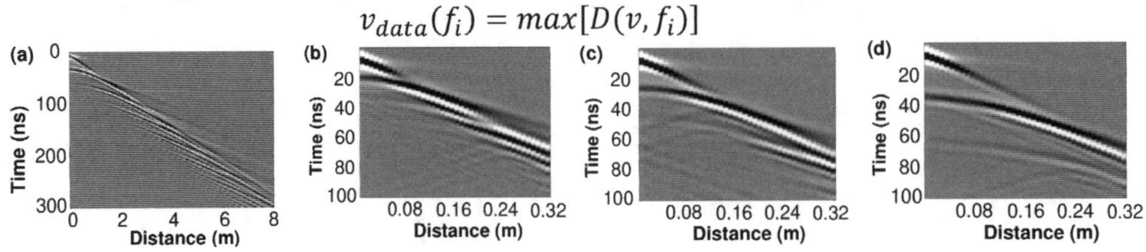

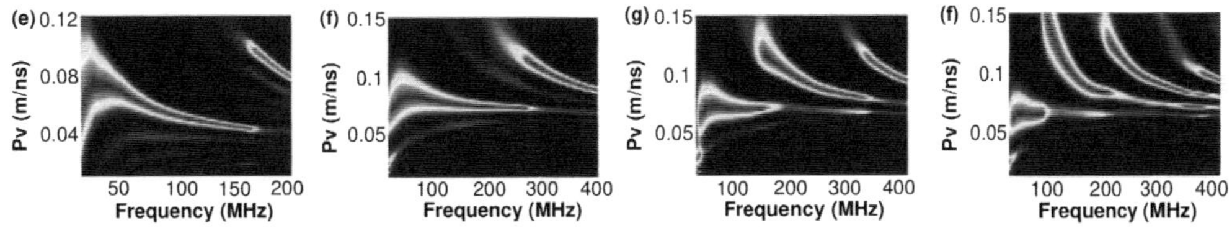

Figure 3. Forward results of two models, (a) CMP data of layered model, (b) (c) (d) CMP data of different positions of dipping interface model, (e) (f) (g) (h) corresponds to phase velocity spectra (a) (b) (c) (d) respectively

It can be seen from the figure that the radar wave is obviously dispersed in the waveguide structure, and the existence of a cut-off frequency can be clearly seen. The energy of the fundamental wave in the high-frequency range decreases with the decrease in the number of receiving antennas, and the dispersion phenomenon of the waveguide gradually decreases with the thickness of the waveguide layer.

4. Dispersion inversion

The maximum value is chosen to extract the dispersion curve in the phase velocity spectrum. Only waves above the cut-off frequency can travel in the waveguide. Dispersion inversion selects only the dispersion curve above the cut-off frequency. Where the following equation calculates the cut-off frequency:

$$f_C^{TE} = \frac{(2n+1)c}{4h\sqrt{\varepsilon_1 - 1}} \quad (4a) \qquad f_C^{TM} = \frac{2nc}{4h\sqrt{\varepsilon_1 - 1}} \quad (4b) \qquad n=0,1,2....$$

In this paper, the models are all layered structures, and the geological conditions are simple, so Reza Dokht proposed the inversion of the dix-type global linear approximation. [5] is chosen here. The inversion results for the two models are shown in Table 1,2.

Table 1. Inversion results of layered model.

Trace	h (m)	ε_1	ε_2
True values	0.32	21	7
Layered model	0.33	21.07	7.83

Table 2. Inversion results of dipping interface model.

Trace	h/h₀ (m)	ε_1	ε_2
Left	0.23/0.24	20.73	8.87
Middle	0.33/0.36	20.83	10.59
Right	0.42/0.48	24.01	10.98

From the inversion results, it can be seen that the permittivity of the waveguide layer and its thickness can be obtained accurately by using the waveguide dispersion inversion of GPR, and there is a large error for the permittivity of the base layer. For the dipping interface inversion, as the thickness of the waveguide layer increases, the inversion results become increasingly critical, especially for the permittivity reconstruction.

All data are processed here to investigate the effect of lateral depth variation on the waveguide dispersion inversion. The two-dimensional phase velocity spectrum of the fundamental wave is extracted from all the data phase velocity spectrum, as shown in figure 4 (a). The inversion result of its two-dimensional subsurface structure diagram is shown in figure 4 (b).

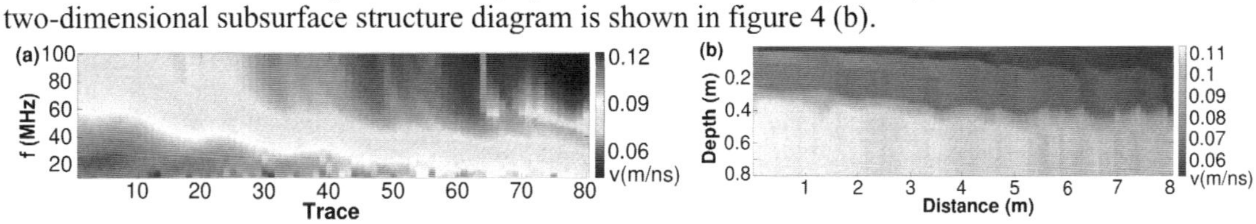

Figure 4. The inversion result (a) two-dimensional phase velocity spectrum, (b) inversion results

It can be seen from the figure that as the thickness of the inclined waveguide layer increases, there are some errors in the phase velocity spectrum and inversion results. However, the overall inversion results are consistent with the model.

5. Experimental data

The experimental data were collected on a specific section of the Suibei railway after rain. The surface layer is about 0.6 m of ballast, then about 0.6 m of gravel, then 1.3 m of non-frost heaving, and the bottom layer is filled soil. The ballast and gravel layers above 1.2 m can be considered as waveguides.

Figure 5 (a) is a CMP data of the experimental data, the dashed box shows the data used for dispersion analysis, and (b) is the phase velocity spectrum obtained from the data. According to this, the subsurface two-dimensional structure is inverted by dispersion, as shown in figure 5 (c).

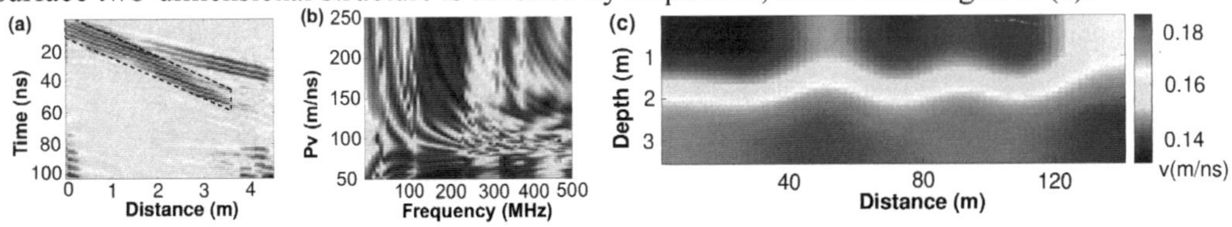

Figure 5. (a) measured data radar data, (b) phase velocity spectrum, (c) inversion results

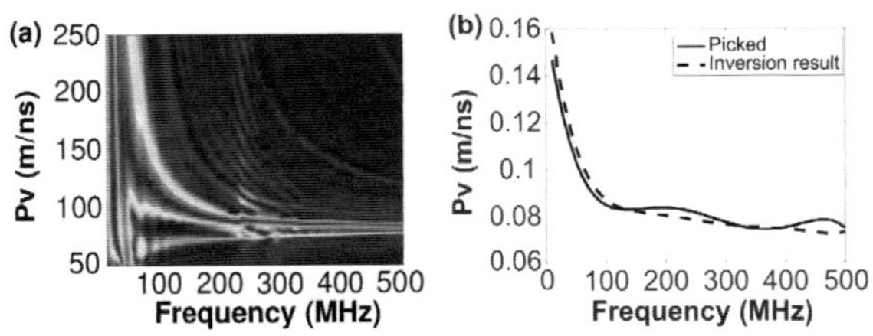

Figure 6 Inversion result for (a) phase velocity spectrum, (b) dispersion curves

The experimental data show that the railway subgrade structure can be considered as a waveguide structure, and the data shows obvious dispersion characteristics. The inversion results also reflect the subsurface structure of the subgrade well. figure 6 is the result of a forward simulation using the inverted model. In Figure. 6b, the dispersion curves of the filed data and the dispersion curves of the inversion are in good agreement, which verifies the accuracy of the inversion results. This can provide some technical support for the safety detection of road subsurface.

6. Summary and conclusions

GPR waveguide dispersion contains a wealth of subsurface information. The thickness and permittivity of the waveguide layer can be determined by dispersion inversion, which provides technical support for detecting the frozen soil and ice layer structure and for the safety detection of subgrade safety after rain. In the future, we will investigate the joint and higher mode inversion to achieve more accurate and stable results.

References

[1] Arcone A S, Peapples R P, Liu L 2003 Propagation of a ground-penetrating radar (GPR) pulse in a thin-surface waveguide *Geophysics* **68(6)** p1922-1933

[2] Kruk D V J, Streich R, Green G A 2006 Properties of surface waveguides derived from separate and joint inversion of dispersive TE and TM GPR data *Geophysics* **71(1)** p19-29

[3] Arcone A S 2012 Field observations of electromagnetic pulse propagation in dielectric slabs *Geophysics* **49(10)** p1763-1773

[4] Kruk D V J, Arcone A S, Liu L 2007 Fundamental and Higher Mode Inversion of Dispersed GPR Waves Propagating in an Ice Layer *IEEE Trans. Geoscience and Remote Sensing* **45(8)** p2483-2491

[5] Dokht D E R, Gholami A, Ohrnberger M 2020 An inexact augmented Lagrangian method for nonlinear dispersion-curve inversion using Dix-type global linear approximation *Geophysics* **85(5)** p77-85

Journal of Physics: Conference Series **2887** (2024) 012061 doi:10.1088/1742-6596/2887/1/012061

Analysis of characteristics of underground pipeline detection by co-polarization and cross-polarization of ground penetrating radar

Zeshan Xu,Min Xiao and Jie Chen

ZOROY Technology and Development Co., Ltd., Dalian,116085,China (2633969556@qq.com)

Abstract: The polarization mode of ground penetrating radar includes co-polarization mode and cross-polarization mode. The co-polarization mode is more commonly used, mainly because of its good polarization matching effect and good detection effect, so the importance of cross-polarization is ignored. In order to achieve better detection effect in underground pipeline detection, this paper compares the co-polarization and cross-polarization experiments, The results show that when the surface roughness of the pipeline is low, the detection effect of VV polarization of the co-polarization mode is better, and when the surface roughness of the pipeline is high, the detection effect of cross polarization mode is better.

1. Introduction

Ground-penetrating radar is currently widely used in municipal roads, Bridges, tunnels, building quality inspection and other fields [1,2,3]. The antenna structure includes separate transmitting antenna and receiving antenna and integrated transmitting antenna and receiving antenna [4]. The most commonly used antenna is integrated transmitting and receiving antenna, which means that a pair of transmitting and receiving antennas are fixed in the same shielding box. The advantages are convenient to carry, simple operation and good shielding effect. A few of them use the transmit and receive separation mode, that is, the transmit and receive antennas can be freely combined. This mode is mainly applied to low and medium frequency antennas, and its advantage is that there are various acquisition methods. In order to study the polarization characteristics, this paper adopts the transmit and receive separation mode, and changes the polarization combination mode of radar by changing the direction of the transmit and receive antennas [5]. The transmitting antenna of ground penetrating radar emits electromagnetic waves in the same direction as its polarization, which are reflected through underground media, The polarization direction of a part of the reflected wave remains unchanged, and the polarization direction of a part of the reflected wave will rotate and change perpendicular to the polarization direction of the incident wave [6], while the receiving antenna will only receive reflected waves in the same direction as its polarization [7]. At present, co-polarization mode is generally used in GPR in engineering applications. The main reason is that co-polarization mode has the best polarization matching effect between transmitting and receiving antennas, and the probability of underground target being detected is large, but the probability of detecting non-target signals is also large, which has great interference to target detection. In most cases, the detection effect of cross-polarization mode is not as good as that of co-polarization mode, this is because only the polarization direction of the reflected wave is changed before it can be received by the receiving antenna, but most of the reflected wave's polarization direction is the same as that of the incident wave, which has the advantage of shielding most background reflections. For underground slender objects, such as pipelines and inclined fractures, the proportion of reflected waves that change the polarization direction will increase due to the appearance surface and other reasons [8], the effect of cross-polarization will be improved. In engineering applications, most of the peers only use VV polarization to detect underground pipelines, and the detection effect is often poor. This paper first introduces the polarization mode of GPR, the detection range of each polarization mode and the influence of the surface roughness of the pipeline on the polarization mode, and finally explains the application scenarios suitable for each polarization mode through underground pipeline experiments.

2. Polarization mode of GPR

2.1 Theory of polarization

The polarization of electromagnetic wave refers to the characteristics of the direction of the field vector at the specified point in space with time, specifically, it refers to the trajectory of the end point of the field intensity vector in front of the wave with time. According to its track shape can be divided into linear polarization, circular polarization, elliptic polarization, GPR often uses electric dipole antenna is produced by linear polarization electromagnetic wave, that is, the direction of the field vector does not change with time, and its intensity changes periodically with time[9] .

At present, GPR generally adopts plane butterfly antenna, the main polarization direction of its electric field is parallel to the long axis of the antenna[10], so the direction of the long axis of the antenna is the direction of polarization. The polarization modes of GPR antenna include V and H, V refers to the vertical direction of antenna polarization, H refers to the parallel direction of antenna polarization. According to the polarization combination of receiving and transmitting antennas, the polarization modes are HH, VV, HV, and VH, as shown in figure 1. If the polarization mode of the transmitting and receiving antennas is the same, it is called the co-polarization mode, including HH and VV, if the polarization mode of the transmitting antenna and the receiving antenna is different, it is called the cross-polarization mode, including VH and HV[11]. The polarization matching effect between transmitting antenna and receiving antenna in co-polarization mode is good, and the reflected wave energy is stronger[12]. At present, HH polarization is mainly used in 3D array ground-penetrating radar, mainly because the HH polarization arrangement is compact, which can improve the density of the measurement line. Single-channel 2D ground penetrating radar mainly adopts VV polarization , mainly because VV polarization has a large effective detection range in the direction of the survey line[13].

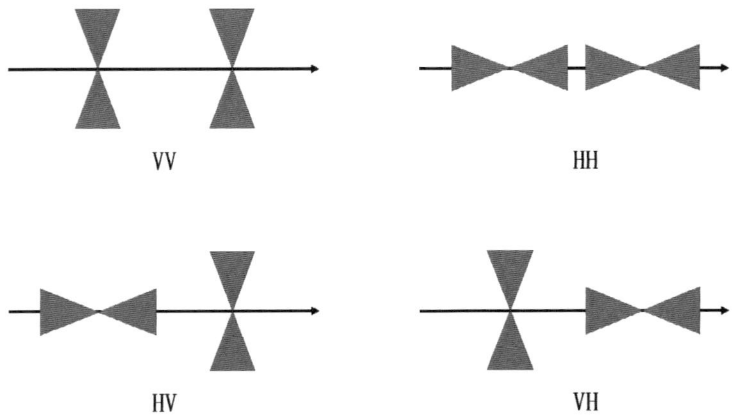

Figure 1. Four polarization modes

2.2 Detection range of different polarizations

Since the electromagnetic wave emitted by the antenna is a spherical wave with uneven energy distribution, the radar antenna has a certain effective radiation range, it is an elliptic cone with an ellipse at the bottom[14,15], as shown in figure. 2. The detection range is large in the direction perpendicular to the long axis of the antenna, and small in the direction parallel to the long axis of the antenna. The effective detection range of GPR is the overlapping area of transmitting and receiving antenna radiation range, as shown in figure 3.

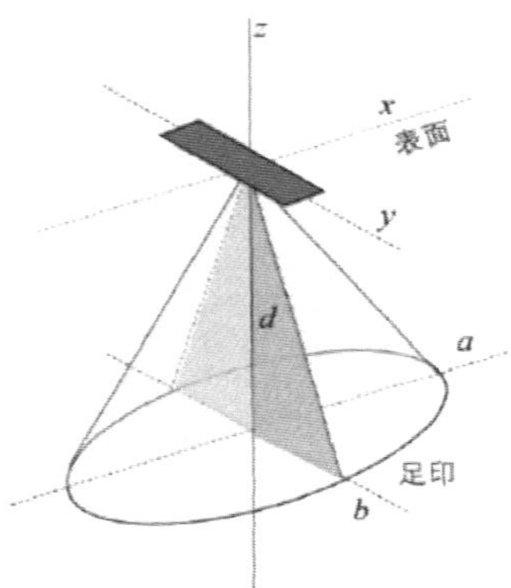

Figure 2. The detection range of the antenna when the depth is d

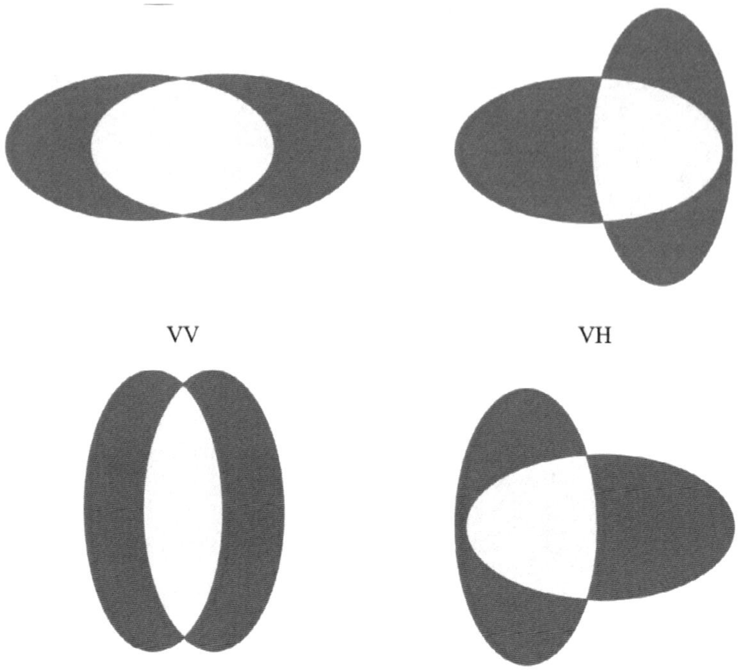

Figure 3. Detection range of different polarization methods

In figure 2, d is the detection depth, and a and b are the major and minor axes of the ellipse when the detection depth is d. The effective radiation range of the ground penetrating radar antenna is related to the detection depth d, the relative dielectric constant ε and the wavelength λ of the electromagnetic wave propagating in the medium, as shown in equations (1) and (2).

$$a = \frac{\lambda}{4} + \frac{d}{\sqrt{\varepsilon - 1}} \quad (1)$$

$$b = \frac{a}{2} \quad (2)$$

As can be seen from the above equations, when VV polarization is adopted, the detection range of the survey line direction is the largest, followed by VH and HV polarization, and the detection range of the survey line direction is the smallest when HH polarization is adopted.

2.3 Influence of surface roughness on polarization

The influence of surface roughness on polarization was studied in detail by Feng Xuan and Zou Lilong[17,18]. For any small surface area of the target body, it is regarded as consisting of many small square grids. First, the roughness degree k_{ij} of any point on the surface is defined as equation (3) :

$$k_{ij} = \frac{\left|a_{i,j-1} - a_{i,j}\right| + \left|a_{i,j+1} - a_{i,j}\right| + \left|a_{i-1,j} - a_{i,j}\right| + \left|a_{i+1,j} - a_{i,j}\right|}{4} \quad (3)$$

In equation (3), a_{ij} is the height value of the previous point (i,j) on the surface.

The greater the roughness degree k_{ij}, the greater the height difference between the point and the surrounding area, and the rougher the surface. Define the overall roughness of the surface as equation (4) :

$$K = \sum_{i=1}^{n}\sum_{j=1}^{m} k_{ij} \tag{4}$$

In equation (4), m and n are the number of grid nodes on the surface along their respective directions. Then the antenna calculates the co-polarization response and cross-polarization response at each measuring point according to a certain line spacing and measuring point spacing. Define the mean amplitude M as equation (5) :

$$M = \frac{1}{xy}\sum_{i=1}^{x}\sum_{j=1}^{y}\frac{\max(A_{VH(i,j)})}{\max(A_{VV(i,j)})} \tag{5}$$

In equation (5), x is the number of measurement lines, and y is the number of measurement points on each measurement line. A_{VV},A_{VH} represents the co-polarization and cross-polarization response amplitudes, respectively. Finally, the average amplitude ratio (VH/VV) of the two polarizations under different roughness is calculated and compared. With the increase of the overall roughness K of the surface, the average amplitude ratio M also increases, indicating that the rougher the surface, the stronger the cross-polarization response amplitude and the weaker the co-polarization response amplitude.

3 Experiment

3.1 Experiment 1

The object of this experiment is two side-by-side cast iron water pipes, both of which are DN800 in size, buried depth of 2m and center distance of 1.8m. A 250MHz antenna was used for field detection, a range-wheel mode was adopted, the channel spacing was 1cm, and the time window was 80ns. VV, VH and HH polarization were used for data acquisition at the same measuring line position, and then the same post-processing was done on the data, including dewow, background removal and bandpass filtering. The radar images were shown in figure. 4, 5 and 6 below.

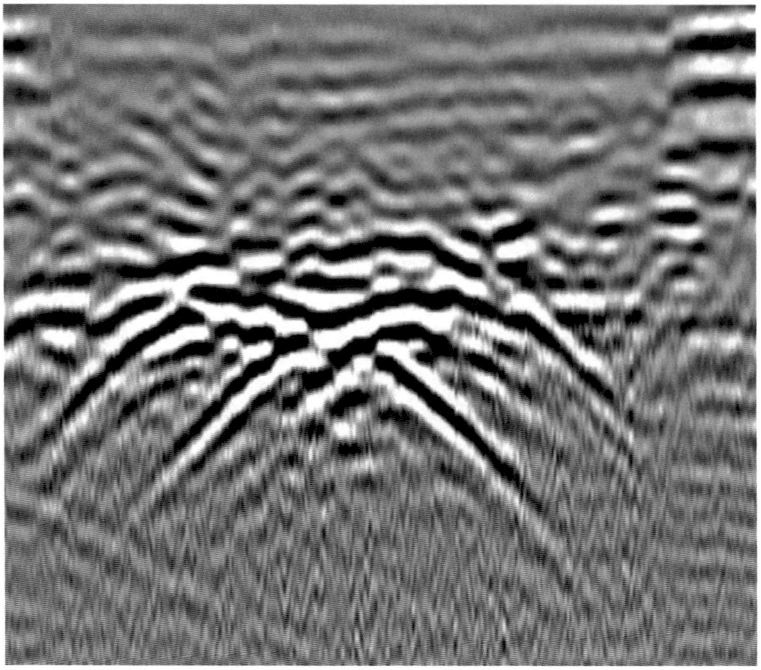

Figure 4. GPR image of VV polarization

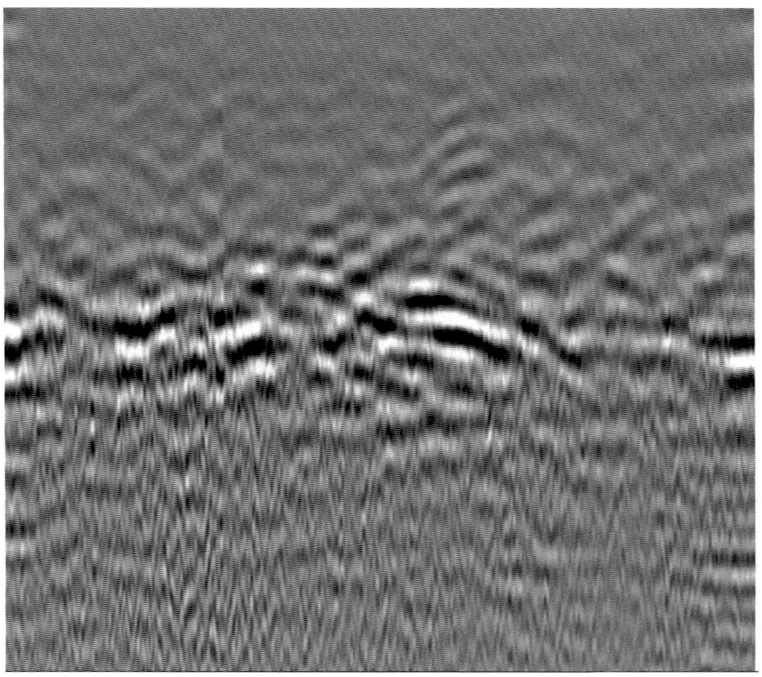

Figure 5. GPRr images of VH polarization

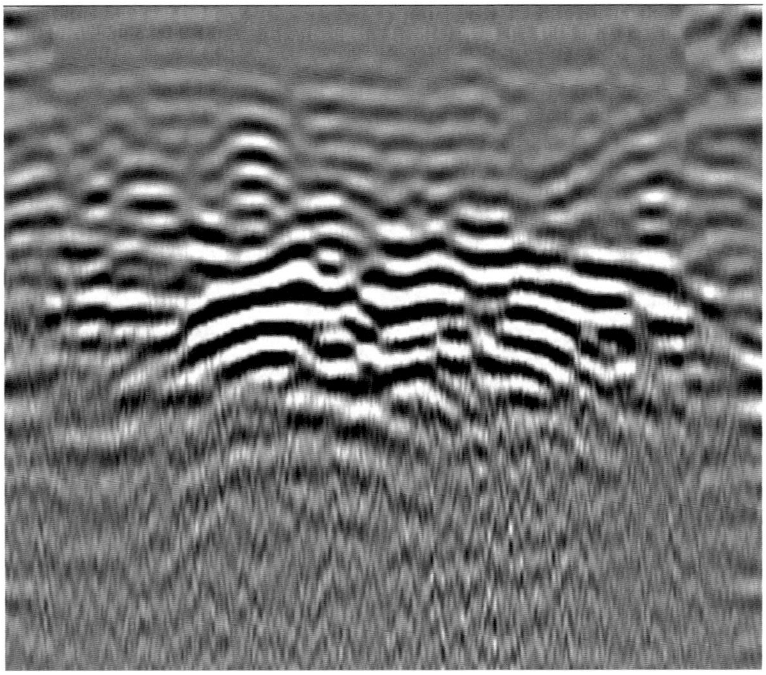

Figure 6. GPR images of HH polarization

As can be seen from the figures above, the amplitude response of the pipeline image in the co-polarization mode is stronger than that in the cross-polarization mode. It is difficult to identify the pipeline in the cross-polarization mode, mainly because the cast iron pipeline is less rough, and the average amplitude of the cross-polarization mode is much smaller than that in the co-polarization mode.

It can be seen from figure. 4 and figure. 6 that the pipe detection effect is better when VV polarization is adopted, the diffracted wave is obvious, and the two pipes can be distinguished. When HH polarization is adopted, the pipe diffraction is small, and the two pipes are not easy to distinguish, mainly because the detection range of HH polarization is small and the diffraction wave is short, which makes the diffraction of the two pipes mixed together.

3.2 Experiment 2

The experiment site is located on an urban road lane with asphalt pavement. Below the experiment area is a concrete pipe, which is used to connect the rain grate and the rain well, the pipe diameter is DN300 and the buried depth is 0.5 meters. The site adopts 250MHz antenna detection, adopts the ranging wheel mode, the track spacing is 1cm, the time window is 50ns, the measurement line layout is perpendicular to the direction of the pipeline, and the site adopts VV, HH, VH and HV polarization for data acquisition at the same position. The same filtering method was adopted for data processing, dewow and bandpass filtering. The display gain of VV and HH polarization was set to 3, and that of VH and HV polarization was set to 5, the radar images were shown in figure. 7, figure 8,figure. 9 and figure. 10.

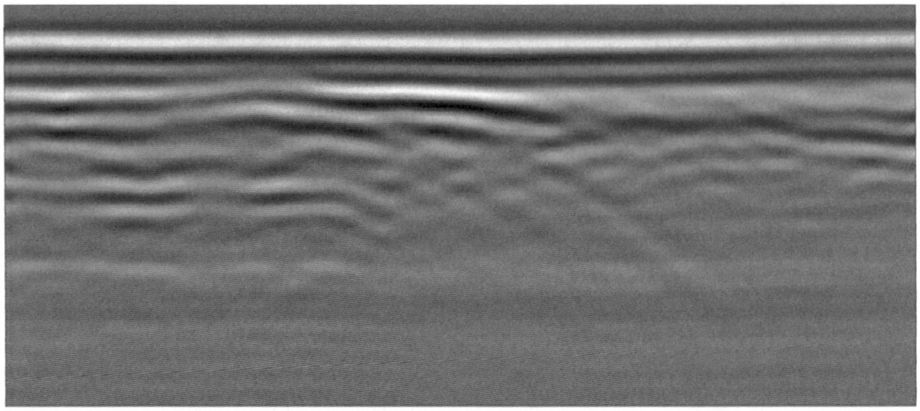

Figure 7. GPR image of pipelines with VV polarization

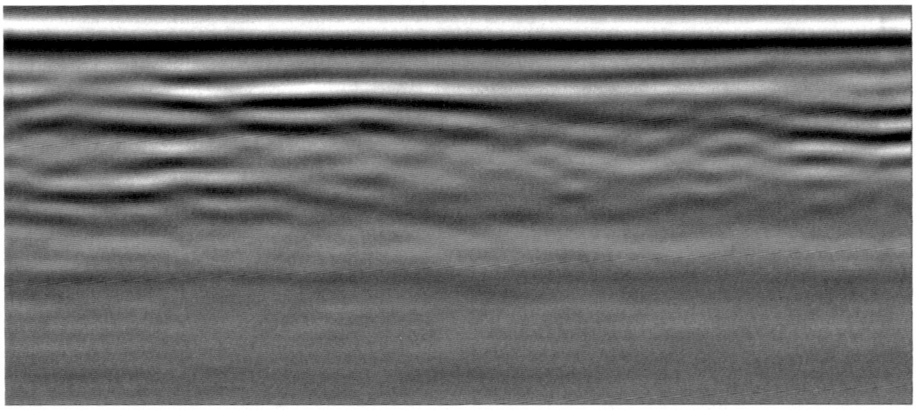

Figure 8. GPRr image of pipelines with HH polarization

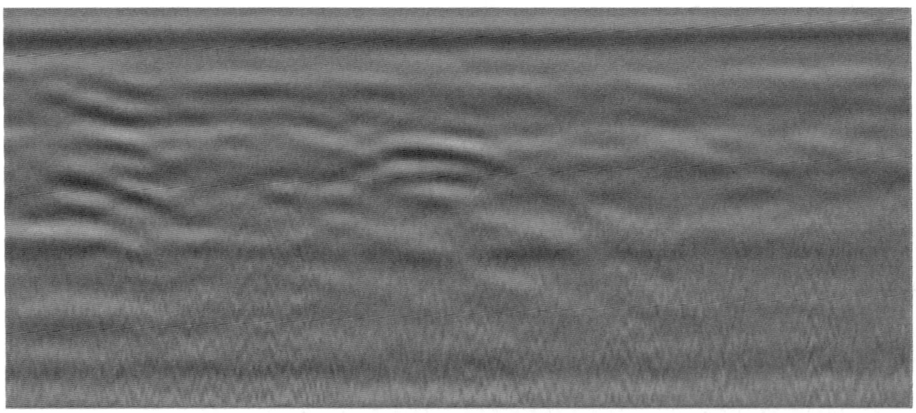

Figure 9. GPR images of pipelines with VH polarization

Figure 10. GPR images of pipelines with HV polarization

As can be seen from the figures above, the overall amplitude response of the radar image is stronger than that of the cross-polarization mode when the co-polarization mode is adopted, mainly because the co-polarization mode has good matching effect and strong reflected wave amplitude. However, for the response of the pipeline, the amplitude response of cross-polarization mode is stronger than that of the co-polarization mode, which is related to the roughness of the concrete pipeline. Because the concrete pipeline has a very rough appearance, the average amplitude of cross-polarization mode is significantly increased, and the average background amplitude of cross-polarization mode is weak, which makes the response characteristics of the pipeline more obvious.

It can be seen from figure 7 and figure 8 that when VV polarization is used, the diffraction of the pipeline image is longer and the pipeline is more obvious. When HH polarization is used, the pipeline image is not obvious and the pipeline is not easy to identify.

4.Conclusion

In this paper, the theory of polarization of GPR is studied and analyzed, and the applicability of each polarization mode in engineering application is proved by experiments. Finally, the following conclusions are obtained:

1. For underground pipelines with small surface roughness, the effect of co-polarization is better, and for underground pipelines with large surface roughness, the effect of cross-polarization is better.

2. The comparison between VV polarization and HH polarization shows that VV polarization diffraction is more obvious, which is beneficial to identify a single underground pipeline or to distinguish several underground pipelines with small spacing.

3. When the situation of underground pipeline in the field is not clear, only the co-polarization method can not guarantee the detection effect, and the co-polarization method and the cross-polarization method should be used respectively.

References

[1] Xu Zeshan, Jian Shikai, Qin Tan,Tang Fawei and Cao Jianlong 2019 *J. Chinese Journal of Engineering Geophysics* **16** 116-25.

[2] ChenWei,GongWenlong and Chang Jun 2020 Application of ground penetrating radar in tunnel lining detection *J. Journal of Suzhou University of Science and Technology (Engineering and Technology)* **33** 54-71.

[3] ZhangWenbo,WeiWenbo,JingJianen and Jia Zhengyuan 2008 Application of Ground Penetrating Radar

Polarization in the Concrete Structure Detection *J. Journal of Jilin University(Earth Science Edition)* **38** 155-59.

[4] Yang Feng, Zhang Quansheng and Wang Pengyue 2009 *Research on geological radar detection technology of Highway Subgrade* (Beijing: Chain Communications Press) pp 3-7.

[5] TongChuanxue,LiuSixin and WangChenhui 2006 The Simulation of GPR Polarization Measurement with Finite-Difference Time-Domain(FDTD) Method *J. Journal of jilin University(Earth Science Edition）* **36** 201-05

[6] LiHang 2016 *Simulation and Analysis of Full-Polarimetric Ground Penetrating Radar Based on FDTD* (Chengdu :Chengdu University of Technology) pp 1-20.

[7] ZhengLongjin,YangXiaolong and ZouLei 2019 Anomaly Detection for Tunnel's Secondary Lining Affected by Linear Polarization Mode of Ground Penetrating Radar *J. Journal of Yangtze River Scientific Research Institute* **36** 164-74.

[8] ChenJun,ChiChangfeng,XuDongsheng and LiangFeng 2020 Application of GPR scattering matrix to estimating azimuths of underground pipelines *J. Carsologica Sinica* **39** 101-9.

[9] GengYuling,JiaXuemin,LiDaming,et al. 2007 *Research on ground penetrating Radar technology in Non-destructive testing of Highway Pavement* (Beijing:Seismological press) pp 5-14.

[10]GuoShili,CaiJianchao,ZhangXueqiang,LiXiuzhong and Dai Gongzheng 2012 *J. Progress in Geophysics* **27** 1812-21.

[11] DongZejun 2022 *Induced Field Rotation Interference from Rough Interfaces and lts Effects on Full-Polarimetric Ground Penetrating Radar Data* (Changchun:Jilin University) pp 1-65.

[12] LiShuangxi,ZengZhaofa,LiJing,SongShiming,ZhangJing and ZhangBaosong 2014 *J. Chinese Journal of Engineering Goephysics* **11** 513-21.

[13] ZhangNing,LiJingxia,XuHang,ZhaoMingrui,LiuLi and Wang Bingjie 2022 Underground Pipeline Distribution Detection by Polarized GroundPenetrating Radar Based on Gray Complementary Code *J. Chinese Joumal of Electron Devices* **45** 767-74.

[14] Bo Huishen 2006 *Practical Handbook of Ground Penetrating Radar Technology* (Beijing: Geological Publishing House) pp 21-31

[15] Wang Jinbiao 2021 *Research and Application of Ground Penetrating Radar in large Depth Detection* (Nanchang: East China University Of Technology) pp 16-20.

[16] Xiao Min, Chen Changyan, Jia Hui,SuZhaofeng,Bai Chaoxu and Zhang Hui 2016 The study of the interference region around metal pipelinein underground disease detection of urban road *J. Geophysical & Geochemical Exploration* **40** 1046-50.

[17] Feng Xuan, Zou Li Long, Liu Cai,LuQi,Liang Wenjing,LiLili and Wang Shiyu 2011 Forward modeling for full-polarimetric ground penetrating radar *J. Chinese Journal of Geophysics* **54** 349-57.

[18] Zou Lilong 2012 *Forward modeling and calibration technique of full Polarimetric GPR* (Changchun: Jilin University) pp 20-39.

Acknowledgements

I would like to thank Mr. Feng from Jilin University for his careful guidance on the content and ideas of the paper, and Engineer Cao Pengtao from Shenzhen Water Planning and Design Institute Co., Ltd. for his patient help in the experiment process.

Application of ground penetrating radar in potential hazards detection of levee

W K Huang[1,2*], S P Li[1,2], C Tan[1,2] and P Y Lu[1,2]
[1]Guangdong Research Institute of Water Resources and Hydropower, Guangzhou 510610, China
[2]State and Local Joint Engineering Laboratory of Estuarine Hydraulic Technology, Guangzhou 510610, China
[*]Corresponding author's e-mail: hwk_gz@163.com

Abstract. Ground-penetrating radar, utilizing electromagnetic wave technology, stands as a crucial geophysical exploration method for analyzing hydraulic engineering defects. In the context of identifying potential hazards in a levee within Yunfu, Guangdong province, GPR technology is deployed. This study delves into the radar's ability to discern characteristics such as inadequate soil compactness, voids, as well as interference from metal pipes and wires. By scrutinizing these radar signatures, the primary causes underlying the development of cracks and uneven settlement across multiple sections of the levee are identified. The insights garnered from this analysis serve as valuable reference points for leveraging ground-penetrating radar technology in the detection of potential hazards within hydraulic engineering contexts.

1. Introduction

Levees constitute a vital component of national infrastructure, with their security and stability holding profound significance in safeguarding lives and fostering social and economic progress[1]. However, the occurrence of cracks, landslides, leakage, and other defects in levees[2], influenced by various natural and human factors, poses a substantial threat to their safety. Consequently, the rapid and precise assessment of levee safety has emerged as an urgent imperative.

Conventional detection techniques like drilling coring and water injection tests often encounter limitations such as high equipment costs, intricate operations, restricted coverage, and other challenges[3]. These shortcomings hinder their ability to meet the demands for comprehensive and accurate monitoring of levee structures and subsurface conditions. Ground-penetrating radar (GPR), as a geophysical exploration method, offers distinct advantages of speed, precision, and non-destructiveness[4]. Consequently, it has found widespread application in geological exploration, resource assessment, archaeological investigations, and beyond. In recent years, the utilization of GPR in hydraulic engineering has garnered increasing attention, particularly in the identification and evaluation of hidden dangers within levees.

2. Theory

The fundamental principle underpinning GPR resembles ultrasonic detection in medical applications. By exploiting variations in wave speed and propagation characteristics of high-frequency electromagnetic waves across different media[5], GPR enables the assessment of underground media properties and formations. GPR operates on the premise of contrasting dielectric constants among various media. Notably, a substantial difference in dielectric constants results in heightened signal energy received by the antenna[3]. Through meticulous analysis of waveform patterns, amplitude intensities, and temporal changes in reflected waves, GPR facilitates the inference of spatial positioning, structural attributes, shapes, and burial depths of the subsurface media[6].

3. GPR application

The levee situated in Yunfu, Guangdong Province, has proven its resilience against numerous floods since its construction, serving as a vital safeguard for the local community's 'life protection line project'. Spanning a total length of 7.8 kilometers, the levee is composed of distinct sections. The upstream segment, measuring 5.7 kilometers (from stake marks 0+000 to 5+700), consists of an earth levee. In the middle, a 1.8-kilometer stretch (from stake marks 5+700 to 7+500) is constructed of stone. The downstream segment, covering 0.3 kilometers, is once again an earth levee.

The primary issue plaguing the levee involves the emergence of cracks and uneven settlement along numerous sections of the earth levee. To investigate the root cause of these defects, ground-penetrating radar (GPR) was deployed along the levee section spanning from 0+000 to 5+600. The GPR equipment utilized was the pulse EKKO PRO, manufactured by Sensors & Software Inc. (SSI), based in Canada. Employing an unshielded antenna operating at a transmission frequency of 100MHz, the GPR system was configured with a step size of 0.25m and a time window of 200ns. The radar was automatically stacked 32 times to collect data in odometer mode which can accurately record the length of the testing line.

The layout of the GPR lines was as follows: Lines 21 to 50 corresponded to stake marks ranging from 0+000 to 4+640; Lines 51 and 52 corresponded to stake marks from 5+400 to 5+600; Lines 53 to 57 corresponded to stake marks from 4+640 to 3+820.

4. Application results

Figure 1 illustrates the findings from Line 24, with the starting and ending stake marks denoted as K0+340 to K0+475. The horizontal axis represents the line's position along the levee's axis, while the left vertical axis indicates depth below the ground, and the right vertical axis signifies time. Notably, within the range of 23 to 40 meters along the line and at depths of 1.8 to 4.0 meters, the reflected signals exhibit greater intensity than the surrounding medium, with disrupted reflection wave patterns and chaotic waveforms observed. This suggests the presence of poorly compacted soil in this area. Further along the line, specifically at positions 98 to 103 meters and depths of 1.0 to 4.0 meters, pronounced reflection signals, and numerous layers of strong reflection indicate the likelihood of shallow buried metal pipes. Additionally, between positions 110 to 135 meters and depths of 1.8 to 7.0

meters, multiple strong reflections are evident. Consistent occurrences of these characteristics across various positions, along with field observations, lead to the presumption of street light interference.

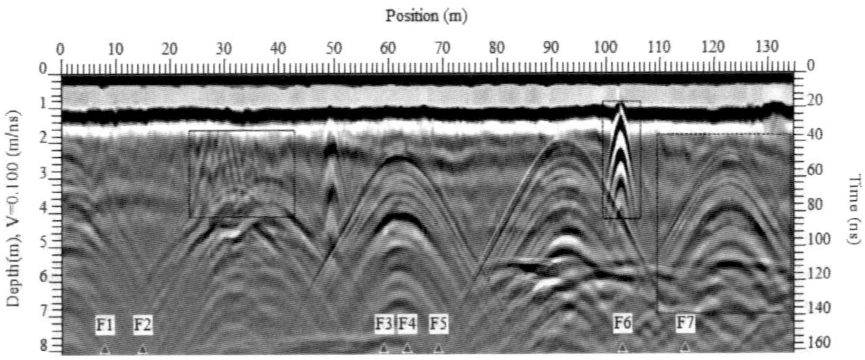

Figure 1. GPR profile of Line 24.

Figure 2 depicts the results obtained from Line 28, spanning from stake marks K0+830 to K0+985. At the position of 100 to 104 meters and depths of 1.0 to 2.8 meters, an anomaly in the form of a hyperbola on the radar profile suggests influence from voids in the area. Similarly, at positions 67 to 71 meters and depths of 0.8 to 2.5 meters, despite chaotic radar profiles, hyperbolic anomalies indicate potential voids. Furthermore, between positions 120 to 145 meters and depths of 2.5 to 6.2 meters, prominent reflections with significant amplitudes point towards wire interference, corroborated by on-site observations.

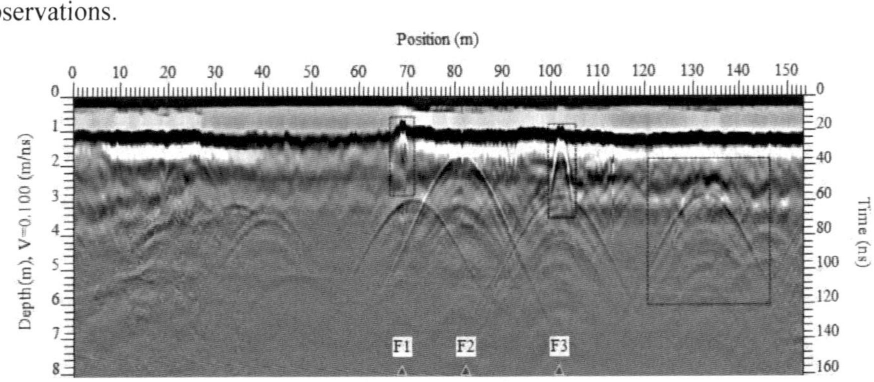

Figure 2. GPR profile of Line 28.

Figure 3 presents the findings from Line 57, encompassing from stake marks K4+090 to K3+820. In this figure, at positions ranging from 12 to 33 meters and depths of 3.2 to 5.2 meters, the reflected signal exhibits greater intensity than the surrounding medium. Discontinuous lineups, along with strong amplitude reflections and localized arc faults, suggest poor densification of the soil in this area. Moving to positions 48 to 54 meters with depths of 2.0 to 6.8 meters, strong reflection signals with isometric arcs indicate the presence of cracks, likely caused by uneven settlement of the soil on the left side. Between positions 94 and 132 meters, with depths of 3.5 to 5.5 meters, a continuous group of homodirectional reflection waves is observed, featuring multiple waves and a strong overall amplitude, indicating a potential water-rich area. At positions 140 to 176 meters and depths of 0.4 to 0.7 meters, a black strip appears in the figure, attributed to defects in the concrete pavement, consistent with field observations. Additionally, at positions 140 to 265 meters and depths of 4.2 to 8.2 meters, multiple regular and strong reflections appear in the figure, which is speculated to be wire interference.

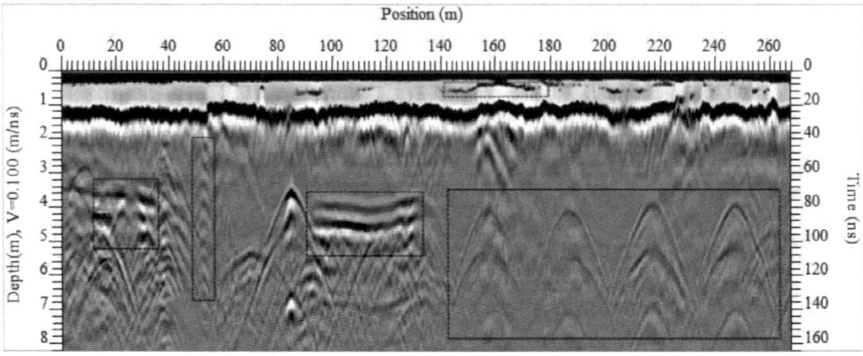

Figure 3. GPR profile of Line 57.

5. Conclusions

GPR was successfully employed in assessing a levee in Yunfu, Guangdong Province, leading to the following conclusions:

(1) The GPR method effectively identifies uncompacted soils, cracks, voids, and other defects, demonstrating its utility in analyzing key factors contributing to hydraulic engineering defects.

(2) The main reasons for the cracks and uneven settlement at the top of levee are the overall poor uniformity of the levee soil and the presence of cavities and non-compact soil in certain areas.

(3) GPR results exhibit multiple solutions and uncertainties. Therefore, interpretation should be complemented by drilling data and a thorough understanding of the field's actual conditions.

Acknowledgment

This work was supported by Science and Technology Innovation Program from Water Resources of Guangdong Province (No. 2022-03, 2023-02).

References

[1] Zhang Y, Zhou L M, Xiao G Q. Analysis and application of wave field of ground enetration radar in defect detection of embankment[J]. *Journal of Changjiang River Scientific Research Institute*, 2019, **36**(10): 151-56.

[2] Xu L Q, Zhang G C, Ma Z K. Development of comprehensive geophysical prospecting technology for hidden danger detection of earth rock dams[J].*Progress in Geophysics*, 2022, **37**(04): 1769-79.

[3] Xu H M, Lin B, Hu J J. Application of ground penetrating radar in detection of dam risks[J]. *Geology of Anhui*, 2017, **27**(01): 39-41.

[4] Zhang J Y, Wang Y, Wu Z Q, et al. Application of ground penetrating radar technology for levee

dngineering hidden trouble detection in cold region[J].*Hydro Science and Cold Zone Engineering*, 2023, **6**(01): 6-9.

[5] Zuo J, Gao Y. Application of Ground penetrating radar technology in detection of dam in the lower Yellow River[J]. *Yellow River*, 2022, **44**(S1): 80-81+87.

[6] Li J., Liu C, Zeng, Z., & Chen, L (2015). GPR Signal Denoising and Target Extraction with the CEEMD Method. *IEEE Geoscience and Remote Sensing Letter,* 12 (8), 1615-1619.

Improved Synchrosqueezing S-Transform and Its Application in Reinforcement Protection Layer Identification

Hongqiang Xiong [1], Jing Li [1, 2*], Tengfei Zhang [3], Rong Hu [1], Tieyu Liu [1]

1 College of Geo-Exploration Science and Technology, Jilin University, 130026, China.

2 Key Laboratory of Applied Geophysics, Jilin University, 130026, China

3 Zhejiang Communications Group Inspection Technology Co.,Ltd., 310000, China

Abstract. Synchrosqueezing is a strategy to enhance the resolution of time-frequency analysis. However, its application in the S-transform leads to complexities due to the variable phase spectrum of the S-transform, resulting in a decrease in resolution and an increase in complexity. This work proposes an Synchrosqueezing Generalized S-Transform (SS-GPST) by introducing phase-shifting and a generalized parameter to simplify the mathematical definition and enhance resolution. Synthetic data trials demonstrate the superiority of SS-GPST over existing synchrosqueezing methods, offering enhanced resolution and minimized inverse transformation errors. In practical ground-penetrating radar applications, we have applied SS-GPST to identify the reinforcement protection layer in tunnel linings. The results show that SS-GPST offers a novel method for thin-layer extraction in ground-penetrating radar applications.

1. Introduction

Time-frequency (TF) analysis is a key means of geophysical data processing. However, traditional TF methods calculate the inner product of the window function (e.g., continuous wavelet transform, S-transform, and short-time Fourier transform) and the analyzed signal, resulting in spectrum energy leakage. Specifically, a certain frequency component will spread in the TF spectrum to the band area with the specific frequency as the center [1].

Daubechies et al. [2] introduced synchrosqueezing transform (SST). SST addresses the issue of TF diffusion near instantaneous frequencies, which can improve the TF resolution of continuous wavelet transform (CWT). S-transform (ST) [3] is widely used in geophysics due to its excellent expression in high-frequency components. Synchrosqueezing can be extended to the ST. Huang et al. [4] proposed synchrosqueezing S-transform (SS-ST) and applied SS-ST to find gas hydrates by the spectral slices of marine seismic data. In the study by [5], the Synchrosqueezing Generalized S-Transform (SS-GST) is defined to improve the TF resolution of the SS-ST further. However, synchrosqueezing depends on the property that the CWT phase spectrum of a harmonic signal is constant over time.

In contrast, the phase spectrum of the ST varies with frequency. Complex mathematical definitions and approximations are utilized in methods such as SS-ST and SS-GST to address this issue. This compromise leads to a decrease in TF resolution and signal reconstruction errors.

This paper introduces a novel method to enhance the SS-ST, termed SS-GPST. This method initially alters the phase spectrum characteristics of the S-transform through phase-shifting. Subsequently, it incorporates a generalized parameter to improve resolution, followed by synchrosqueezing. The SS-GPST simplifies the definitions of forward and inverse transformations, enhancing TF resolution and signal reconstruction accuracy. Identifying thin layers has always been a critical challenge in ground-penetrating radar (GPR) engineering detection tasks. We applied SS-GPST to GPR data for thin-layer identification. The results demonstrate that SS-GPST slices reveal the reinforcement protection layer.

2. Theory

The ST of a time series signal $h(t)$ is

$$ST(\tau,f)=\int_{-\infty}^{\infty} h(t)\frac{|f|}{\sqrt{2\pi}}e^{-\frac{(t-\tau)^2 f^2}{2}}e^{-i2\pi ft}dt, \tag{1}$$

where t and f represent time and frequency variables, respectively. τ is a time variable defining the window function position over time. On the ST, the corresponding instantaneous frequency is $\tilde{f}$.

$$\tilde{f}(\tau,f)=f+\frac{1}{2\pi i ST(\tau,f)}\frac{\partial ST(\tau,f)}{\partial\tau}. \tag{2}$$

The synchrosqueezing involves compressing the ST coefficient value towards the nearest instantaneous frequency. In eqn. 2, the instantaneous frequency is related to f, indicating that the ST phase is dependent on f. Because the sum of the complex numbers $ST(\tau,f)$ can cancel each other, resulting in reduced energy. To address this limitation, we involve phase-shifting ST (PST).

$$PST(\tau,f)=\int_{-\infty}^{\infty} h(t)\frac{|f|}{\sqrt{2\pi}}e^{\frac{-(t-\tau)^2(f)^2}{2}}e^{-i2\pi f(t-\tau)}dt. \tag{3}$$

The PST and ST have the same TF characteristics because the ST multiplied by $e^{i2\pi f\tau}$ does not change amplitude. To optimize the TF resolution of the PST, we introduce a parameter to the PST to control the shape of the Gaussian window. The generalized phase-shifting S-transform (GPST) is

$$GPST(\tau,f)=\int_{-\infty}^{\infty} h(t)\frac{|Af|}{\sqrt{2\pi}}e^{\frac{-(t-\tau)^2(Af)^2}{2}}e^{-i2\pi f(t-\tau)}dt, \tag{4}$$

In the TF analysis of GPST, A is positively correlated with time resolution and negatively correlated with frequency resolution. The instantaneous frequency $\bar{f}$ of GPST is

$$\bar{f}(\tau,f)=\frac{1}{2\pi i GPST(\tau,f)}\frac{\partial GPST(\tau,f)}{\partial\tau} \tag{5}$$

Based on the instantaneous frequency $\bar{f}$, the synchrosqueezing transform can be written as

$$SSGPST(\tau,\bar{f}_l)=\sum_{f_k:|\bar{f}(\tau,f_k)-\bar{f}_l|\le\Delta\bar{f}/2} GPST(\tau,f_k)f_k^{-1}, \tag{6}$$

where f_k is the discrete frequency, $\bar{f}_l$ and $\Delta\bar{f}$ are the central frequency and the frequency interval of the SS-GPST. The signal can be reconstructed from SS-GPST by

$$h(t)=\mathrm{Re}\left[C^{-1}\sum_l SSGPST(\tau,\bar{f}_l)\Delta\bar{f}\right], \tag{7}$$

where $C=-A/2\int_0^{\infty}\hat{\omega}(\xi)\xi^{-1}d\xi$, $\omega(t)=(1/\sqrt{2\pi})e^{(-t^2A^2/2)}e^{i2\pi t}$ and $\hat{\omega}(\xi)$ is the Fourier Transform of $\omega(t)$.

3. Synthetic data experiment

Figure 1(c) shows a composite signal with three frequency components with a time range $t\in[0,10]$. Each component exhibits unique characteristics: $s_1(t)$ it features the lowest frequency and largest amplitude, whereas the amplitude $s_2(t)$ reduces with increasing frequency $s_3(t)$ is a 9 Hz cosine signal with the smallest amplitude. This composite signal represents the typical feature of stronger attenuation of high-frequency components in GPR signals.

Figure 1(a)-(d) shows the TF spectrum of the synthetic signal obtained using traditional TF methods. Figure 1(a)-(d) reveals that the energy appears "blurred" on the TF spectrum of WT, ST, GST, and GPST. Conversely, synchrosqueezing methods significantly improve the TF resolution in the SST, SS-ST, SS-GST, and SS-GPST, as depicted in Figure 1(e)-(h). SST excels in focusing on the low-frequency component $s_1(t)$. In contrast, ST-based methods typically excel in high-frequency components interested in GPR. However, inherent limitations of ST in synchrosqueezing result in energy dispersion and lower resolution in both SS-ST and SS-GST. For example, with the fixed 9Hz signal $s_3(t)$, SS-ST and SS-GST yield imprecise results, in contrast to the superior accuracy of SS-GPST.

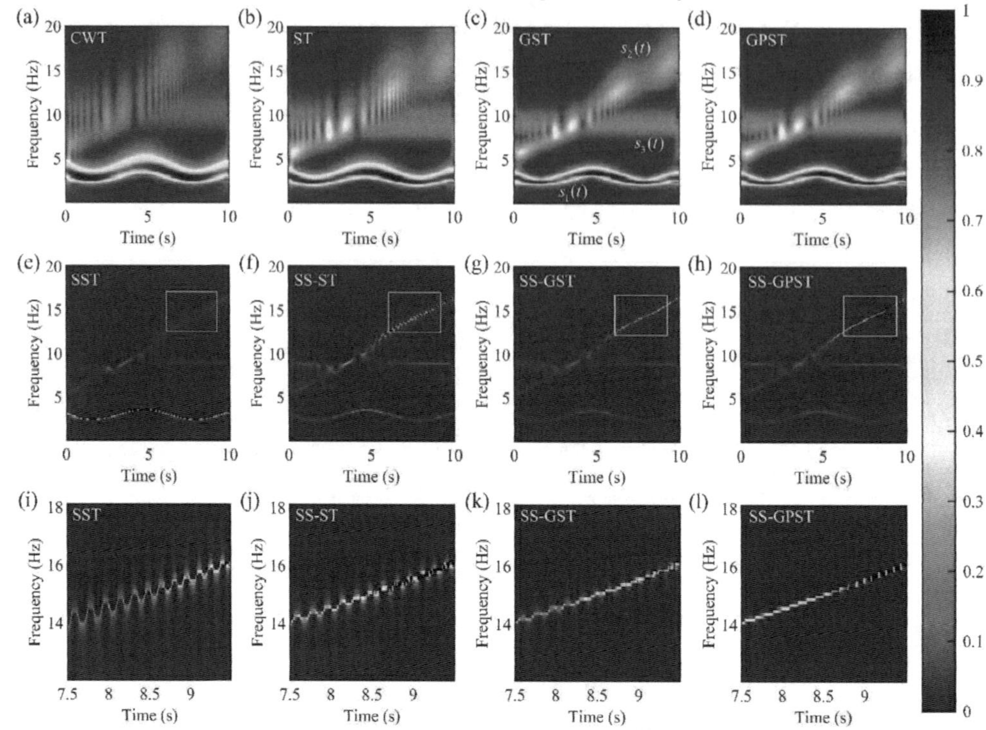

Figure 1. The results of different TF analysis methods. Where (i) - (l) represents local amplification.

In Table 1 maximum error E_{max} and mean squared error (MSE) associated with SS-ST and SS-GST are significant, suggesting that these methods might lead to information loss during signal reconstruction. In contrast, SS-GPST demonstrates more precise reconstruction capabilities.

Table 1. Signal reconstruction error of different methods.

Indexs	SST	SS-ST	SS-GST	SS-GPST
E_{max}	0.03	0.51	0.70	0.04
MSE	1.97e-4	0.04	0.10	6.00e-4

4. Real data testing

As depicted in Figure 2, the raw GPR data were recorded by a vehicle-mounted GPR system [6] equipped with 300 MHz air-coupled antennas in a highway tunnel. Identifying the thickness of the reinforcement protection layer is a critical task in tunnel inspections. However, with a 300 MHz antenna, the resolution limit is exceeded when detecting a design thickness of approximately 3cm.

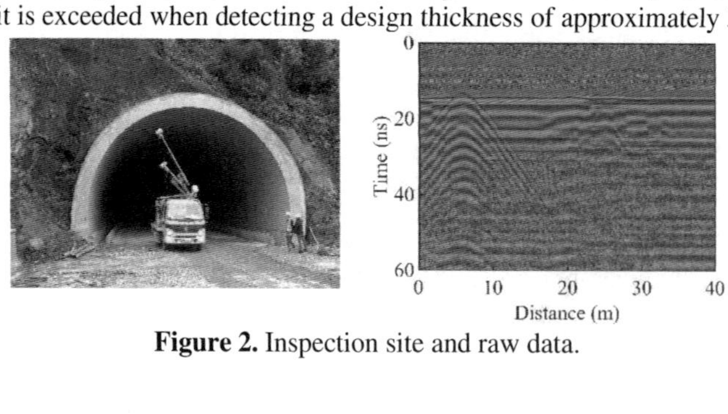

Figure 2. Inspection site and raw data.

doi:10.1088/1742-6596/2887/1/012063

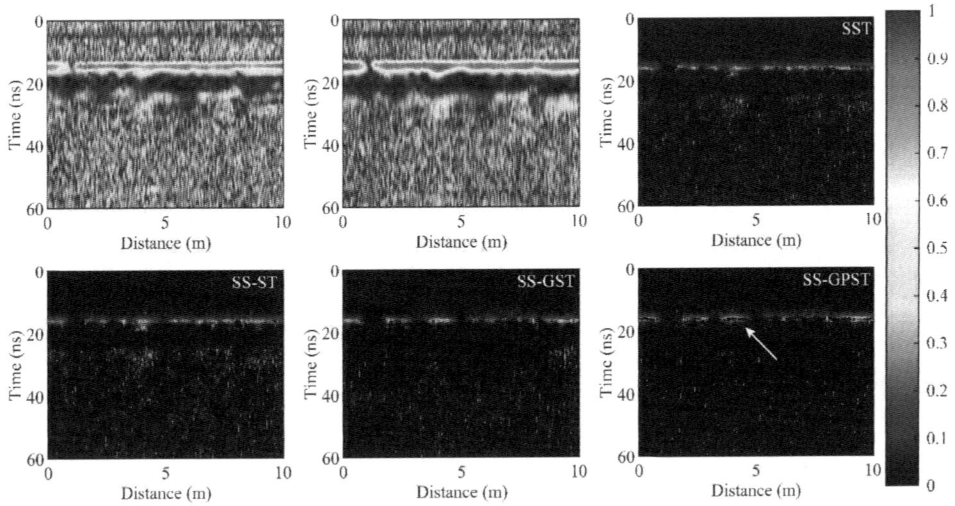

Figure 3. Normalized 900 MHz TF slice of different methods.

Figure 3 presents the results of different TF analysis methods at a 900 MHz slice. Figure 3 demonstrates that ST, CWT, SST, SS-ST, and SS-GST face challenges in defining layer thickness due to limited resolution. However, SS-GPST distinctly delineates the upper and lower boundaries of the reinforcement protection layer. SS-GPST results indicate a reinforcement protection layer thickness of 3.0-3.5 cm, aligning with the actual design specifications.

5. Conclusion

This paper presents an improved synchrosqueezing S-transform, which enhances TF resolution and reconstruction accuracy through phase-shifting and generalized parameters. Results from simulated tests indicate that the SS-GPST outperforms other methods in TF resolution and signal reconstruction precision. In practical data, we applied SS-GPST to identify the thin reinforcement protection layer in tunnel detection using GPR. The findings show that the high-frequency SS-GPST slices accurately reveal the reinforcement protection layer's upper and lower interfaces.

Acknowledgments

This work was supported in part by the Natural Science Foundation of China under Grant 42222407 and Grant 42174065

References

[1] Auger F, Flandrin P, Lin YT, et al. Time-frequency reassignment and synchrosqueezing: an overview. *IEEE Signal Process. Mag.* 2013; 30(6): 32–41.
[2] Daubechies I, Lu J, Wu HT. Synchrosqueezed wavelet transforms an empirical mode decomposition-like tool. *Appl. Comput. Harmon. Anal.* 2011; 30(2): 243–61.
[3] Stockwell RG, Mansinha L, Lowe RP. Localization of the complex spectrum: the S transform. *IEEE Trans. Signal Process.* 1996; 44(4): 998–1001.
[4] Huang Z, Zhang J, Zhao T, et al. Synchrosqueezing S-transform and its application in seismic spectral decomposition. *IEEE Trans. Geosci. Remote Sens.* 2015; 54(2): 817–25.
[5] Li J., Liu C, Zeng, Z., & Chen, L (2015). GPR Signal Denoising and Target Extraction with the CEEMD Method. *IEEE Geoscience and Remote Sensing Letter,* 12 (8), 1615-1619.
[6] Xiong H, Li J, Su G, et al. Automatic defect detection in operational high-speed railway tunnels guided by train-mounted ground penetrating radar data. *J. Appl. Geophys.* 2023; 219: 105219.

Polarimetric Attributes of Martian Regolith at the Tianwen-1 Landing Site

Jiarun Yang[1], Xuan Feng[1], Tseeulam Khuut[2], Qi Lu[1], Haoqiu Zhou[1], Zejun Dong[1], Minghe Zhang[1], Yafei An[1]

[1] College of Geo-Exploration Science and Technology, Jilin University, Changchun 130026, China (jryang22@mails.jlu.edu.cn)
[2] Mongolian University of Science and Technology, Ulaanbaatar 17025, Mongolia

Abstract. We designed a data processing procedure for polarization analysis of the Martian regolith and applied it to the RoPeR CH2 channel data of the Zhurong rover, obtaining the structure distribution and weathering degree distribution of the Martian regolith of about 1900 m (325 sols). Three typical subsurface structures are studied in detail, and the results show that the structures in the Martian regolith all have relatively rough surfaces. This indicates that the structures within the regolith have suffered a high degree of weathering.

1. Introduction

Martian regolith's structure, density, grain size, and water ice content reflect the history of geological activities, space weathering modification, and weak impacts [1, 2]. This information will deepen our understanding of the history of Martian geological and climatic evolution and help explore in-situ resources and the construction of settlements on Mars. Full-polarimetric ground penetrating radar (FP-GPR) is an effective technique to detect the fine-scale structure of Martian regolith.

On May 15, 2021, the Chinese Tianwen-1 (TW-1) lander carrying the Zhurong rover landed on the Utopian Plain (109.925°E, 25.066°N) in the northern lowlands of Mars [3]. The rover is equipped with the Mars Rover Penetrating Radar (RoPeR) for nondestructive investigation of the Martian subsurface structure [4]. The TW-1 landing area is an ancient impact basin with a more than 3,300 km diameter. The surface is covered with a thick regolith layer after a long geological stage of weathering, deposition, and migration. The overall surface of the Utopian plain is smooth, but it also has distinct geological features, such as crenulated ridges, lobate margins, and troughs [2].

RoPeR on the Zhurong Rover can use two different frequency channels to make in-situ measurements of the fine-scale subsurface structure and dielectric properties of subsurface materials [5]. Channel 1 (CH1) operates at a frequency of 15 ~ 95 MHz, and Channel 2 (CH2) operates at a frequency of 0.45 ~ 2.15GHz [5]. The CH2 channel is an FP-GPR using an antenna array composed of two orthogonal Vivaldi antennas, which can transmit and receive horizontally (H) or vertically (V) polarized electromagnetic waves, thus forming four sub-channels: HH, HV, VH, VV [4]. The CH2 channel, with a depth range of 3 to 10 meters and a resolution in centimeters, has the ability to detect the fine structure of Martian regolith.

In this study, we investigated the polarimetric attributes of the Martian regolith in the landing area using RoPeR CH2 data from the first 325 sols (approximately 1900m travel distance). Combined with

The Navigation and Terrain Camera (NaTeCam), the strong reflective structures in the Martian regolith at a depth of 6 meters were identified, and the weathering degree of the structures in the regolith was analyzed.

2. Methodology

2.1. RoPeR Data Processing Procedure

We studied the initial ~1900 m radar data of the RoPeR CH2 channel. The specific processing process is as follows:

1. Delete the self-check traces. During the data acquisition process, the RoPeR on the Zhurong rover self-checked the performance of the instruments. No radar signal is sent during the CH2 channel self-check; only signal reception is performed. These self-detection paths can affect radar images. This step deletes all self-test traces.
2. Delete duplicate traces. It is noted that part of the CH2 channel data is collected at the same location, and these duplications will affect the radar image. Only one trace of data is retained at the same position.
3. Time zero correction. CH2 channel data has 2048 data points per record, corresponding to a time window of 235ns. Data zero is set to 18ns.
4. DC shift correction. This step removes the DC shift. For each trace of data, subtract the average of the last third of the sample points.
5. Remove environment noise. No radar signal was sent during the CH2 channel self-check; only the receiver worked. So, the self-check can represent the noise and interference of the environment and the instruments. This step subtracts the average of the self-checking traces for all normal traces.
6. Band-stop filtering. We found strong interference between 1.0 and 1.5 GHz. This step suppresses this part of the noise by band-stop filtering.
7. Remove background noise. Strong reflections from the surface and lateral noise from the instrument can adversely affect the data. This step is performed by subtracting the average of all traces.
8. Amplitude compensation. We use the average of AGC gain function to compensate for and correct the attenuation of radar electromagnetic wave energy.
9. Bandpass filtering. The bandwidth of the signal transmitted by the high-frequency channel is limited to 0.45-2.15 GHz.
10. Time-depth conversion. $\varepsilon_r = 3.6$ is used for time-depth conversion.

By processing RoPeR CH2 data from the first 325 sols, the radargram of 1,900 meters was obtained.

2.2. Polarization Analysis

We use Freeman decomposition (FD) and H-Alpha decomposition to study the structure and weathering of Martian regolith. Because the spectrum of the VH channel differs from the other three channels, we use the data of HH, HV, and VV channels for polarization analysis. P_S is used to characterize the internal structure of Martian regolith, and polarization-scattering entropy H is used to describe the degree of weathering of shallow materials.

3. Results and Discussion

No significant layers were observed in the processed radar images (Fig. 1). Only the strong reflections appear as banding on the radargram, which is caused by subsurface structures. There are obvious anomalies in the radar image of the VH channel, which are reflected in its power spectrum. So, we use the radar data of HH, HV, and VV to obtain P_S and H for polarization analysis. P_S image is similar to the radargrams but has a lower noise level, and the reflection from the subsurface structure is more prominent (Fig. 1 (a)). H greater than 0.5 indicates that the scattering mechanism of subsurface reflections is highly random (Fig. 1 (b)). This shows a high degree of weathering in the Martian regolith.

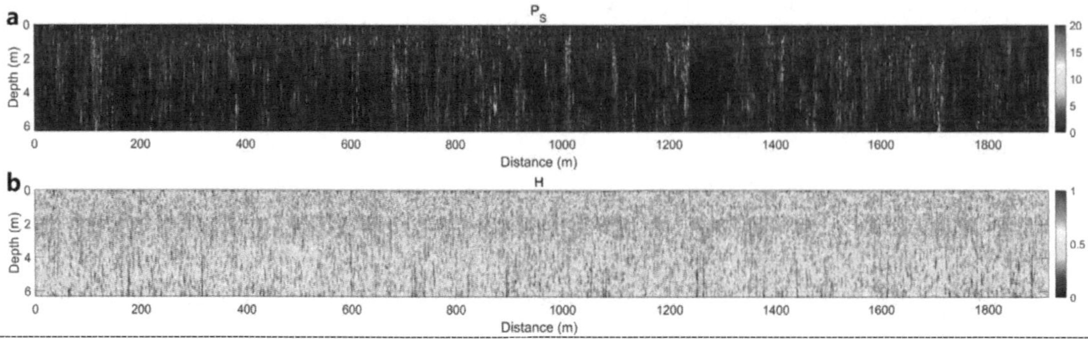

Figure 1. P_S and H are calculated from HH, HV, and VH channel data.

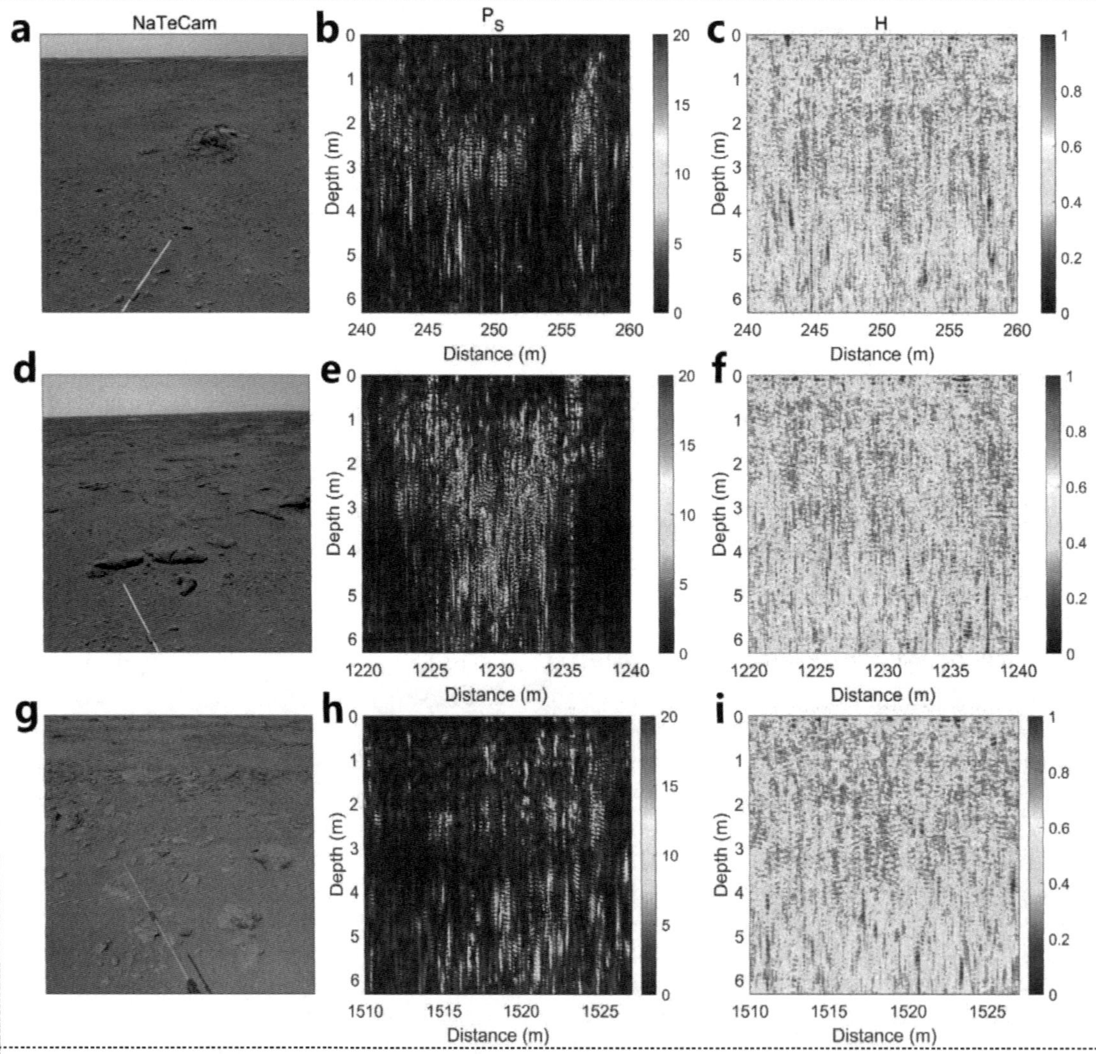

Figure 2. Three typical subsurface structures. (a), (d) and (g) are the NaTeCam images. (a) NaTeCam image of site 1. (d) NaTeCam image of site 2. (g) NaTeCam image of site 3. (b), (e), and (h) are P_S profiles. (c), (f) and (i) are H profiles.

We combined NaTeCam images to analyze three typical subsurface structures. A rock outcrop can be seen at site 1 (Fig. 2 (a)), which is reflected as an inclined structure on the $\boldsymbol{P_S}$ profile (Fig. 2 (b)). It is difficult to distinguish this structure on the $\boldsymbol{H}$ profile (Fig. 2 (c)), indicating that the scattering randomness of the reflection from the rock surface is similar to that of the surrounding medium. This shows that the rock's surface is highly rough and highly weathered. Site 2 is a large slab rock outcrop (Fig. 2 (d)), which shows a high-energy structure on the $\boldsymbol{P_S}$ profile (Fig. 2 (e)). It extends more than 6 meters subsurface. It can be seen from the $\boldsymbol{H}$ profile (Fig. 2 (f)) that the $\boldsymbol{H}$ value of the subsurface structure is slightly lower than that of the surrounding medium. However, the overall fragmentation is also high. Site 3 is located on the edge of a rough terrain area, with a large rocky outcrop visible in the NaTeCam image (Fig. 2 (g)). Strong reflection structures at different depths can be observed on the $\boldsymbol{P_S}$ profile (Fig. 2 (h)), indicating that the structure is relatively broken and extends down more than 6m. It can be seen from the $\boldsymbol{H}$ profile (Fig. 2 (i)) that the lower part of the structure has a slightly lower value of polarization scattering entropy $\boldsymbol{H}$ than the surrounding medium. In contrast, the upper part is difficult to distinguish from the surrounding medium. This indicates that the upper part is more weathered than the lower part.
Intense winds with sand may be the reason for Martian regolith erosion, as transverse aeolian ridges (TARs) and ventifacts are ubiquitous in the landing zone [3]. Meanwhile, Some rocks with pitted surfaces, similar to igneous rocks observed in previous missions, show the possibility of brine-related dissolution processes under cold environments [6].

4. Conclusion
The data processing procedure for polarization analysis of Martian regolith is designed. We applied it to the RoPeR CH2 channel data of the Zhurong rover and obtained the structure distribution and weathering degree distribution of about 1900 meters of Martian regolith. P_S was used to image the underground structure, and polarization scattering entropy H was used to characterize the degree of weathering. Three typical underground structures are studied in detail, and the results show that the structures in the Martian regolith all have relatively rough surfaces. This indicates that the structures within the regolith have suffered a high degree of weathering.

Acknowledgments
This work was supported by the National Key Research and Development Program of China under grant 2022YFC3003402, the Science and Technique Development Program of Jilin Province under grant 20220101148JC, the Graduate Innovation Fund of Jilin University under grant 2024CX234. (Corresponding author: Xuan Feng.)

References
[1] C. Atwood-Stone and A. S. McEwen, *Geophysical Research Letters,* vol. 40, no. 12, pp. 2929-2934, 2013.
[2] C. M. Stuurman *et al.*, *Geophysical Research Letters,* vol. 43, no. 18, pp. 9484-9491, 2016, doi: 10.1002/2016GL070138.
[3] L. Ding *et al.*, "Surface characteristics of the Zhurong Mars rover traverse at Utopia Planitia," *Nature Geoscience,* vol. 15, no. 3, pp. 171-176, Mar 2022, doi: 10.1038/s41561-022-00905-6.
[4] H. Q. Zhou *et al.*, "Polarized Orientation Calibration and Processing Strategies for Tianwen-1 Full-Polarimetric Mars Rover Penetrating Radar Data," *Ieee Transactions on Geoscience and Remote Sensing,* Article vol. 60, p. 14, 2022, Art no. 5119914, doi: 10.1109/Tgrs.2022.3219932.
[5] S. Liu *et al.*, "Data Pre-Processing and Signal Analysis of Tianwen-1 Rover Penetrating Radar," *Remote Sensing,* vol. 15, no. 4, p. 966, 2023, doi: 10.3390/rs15040966.
[6] J. W. Head, M. A. Kreslavsky, and D. R. Marchant, "Pitted rock surfaces on Mars: A mechanism of formation by transient melting of snow and ice," (in English), *J. Geophys. Res.-Planets,* Article vol. 116, p. 11, Sep 2011, Art no. E09007, doi: 10.1029/2011je003826.

Journal of Physics: Conference Series **2887** (2024) 012065 doi:10.1088/1742-6596/2887/1/012065

Time–lapse monitoring of soil water content by full GPR waveform inversion with Grey Wolf Optimizer

M H Zhang[1,2], M Bano[1] and X Feng[2]

[1]ITES UMR-7063, EOST, University of Strasbourg, F-67000 Strasbourg, France
[2]College of Geo-Exploration Science and Technology, Jilin University, Changchun 130026, China

Abstract. Ground-Penetrating Radar (GPR) has become a powerful geophysical technique for investigating soil water dynamics. In this study, we proposed an inversion scheme based on Grey Wolf Optimizer (GWO) to monitor the change of SWC from the time-lapse of the full waveform of GPR data. The scheme was applied to GPR data acquired during a drainage experiment. Five traces at different times were selected in the experiment. The results show that the proposed scheme is very promising for direct~~ly~~ and non-invasive~~ly~~ monitoring of SWC change from the time-lapse of the full waveform inversion of GPR data based on field scale GWO algorithm.

1. Introduction

Monitoring soil water content (SWC) in the vadose zone is important in many fields including hydrology, agriculture, soil science and environmental science (Binley et al., 2001). In recent year, geophysical methods have become a promising technique for measuring SWC because they are non-invasive to the soil and easy to use. Among geophysical methods, ground penetrating radar (GPR) have the advantages of high-precision, large detection depth and can give the dielectric constant directly related to the SWC (Huisman et al., 2003; Klotzsche et al., 2018). Therefore, GPR is a suitable and popular geophysical method for SWC investigation. Most research to estimate SWC is based on traveltime information from GPR data (Klotzsche et al., 2018; M. Zhang et al., 2022). Compared to inversion with traveltime data, full GPR waveform inversion can provide more accurate and high-resolution results because the full waveform data includes more information (Meles et al., 2010; Klotzsche et al., 2019).

At present, inversion techniques are divided into two directions: local and global optimization. The disadvantages of the local inversion are that it is easily trapped by the local minimum and relies on the initial model. The global optimization helps overcome these problems. In this paper, we use a grey wolf optimization (GWO) algorithm to optimize the inversion problem. The GWO is inspired by the social hierarchy and hunting behaviour of the grey wolves and has begun to be applied to the geophysical inversion such as Rayleigh wave dispersion curve inversion (Song et al., 2015) and denoising magnetotelluric data (X. Zhang et al., 2022). Features of the GWO include simplicity, flexibility and fewer parameters needed to adjust.

The aim of this study was to applied the proposed inversion scheme to time-lapse GPR full waveform data measured at different times to monitor water flow during a drainage experiment. The proposed scheme was applied to the GPR full waveform data collected at five different time points during the drainage experiment and the SWC were directly estimated from the GPR data on each measurement time points.

2. Methods

2.1. Petrophysical relationships
In this study, the Topp equation (Topp et al., 1980) is used to relate the relative dielectric permittivity with the water content. The Topp equation is expressed as follows:

$$\kappa = 3.03 + 9.30\theta + 146.00\theta^2 - 76.70\theta^3$$

where $\kappa$is the relative dielectric permittivity and θrepresent the water content.

2.2. GPR forward modeling
After obtaining the relative dielectric permittivity from the water content, we realize the GPR forward modelling based on a MATLAB code generated by Bano (2004).

2.3. GWO inversion scheme
In this study, the objective function for inversion is defined as follows:

$$fitness(i) = \sum_{i=1}^{N} \left(E_{obs}^{i} - E_{cal}^{i}\right)^2$$

where N is the number of samples, $E_{obs}^{i}$and E_{cal}^{i} represent the observed data and the calculated data at ith sampling point, respectively. The GWO algorithm was used to optimize the objective function.
For GWO algorithm, the grey wolves are divided into three groups α, β, δ in the grey groups. The αwolves represent the best solution. The process can be expressed by the following equations:

$$\vec{D} = \left|\vec{C} \cdot \vec{X}_P(t) - \vec{X}(t)\right|$$
$$\vec{X}(t+1) = \vec{X}_P(t) - \vec{A} \cdot \vec{D}$$

where t indicates the current iteration, $\vec{A}$ and $\vec{C}$ stand for the coefficient vectors, $\vec{X}_P$ represents the position vector of the prey, and $\vec{X}$ is the position vector of a grey wolf.
The vectors $\vec{A}$ and $\vec{C}$ are calculated as follows:

$$\vec{A} = 2\vec{a} \cdot \vec{r}_1 - \vec{a}$$
$$\vec{C} = 2 \cdot \vec{r}_2$$

where $\vec{a}$ is linearly decreased from 2 to 0 over the course of iterations and $\vec{r}_1$, $\vec{r}_2$ are random vectors in [0, 1].

3. Results and discussion
Figure 1 shows the profiles at five different time points during the drainage experiment. We can see that the reflection from the two sandy layers boundary (around 10 ns showed by red arrow) becomes clear with the decreasing of the overall water content during the drainage experiment. In order to see the phenomenon clearly, we picked up one trace from every profile and showed in Figure 2. From figure 2, the reflection signals appear and gradually becomes stronger when the water level moves down the two layers boundary (the trace after 10:30) and the arrival (around 10 ns) moves to shorter travel times.
Figure 2 also show the inversion results. In Figure 2, all the calculated traces match the observed traces perfectly. Figure 3 presents the inverted SWC and the measured SWC at different time points during the drainage experiment. Firstly, it is noted that the inverted SWC have the similar trend with the measured SWC. The move down of the water level can be identified with the increase of time points. We did not observe the change of SWC at the boundary (around 0.5 m) from the inverted SWC and measured SWC. After, with the descent of the water level, the SWC also decreases. From all the profiles after 12:30, the change of SWC at the boundary can be observed from both inverted SWC profiles and measured SWC profiles.

Journal of Physics: Conference Series **2887** (2024) 012065 doi:10.1088/1742-6596/2887/1/012065

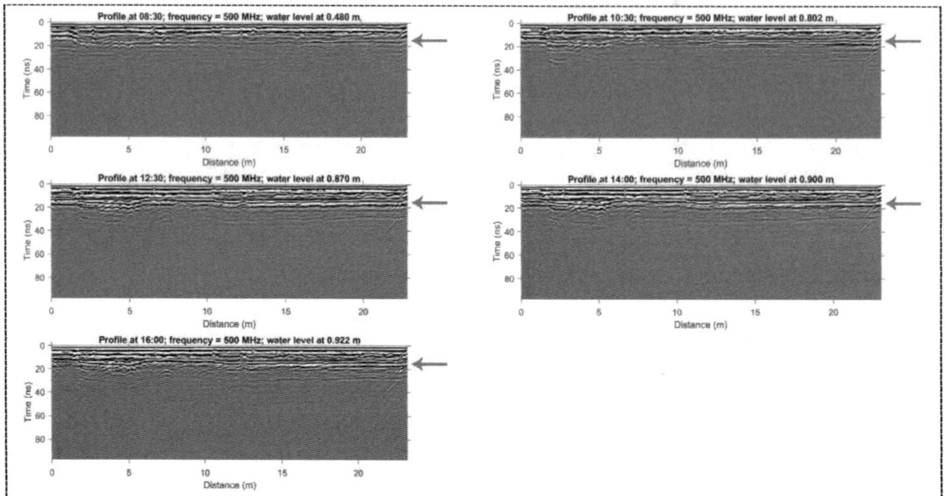

Figure 1. Processed observed GPR profiles during drainage experiment at different time points. The water tables are 0.480 m (8 h 30 min), 0.802 m (10 h 30 min), 0.870 m (12 h 30 min), 0.900 m (14 h) and 0.922 m (16 h) for all the five profiles.

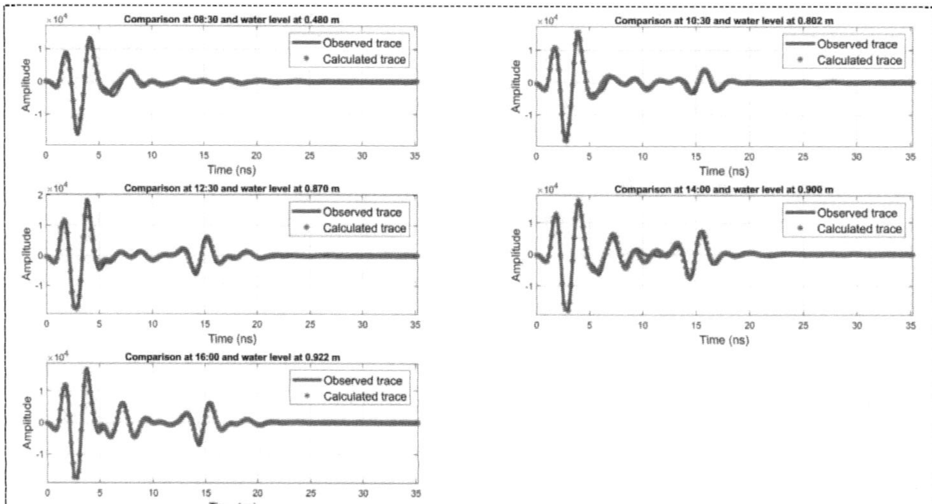

Figure 2. The fitness between observed and calculated traces at different times during the drainage experiment. The solid red lines represent the observed data and the mark dotted lines represent the calculated data.

4. Conclusion

In this paper, an inversion technique was applied to the time-lapse GPR to study soil water dynamics. The spatiotemporal variability of SWC was estimated by the inversion technique in unsaturated sandy soil during a drainage field experiment. The results show that the inversion technique based on GPR full waveform data with GWO algorithm is a promising method to non-invasively estimate and monitor the change of the SWC. In the future, the proposed scheme can be extended to high dimensional problems and to invert the hydraulic properties of the soil.

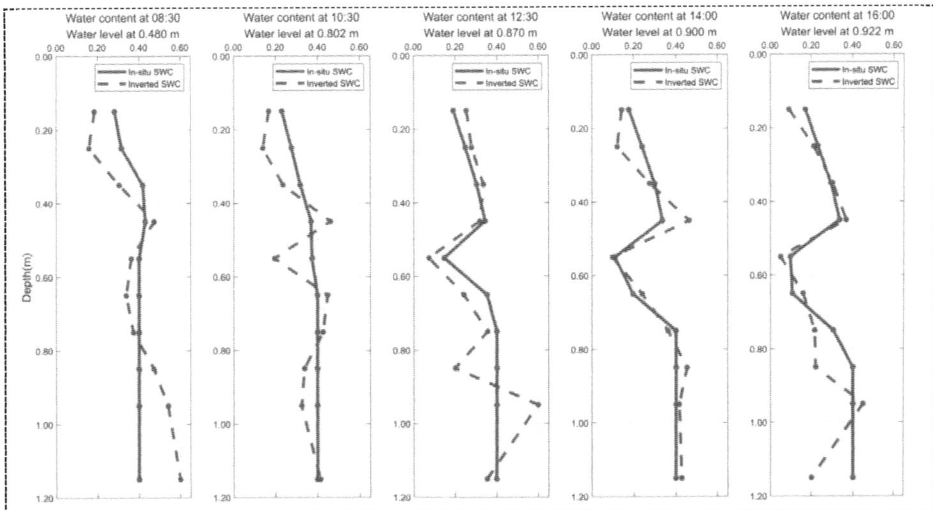

Figure 3. The SWC profiles at different times during the drainage experiment. The solid red lines represent the SWC measured by Sentek sensors. The dashed blue lines represent the inverted SWC.

References

[1] Binley A, Winship P, Middleton R, Pokar M and West J 2001 High-resolution characterization of vadose zone dynamics using cross-borehole radar *Water Resour. Res.* **37** pp 2639-2652

[2] Huisman J A, Hubbard S S, Redman J D and Annan A P 2003 Measuring Soil Water Content with Ground Penetrating Radar: A Review *Vadose Zone J.* **2** pp 476-491

[3] Klotzsche A, Jonard F, Looms M C, van der Kruk J and Huisman J A 2018 Measuring Soil Water Content with Ground Penetrating Radar: A Decade of Progress *Vadose Zone J.* **17** p 180052

[4] Zhang M, Feng X, Bano M, Xing H, Wang T, Liang W, Zhou H, Dong Z, An Y and Zhang Y 2022 Review of Ground Penetrating Radar Applications for Water Dynamics Studies in Unsaturated Zone *Remote Sens.* **14** p 5993

[5] Meles G A, van der Kruk J, Greenhalgh S A, Ernst J R, Maurer Hand Green A G 2010 A New Vector Waveform Inversion Algorithm for Simultaneous Updating of Conductivity and Permittivity Parameters from Combination Crosshole/Borehole-to-Surface GPR Data *IEEE Trans. Geosci. Remote Sens.* **48** pp 3391-3407

[6] Klotzsche A, Vereecken H and van der Kruk J 2019 Review of crosshole ground-penetrating radar full-waveform inversion of experimental data: Recent developments, challenges, and pitfalls *Geophysics* **84** pp H13-H28

[7] Song X, Tang L, Zhao S, Zhang X, Li L, Huang J and Cai W 2015 Grey Wolf Optimizer for parameter estimation in surface waves *Soil Dyn. Earthq. Eng.* **75** pp 147-157

[8] Zhang X, Li D, Li J and Li Y 2022 Grey wolf optimization-based variational mode decomposition for magnetotelluric data combined with detrended fluctuation analysis *Acta Geophys.* **70** pp 111-120

[9] Topp G C, Davis J L and Annan A P 1980 Electromagnetic determination of soil water content: Measurements in coaxial transmission lines *Water Resour. Res.* **16** pp 574-582

[10] Bano M 2004 Modelling of GPR waves for lossy media obeying a complex power law of frequency for dielectric permittivity *Geophys. Prospect.* **52** pp 11-26

Acknowledgments

We thank Nolwenn Lesparre, Benjamin Belfort, Pascal Friedmann, Jean-François Girard, François Lehmann and Oliver Razakarisoa for helping us acquire GPR data. The experiment was funded by the ITES laboratory via an internal call.

Journal of Physics: Conference Series **2887** (2024) 012066 doi:10.1088/1742-6596/2887/1/012066

Estimation of soil water content based on full waveform inversion technique of GPR data with a Grey Wolf Optimizer

M. H. Zhang[1,2], M. Bano[1], X. Feng[2], T. Khuut[3] and Q. Lu[2]

[1]ITES UMR-7063, EOST, University of Strasbourg, F-67000 Strasbourg, France
[2]College of Geo-Exploration Science and Technology, Jilin University, Changchun 130026, China
[3]Mongolian University of Science and Technology, Ulaanbaatar 17025, Mongolia
E-mail: minghe.zhang@etu.unistra.fr

Abstract. This paper proposes a novel inversion scheme for estimating soil water content (SWC) based on full waveform inversion technique ground penetrating radar (GPR) using the Grey Wolf algorithm Optimizer. In order to evaluate the performance of the proposed inversion scheme, a numerical experiment was carried out. Then, the proposed inversion scheme was applied ~~on~~ to the data with and without noise. The results show a good match between the observed and calculated GPR data. And the SWC can still be perfectly inverted even if there is noise in the data. These demonstrate the efficiency and applicability of the proposed inversion scheme.

1. Introduction

In many disciplines such as hydrology, agriculture, soil science, soil water content (SWC) plays an important role (Blonquist et al., 2006; Vereecken et al., 2008). Traditional techniques including the neutron method, the γ-ray method and time domain reflectometry (TDR), are time consuming and destructive for the soil. Regarding scale, most traditional techniques are point-scale techniques. For large scale measurement, remote sensing technique is a suitable method, but the detection depth is shallow.

In the last decades, ground penetrating radar (GPR) method has become a promising geophysical method for estimating the SWC because the GPR wave velocity is very sensitive to the change of SWC and GPR is non-invasive to the soil. Additionally, GPR fills the gap between point-scale and large-scale measurements due to high-precision measurements and large detection depth (Anbazhagan et al., 2020; Cao et al., 2020). Several reviews have been published to summary the development for SWC estimation ~~by~~ using GPR (Huisman et al., 2003a; Klotzsche et al., 2018; Liu et al., 2019; M. Zhang et al., 2022). However, most of the method relies on the travel time data to estimate the SWC.

In geophysics, full waveform inversion (FWI) has shown its advantages over the traveltime-based method. FWI uses the entire waveform information and can provide higher resolution results. Several studies have applied the GPR FWI to estimate the SWC by constructing the objective function based on the Green's function from off-ground GPR reflection data (Lambot et al., 2008; Jonard et al., 2012). However, SWC estimation based on GPR FWI is still a relatively new technique.

Inversion techniques mainly include local optimization and global optimization methods. The local inversion methods have been widely used in GPR FWI (Busch et al., 2012; Lavoué et al., 2014; Li et al., 2017). However, the local inversion techniques are easily to trapped by local minima and rely on initial model selection. The global optimization techniques can overcome the disadvantages of local inversion techniques. Many global optimization methods have been applied in many fields, including genetic algorithm (GA) (Bijani et al., 2012) and particle swarm optimization (PSO) (Kaplanvural et al., 2020). As a newcomer in the population-based swarm intelligence optimization method. The grey wolf optimizer (GWO) method is simple, flexibile, derivation-free and avoids local optima. Therefore, the GWO method has become one of the popular optimization methods in recent years and has begun to apply to geophysical inversion (Song et al., 2015; X. Zhang et al., 2022).

In this study, we proposed a new efficient inversion scheme to estimate the SWC from GPR full waveform data based on optimization by GWO. In order to demonstrate the applicability of the proposed inversion scheme, a numerical example was carried out and the proposed inversion scheme was applied to the numerical example. The numerical experiment results demonstrate that the proposed inversion scheme can successfully estimate the SWC.

2. Methods

2.1. Petrophysical relationships

In this study, the Topp equation (Topp et al., 1980) is used to relate the relative dielectric permittivity with the water content. The Topp equation is expressed as follows:

$$\kappa = 3.03 + 9.30\theta + 146.00\theta^2 - 76.70\theta^3$$

where $\kappa$is the relative dielectric permittivity and θrepresent the water content.

2.2. GPR forward modeling

After obtaining the relative dielectric permittivity from the water content, we realize the GPR forward modelling based on the following equation (Bano, 2004):

$$E(\omega, z_1) = G(z_1)R(\omega)E_0(\omega)exp\left[i\frac{\omega}{V}2z_1\right]exp[-\alpha 2z_1]$$

where $R(\omega)$ is the reflection coefficient at $z = z_1$ and $G(z_1)$ is the geometrical spreading. According to Bano (2004), α is equal to $\omega/\,2VQ$. Finally, a GPR response can be gained by inverse Fourier transform. In this paper, we realize the 1D GPR forward modeling by using a MATLAB code generated by Bano (2004).

2.3. GWO inversion scheme

For an inversion process, the main goal is to invert the parameters in a given solution space from observed data. In the inversion process, an objective function is necessary. In this study, the objective function is defined as follows:

$$fitness(i) = \sum_{i=1}^{N}\left(E_{obs}^{i} - E_{cal}^{i}\right)^2$$

where N is the number of samples, $E_{obs}^{i}$and E_{cal}^{i} represent the observed data and the calculated data at ith sampling point, respectively. In this study, the GWO algorithm was used to optimize the objective function for finding the global minimum and finally invert the soil water content from GPR full waveform data.

The GWO is one of the swarm intelligence optimization algorithms, which is inspired by the biological habits of the grey wolf and simulating the social hierarchy and the hunting behaviour of grey wolves. In the grey groups, the grey wolves are divided into three groups α, β, δ. The αwolves represent the best solution. The process can be expressed by the following equations:

$$\vec{D} = \left|\vec{C}\cdot\vec{X}_P(t) - \vec{X}(t)\right|$$

$$\vec{X}(t+1) = \vec{X}_P(t) - \vec{A}\cdot\vec{D}$$

where t indicates the current iteration, $\vec{A}$ and $\vec{C}$ stand for the coefficient vectors, $\vec{X}_P$ represents the position vector of the prey, and $\vec{X}$ is the position vector of a grey wolf.

The vectors $\vec{A}$ and $\vec{C}$ are calculated as follows:

$$\vec{A} = 2\vec{a}\cdot\vec{r}_1 - \vec{a}$$

Journal of Physics: Conference Series **2887** (2024) 012066 doi:10.1088/1742-6596/2887/1/012066

$$\vec{C} = 2 \cdot \vec{r}_2$$

where $\vec{a}$ is linearly decreased from 2 to 0 over the course of iterations and $\vec{r}_1$, $\vec{r}_2$ are random vectors in [0, 1].

3. Results and discussion

In order to evaluate the performance of the proposed inversion scheme, we built a synthetic model which is showed in Figure 2 (blue dotted line). Figure 2 also shows the inverted SWC (red solid line). Figure 1 presents the corresponding GPR forward response from the true SWC (blue star dotted line) and the inverted SWC (red solid line). From the Figure 1, it is noted that the observed and calculated GPR response match perfectly. In the Figure 2, the inverted SWC and the true SWC fit very well. Figure 3 shows the fitness of the iteration for synthetic model. In the early stage, the curve has an abrupt decrease. Then the change gradually converges to zero after around 1200th iteration. These demonstrate that the proposed inversion scheme can be used for estimating the SWC from the GPR full waveform data and the algorithm has the ability to achieve fast convergence.

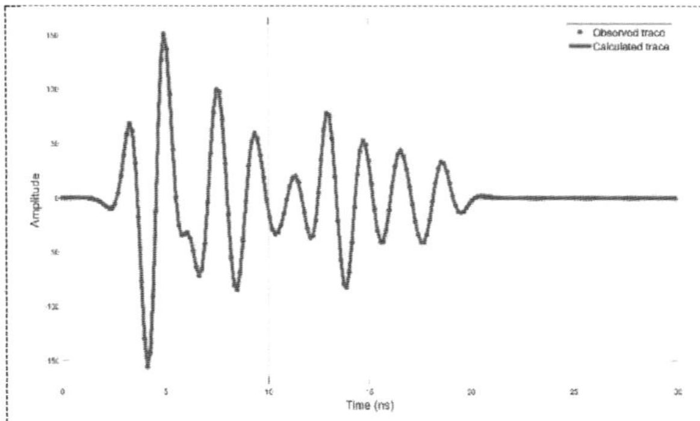

Figure 1. Observed data (blue star dotted line) and calculated GPR response (red solid line).

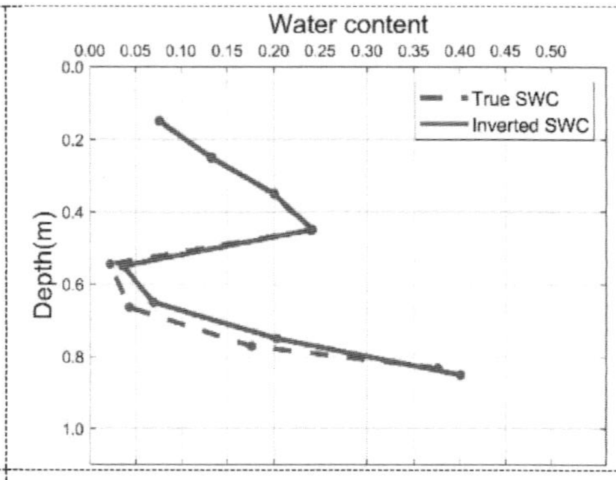

Figure 2. True SWC (blue dotted line) and inverted SWC (red solid line).

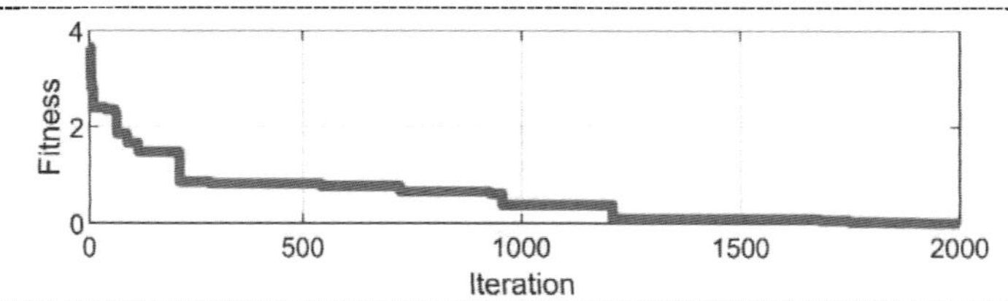

Figure 3. Fitness curve of iteration for synthetic model.

4. Conclusion

In this paper, a new efficient SWC estimation inversion scheme from GPR full waveform data based on GWO is proposed. The proposed inversion scheme was tested on a numerical experiment. And the scheme is applied to data with and without noise, respectively. In the numerical experiment, the SWC were successfully inverted using the proposed inversion scheme. The results demonstrate the ability of the proposed inversion scheme for to estimate the SWC from the GPR full waveform data. This study provides the opportunity for non-invasively estimating the SWC with acceptable accuracy.

References

[1] Blonquist J M, Jones S B and Robinson D A 2006 Precise irrigation scheduling for turfgrass using a subsurface electromagnetic soil moisture sensor *Agric. Water Manage.* 84 pp 153-165

[2] Vereecken H, Huisman J A, Bogena H, Vanderborght J, Vrugt J A and Hopmans J W 2008 On the value of soil moisture measurements in vadose zone hydrology: A review *Water Resour. Res.* 44 W00D06

[3] Anbazhagan P, Bittelli M, Pallepati R R and Mahajan P 2020 Comparison of soil water content estimation equations using ground penetrating radar *J. Hydrol.* **588** p 125039

[4] Cao Q, Song X, Wu H, Gao L, Liu F, Yang S and Zhang G 2020 Mapping the response of volumetric soil water content to an intense rainfall event at the field scale using GPR *J. Hydrol.* **583** p 124605

[5] Huisman J A, Hubbard S S, Redman J D and Annan A P 2003 Measuring Soil Water Content with Ground Penetrating Radar: A Review *Vadose Zone J.* **2** pp 476-491

[6] Klotzsche A, Jonard F, Looms M C, van der Kruk J and Huisman J A 2018 Measuring Soil Water Content with Ground Penetrating Radar: A Decade of Progress *Vadose Zone J.* **17** p 180052

[7] Liu X, Chen J, Cui X, Liu Q, Cao X and Chen X 2019 Measurement of soil water content using ground-penetrating radar: a review of current methods *Int. J. Digit. Earth***12** pp 95-118

[8] Zhang M, Feng X, Bano M, Xing H, Wang T, Liang W, Zhou H, Dong Z, An Y and Zhang Y 2022Review of Ground Penetrating Radar Applications for Water Dynamics Studies in Unsaturated Zone *Remote Sens.* 14 p 5993

[9] Lambot S, Slob E, Chavarro D, Lubczynski M and Vereecken H 2008 Measuring soil surface water content in irrigated areas of southern Tunisia using full-waveform inversion of proximal GPR data *Near Surf. Geophys.***6** pp 403-410

[10] Jonard F, Weihermüller L, Vereecken H and Lambot S 2012 Accounting for soil surface roughness in the inversion of ultrawideband off-ground GPR signal for soil moisture retrieval *Geophysics***77** pp H1-H7

[11] Busch S, van der Kruk J, Bikowski J and Vereecken H 2012 Quantitative conductivity and permittivity estimation using full-waveform inversion of on-ground GPR data *Geophysics* 77 pp H79-H91

[12] Lavoué F, Brossier R, Métivier L, Garambois S and Virieux J 2014 Two-dimensional permittivity and conductivity imaging by full waveform inversion of multioffset GPR data: a frequency-domain quasi-Newton approach *Geophys. J. Int.* 197 **pp** 248-268

[13] Bijani R, Neto C F P, Martins S S and Travassos J M 2012 2-D tomography of first-arrivals using the Genetic Algorithm with Elitism in *SEG Technical Program Expanded Abstracts 2012* pp 1-6

[14] Kaplanvural İ, Pekşen and Özkap K 2020 1D waveform inversion of GPR trace by particle swarm optimization *J. Appl. Geophys.***181** p 104157

[15] Song X, Tang L, Zhao S, Zhang X, Li L, Huang J and Cai W 2015 Grey Wolf Optimizer for parameter estimation in surface waves *Soil Dyn. Earthq. Eng.***75** pp 147-157

[16] Zhang X, Li D, Li J and Li Y 2022 Grey wolf optimization-based variational mode decomposition for magnetotelluric data combined with detrended fluctuation analysis *Acta Geophys.***70** pp 111-120

[17] Topp G C, Davis J L and Annan A P 1980 Electromagnetic determination of soil water content: Measurements in coaxial transmission lines *Water Resour. Res.* **16** pp 574-582

[18] Bano M 2004 Modelling of GPR waves for lossy media obeying a complex power law of frequency for dielectric permittivity *Geophys. Prospect.***52** pp 11-26

[19] Li J., Feng Z., & Schuster G., (2017) Wave-equation dispersion inversion, *Geophysical Journal International*, 208 (3), 1567–1578,

Acknowledgements

The research was funded by the National Key Research and Development Program of China (No. 2022YFC3003402). We thank Nolwenn Lesparre, Benjamin Belfort, Pascal Friedmann, Jean-François

Girard, François Lehmann and Oliver Razakarisoa for helping us to acquire GPR data. This experiment was funded by the ITES laboratory via an internal call.

Journal of Physics: Conference Series **2887** (2024) 012067 doi:10.1088/1742-6596/2887/1/012067

Estimation of soil hydraulic properties based on full waveform inversion of ground-penetrating radar data with Grey Wolf optimization algorithm

M. H. Zhang[1,2], M. Bano[1], X. Feng[2], T. Khuut[3] and Q. Lu[2]

[1]ITES UMR-7063, EOST, University of Strasbourg, F-67000 Strasbourg, France
[2]College of Geo-Exploration Science and Technology, Jilin University, Changchun 130026, China
[3]Mongolian University of Science and Technology, Ulaanbaatar 17025, Mongolia
E-mail: minghe.zhang@etu.unistra.fr

Abstract. In this paper, we propose an inversion scheme to directly and non-invasively determine soil hydraulic properties. The scheme is based on hydrogeophysical inversion full waveform ground penetrating radar (GPR) data with Grey Wolf Optimizer algorithm. This inversion scheme includes a soil hydraulic model, a petrophysical relationship, a ground penetrating radar forward modelling and a Grey Wolf Optimizer algorithm. In order to evaluate its performance, a numerical experiment was carried out and the proposed scheme was applied to the experimental data. The results indicate that the proposed inversion scheme is promising for determining soil hydraulic properties with GPR.

1. Introduction

In many fields such as hydrology, agriculture, soil and environment science, soil hydraulic properties (SHP) are important for accurately modelling the water flow and contaminant transport as well as to manage water and soil resources. In recent decades, among the geophysical methods, ground penetrating radar (GPR) has become one of the most promising methods for the estimation and monitoring of soil water content (SWC) because the GPR wave velocity is very sensitive to the change of SWC (Huisman et al., 2003; Klotzsche et al., 2018; Zhang et al., 2022). And more and more researchers began to be interested in determining SHP from GPR data with the development of the interdisciplinary research (Scholer et al., 2012; Léger et al., 2014; Jaumann et al., 2018). However, estimating SHP using GPR data is still a new technique. Most methods for determining SHP with GPR are based only on travel time information from GPR data.

Full waveform inversion (FWI) has been widely used in GPR inversion. Compared to traveltime-based inversion, FWI can provide higher resolution results because the entire waveform information is used in FWI. Several authors have applied FWI to the GPR data to determine the SHP by building the objective function based on Green's functions (Lambot et al., 2006; Jadoon et al., 2008). In general, the inversion techniques are divided into local and global optimization methods. Many local inversion methods have been used for GPR FWI. However, the local inversion is easily trapped by local minima and also depends on the initial model. Global optimization methods can overcome local inversion problems. As a newcomer swarm intelligence optimization algorithm, the grey wolf optimizer (GWO) algorithm has become a popular technique for solving the optimization problem in recent years. The

GWO is inspired by the social leadership hierarchy and hunting behaviour of grey wolves and has been successfully applied to the geophysical inversion (Song et al., 2015). However, the GWO was not used for inversion of GPR data.

In this study, an inversion scheme was proposed to determine the SHP based on FWI of GPR data with the GWO algorithm. The proposed scheme was tested on numerical experiment data. The results demonstrate that the proposed inversion scheme can successfully determine the SHP from the full waveform of GPR data with GWO algorithm.

2. Methods

2.1. Soil hydraulic model

The SWC profile can be described by the soil water retention curve when the soil was in hydrostatic equilibrium. In this study, we selected the van Genuchten model to express the soil water retention curve which is expressed as (Jonard et al., 2015):

$$\theta(h) = \begin{cases} \theta_r + (\theta_s - \theta_r)[1 + |\alpha h|^n]^{-m}, & h < 0 \\ \theta_s \qquad , & h \geq 0 \end{cases}$$

where θ $[\mathrm{m}^3 \cdot \mathrm{m}^{-3}]$ is the volumetric water content of the soil, $\theta_r[\mathrm{m}^3 \cdot \mathrm{m}^{-3}]$ and $\theta_s[\mathrm{m}^3 \cdot \mathrm{m}^{-3}]$ represent the residual and saturated water contents of the soil, respectively, h [m] stands for the pressure head, α $[\mathrm{m}^{-1}]$and n [−]determine the shape of curve, which are related to the inverse of the air entry value and the pore size width distribution, respectively, and m [−]is expressed as follows according to the Mualem condition:

$$m = 1 - (1/n), \quad n > 1.$$

When a water table is located at position z_w[m], the pressure head $h$can be related to the vertical position z [m]by:

$$h = z - z_w.$$

2.2. Petrophysical relationships

In this study, the Topp equation (Topp et al., 1980) is used to relate the relative dielectric permittivity with the water content. The Topp equation is expressed as follows:

$$\kappa = 3.03 + 9.30\theta + 146.00\theta^2 - 76.70\theta^3$$

where $\kappa$is the relative dielectric permittivity and θrepresent the water content.

2.3. GPR forward modeling

After obtaining the relative dielectric permittivity from the water content, we realize the GPR forward modelling based on the following equation (Bano, 2004):

$$E(\omega, z_1) = G(z_1)R(\omega)E_0(\omega)exp\left[i\frac{\omega}{V}2z_1\right]exp[-\alpha 2z_1]$$

where $R(\omega)$ is the reflection coefficient at $z = z_1$ and $G(z_1)$ is the geometrical spreading. According to Bano (2004), α is equal to $\omega/\, 2VQ$. Finally, a GPR response can be gained by inverse Fourier transform. In this paper, we realize the 1D GPR forward modeling by using a MATLAB code generated by Bano (2004).

2.4. GWO inversion scheme

In this study, the objective function for inversion is defined as follows:

$$fitness(i) = \sum_{i=1}^{N} \left(E_{obs}^i - E_{cal}^i\right)^2$$

where N is the number of samples, $E_{obs}^i$and E_{cal}^i represent the observed data and the calculated data at ith sampling point, respectively.

The GWO is one of the swarm intelligence optimization algorithms, which is inspired by the biological habits of the grey wolf and simulating the social hierarchy and the hunting behaviour of grey wolves. In the grey groups, the grey wolves are divided into three groups α, β, δ. The α wolves represent the best solution. The process can be expressed by the following equations:

$$\vec{D} = \left|\vec{C} \cdot \vec{X}_P(t) - \vec{X}(t)\right|$$

$$\vec{X}(t+1) = \vec{X}_P(t) - \vec{A} \cdot \vec{D}$$

where t indicates the current iteration, $\vec{A}$ and $\vec{C}$ stand for the coefficient vectors, $\vec{X}_P$ represents the position vector of the prey, and $\vec{X}$ is the position vector of a grey wolf.

The vectors $\vec{A}$ and $\vec{C}$ are calculated as follows:

$$\vec{A} = 2\vec{a} \cdot \vec{r}_1 - \vec{a}$$

$$\vec{C} = 2 \cdot \vec{r}_2$$

where $\vec{a}$ is linearly decreased from 2 to 0 over the iterations and $\vec{r}_1$, $\vec{r}_2$ are random vectors in [0, 1].

3. Results and discussion

A synthetic model is built to evaluate the performance of the proposed inversion scheme. The SHP of the model is defined according to the HYDRUS-1D software. Table 1 lists the input SHP and search space for synthetic model inversion. Figure 2 shows the soil water retention curve (SWRC) calculated from the input SHP and Figure 1 shows the corresponding forward GPR response.

The inversion results are listed in Table 1. The SWRC calculated from the inversion results is shown in Figure 2 and the corresponding forward GPR response is shown in Figure 1. From Figure 1, the observed and calculated GPR response fit very well. In Figure 2, the calculated SWRC based on the inversion results perfectly matches the SWRC calculated based on the actual input value.

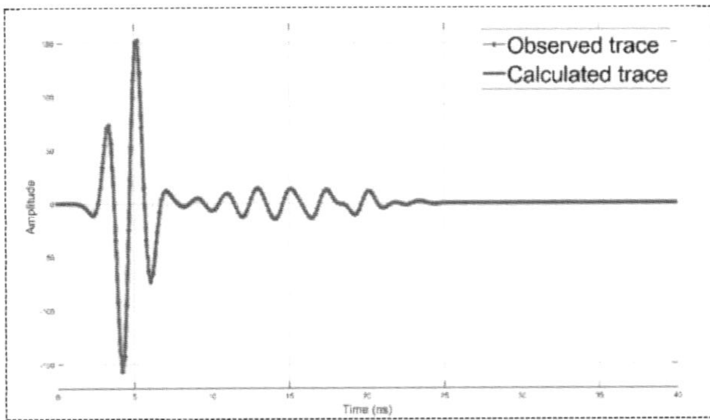

Figure 1. Observed data (blue star dotted line) and calculated GPR response (red solid line).

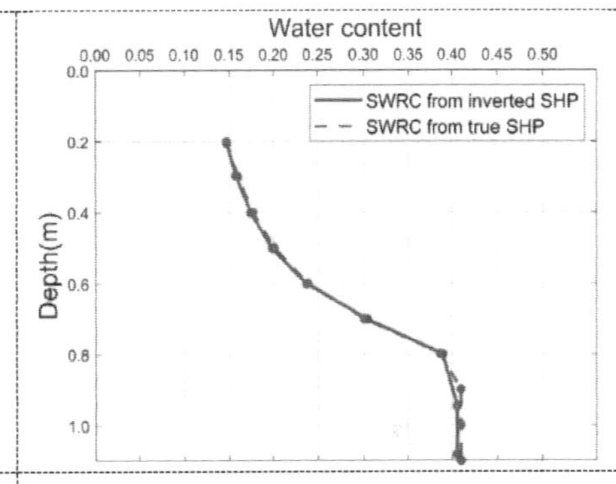

Figure 2. SWRC from true SHP (blue dotted line) and inverted SHP (red solid line).

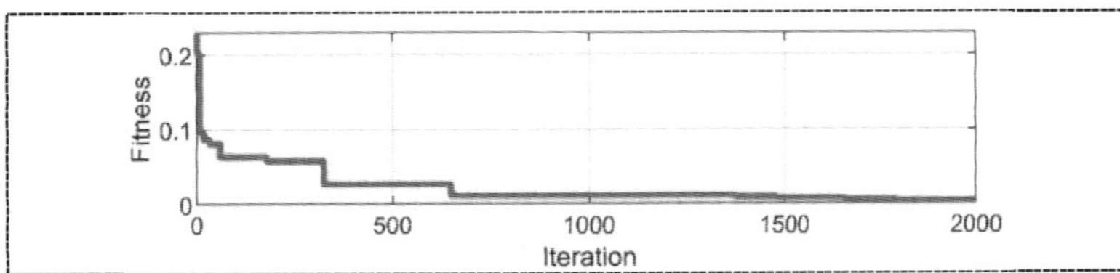

Figure 3. Fitness curve of iteration.

Figure 3 shows the iterative fitness and we notice a sharp decrease at the beginning. Then the change gradually becomes stable and converges to zero after about 600th iteration. These demonstrate that the GWO shows a good balance between local exploitation and global exploration. The algorithm has the ability to avoid the local optimum and achieve rapid convergence.

Table 1. The search space, actual value, and inversion results of the SHP for the synthetic model.

SHP	θ_s	θ_r	α	n
Search space	0.205-0.615	0.033-0.098	3.750-11.250	0.945-2.835
Actual value	0.410	0.065	7.500	1.890
Inversion results	0.406	0.090	7.008	2.113

4. Conclusion

This paper proposes a new SHP estimation scheme from the full waveform of GPR data with GWO algorithm. The proposed inversion scheme was tested by a numerical experiment. The results show that the SHP were successfully determined non-invasively by the proposed inversion scheme. In the future, this technique could be extended to higher dimensional data.

References

[1] Huisman J A, Hubbard S S, Redman J D and Annan A P 2003 Measuring Soil Water Content with Ground Penetrating Radar: A Review *Vadose Zone J.*2 pp 476-491

[2] Klotzsche A, Jonard F, Looms M C, van der Kruk J and Huisman J A 2018 Measuring Soil Water Content with Ground Penetrating Radar: A Decade of Progress *Vadose Zone J.* **17** p 180052

[3] Zhang M, Feng X, Bano M, Xing H, Wang T, Liang W, Zhou H, Dong Z, An Y and Zhang Y 2022 Review of Ground Penetrating Radar Applications for Water Dynamics Studies in Unsaturated Zone *Remote Sens.* 14 p 5993

[4] Scholer M, Irving J, Looms M C, Nielsen L and Holliger K 2012 Bayesian Markov-Chain-Monte-Carlo Inversion of Time-Lapse Crosshole GPR Data to Characterize the Vadose Zone at the Arrenaes Site, Denmark *Vadose Zone J.* **11** p vzj2011.0153

[5] Léger E, Saintenoy A andCoquet Y 2014 Hydrodynamic parameters of a sandy soil determined by ground-penetrating radar inside a single ring infiltrometer*WaterResour. Res.***50**pp5459-5474

[6] Jaumann S and Roth K 2018 Soil hydraulic material properties and layered architecture from time-lapse GPR *Hydrol. Earth Syst. Sci.* **22** pp 2551-2573

[7] Lambot S, Slob E C, Vanclooster M and Vereecken H 2006 Closed loop GPR data inversion for soil hydraulic and electric property determination *Geophys. Res. Lett.* **33** L21405

[8] Jadoon K Z, Slob E, Vanclooster M, Vereecken H and Lambot S 2008 Uniqueness and stability analysis of hydrogeophysical inversion for time-lapse ground-penetrating radar estimates of shallow soil hydraulic properties *Water Resour. Res.* 44 W09421

[9] Song X, Tang L, Zhao S, Zhang X, Li L, Huang J and Cai W 2015 Grey Wolf Optimizer for parameter estimation in surface waves *Soil Dyn. Earthq. Eng.***75** pp 147-157

[10] Jonard F, Weihermüller L, Schwank M, Jadoon K Z, Vereecken H and Lambot S 2015 Estimation of Hydraulic Properties of a Sandy Soil Using Ground-Based Active and Passive Microwave Remote Sensing *IEEE Trans. Geosci. Remote Sens.* **53** pp 3095-3109

[11] Topp G C, Davis J L and Annan A P 1980 Electromagnetic determination of soil water content: Measurements in coaxial transmission lines *Water Resour. Res.* **16** pp 574-582

[12] Bano M 2004 Modelling of GPR waves for lossy media obeying a complex power law of frequency for dielectric permittivity *Geophys. Prospect.***52** pp 11-26

Acknowledgements

The research was funded by the National Key Research and Development Program of China (No. 2022YFC3003402). We thank Nolwenn Lesparre, Benjamin Belfort, Pascal Friedmann, Jean-François Girard, François Lehmann and Oliver Razakarisoa for helping us to acquire GPR data. This experiment was funded by the ITES laboratory via an internal call.

Journal of Physics: Conference Series **2887** (2024) 012068 doi:10.1088/1742-6596/2887/1/012068

Mapping engineering communications based on radiovision and augmented reality technologies

Vitaly Khmelev[1], Ilya Tseplyaev[1], Sergey Shipilov[1,4], Ivan Fedyanin[1], Maxim Dimov[1], Rail Satarov[2] and Dmitry Romanov[3,4]

[1] Radiophysics Faculty, Tomsk State University, 634050 Tomsk, Russia
[2] Limited Liability Company "Radiovision", Russia
[3] Research and Production Association "Terrazond", Moscow, Russia
[4] Author to whom any correspondence should be addressed.

E-mail: info@terrazond.ru

Abstract
The article presents a mathematical apparatus of radio wave tomosynthesis for obtaining a radar image using an ultra-wideband multi-channel antenna array. Also presented are the results of processing radar data and the resulting three-dimensional radio tomogram of inhomogeneities in a homogeneous layer. Using the example of restoring the shape of test objects, the possibility of isolating individual layers at the required depth is shown. By combining radio wave and video information, the possibility of implementing augmented reality is shown. This combination provides a more visual interpretation of GPR data in construction and repair tasks.

1. Introduction

One of the tasks of subsurface exploration is to develop effective methods for detecting and visualizing hidden engineering structures. We propose to use the method of radio wave tomosynthesis for these and similar purposes [1]. The use of the radio tomosynthesis method for constructing 3D radio images significantly reduces the influence of multiple interactions and diffraction effects. Thus, a decent description of the dominant wave interactions can be described in the approximation of a single scatter.

2. The method of radio wave tomosynthesis

The proposed method is based on the Stolt method from geophysics [2]. Let us write down the expression for the once scattered field, using the so-called Born approximation. This is when a point source located at point $\mathbf{r}_0$ creates secondary currents in an inhomogeneous medium, of the form bellow:

$$j(\mathbf{r}_1) = k^2 \Delta\varepsilon(\mathbf{r}_1) G_0(\mathbf{r}_1 - \mathbf{r}_0).$$

Here $G_0(\mathbf{r}) = \exp(ik|\mathbf{r}|)/4\pi|\mathbf{r}|$ – Green's function of free space.

Value $k = 2\pi\sqrt{\varepsilon}/\lambda = 2\pi f\sqrt{\varepsilon}/c$ – the wave number corresponding to the background environment, $\varepsilon = \varepsilon' + i\varepsilon''$ – the complex dielectric constant of the background medium and f – the frequency of monochromatic radiation, or the spectral frequency in the case of pulsed radiation. Let's make an assumption and consider objects with a dielectric constant that differs slightly from the background environment: $\varepsilon = \varepsilon_f(1 + \Delta\varepsilon)$, where $\Delta\varepsilon << 1$. And now it is possible to use the Born approximation for this task based on these conditions.

If the radiation is observed at a point lying in free space, at a certain height H above the interface of the medium, then its value is related to the surface distribution of the field by a simple ratio.:

$$\mathrm{E}(\mathbf{r}) = (2\pi)^{-2} \iint \hat{\mathrm{E}}_s(\boldsymbol{\kappa}_\perp) \exp(i\boldsymbol{\kappa}_\perp \mathbf{r} + i\kappa_z H) d^2\boldsymbol{\kappa}_\perp.$$

Here $\hat{\mathrm{E}}_s(\boldsymbol{\kappa}_\perp)$ means a two-dimensional spatial spectrum of the surface distribution of the field.

As a result, solving the inverse problem of tomography for inhomogeneities hidden behind the interface of medium reduces to a sequence of operations:

1. Calculating of the spatial frequency spectrum of the received field
$$\hat{\mathrm{E}}(\boldsymbol{\kappa}_\perp) = \iint \mathrm{E}(\mathbf{r}) \exp(-i\boldsymbol{\kappa}_\perp \mathbf{r}) d^2\boldsymbol{\kappa}_\perp.$$
2. Finding the spectrum at the level of the medium interface
$$\hat{\mathrm{E}}_s(\boldsymbol{\kappa}_\perp) = \mathrm{E}(\boldsymbol{\kappa}_\perp) \exp(-i\kappa_z H).$$
3. Focusing the field into the background medium and finding the spectrum of spatial frequencies of currents at depth
$$\hat{j}(\boldsymbol{\kappa}_\perp) = -\exp(-i\kappa_z h) 2i\kappa_z \hat{\mathrm{E}}_s(\boldsymbol{\kappa}_\perp).$$
4. Restoration of an equivalent spatial current distribution
$$j(\mathbf{r}) = (2\pi)^{-2} \iint \hat{j}(\boldsymbol{\kappa}_\perp) \exp(-i\boldsymbol{\kappa}_\perp \mathbf{r}) d^2\boldsymbol{\kappa}_\perp.$$

The last operation is equivalent to restoring the distribution of inhomogeneities. All this sequence of operations physically means using two-stage reverse focusing of radiation. Certain approximations are, of course, allowed, and the applicability of which can be easily tested experimentally.

Journal of Physics: Conference Series **2887** (2024) 012068 doi:10.1088/1742-6596/2887/1/012068

3. Processing of experimental data

The practical use of the proposed method for tomography of building structures is interesting [3]. One of the main goals of subsurface exploration is the detection and visualization of hidden engineering structures. The task becomes more complicated when the object of detection is extended cables, pipes or fittings located at different depths. When radiowave radiation penetrates a medium, we can talk about restoring its internal structure using the transmitted or scattered field. This structure represents the spatial distribution of permittivity values. Sharp gradients of changes in permeability are characteristic of boundaries of medium or objects immersed in them. Although these tasks are not easy, a number of effective solutions have been developed based on the use of radiation focusing phenomena for them.

We use a developed ground-penetrating radar with a linear active antenna array to obtain experimental data. Its called "Terrazond" (Multi channel GPRS "GRT-32S"). This radar is designed to penetrate concrete and other materials (see figure 1). Here, the antenna elements are coordinated with concrete for better penetration into the depth of building structures.

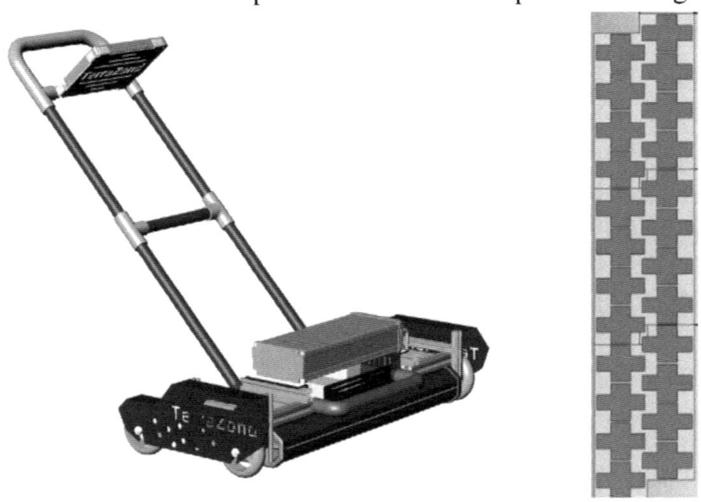

Figure 1. Concrete GPR and its antenna array.

The concrete microscope has the following characteristics: frequency range: 0.5 – 4 GHz; penetration depth up to 1.5 m; width resolution up to 4 cm; width of 16 element antenna array 0.7 m; scanning speed for 1cm traces step 10 m\s.

We performed tomography on test objects in the form of metalized strips to verify the feasibility of the radio wave tomosynthesis method. Each strip was 2 cm wide and was arranged like a snake between two planes of aerated concrete blocks, with a refractive index of n=1.6. The first metal strip was located at a depth of 10 cm and the second was at a depth of 20 cm below the surface. The height of the antenna array above the plane was 5 cm (figure 2).

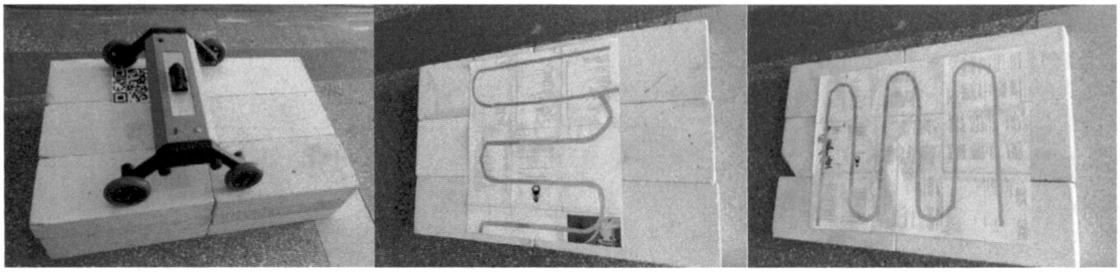

Figure 2. Test scene of two-layer heterogeneity in aerated concrete.

As a result of the scan, an oscillogram was obtained. Part of it is shown in figure 3. Previously, the initial data was processed using an algorithm proposed by the authors to remove common-mode interference [4]. The oscillogram shows hyperbolic patterns characteristic of point-like objects, which are located at different depths. Extended objects also produce hyperbolas on the oscilloscope, provided that ground-penetrating radar moves perpendicular to their extent.

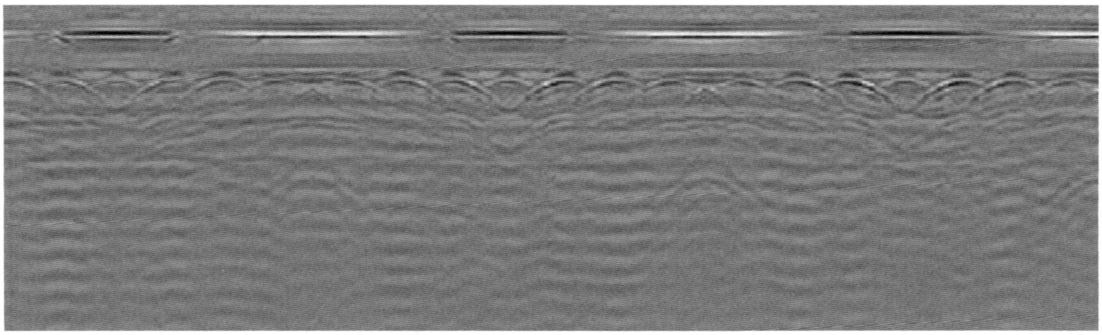

Figure 3. The waveform of the signal obtained during the experiment.

We have reconstructed a three-dimensional radio tomographic image of the studied area using the tomographic technique based on tomosynthesis. The result is shown in figure 4. Slices of this image are displayed at depths corresponding to the positions of the test objects: 10 and 20 centimeters.

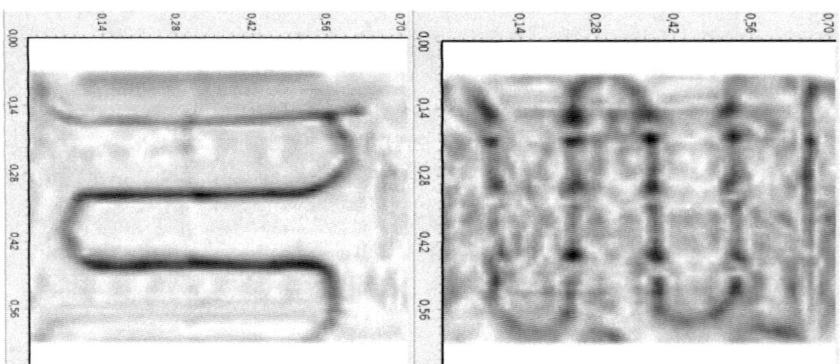

Figure 4. Radiotomogram slices at depths of 10 and 20 cm.

Combining two digital images can create an augmented reality effect [5]. Each of these has a different nature. Therefore, it is necessary to translate one image into another. In this article, we will focus on the image area captured by the camera and calculate its geometric dimensions in meters. This information will allow us to overlay radio images over the classical optical image. We should immediately note that the digital image captured by the camera is an angular image. The pixels in the image are the discrete units of the vertical and horizontal angles of the camera's field of view. And without information about the distance to the object of study or information about the sizes of objects inside the scene, angular dimensions cannot be converted into meters [6]. We solved this problem by adding a predefined object with known geometric dimensions to the scene.

In our experiment, a piece of paper with a printed QR code was used as an optical beacon, which provides invariance to planar rotation. The structure of the QR code also allows us to estimate camera tilt and affine distortions. So that they can be compensated. The scanning process begins with setting the ground-penetrating radar at the QR code, then the operator moved the ground-penetrating radar along a free trajectory synthesizing a virtual aperture. This made it possible to obtain a radio image with sufficient quality. The operator also filmed the test scene without releasing the QR code from the scene. In each frame where the QR code was recognized, a plane was completed from its edge to the sizes of reconstructed radio image. The angular dimensions were calculated based on the decreasing size of the QR code elements. As a result of the combination of a radio tomogram and a real image, the results presented in figure 5 were obtained. The authors also suggest using color differentiation of objects by depth.

Figure 5. The result of overlaying a radio tomogram onto a real image of an object.

4. Conclusion

Thus, in the course of the work, a three-dimensional radio tomogram of inhomogeneities was obtained in a homogeneous layer using radio wave sensing data obtained using a concrete GPR. Thus, in the course of the work, a three-dimensional radio tomogram of inhomogeneities was obtained in a homogeneous layer using radio wave sensing data obtained using a concrete GPR. The authors showed the possibility of selecting individual layers at the required depth, using the example of restoring the shape of test objects. We also showed the possibility of implementing augmented reality by combining radio wave and video information. This combination provides a more visual interpretation of GPR data in construction and repair tasks.

Acknowledgments

This study was supported by the Tomsk State University Development Programme («Priority-2030»).

References

[1] Shipilov S, Satarov R, Yakubov V, Minin O, Minin I and Yurchenko A 2017 *Opt. Quant. Electron.* **49** 339
[2] Stolt R 1978 *Geophysics*, **43** 23–48.
[3] Wallace W, Xavier D, Peter A, 2018 *NDT & E International.* **96** 58-78
[4] Gorst A, Tseplyaev I, Eremeev A, Shipilov S, Romanov D, Eremin R 2023 *Remote Sensing* **15** 2751
[5] Dickmann F, Keil J, Dickmann P 2021 *KN J. Cartogr. Geogr.* **71** 285–295

[6] Li J., Zeng Z., Sun J., Liu F., (2012) Through-wall detection of human being's movement by UWB radar. *IEEE Geoscience and Remote Sensing Letter*, 9(6), 1079-1083.

Lunar surface permittivity distribution along route of Yutu-2 rover

Du W[1,2], Su Y[1,2], Dai S[1], Ding C Y[3]

[1]*Key Laboratory of Lunar and Deep Space Exploration, National Astronomical Observatories, Chinese Academy of Sciences, Beijing 100101, China*
[2]*School of Astronomy and Space Science, University of Chinese Academy of Sciences, Beijing 100049, China*
[3]*Institute of Moon-base Exploration and Observation, Shenzhen University, Shenzhen 518060, China.*

Abstract. On January 3, 2019, Chang'E-4 (CE-4) landed in the Von Kármán crater within the South pole-Aitken Basin on the Moon's farside. The lunar penetrating radar (LPR) onboard the Yutu-2 rover is the same as the Chang 'E-3 mission, which can obtain information about the thickness and strutures of lunar regolith as well as upper crust along the excursion path. The permittivity of lunar material is one of the important parameters for interpreting radar data and can also be obtained by LPR. In this paper, we use the metal plate reflection method based on LPR data and panoramic camera (PC) data to estimate the relative permittivity distribution of 1373m route during the first 53 lunar days. We estimated an average relative permittivity of ~2.78, which is less than ~2.9 of Chang'E-3 (CE-3).

1. Introduction

At present, the permittivity of lunar surface can be obtained by remote sensing inversion, sample measurement and ground penetrating radar inversion [1]. In this paper, we use the CE-4 LPR data. LPR is a dual-frequency dual-channel ground penetrating radar system. Channel 1 operates at a center frequency of 60 MHz (low frequency) with a bandwidth of 40-80MHz, while the Channel 2 operates at a center frequency of 500 MHz (low frequency) with a bandwidth of 250-750MHz [2]. The radar emits electromagnetic waves into the lunar surface, which will be reflected and scattered when discontinuous interface of different mediums occur. The detection data can be obtained after the receiver amplifies and samples the reflection signals. Figure 1 shows the traverse during the first 53 lunar days of Yutu-2, and Figure 1 shows the obtained radar profile along the traverse. In this paper, we first proposed a radar signal correlation alignment method for zero-time correction. Then, we use the LPR high frequency data and metal plate test data to calculate a relative permittivity distribution map along the exploration path of the Yutu-2 rover.

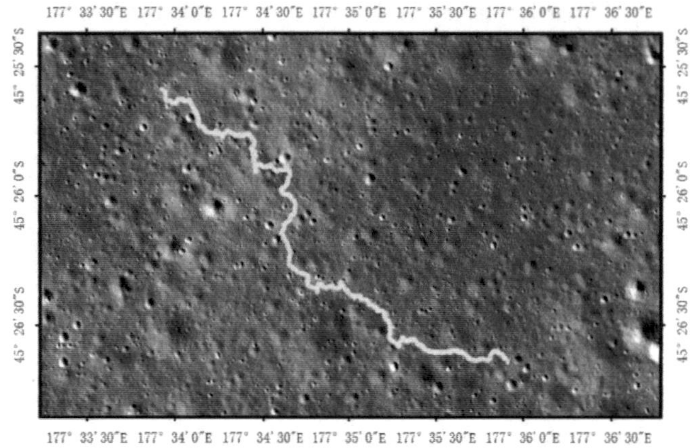

Figure 1. The Yutu-2 route within the first 53 lunar days. The red star indicates the landing site of rover. The LPR performed observation along the ~1373m route. The background image comes from Lunar Reconnaissance Orbiter (LRO) Narrow Angle Camera (NAC) (File ID: M1303619844).

2. Estimation method of lunar surface permittivity

The metal plate reflection method has been widely used to estimate the Lunar or Mars surface material permittivity [3, 4, 5]. In the ground test experiment, the reflected signal of the metal plate is obtained by placing a large metal plate under the antenna for measurement, and the antenna is pointed upward towards the sky to obtain the received waveform in free space (direct coupled wave). We need to subtract the direct coupled wave from the plate/lunar reflection signal to obtain a corrected waveform. Before that, a zero-time adjustment should be adopted. In this paper, a correlation alignment method is used to align the troughs of trace data efficiently and accurately. Specifically, the signals are interpolated first, and then the cross-correlation sequence $R_{xy}(l)$ of n_{th} column lunar surface reflection signal $x(n)$ and the directly coupled wave $y(n)$ is calculated using formula (1), then align $x(n)$ to $y(n)$ according to the maximum correlation value. Figure 2 shows the effect of a lunar signal aligned with a direct coupled wave.

$$R_{xy}(l) = y(n) * x^*(-n) \tag{1}$$

Next, using the amplitude of metal plate reflection signal A_m to represent the incident signal amplitude, A represents the lunar surface reflection amplitude. The permittivity of lunar surface can be calculated from [3]:

$$\varepsilon_r = \left(\frac{A_m + A}{A_m - A}\right)^2 \tag{2}$$

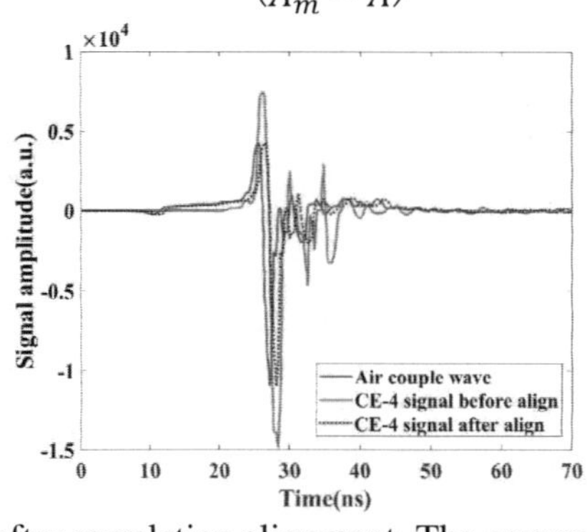

Figure 2. signals before and after correlation alignment. The orange solid line represents the direct coupled wave, and the blue solid line and dashed lines represent the surface reflection signal before and after alignment respectively.

3. Results

Figure 3 shows the estimated results of the relative permittivity of lunar surface using the method in Section 2, with values ranging from ~2~6, a mean value $\mu \approx 2.78$, standard deviation $\sigma \approx 0.26$. We can see that the values fluctuate along the excursion path, together with some high values. This may be related to variations in the antenna height from the ground, inhomogeneity of the lunar surface material, etc. Lunar surface permittivity distribution along the route shown by Figure 4 can be drawn using the located the radar data. We identified a total of 7 outliers by combining the PC images and velocity information from the rover, found that the corresponding locations of these anomalies have gravels underneath the antenna or undergo an undulation of terrain. It is likely that these factors could have caused the anomalies, and the specific effect of debris size or distance on the dielectric constant can be further analysed in future studies.

In summary, the average relative permittivity of lunar surface of Chang'E-4 landing site is ~2.78, which is lower than ~2.9 of Chang'E-3 [4], and its surface is relatively flatter and smoother, scattered with small-size gravels. In addition, Combined with Figure 3 and Figure 4, it can also be found that the permittivity value increases slightly after ~1000m path, and the distribution of rocks around the corresponding track increases. However, there is still no large-size rocks or topographic fluctuation similar to the landing site of Chang'E-3, so the reason for the change of the permittivity value of this segment may be the difference of regolith material.

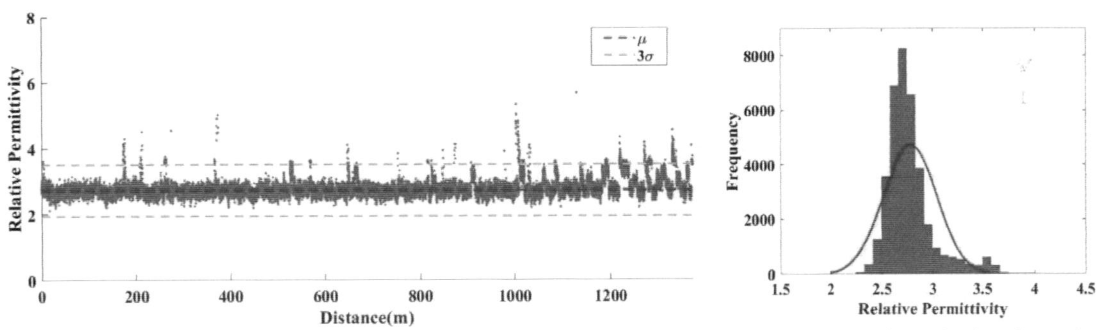

Figure 3. Calculated value of lunar surface permittivity. The blue points indicate the calculated results of each trace.

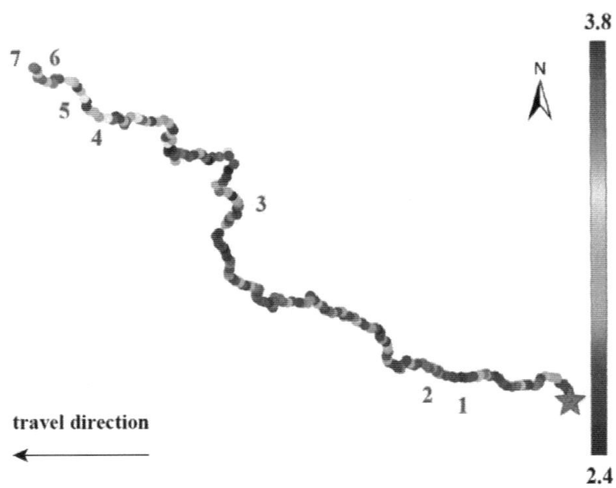

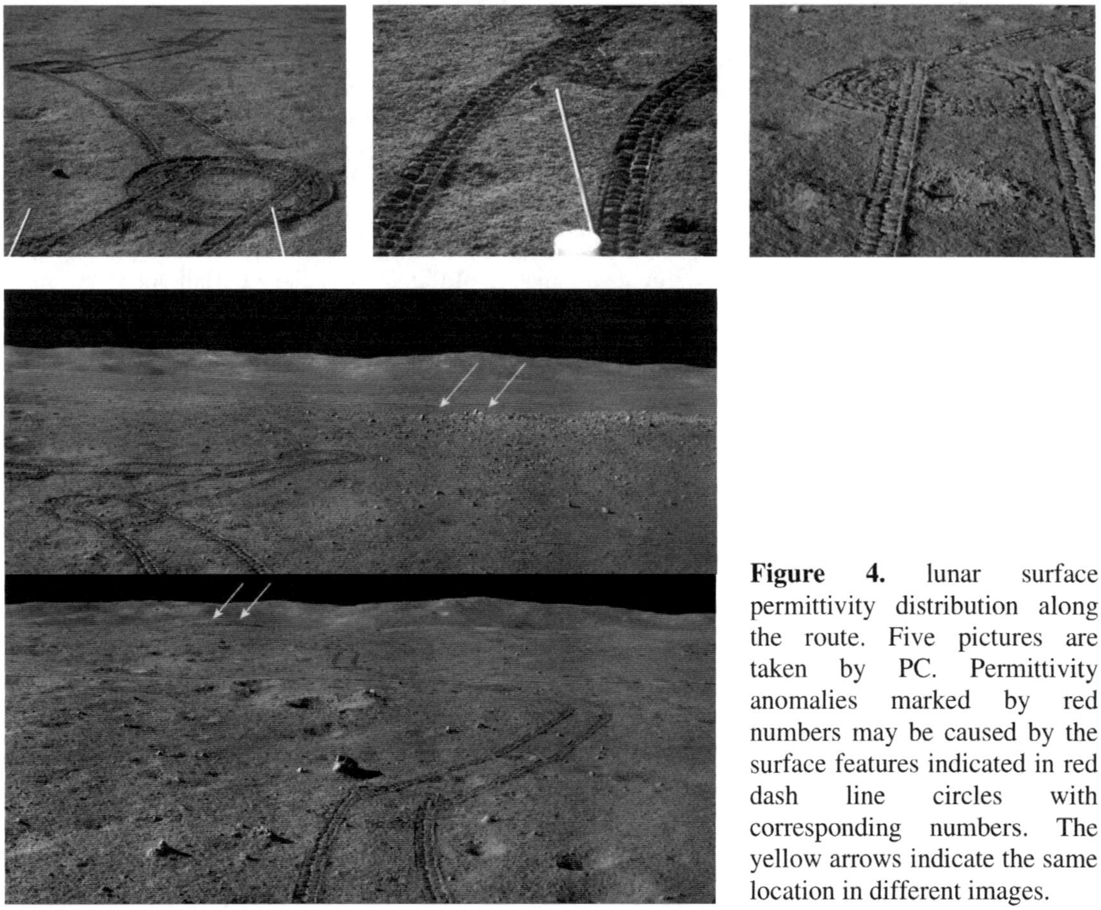

Figure 4. lunar surface permittivity distribution along the route. Five pictures are taken by PC. Permittivity anomalies marked by red numbers may be caused by the surface features indicated in red dash line circles with corresponding numbers. The yellow arrows indicate the same location in different images.

Reference

[1] Fa W Z. Overview of the dielectric permittivity of lunar surface materials and implications for Chang'E-5 sample measurements. (in Chinese) 2022 *Sci Sin-Phys Mech Astron* **53** 239602

[2] Fang G Y et al., Lunar penetrating radar onboard the Chang'e-3 mission. 2014 *Res. Astron. Astrophys.* **14** 1607

[3] Su Y et al., Hyperfine structure of regolith Unveiled by Chang'E-5 Lunar Regolith Penetrating Radar. 2022 *IEEE Transactions on geoscience and remote sensing* **60** 5110414

[4] Dong Z H, Fang G Y, Ji Y C, Gao Y Z, Wu C, Zhang X J. Parameters and structure of lunar regolith in Chang'E-3 landing area from lunar penetrating radar (LPR) data. 2017 *Icarus* **282** p 40-46

[5] Dong Z H, Fang G Y, Zhou B, Zhao D, Gao Y Z and Ji Y C. Properties of lunar regolith on the Moon's farside unveiled by Chang'E-4 Lunar Penetrating Radar. 2021 *Journal of Geophysical Research: Planets* **126** e2020JE006564

Acknowledgments

The authors are grateful to the Ground Research and Application System team members for their contribution to data receive and release. The Chang'E-4 data can access from the link of https://moon.bao.ac.cn/.

Journal of Physics: Conference Series **2887** (2024) 012070 doi:10.1088/1742-6596/2887/1/012070

Reverse time migration of GPR data based on GPU parallel Symplectic Euler Algorithm and Total Variation Regularization

Xin Gao[1], Jianwei Lei[1*], Zhiqing Chen[1], Hongyuan Fang[1]

[1]School of Water Conservancy and Transportation, Zhengzhou University, Zhengzhou, China
*email: 15333827823@163.com

Abstract. Reverse time migration is important in GPR image processing, but its efficiency and accuracy need to be improved. In terms of efficiency, reverse time migration uses the Finite Difference Time Domain (FDTD) algorithm for parallel CPU computation, which is inefficient. In terms of accuracy, it usually uses cross-correlation conditions for imaging, which reduces the imaging accuracy. To solve the above problems, this paper proposes the solution: using symplectic algorithm parallel GPU acceleration technology to improve the migration efficiency and using TV regularization technology to improve imaging accuracy. Finally, the proposed algorithm is verified by the road cavity model and the underground pipeline model.

1. Introduction

Ground penetrating radar (GPR) is a non-destructive testing method that is widely used in detecting and locating abnormal bodies. Migration is widely used in detecting and locating abnormal bodies in shallow underground structures. Now the most commonly used and accurate migration algorithm is reverse time migration (RTM)[1], but traditional RTM still faces efficiency and accuracy challenges.

To improve computing efficiency, this paper introduces the symplectic Euler algorithm, applies it to reverse time migration (RTM) and combines it with GPU acceleration to enhance its computational efficiency [2-3]. This paper applies the L1-TV regularization denoising strategy using the alternating direction method of multipliers (ADMM) to GPR RTM profiles, creating high-quality RTM images.

2. Methodology

2.1 RTM based on GPU parallel symplectic Euler algorithm.

In two-dimensional isotropic media, Maxwell equation can be expressed as follows:

$$\begin{aligned} \frac{\partial E}{\partial t} &= \frac{1}{\varepsilon}\nabla \times H - \frac{\sigma}{\varepsilon}E \\ \frac{\partial H}{\partial t} &= -\frac{1}{\mu}\nabla \times E \end{aligned} \tag{1}$$

where H and E respectively represent the magnetic and electric field vector; σ is the electrical conductivity, ε is the dielectric permittivity, and μ is the magnetic conductivity.
By introducing the vector magnetic potential $H=\nabla\times A$ and setting $E=-U$, applying symplectic Euler algorithm to discretise Equation, the following iterative calculation formula is obtained [4]:

$$
\begin{aligned}
A_{i,j}^{n+1} &= A_{i,j}^{n} + \frac{dt}{\mu} U_{i,j}^{n} \\
U_{i,j}^{n+1} &= \frac{\varepsilon - dt\sigma}{\varepsilon} U_{i,j}^{n} + \frac{dt}{\varepsilon} \nabla^2 A_{i,j}^{n+1}
\end{aligned}
\quad (2)
$$

where $A_{i,j}^{n}$ and $U_{i,j}^{n}$ represent the discrete values of the field components A and U at time ndt, space grid nodes (i, j), and ∇^2 is he laplace operator.

2.2 Total variation regularization
In this paper, the alternating direction method of multipliers (ADMM) is adopted to solve the TV problem [5]. Suppose x is the clear RTM profile after noise removal, and y is the original RTM profile with noise, which can be expressed as:

$$y = x + n \quad (3)$$

where n is noise. Introducing the L1-TV regularization and the expression for the constrained optimization problem:

$$
\begin{aligned}
\hat{x} &= \min_{x,z} \left\{ \frac{1}{2} \| y - x \|_2^2 + \lambda \xi(z) \right\} \\
&s.t.\ \ x = z
\end{aligned}
\quad (4)
$$

where λ is the regularization parameter; z is the slack variable, $\xi(z) = \|z\|_{TV}$.

3. Numerical Simulation

3.1 Road cavity model
The efficiency of the migration algorithm was verified by the road cavity model, as shown in Fig. 1. The depth of the simulated area is 4m, the horizontal distance is 8m, and the spatial step size $dx=dz=0.01$m. The model is divided into two layers, the upper dielectric relative permittivity $\varepsilon_r=5$, conductivity $\sigma=0.002$S/m. And the lower $\varepsilon_r=9$, $\sigma=0.005$S/m. The model consists of four circular voids with radii of 0.15m, 0.2m, 0.13m and 0.18m, whose relative dielectric constant $\varepsilon_r=1$ and electrical conductivity $\sigma=0$S/m. The excitation source selects the Ricker wavelet with the centre frequency of 400 MHz, the time step is 0.02 ns and the total recording length is 70 ns. It adopts the sending and receiving mode of self-excitation and self-receiving to record a total of 400 waveforms.
The forward simulation results are shown in Fig. 2. Further migration processing of the forward profile is required. The profile processed by the reverse time migration algorithm is shown in Fig. 3. Most of the models were restored. But some low-frequency noise and artifacts can still be seen. The position of the red dotted line in Fig. 3 is selected to draw the single-channel migration data. The results are shown in Fig. 4. The voids and interfaces are effectively restored, but some low-frequency noise and artifacts can still be seen. The calculation results of different algorithms are shown in Table 1.

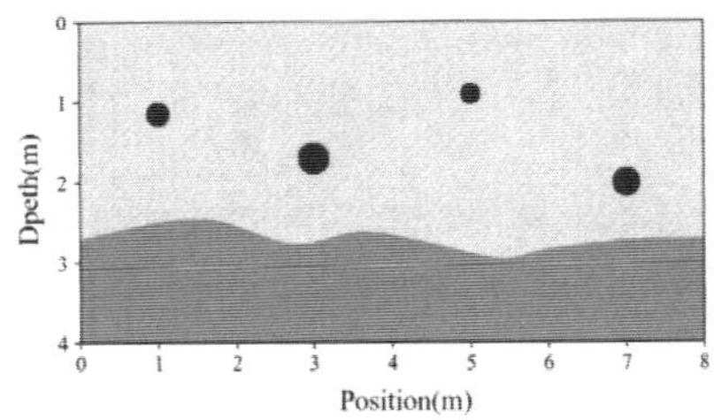

Fig. 1 The diagram of road cavity model

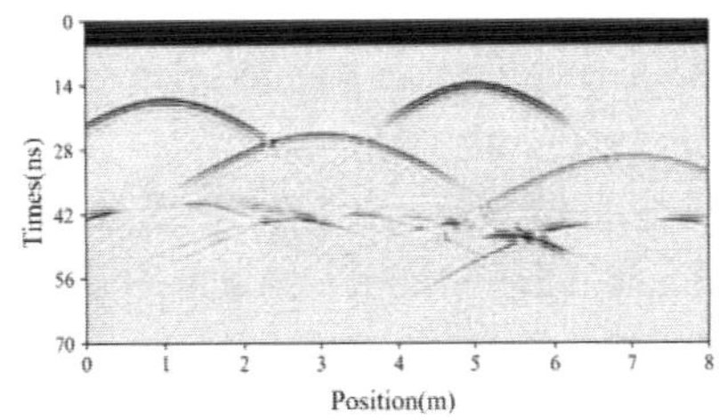

Fig. 2 The forward profile of road cavity model

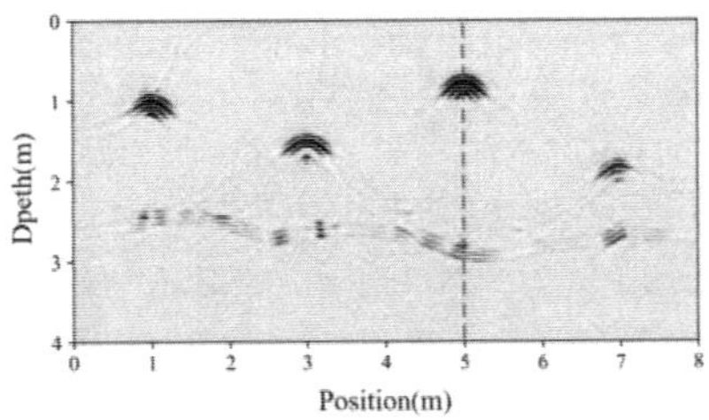

(a) CPU serial FDTD

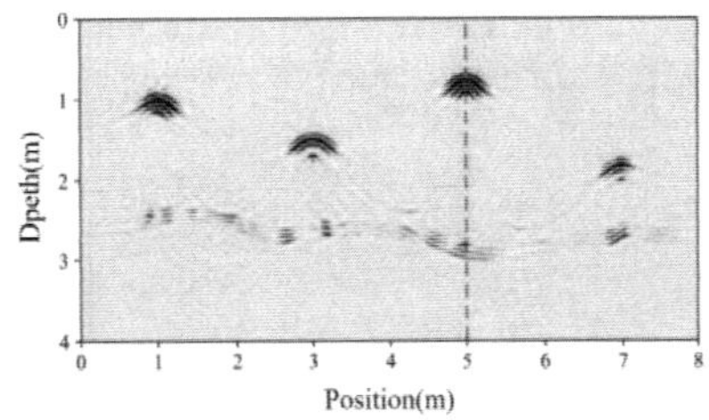

(b) CPU serial symplectic Euler algorithm

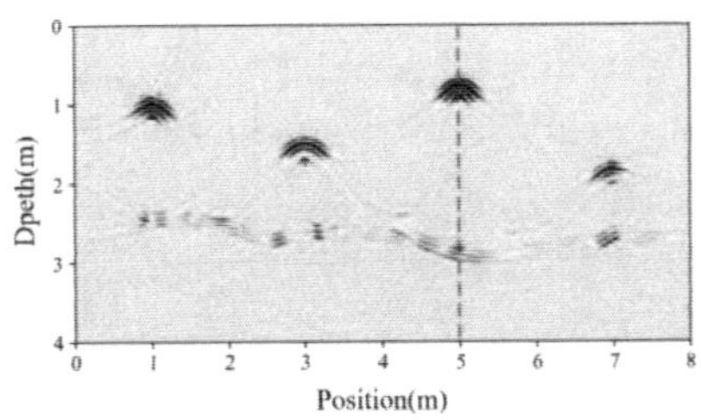

(c) GPU parallel symplectic Euler algorithm

Fig. 3 The processing results of RTM algorithms

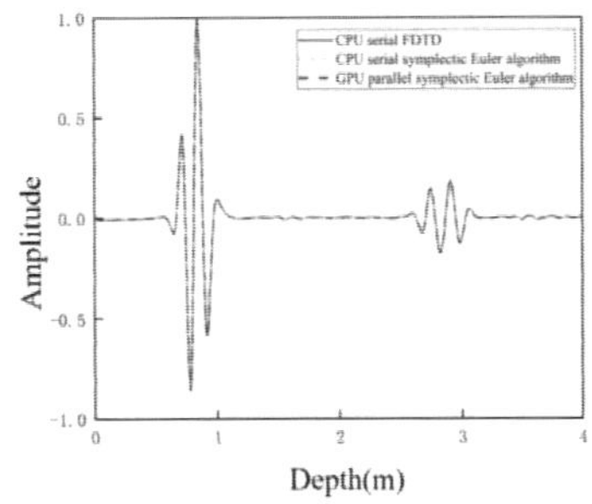

Fig. 4 The comparison diagram of RTM singledata obtained by different algorithms

Table 1 Simulation time of different algorithms

Method	Grid	Time /s
CPU serial FDTD	400×800	35169.32
CPU serial symplectic Euler algorithm	400×800	25013.87
GPU parallel symplectic Euler algorithm	400×800	1620.46

3.2 Underground pipeline model

The high accuracy of the migration algorithm was verified by a underground pipeline model, as shown in Fig. 5. The depth of the simulated area is 4m, the horizontal distance is 8m, and the spatial step size is dx =dz= 0.01m. The model comprises two layers, the upper dielectric relative permittivityε_r=5, conductivity σ=0.003S/m, and it includes an uncompact area and an irregular void (ε_r=1, σ=0S/m), and PVC pipes of different radii (ε_r=3,σ=0S/m) and concrete pipes (ε_r=8, σ=0.002S/m), and air (ε_r=1, σ=0S/m). The lower ε_r=12, σ= 0.005S/m. Two irregularly elliptical objects are arranged in the lower layer, one is an inflated cavity (ε_r=1, σ=0S/m), and the other is a water-filled cavity (ε_r=81, σ=0.001S/m). The excitation source selects the Ricker wavelet with the centre frequency of 900 MHz, other parameters are the same as previous model. Forward simulation results are shown in Fig. 6. The profiles processed by traditional RTM migration imaging and TV regularization technology are shown in Figure 7 and 8. Clutter is reduced and the location of the void can be roughly judged in Fig 8.

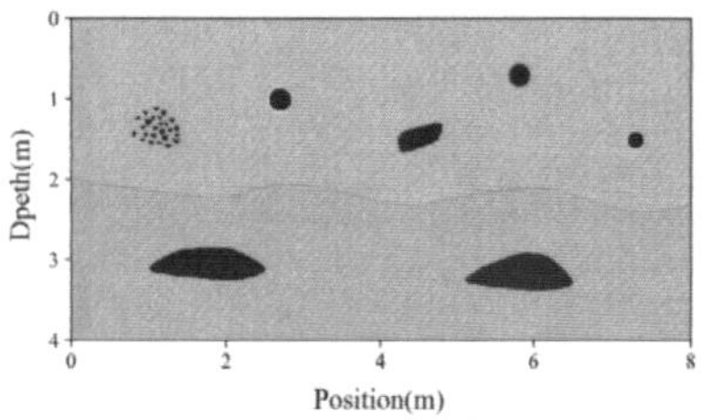

Fig. 5 The schematic diagram of underground pipeline model

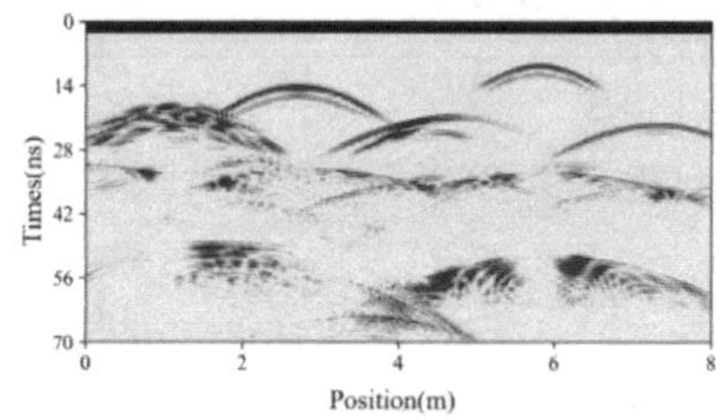

Fig.6 The forward profile of underground pipeline model

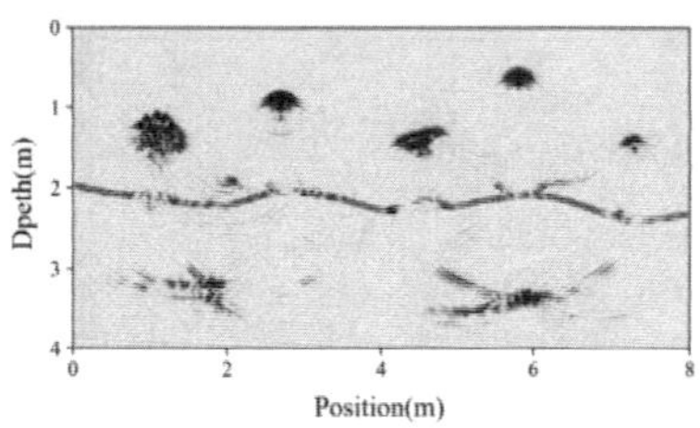

Fig. 7 Traditional RTM processing results

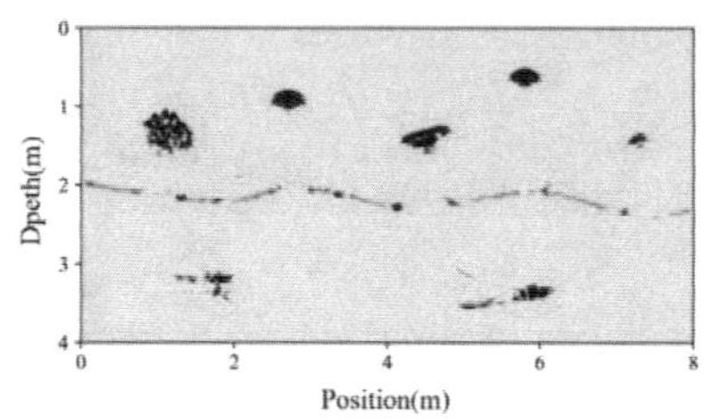

Fig. 8 Results of TV regularization RTM processing

4. Conclusion

In this paper, the RTM algorithm of GPU parallel symplectic algorithm is used for efficient migration imaging of GPR profile, and TV regularization technology is introduced to improve the accuracy. Simulation data show that The efficiency of GPU parallel symplectic algorithm can be improved by more than 20 times. TV regularization can remove low frequency noise and artifacts from RTM profile and improve imaging quality. In summary, the proposed algorithm improves the efficiency and accuracy of RTM, and can obtain the information of abnormal bodies of underground structures accurately.

5 References

[1] Shen, R.Q., Zhao, Y.H., Hu, S.F. et al.: 'Reverse-time migration imaging of ground-penetrating radar in NDT of reinforced concrete structures', Remote Sensing, 2021, 13, (10).

[2] Liu, H.W., Li, B., Liu, H., et al.: 'The issues of prestack reverse time migration and solutions with graphic processing unit implementation', Geophysical Prospecting, 2012, 60, (5), pp. 906–918.

[3] Fang, H.Y., Lin, G.: 'Symplectic partitioned Runge–Kutta methods for two-dimensional numerical model of ground penetrating radar', Computers & Geosciences, 2012, 49, pp. 323–329.

[4] Li J., Liu C, Zeng, Z., & Chen, L (2015). GPR Signal Denoising and Target Extraction with the CEEMD Method. *IEEE Geoscience and Remote Sensing Letter,* 12 (8), 1615-1619.

[5] Yang, S., Wang, J., Fan, W., et al.: 'An efficient ADMM algorithm for multidimensional anisotropic total variation regularization problems', Proceedings of the 19th ACM SIGKDD International Conference on Knowledge Discovery and Data Mining, Chicago, USA, 2013, pp. 641–649.

Journal of Physics: Conference Series **2887** (2024) 012071 doi:10.1088/1742-6596/2887/1/012071

Subsurface characterization of Removable Urban Pavements (RUP) using Ground Penetrating Radar (GPR)

G Andreoli[1], F Schmidt[2], A Ihamouten[3], A Cothenet[3], T Dezert[4], M L Nguyen[3] and T Sedran[1]

[1] MAST/MIT, Nantes Gustave Eiffel University – France (gregory.andreoli@univ-eiffel.fr)
[2] MAST/EMGCU, Marne-la-Vallée Gustave Eiffel University – France
[3] MAST/LAMES, Nantes Gustave Eiffel University – France
[4] FI-NDT, Nantes – France

Abstract. In the pursuit of improving transport infrastructure as part of the overall sustainable development policy, new coatings and emerging structures are coming to the forefront. This is the case with Removable Urban Pavements (RUP). This new type of structure, thanks to lightweight equipment, is removable in parts and allows quick access to underground networks. To assess the design and potential degradation of RUP, the use of multi-technique tools for Non-Destructive Techniques (NDT) is implemented as part of the European HERON project. These methods are increasingly sought after by transport infrastructure managers, both for corrective and preventive/predictive maintenance. Among them, Ground Penetrating Radar (GPR) is gaining popularity among pavement professionals for multi-object detection/characterization in the subsurface, as well as for assessing the geometry of structures. In this context, a RUP structure, installed on the fatigue carousel at Gustave Eiffel University (Nantes, France) and subjected to various mechanical and hydraulic solicitations, has been studied using GPR.

1. Introduction

In a policy aiming for zero risk, it is essential to control hazards and enhance our knowledge of transportation infrastructures. The objective is to implement corrective, preventive, and/or predictive maintenance that minimizes risks below a certain threshold of acceptability. Therefore, the instrumentation of pavements, particularly in the context of preventive monitoring, is a crucial tool in the risk analysis approach. It is important in this case to better understand the behavior of structures and provide reliable data for a robust risk analysis. For this purpose, the European project HERON [1] aims to combine the principles of repair and/or replacement action with automatic detection of areas that may exhibit defects. In this context, the Forever Open Road (FOR) [2] principle with its national variations (such as R5G in France) has been developed by the Forum of European National Highway Research Laboratories (FEHRL). Thus, HERON addresses three FOR principles: resilience, automation, and adaptability. In terms of technological objectives, the HERON project aims to develop an integrated automated system for the maintenance and renovation of the road network, including crack sealing, pothole patching, rejuvenation of asphalt pavement, detection/removal of traffic cones and other potential obstacles, detection of marking defects, autonomous replacement of elements of Removable Urban Pavements (RUP) [3], etc. We focus here on this latter innovative technology. To accomplish

Journal of Physics: Conference Series **2887** (2024) 012071 doi:10.1088/1742-6596/2887/1/012071

these tasks, the HERON system consists of an instrumented autonomous terrestrial robotic vehicle and drones for coordinating maintenance and intervention phases, a detection interface allowing enhanced monitoring of structural conditions, augmented reality visualization tools, data processing of collected information, and workflow coordination through deep learning methods (YOLO method). Finally, HERON enables a modular design of the operational system, maximizing capabilities and adaptability for transportation infrastructures, while reducing human risks, maintenance costs, traffic disruptions, and thereby contributing to the improvement of network efficiency.
Among the innovative structures, RUP have been selected for evaluation within the project. Gustave Eiffel University is involved in the design and deployment of RUP on various test sites in France [4]. This enables full-scale trials, where the HERON project can now fit into a multi-technique Non-Destructive Techniques (NDT) framework, whether on the surface (LiDAR, photogrammetry, imaging, etc.) and subsurface (GPR). In the frame of this research work, we first introduce the principle of RUP and highlight the relevance of using GPR for subsurface characterization. Next, we present the fatigue carousel at the university on which a test structure has been implemented. Finally, we conclude with the initial results obtained using GPR, followed by a conclusion and research outlook.

2. Presentation of RUP

In 2008, Gustave Eiffel University proposed an original concept of Removable Urban Pavement (RUP) whose purpose is to ensure an easy and fast access to the sub-base and to the networks underneath, for maintenance or extension needs. The concept is based on interlocked precast hexagonal concrete slabs. The slabs were designed with keys to avoid slab rocking risk under heavy traffic. A first prototype confirmed the proof of concept [3]. Within a collaborative project Gustave Eiffel University has been developing since 2017 a new type of permeable RUP as this property is becoming more and more pregnant in a context of climate change and violent climatic events [4]. The presence of vertical holes together with a wearing course made of porous concrete allows the water drainage. The slabs are laid on an original excavatable and draining cement treated base (figure 1 and figure 2).

Figure 1. RUP (hexagonal slab).

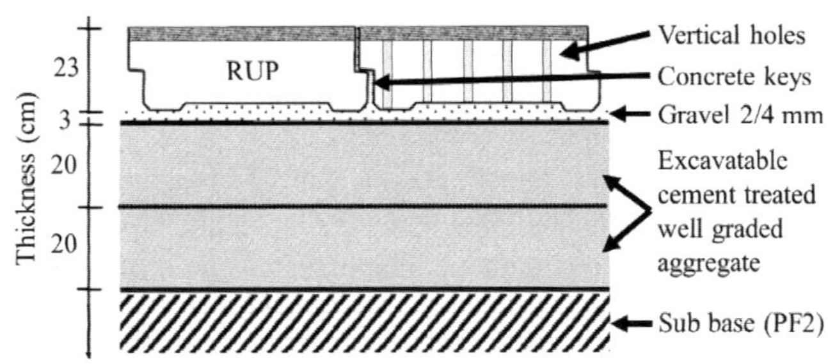

Figure 2. Implementation of RUP on a structure.

3. Experiment and result

The Gustave Eiffel University (Nantes, France), has been equipped for over 45 years with an accelerated pavement testing system under the loads of a heavy vehicle at speeds up to 100 km/h [5] (figure 3). This allows the validation of new structures through fatigue testing before scaling up to demonstrator testing on the road network. Thus, for the execution of a mechanical and hydraulic design campaign on RUP conducted in 2020/2021 for 2 months, the University constructed a section measuring 8.35 meters in length by 2.31 meters in width. The section consists of slabs with a 46 cm edge and 23 cm thickness (including 4 cm of porous concrete on the surface for the 22 double-layer hexagonal slabs, 4 cm of hydraulic concrete for the 15 half-edge slabs, and 2 single-layer quarter-slabs) (figure 4). This section was subjected to 200,000 cycles of loading by a half-axle with dual wheels (which corresponds on average to traffic between 45 and 91 buses/day/direction for 20 years) (figure 5).

Journal of Physics: Conference Series **2887** (2024) 012071 doi:10.1088/1742-6596/2887/1/012071

Figure 3. Fatigue carousel (Gustave Eiffel University – Nantes, France).

Figure 4. Studied RUP Structure.

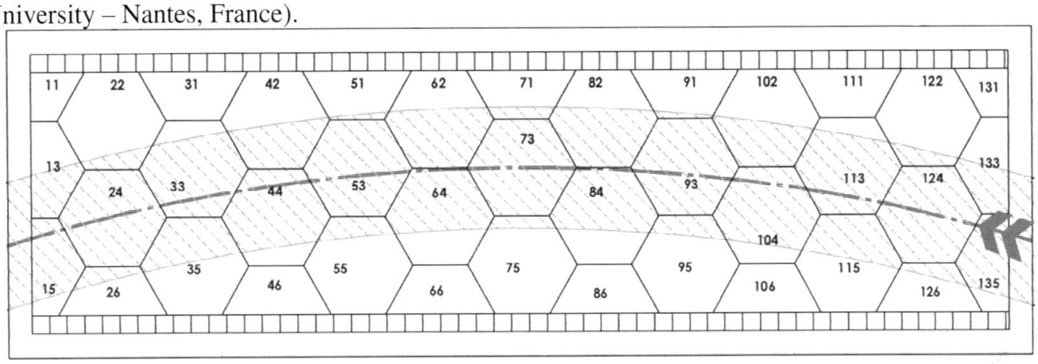

Figure 5. Layout of the RUP structure on the fatigue carousel (the hatched surface represents the half-axle wandering).

As slabs interlocking is mandatory for the concept, it is important to verify its integrity. As this information is not reachable from a visual inspection, Ground Penetrating Radar (GPR) appears as a potential alternative. Using multi-channel GPR ground-coupled antenna, we have inspected the RUP structure after 200,000 loading cycles on a large frequency bandwidth [1 – 4.5] GHz. The goal here is to observe the subsurface structure and collect a large dataset on continuous acquisition to identify some anomalies. A global approach (raw signals without pre-processing) is initially employed, followed by a local approach (pre-processing with the extraction of key features) to enable data processing through artificial intelligence methods. We perform automatic anomaly detection using a YOLO-type deep learning algorithm combined with an SVM-type machine learning algorithm for effective classification. Building upon the key features observed in figure 6, the first results show excellent automatic anomaly detection in B-scans (figures 6a and 6d), cross-scans (figures 6b and 6e), and A-scans (figures 6c and 6f). The objective now is to enhance the classification of these anomalies and integrate them into a multi-technique approach (surface and subsurface), helping decision-making, as advocated in the HERON project.

4. Conclusion et perspectives

In conclusion, it seems to be possible to characterize RUP subsurface, thereby extracting data/images that can serve as features for building a database complementary to the surface database obtained through imaging, LiDAR, and/or photogrammetry techniques. These features are used as inputs to deep learning algorithms for pattern recognition (as YOLO v4), able to identify potential structural anomalies and SVM for the classification [6]. In addition, thanks to Full Waveform Inversion (FWI) methods, it is possible to obtain observable physical quantities such as the dielectric permittivity of the examined medium. A hybrid data processing approach will then enhance algorithm performance while reducing computation times [7]. With the creation of a comprehensive multi-technique database, the possibility of decision-making and intervention in maintenance operations becomes feasible.

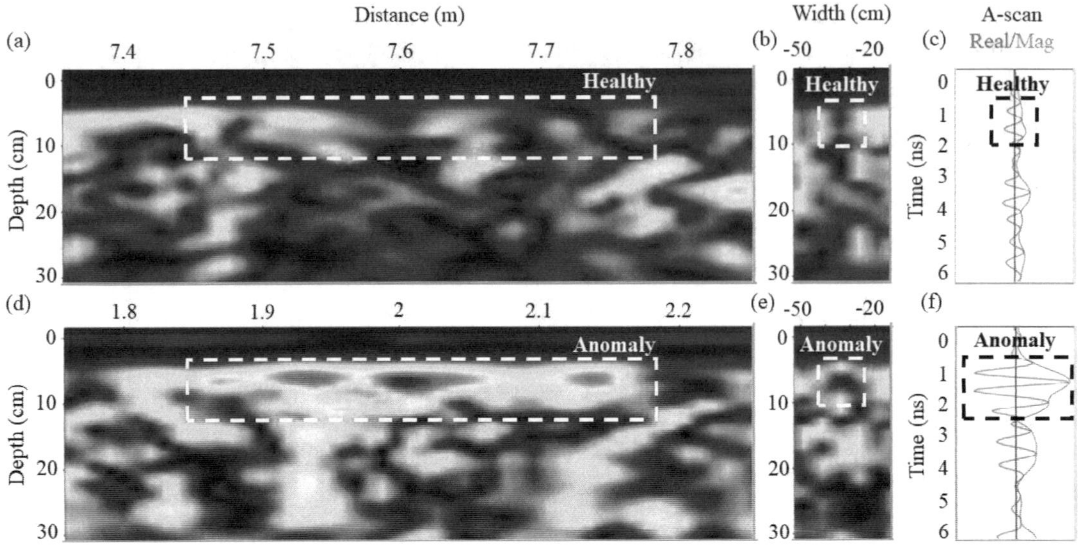

Figure 6. RUP Data (in magnitude) from the fatigue carousel after 200 kcycles of loading (a) Slab n°42 B-scan (b) Slab n°42 Cross-scan (c) Slab n°42 A-scan (d) Slab n°122 B-scan (e) Slab n°122 Cross-scan (f) Slab n°122 A-scan

5. References

[1] Katsamenis I, Bimpas M, Protopapadakis E, Zafeiropoulos C, Dimitris D, Doulamis A, Doulamis N, Martín-Portugués Montoliu C, Handanos Y, Schmidt F, Ott L, Cantero M, and Lopez R 2022 Robotic Maintenance of Road Infrastructures: The HERON Project *15th International Conference on PErvasive Technologies Related to Assistive Environments (PETRA '22)* Association for Computing Machinery, New York, USA, 628–35.

[2] Forever Open Road (FOR), www.foreveropenroad.eu/

[3] de Larrard F, Sedran T and Balay J-M 2013 Removable urban pavements: an innovative, sustainable technology *International Journal of Pavement Engineering* **14**(1), ed Taylor & Francis pp 1–11.

[4] Sedran T, Gennesseaux E, Waligora J, Klein P, Mai Lan Nguyen, Cesbron J, Ropert C and Monnier L 2023 Development of a Permeable Removable Urban Pavement *14th International Symposium on Concrete Roads, Karkow, Poland.* (https://hal.science/hal-04354455).

[5] Blanc J, Hornych P, Sotoodeh-Nia Z, Williams C, Porot L, Pouget, Boysen R, Planche J-P, Lo Presti D, Jimenez A and Chailleux E 2019 Full-scale validation of bio-recycled asphalt mixtures for road pavement *Journal of Cleaner Production* **227**, pp 1068–78.

[6] Andreoli G, Ihamouten A, Jaufer R, Todkar S S, Guilbert D, Nguyen M L, Simonin J-M, Tissier J and Derobert X 2022 The use of machine learning method to classify electromagnetic characteristics of pavement tack coat *19th International Conference on Ground Penetrating Radar (GPR 2022)*, Golden Colorado, USA.

[7] Andreoli G, Ihamouten A and Derobert X 2024 Hybridization of Machine Learning / Full Waveform Inversion Using Permittivity in SVM/SVR Models to Improve Tack coat Characterization *NDT & E International* (Under review).

Acknowledgements
The authors would like to thank the European Union's Horizon 2020 research and innovation program which, through the HERON project, allowed us to conduct this study at Gustave Eiffel University (Nantes/France).

Journal of Physics: Conference Series **2887** (2024) 012072 doi:10.1088/1742-6596/2887/1/012072

Multi-frequency electromagnetic sensor for detecting NaCl concentration in water - experimental study

Haoyu Wang[1], Zhaofa Zeng[1*] and Zhilian Li[1]
[1]College of Geo-Exploration Science and Technology, Jilin University, 938 Ximinzhu Street, Changchun 130026, China
E-mail: haoyuw@mails.jlu.edu.cn

Abstract. Clean water resources have become the focus of global attention. The rapid detection of water pollution and the rapid evaluation of NaCl content in water is an important evolution the quality of water or soil in environment and agriculture. The main method analysis the content in water or soil is based on laboratory chemical tests at present. In this paper, we used a method based on a non-uniform microwave sensor array to determine the concentration of NaCl in water. Based on the experiment data, empirical formulas obtained between the concentration of NaCl in water and different electromagnetic frequency which can be used for NaCl content estimation. The results of the study demonstrated the feasibility of microwave sensors in detecting chemical content in water, and as basic experiment for further study for water pollution.

1. Introduction

The importance of irrigation and fertilization in ensuring food production has increased significantly due to the impact of global climate change and population growth. Furthermore, these practices will have a profound effect on the processes of the water cycle, energy flow, and material cycling in terrestrial ecosystems. The concentration of NaCl in water has a significant impact on agricultural irrigation or the environmental problem of seawater intrusion. High NaCl concentrations can affect crop growth and production, and even lead to soil salinization. Therefore, rapid detection of NaCl concentration in water is an important issue in the field of environmental and agricultural sciences. At present, the primary measurement methods include automatic titrator method, spectrophotometer method and other techniques. These methods are time-consuming, expensive, and difficult to expand to large-scale water bodies and frequent detection scenes. Therefore, it is of greatly significant to study the rapid and low-cost detection method for NaCl concentration in water.

Planar structure-based microwave sensing is a new technology, which has the advantages of lightweight, compact, low cost and flexibility. This technology enables rapid measurements compared to complex measurement procedures in laboratories [1]. It has been successfully used in the fields of glucose detection [2], blood glucose detection [3], biosensors [4] and food quality assessment [5].

Maryam et al. (2019) proposed a new sensor based on complementary split-ring resonator (CSRR) which sensor was fabricated and used to measure the permittivity and permeability of several materials. [6]. Zhang et al. (2019) proposed the concept of microwave array sensor for water quality detection [7], where sensor array is presented along with the results of testing that with water samples with heavy metal pollutants including Chromium (Cr), Lead (Pb), and Mercury (Hg). Yee et al. (2021) proposed a microwave sensing technology for ammonia and iron concentration in water based on complementary split-ring resonator, which provided a new way to determine ion concentration in water [1].

In this paper, we proposed a new method for water quality measurement based on non-uniform microwave sensor arrays. This method employs multi-frequency electromagnetic measurement techniques to accurately determine the concentration of NaCl in water, achieving high precision.

2. Microwave array sensor and measurement system

The design of the sensor array is carried out in Altair FEKO software based on the reference [7]. The substrate of sensor is chosen to be Rogers RO4350 with dielectric properties of ε_r =3.66 and tanδ =0.0031[7], we can monitor the change of the intermediate electrical properties of the solution by measuring the resonant frequency shift and the amplitude change of the response at the resonant frequency.

We design and build a microwave sensor array for water quality testing (Figure 1a) which consists of five elements, each based on a microstrip transmission line design equipped with a complementary split-ring resonator (CSRR) resonating at a different frequency in the range of 1 GHz to 10 GHz [7]. When the CSRR sensor is exposed to water samples with different NaCl concentrations, peaks of the sensor at 1.5, 3.5, 5.5 and 7.5GHz changed and the microwave parameters such as resonance frequency, quality factor, and reflection coefficient also changed.

In the experiment, four peaks in the responses were observed at resonance frequencies. A method for determining the concentration of NaCl in water samples was established by simulating the variation of peak values. The results of measuring the responses of the proposed microwave sensor array (| S 21 |) when the sensor was exposed to water solutions discussed in section II are presented here. The response of the sensor is measured with a N9914A FieldFox Vector Network Analyzer (VNA). In order to reduce the effect of random noise on the measurements, an intermediate frequency (IF) bandwidth of 100 Hz is applied on Vector Network Analyzer (VNA). Each time, the sensor container is filled with one of the water solutions with NaCl content. After each solution, the container is cleaned and dried to make it ready for testing the next water solution without any mixing of the pollutants.

Figure 1. Diagram of experimental instruments (a) Microwave array sensors (b) Network Analyzer.

3. Numerical simulation experiment

3.1. Numerical simulation of responses to different concentrations of NaCl in water

A numerical simulation experiment was conducted. GPRMax software was utilized to simulate four different responses under various concentrations, as illustrated in Figure 2. It can be observed that the amplitude of peaks decreases with an increase in NaCl concentration, as illustrated in Figure 2. Linear fitting was performed between amplitude data and NaCl content, resulting in the linear formulas shown in Figure 2.

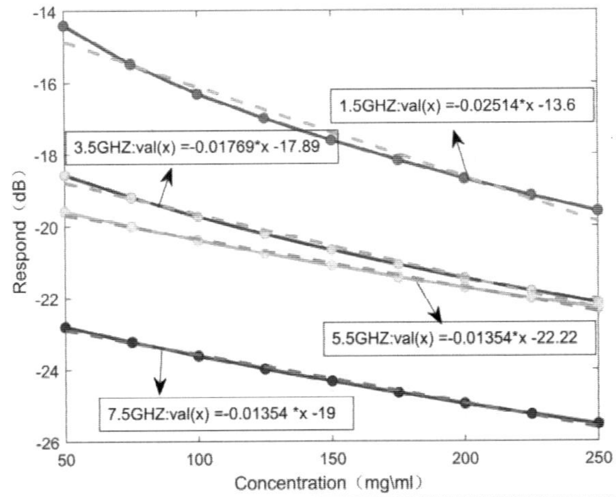

Figure 2. Simulation diagram of NaCl concentration measured by microwave array sensor in water, where the solid line represents the simulated data, and the dashed line represents the empirical formula of data fitting for the simulated data. The text in the text box represents the fit formula.

3.2. Experiment and data analysis of responses to concentrations of NaCl solution

We prepared 200 ml NaCl solutions with the concentrations from 50mg/ml to 250mg/ml with 25mg/ml interval. Solutions of different concentrations of place to the sensing part, measuring its response, and save the data. To enhance the reliability of the experimental data, each solution was measured three times, and the average of the three measurements was calculated. This can reduce the impact of random errors on the experimental results, and improve the accuracy and credibility.

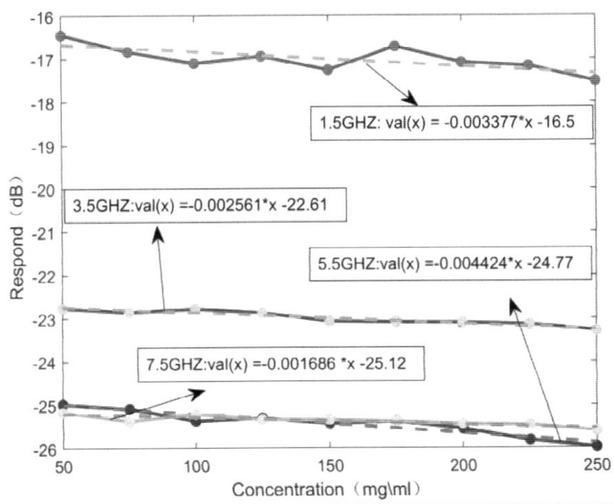

Figure 3. Laboratory measured data of NaCl concentration in water measured by microwave array sensor, where the solid line represents the measured data, and the dashed line represents the empirical formula of data fitting of the measured data. The text in the text box represents the fit formula.

Figure 3 illustrates the responses at four different frequencies obtained in laboratory under varying concentrations of NaCl. We perform linear fitting between amplitude data and NaCl content, and get the linear formulas as shown in Figure 3.

By comparing the fitting formula obtained from the measured data and the simulated data, we can find that the discrepancy between the measured data and the simulated data of the fourth peak value represented by 7.5GHz is too large, which may be caused by welding, no container in the simulation experiment and other influences. However, the gap between the measured data and the simulated data represented by the other three peaks can be said to be very small. Therefore, we can use the fitting formula of these three peaks to judge the concentration of NaCl in water.

We conducted data analysis on the fitting formulas of the three peaks represented by 1.5, 3.5 and 5.5 GHz, in which the average relative error is 1.0379%, 0.2090% and 0.2677%. It can be seen that although the average relative error of the first peak was greater than that of the other two peaks, but in general, the average relative error of the three peaks is less than 1.5%, which is negligible.

4. Conclusion and prospect

In this paper, we designed and manufactured a microwave array sensor, which has the characteristics of high precision, strong anti-interference, long life, low power consumption, etc. We used this sensor to perform experiments on different concentrations of NaCl in water. We get three empirical formulas for determining the concentration of NaCl in water through measured data, offering a method to address the challenge of rapidly measuring NaCl in water. The experimental results demonstrated that by combining the use of a network analyzer and sensor array, it is possible to effectively measure the response of NaCl solutions at different concentrations and establish the relationship between concentration and response through formula fitting.

This study demonstrates the advantages and potential of microwave sensors for the efficient measurement of NaCl content in water. Future research will focus on the response of microwave sensors to other pollutants, including Cr, Pb, and Hg. The proposed sensing technology can be adopted as portable water quality systems, promoting in situ water quality measurements.

5. Acknowledgements

This study was supported by 2022YFC3003402.

6. References

[1] YeeS K, Ong NTJ, Lim SCJ, Mohd Zin NS , SH Dahlan , AYI Ashyap and CF Soon 2021 Micorowave sensing of ammonia and iron concentration in water based on complementary double split-ring resonator *Sensors and Actuators Reports* Volume **3**.

[2] Bababjanyan A, Melikyan H, Kim S, Kim J, Lee K and Friedman B 2010 Real-time noninvasive measurement of glucose concentration using a microwave biosensor *J. Sensors* Volume **2019.**

[3] Choi H, Luzio S, Beutler J and Porch A 2018 Microwave noninvasive blood glucose monitoring sensor: Penetration depth and sensitivity analysis *2018 IEEE International Microwave Biomedical Conf* pp 52-54.

[4] Li J., Zeng Z., Sun J., Liu F., (2012) Through-wall detection of human being's movement by UWB radar. *IEEE Geoscience and Remote Sensing Letter,* 9(6), 1079-1083.

[5] Daniel A, Ihar R, Paul B I, Gil K , Dror B and Yuri F 2016 A microwave sensor for the characterization of bovine milk *Food Control* Vol **63** pp 195–200.

[6] Maryam S S, Vahid N, Mostafa K, Mohammad S and Omar M R 2019 A CSRR-Based sensor for full characterization of magneto-dielectric materials *IEEE Trans Microw Theory Tech* Volume **67** pp 806 - 814

[7] Zhang K Y, Amineh R K, Dong Z Q and David N 2019 A microwave sensor array for water quality testing *2019 IEEE 20th Wireless and Microwave Technology Conference*

Journal of Physics: Conference Series **2887** (2024) 012073 doi:10.1088/1742-6596/2887/1/012073

Application of Geo-Radar in the Detection of Various Adverse Geology in Tunnel Over-advance Geological Forecasting

Wenlong Zhao,Yuhao Tang, Siwei Zhao

972722766@qq.com

Abstract: Tunnel excavation process will encounter a variety of adverse geological problems, different adverse geologic bodies on the safety of tunneling projects will have more or less impact. This paper summarizes and analyzes several kinds of common adverse geological bodies based on previous engineering examples, and compares and analyzes the prediction results of geo-radar with the actual excavation characteristics in the preliminary over-advanced geological prediction, which shows that different adverse geological bodies such as fracture zones (faults), water-bearing bodies, cavities, joints and fissures development zones, etc., will have an impact on the frequency and amplitude (energy) of radar waves, which will confirm the frequency and amplitude changes of radar waves in the face of the palm. By analyzing the frequency and amplitude changes of radar waves, we can confirm the change of surrounding rocks in front of the palm face. The application of geo-radar in over-advanced geological forecasting and detection of adverse geology has solved the problems encountered in the process of tunnel construction and excavation, and has provided reliable technical support for safe tunnel construction.

Keywords: geo-radar; Undesirable Geological Bodies; overcasting; tunnel construction

1 Introduction

With the continuous upgrading and development of China's economy, more and more infrastructure measures are under construction and development, especially the construction of highways and high-speed railroads in mountainous areas of China is further extended to all corners, and the construction of mountain tunnels encounters various undesirable geologic bodies such as karsts, fracture zones, fissures, etc., which seriously affects the safety and progress of the tunneling project, therefore, the over-advance geological forecast has an important guiding significance during the excavation process of the tunnel. Therefore, advance geological prediction has an important guiding significance in the process of tunnel excavation, and geo-radar, as an

Journal of Physics: Conference Series **2887** (2024) 012073 doi:10.1088/1742-6596/2887/1/012073

important detection means of advance geological prediction, plays a crucial role in the prediction of the undesirable geological bodies in front of the tunnel palm face.

Geological radar has the advantages of high resolution, fast and efficient in the selection of methods for overpass geological forecasting [1] [2], through the data acquisition of the tunnel palm face, combined with the overpass horizontal drilling holes, deepening the gun holes as well as the relevant survey data in the early stage of the orthogonal simulation, data image analysis, you can make a fast and efficient judgment of the type, scale and approximate scope of the influence of the undesirable geologic body in the front of the palm face, which can provide the data support for the further excavation of the tunnel. Tunnel further excavation to provide data support.

2 Geological Radar Fundamentals

In the working process of geo-radar, the transmitter antenna transmits high-frequency electromagnetic waves to the detected body, when the high-frequency electromagnetic waves are transmitted to the interface of two different media in the detected body, due to the different dielectric constants of the two media and the electromagnetic waves are reflected and refracted, and the propagation of the incident wave, reflected wave and refracted wave follows the law of reflection and the law of refraction, and the reflected wave is returned to the surface of the detected body and is received by the receiving antenna of geo-radar, forming the radar image. The reflected wave returns to the surface of the detected body and is received by the antenna of the geo-radar, forming the radar image. The schematic diagram of the detection principle is shown in Figure 1.

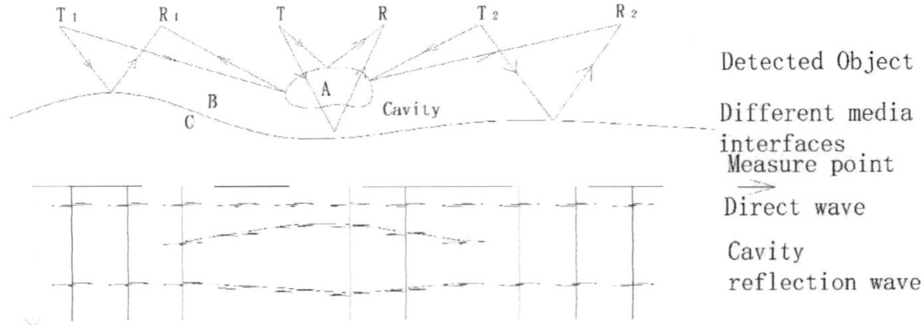

Figure 1 Schematic diagram of the working principle of geological radar

Radar image contains rich information of the detected body, according to the radar image characteristics of the detected body (e.g., dense zone of nodules, cavities, reflective interfaces, etc.) for qualitative interpretation, and then according to the formula (1) can be quantitatively interpreted for the anomalous parts of the detected body.

$$H = \frac{tc}{2\sqrt{E_r}} \ldots\ldots\ldots\ldots\ldots\ldots (1)$$

Where: H is the burial depth of the target body, t is the two-way time for the electromagnetic wave propagating from the surface of the detector body to the anomalous part of the detector body (or the interface of different media) and then reflecting back to the surface, c is the propagation speed of radar wave in vacuum, and Er is the relative permittivity of the medium of the target body. The relative dielectric constants of common surrounding rocks and media are shown in the following Table 1. The relative dielectric constants are related to a variety of substances such as lithology, voids and void fillers, and groundwater.

Table 1 Relative dielectric constants of common media

medium	relative dielectric constants	medium	relative dielectric constants
air	1	shale	7
waters	81	limestone	7~8
Ice	4	concrete	6.4
clays	8~12	granite	5~7

Ground Penetrating Radar image mainly reacts to the electrical distribution characteristics of the probing medium, and through statistical analysis and excavation verification, the radar graphic characteristics of typical undesirable geologic bodies such as relatively intact rock bodies, fracture zones, water-bearing bodies, rock vein bodies, nodal fissure dense zones and lithological transition zones have been summarized, as shown in Table 2.

Table 2 Relationships between radar images, waveform characteristics of typical bad geologic bodies [3]

Classification of Undesirable Geological Bodies	Radar image characterization			
	Energy Characteristics	frequency change	In-phase axis characteristics	waveform variation
Relatively intact rock mass	Weak, slow decay	High	sequential	homogeneous
crushing belt	Stronger, faster decay	Low, homogeneous~messy	Discontinuous	Less regular
dendrite (geology)	Stronger, Faster Decay	Low, Cluttered	Somewhat directional	More heterogeneous
water-bearing body	Stronger, Faster Decay	Low, Cluttered	Discontinuous, concentrated	Poorly regular
jointed fissure zone (geology)	Stronger, Faster Decay	Lower, More Cluttered	Discontinuous	More chaotic
rock transition zone	Stronger, Faster Decay	Lower, More Cluttered	Continuous to discontinuous	More heterogeneous
Karst (caves)	Stronger, Faster	Lower, Cluttered	Certain direction	More homogeneous

3 Engineering examples

3.1 Fractured zone detection

After excavating a slate sandstone tunnel to a certain mileage, the surrounding rock changed from weak weathering to strong weathering folder (full weathering), and there is a twisting phenomenon, the rock body is extremely broken, for this reason, in the palm face to carry out the geo-radar over-the-horizon geologic prediction detection, through the digital filtering, cheaper processing, energy equalization, time-depth conversion, and other processing, and ultimately generated radar generated profiles (Figure 2-a). Combined with the site surrounding rock situation and the analysis of the previous investigation data, it is believed that there is a fault fragmentation zone behind the palm face, and the support measures should be strengthened in the later excavation; the later excavation also verified the existence of the fault fragmentation zone.

3.2 Detection of dyke bodies

A granite tunnel excavation to a palm face after the emergence of gabbro strips, gabbro strips and granite contact zone surrounding rock fragmentation, high risk of falling blocks, in order to ensure the construction safety, temporarily increase the geo-radar method to detect the distribution range and size of the front gabbro strips. After various processing means, the radar profile was finally generated (Figure 2-b), and it was judged that there were several gabbro vein intrusions in the ranges of 0-5m, 7-8m, 10-11m and 17-22m in front of the palm face, the surrounding rock was broken, and the joints and fissures were developed intensively, and the risk of dropping blocks was high, and the over-head support measures should be strengthened in the late excavation; and the excavation process at the later stage also revealed the existence of multiple vein bodies, which avoided the disaster in time successfully. The disaster was successfully avoided in time.

3.3 Detection of water-bearing bodies

A granite tunnel excavation to a palm face, according to the preliminary investigation and design data to judge the palm face behind the water-rich area, in order to grasp the water-rich conditions in front of the palm face in advance, to carry out geological radar exploration. After processing, it was judged that there was a water-bearing body 6-19m behind the palm face to the right of the palm face. In order to prevent the occurrence of water influx in the palm face to the right of the position of the arrangement of overpass horizontal drilling, drilling to 8m there is a strand of water flow, and with pressure (Figure 2-c); this geo-radar prediction will be successful in the front of the palm face pressurized water leakage, to ensure the safety of the construction.

3.4 Karst Detection

A gray rock tunnel in the excavation process to do geological radar forecast, through the interpretation analysis inferred that the palm face in front of the 5m may exist within the cavern or peripheral rock dissolution fissure development; in the follow-up process of footing found in the palm face on the right side of the existence of a small and medium-sized cavern (Figure 2-d).

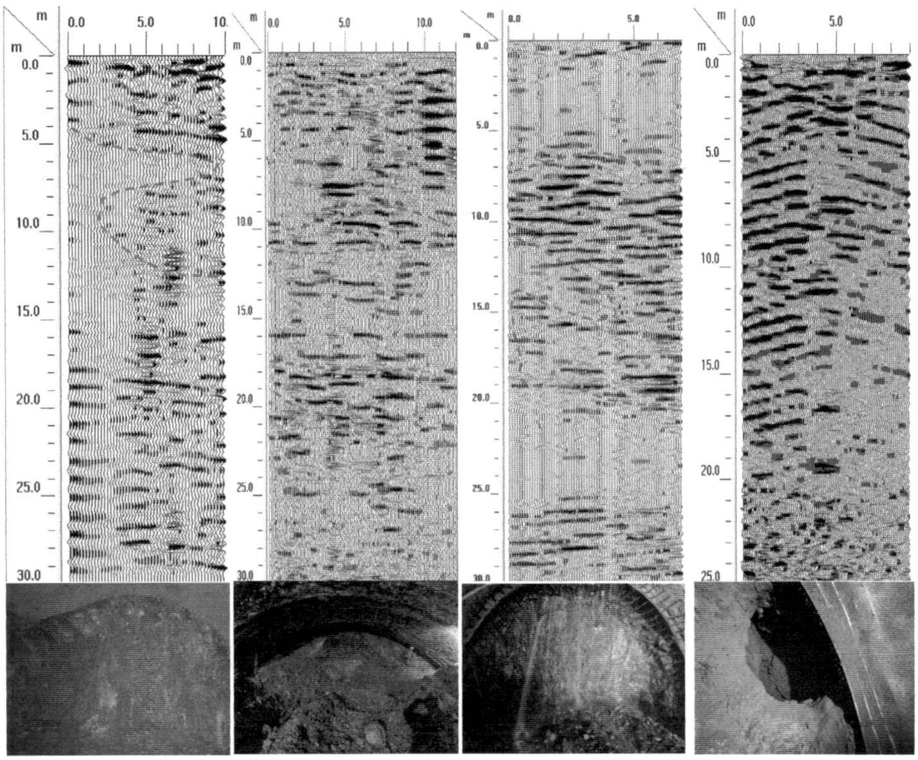

a - Crush zone b - dyke body c - Water-bearing bodies d - Caves

Fig. 2 Georadar waveform characteristics of different undesirable geologic bodies

4. Conclusion

Geological radar has good application in the detection of different bad geology in the over-advance geological prediction. Combined with the preliminary investigation and design data and the lithological characteristics of the ground, it can realize the accurate detection of many kinds of bad geology bodies. At the same time, combined with the characteristics of geological radar waveforms in tunnels of different lithologies, analyzing the energy characteristics, waveform characteristics, frequency characteristics and isotropic axial continuity of the radar waveforms can

Journal of Physics: Conference Series **2887** (2024) 012073 doi:10.1088/1742-6596/2887/1/012073

further explain the type of adverse geological bodies in front of the palm face, and provide a scientific basis for the interpretation of the safe construction of tunnels.

Reference

[1]. ZHANG Chongmin,ZHANG Fengkai,et al. AStudy of Full Waveform Inversion of Advance Tunnel Geological Prediction by Ground Penetrating Radar [J]. Tunnel Construction, 2019, 39(1): 0102-08.

[2]. ZENG Zhaofa,LIU Sixin,FENG Xuan.Principle and application of ground penetrating radar [M] .Beijing:Publishing House of Electronics Industry,2010.Publishing House of Electronics Industry.

[3]. SHI Pengchao,CHI Changfeng, et al. Application of ground penetrating radar in advanced geological prediction of tunnels [J]. Tunnel and Construction Engineering, 2019, 03: 0170-04.

AUTHOR INDEX